Geophysical Monograph Series

Including
IUGG Volumes
Maurice Ewing Volumes
Mineral Physics Volumes

Geophysical Monograph 184

Carbon Cycling in Northern Peatlands

Andrew J. Baird
Lisa R. Belyea
Xavier Comas
A. S. Reeve
Lee D. Slater
Editors

American Geophysical Union
Washington, DC

Library of Congress Cataloging-in-Publication Data

Carbon cycling in northern peatlands / Andrew J. Baird ... [et al.].
     p. cm. — (Geophysical monograph ; 184)
    Includes bibliographical references and index.
    ISBN 978-0-87590-449-8
  1. Carbon cycle (Biogeochemistry)—Northern Hemisphere. 2. Peatlands—Environmental aspects—Northern Hemisphere. 3. Carbon sequestration—Northern Hemisphere. 4. Greenhouse gases—Northern Hemisphere. I. Baird, Andrew J., 1969-
    QH344.C385 2009
    577'.144—dc22

2009024462

ISBN: 978-0-87590-449-8
ISSN: 0065-8448

Cover Photo: Open pools in the central unit of Caribou Bog, Orono, Maine. Modified from R. B. Davis original (taken June 2006).

CONTENTS

PREFACE

Even though they cover only between 2 and 3% of its land mass, peatlands are a major component of the Earth's carbon cycle, containing about one third of the carbon in the pedosphere. These large carbon stores remove carbon from, and release it to, adjacent systems (most significantly the atmosphere) in a complex cycle. Although large peatlands are found in the tropics, this monograph focuses on recent developments in our understanding of carbon dynamics in northern peatlands, that is, those peatlands that occur at latitudes higher than 45°N. We focus exclusively on northern peatlands because of their significance in terms of surface areal extent and their importance as carbon stores. Northern peatlands are also most likely to be affected by climate change and warming.

Peatlands science cuts across disciplines, and it can be difficult for peatland researchers in one discipline to find advances made by peatland scientists in other disciplines. Recognition of this problem was partly behind our decision to propose the monograph; in other words, we wanted to produce a collection of papers that brought together the state of knowledge on peatland science. Obviously, a monograph that considered all areas of peatland science would have been a huge, if not impossible, undertaking. We chose to focus on carbon cycling and climate, and we did so for two reasons. We felt that much previous work has looked at peatlands primarily as archives of climate change, with less emphasis on the processes that control how a peatland responds to variations in climate and how it may itself influence climate. We were also aware of the need to include peatlands in climate models and the need to communicate current understanding of the role of peatlands in the global carbon cycle to a larger audience, especially, but not exclusively, climate modelers.

Many individuals helped with the production of this monograph. Each chapter was independently reviewed, and we are indebted to those academics who undertook reviews, and in some cases rereviews, on very tight deadlines. Finally, this effort was inspired by a National Science Foundation funded "Peatlands Geophysics Workshop" held at the University of Maine in June 2007. The purpose of the workshop was to bring together peatland scientists from a range of disciplines to consider novel ways of mapping the subsurface structure of peatlands. We hope some of that ambition and novelty is reflected in the current collection.

Lee D. Slater
Rutgers, State University of New Jersey

Andrew J. Baird
University of Leeds

Lisa R. Belyea
Queen Mary University of London

Xavier Comas
Florida Atlantic University

A. S. Reeve
University of Maine

Carbon Cycling in Northern Peatlands
Geophysical Monograph Series 184
Copyright 2009 by the American Geophysical Union.
10.1029/2008GM000872

Understanding Carbon Cycling in Northern Peatlands: Recent Developments and Future Prospects

Andrew J. Baird,[1] Xavier Comas,[2] Lee D. Slater,[3] Lisa R. Belyea,[4] and A. S. Reeve[5]

Although one of the earliest recorded investigations of peatlands is attributed to *King* [1685] more than 3 centuries ago [*Gorham*, 1953], *Weber*'s [1902] treatise on the Aukštumala Raised Bog in Lithuania is still considered the first comprehensive ecohydrological study of peatlands and the foundation for modern peatland science. Weber's monograph was pioneering for different reasons: (1) it integrated disciplines such as stratigraphy, hydrology, chemistry, and ecology to describe, classify, and model peatlands forms and their development (e.g., internal processes), and (2) it investigated potential interactions between peatland (mostly hydrology) and changes in climate and sea level (e.g., external forcing). Since then, most peatlands science has focused on the peatland archive (e.g., pollen and macrofossils) for environmental and climate reconstruction over the Holocene, whereas processes controlling the response of peatlands to climate change have tended to be overlooked.

The effect of peatlands on global climate is currently unclear. Peatlands influence climate by sequestering CO_2 from the atmosphere and storing it in living and dead biomass. They return some of this CO_2 via the decay of plant litter and peat and are the largest natural terrestrial source of atmospheric methane (CH_4), which is produced during anaerobic decomposition. Through the Holocene, peatlands have been a persistent sink for atmospheric CO_2 and a persistent source of atmospheric CH_4. Although CH_4 is a much more potent greenhouse gas than CO_2, modeling work by *Frolking et al.* [2006] suggests that peatlands have had a negative radiative forcing (cooling) effect on climate through the Holocene. However, that may change in the coming decades as peatlands respond to climate change. For example, existing peatlands may become net emitters of CO_2 as peat warms and rates of decomposition increase, while in areas of permafrost the formation of thaw lakes may lead to higher rates of CH_4 loss. On the other hand, new peatlands may develop in areas that are currently tundra and become large sinks of atmospheric CO_2 in the next 100–200 years, thus offsetting, at least in part, greater losses of CO_2 and CH_4 from northern peatlands at lower latitudes. To understand how peatlands affect global climate, we need to represent them as land surface schemes coupled to global climate models (GCMs). Before we can do so, there is a need to understand better how peatlands function as ecohydrological entities. Part of this understanding can come from the paleorecord, but part must come from process and modeling studies at a range of scales. Where possible, peatland models should be process-based and applicable to a range of climatic, geomorphological, and geological settings; that is, they should be transportable.

Scale is an important factor to consider in studies of carbon cycling in northern peatlands for at least two reasons. First, when considering processes operating across the whole peatland or at the regional scale, it is uncertain whether small-scale processes (those at plan scales of about 1 m) can be ignored or simplified adequately. Second, it is becoming clear that little is known about the off-site components of the carbon budget of peatlands; carbon exchanges are not just between peatlands and the atmosphere; dissolved carbon can also be exported from peatlands, where its fate remains uncertain. As part of this second reason, other off-site factors such as topographic setting are also important controls on carbon cycling in individual peatlands. Although this monograph

[1]School of Geography, University of Leeds, Leeds, UK.

[2]Department of Geosciences, Florida Atlantic University, Boca Raton, Florida, USA.

[3]Department of Earth and Environmental Sciences, Rutgers, State University of New Jersey, Newark, New Jersey, USA.

[4]Department of Geography, Queen Mary University of London, London, UK.

[5]Department of Earth Sciences, University of Maine, Orono, Maine, USA.

Carbon Cycling in Northern Peatlands
Geophysical Monograph Series 184
Copyright 2009 by the American Geophysical Union.
10.1029/2008GM000875

focuses on peatlands occurring across a limited latitudinal range, study sites are included across geographical longitudes spanning three different continents and a total of seven countries.

The chapters consider some of these scale and geographical issues and also how we can improve our understanding of key ecohydrological processes in peatlands and how they affect carbon cycling. The monograph is divided into four sections, and the content of each is discussed briefly below.

Section 1 considers the role of peatlands in the global carbon cycle. It does so from a variety of perspectives. The first chapter considers how peatlands respond to climate change. Many workers still consider peatlands to respond in simple linear ways to climate. However, it is becoming clear that peatlands are complex adaptive systems that do not show simple responses to climate. Sometimes peatlands undergo large ecohydrological changes in the absence of a climate driver, and sometimes they show extreme sensitivity to apparently small climatic changes. If we are to model peatlands adequately, then we need to know the reasons behind such nonlinear behavior. Climate modelers are only now starting to recognize the importance of the peatland carbon store and the need to include peatlands in GCMs. How peatlands should be represented in climate models is the focus of the second and third chapters. The second looks at how peatlands differ from other land surface types and the challenges these differences present when trying to incorporate peatlands into land surface schemes. The third looks at the problem of what it means to ignore small-scale variability when modeling CH_4 losses from peatlands and the importance of such small-scale variability when trying to represent peatlands in GCMs. The fourth chapter provides a broad temporal and spatial perspective and uses meta-analyses of data from previous studies to investigate the factors that influenced peatland initiation and carbon accumulation during the Holocene. Finally, there is a chapter on direct human impacts on the peatland carbon store. Much of the current concern is with the indirect human impacts (climate change) on peatlands; therefore, it is useful also to consider how direct impacts such as changes in land use to forestry and agriculture affect carbon cycling.

Section 2 focuses on processes operating at and near the peatland surface, where climate change is likely to have the greatest impact through disturbance (e.g., fire and permafrost thaw) and changes in water table and temperature regimes. One of the unique characteristics of peatland land surfaces is the prevalence of mosses, which have distinct biophysical and biochemical properties compared with vascular plants. The relative abundance of these plant types has a profound effect on carbon cycling because the biochemistry of plant-derived substrates is a key control on rates and pathways of decomposition. The other major control on carbon cycling relates to vertical and horizontal heterogeneity in water table regime, which, in turn, controls oxygenation, the distribution of plants and microbes, and microbial metabolic pathways. The first chapter in this section reviews remote sensing approaches to obtaining land surface data relevant to the carbon cycle, both for generating land surface classifications and for retrieving biophysical properties that can be used to parameterize process-based models. The second chapter considers mass loss and nutrient release from fresh plant litter, noting that changes that affect the relative production of *Sphagnum* versus vascular plant litter are likely to have feedbacks on carbon and nutrient cycling. The third chapter examines carbon flow from a microbial perspective, identifying the main microbial players, metabolic pathways, and factors controlling substrate use in the oxic, periodically oxic/anoxic, and permanently anoxic zones. The final chapter considers the relative amounts of CO_2 and CH_4 produced during terminal mineralization of carbon under anoxic conditions, noting the importance of both substrate characteristics and physical factors and making a plea for further studies using consistent methodologies.

Section 3 describes the state of knowledge on CH_4 accumulation in, and release from, peatlands by considering both deep and shallow sources of the gas at a wide range of scales. Methane is lost to the atmosphere through three main mechanisms: diffusion through the peat matrix, transport through vascular plants, and ebullition (as bubbles). Until recently, most studies considered the first two mechanisms almost exclusively. However, there is burgeoning interest in the significance and causes of ebullition. Nonsteady or episodic ebullition events have generated particular interest because of the potentially large amounts of CH_4 involved. Ebullition fluxes in northern peatlands typically exceed average diffusive fluxes on a per-event basis and often on a seasonal basis as well. Current ebullition estimates are unclear for several reasons: (1) our poor understanding of gas spatial variability related to the heterogeneous nature of the peat matrix; (2) uncertainties related to contrasting models of gas accumulation (e.g., shallow entrapment in poorly decomposed peat versus deep entrapment below confining layers of woody peat); and (3) factors affecting ebullition dynamics, often related to environmental parameters such as soil chemistry, substrate quality, or plant community structure. To further complicate estimates, biogenic gas emissions from wetlands are often related to changes in temperature, atmospheric pressure, and/or water table elevation. The first and second chapters in this section emphasize the importance of CH_4 accumulation in deep peats (i.e., >2 m) and describe the use of minimally invasive techniques such as global positioning systems (GPS) and ground-penetrating radar to investigate deep free-phase gas accumulations in peatlands. The third and fourth chapters examine the role of shallow peat soils (i.e., <1 m) as both zones of CH_4 production and zones from which CH_4 is lost to the atmosphere. The third chapter proposes a new conceptual model for bubble buildup and release in shallow peat soils, while the fourth identifies key zones of enhanced methanogenesis at shallow depths based on carbon isotope composition. The fifth chapter presents an overview of an experimental design that can be used for investigating the accumulation and release of CH_4 from shallow peats under controlled laboratory conditions. Finally, the last chapter looks at some of the controls that may induce losses of CH_4 gas from peatlands such as atmospheric pressure, peat temperature, and water table position for both deep and shallow peats.

All of the topics presented in this monograph reveal the importance of the physical and chemical processes related to water supply to and movement within peatlands. Section 4 focuses on peatland hydrology and its role in carbon dynamics. Efforts to understand peatland hydrological processes typically focus on (1) saturation state and water table position and (2) rates and directions of water movement. Water in peatlands isolates organic matter from the atmosphere, altering the redox state and slowing the oxidation of organic matter while creating an environment favorable for methane production. This mixture of organic matter and water indirectly results in high concentrations of dissolved organic carbon (DOC) in peat pore waters. The production of DOC within and export from peatlands is discussed in the first two chapters of section 4. The rate and direction of surface and groundwater flow within a peatland regulate the export of DOC from the peatland and influence the supply of nutrients to it. The third chapter describes the hydrological and hydrochemical importance of natural soil pipes in the peatlands in which they occur. Chapter five focuses on the hydrodynamics of the unsaturated zone in harvested peatlands.

Chapter six discusses the role of subsurface heterogeneity on groundwater flow patterns. There are a variety of feedback mechanisms between the hydrology of a peatland and associated carbon dynamics that complicate this relationship. The relationship between hydrology and biogenic gas dynamics is one of these feedback systems and is discussed in chapter four. While there are many similarities among the peatland systems discussed in this section, it is important to note the differences between individual systems and to use caution when generalizing processes observed in one peatland to another.

REFERENCES

Frolking, S., N. Roulet, and J. Fuglestvedt (2006), How northern peatlands influence the Earth's radiative budget: Sustained methane emission versus sustained carbon sequestration, *J. Geophys. Res.*, *111*, G01008, doi:10.1029/2005JG000091.

Gorham, E. (1953), Some early ideas concerning the nature, origin, and development of peat lands, *J. Ecol.*, *41*(2), 257–274.

King, W. (1685), On the bogs and loughs of Ireland, *Philos. Trans. R. Soc. London*, *15*, 948–960.

Weber, C. A. (1902), *Über die Vegetation und Entstehung des Hochmoors von Augstumal im Memeldelta mit vergleichenden Ausblicken auf andere Hochmoore der Erde*, Paul Parey, Berlin.

A. J. Baird, School of Geography, University of Leeds, Leeds LS2 9JT, UK.

L. R. Belyea, Department of Geography, Queen Mary University of London, Mile End Road, London E1 4NS, UK.

X. Comas, Department of Geosciences, Florida Atlantic University, Boca Raton, FL 33431, USA.

A. S. Reeve, Department of Earth Sciences, University of Maine, Orono, ME 04469, USA.

L. D. Slater, Department of Earth and Environmental Sciences, Rutgers, State University of New Jersey, Newark, NJ 07102, USA. (lslater@andromeda.rutgers.edu)

Nonlinear Dynamics of Peatlands and Potential Feedbacks on the Climate System

Lisa R. Belyea

Department of Geography, Queen Mary University of London, London, UK

Peatlands have potential for strong feedback on the global climate system, but their response to future climate change is highly uncertain. In this chapter, I review a range of evidence demonstrating that peatland dynamics are nonlinear. Rather than gradual change that converges on a single dominant pathway and matches the frequency of external forcing, peatlands show (1) sensitivity to initial conditions and divergence onto multiple pathways of development, (2) long periods of little change, punctuated by abrupt transitions of state even under weak or steady environmental forcing, and (3) responses to external forcing at unexpected frequencies. Nonlinear systems exhibit persistence when stabilizing forces (i.e., negative feedback mechanisms) dominate and undergo rapid transformation when destabilizing forces (i.e., positive feedback mechanisms) dominate. In peatlands, stabilizing and destabilizing forces result from interactions among hydrological processes, organic matter dynamics, and energy exchanges. The depth dependence of peat hydraulic conductivity tends to stabilize hydrological conditions, whereas local flow networks may amplify water losses when vascular plant transpiration is high. Peat formation rate is generally constrained by water storage change but occasionally can trigger a rapid increase or decrease in thickness of the unsaturated zone. Regionally, increases in evapotranspiration may be counteracted by recycling and precipitation of evaporated water over peatlands, whereas contrasts in albedo and energy partitioning across peatlands and surrounding forests may promote rapid spring thaw. In order to predict feedbacks on the climate system, it will be essential to reduce the complexity of peatlands by identifying the key variables and interactions that control nonlinear behavior.

1. INTRODUCTION

Throughout the Holocene, the radiative forcing function of peatlands has shifted from a net warming to a net cooling [*Smith et al.*, 2004; *Frolking et al.*, 2006], but the future impact of peatlands on a changing climate system is highly uncertain. Of central concern is the vulnerability of the large peatland carbon pool to processes that might release CO_2 and CH_4 to the atmosphere, thereby amplifying human-induced changes to atmospheric chemistry. Peatlands are complex systems [*Belyea and Baird*, 2006], and as I will show in this chapter, their dynamics are often nonlinear. Abrupt, step-like changes in peatland structure (e.g., the distribution of vegetated and nonvegetated land surface types) and function (e.g., hydrological processes, organic matter dynamics, and energy exchanges) may be linked only weakly to climate forcing [e.g., *Belyea and Malmer*, 2004; *Yu*, 2006a].

Carbon Cycling in Northern Peatlands
Geophysical Monograph Series 184
Copyright 2009 by the American Geophysical Union.
10.1029/2008GM000829

If these nonlinear changes involve the peatland carbon pool, they may have profound effects on the climate system.

A key question, then, is "What controls nonlinear dynamics in northern peatlands and what are the potential feedbacks on the climate system?" I will approach this question by reviewing some of the evidence for nonlinear behavior in peatland ecosystems, identifying some of the stabilizing (negative feedback) and destabilizing (positive feedback) forces that operate within them, and discussing possible biogeochemical and biogeophysical feedbacks on the global climate system for two "what if?" scenarios of future environmental change.

2. EVIDENCE OF NONLINEAR DYNAMICS IN NORTHERN PEATLANDS

The nonlinear behavior of some complex systems is characterized by long periods of stasis punctuated by occasional, abrupt shifts to alternative regimes, which differ in fundamental structure and processes from the previous regime. Such threshold (step-like) dynamics occur in many physical and biological systems and can be classified into three basic types [*Andersen et al.*, 2009]: (1) "driver threshold," the system state responds linearly to an environmental driver, which itself undergoes a step-like change; (2) "state threshold," the system state responds in a step-like way after a slowly varying environmental driver exceeds a threshold value; (3) "driver-state hysteresis," the threshold value for step-like response of the system state differs, depending on whether the environmental driver is increasing or decreasing. The last two types are of particular concern in the context of climate change and the carbon cycle, because abrupt shifts in fundamental structure and function can occur unexpectedly under weak external forcing.

Paleorecords and long-term instrumental records can provide evidence of three types of behavior that are characteristic of nonlinear systems [*Rial et al.*, 2004]: (1) Nonlinear systems are highly sensitive to initial conditions and show divergence onto multiple pathways rather than convergence onto a single dominant pathway of evolution. (2) Even under weak or steady external forcing, nonlinear systems show rapid, abrupt transitions of state rather than slow, gradual changes proportional to external forcing. (3) Nonlinear systems respond to oscillations at unexpected frequencies rather than matching the frequency of the external forcing. All of the following examples from peatlands suggest at least one of these distinctively nonlinear behaviors.

2.1. Peatland Initiation and Expansion

Peatlands initiate by three mechanisms [*Rydin and Jeglum*, 2006]: terrestrialization by infilling of a lake or pond, pri-

mary peat formation on newly exposed mineral soil, and paludification by "swamping" of mineral soil that was previously covered by forest, grassland, heathland, or tundra. In previously glaciated landscapes, peat formation often begins by terrestrialization of small depressions, and these initially small peatlands subsequently expand across the landscape by primary peat formation or paludification [*Korhola et al.*, 1996; *Anderson et al.*, 2003]. Very few studies have examined the processes directly involved in lateral expansion by primary peat formation and paludification, but it seems clear that local positive feedback must be involved. Specifically, the groundwater mound must grow with the peat deposit, with the result that surrounding mineral soils become "swamped." A range of pedogenic processes that decrease soil permeability may also be involved [*Rydin and Jeglum*, 2006]. The increase in wetness of the mineral soil allows establishment of peat-forming plants such as *Sphagnum*, triggering a further network of positive feedbacks, involving water retention, changes in soil water chemistry, and organic matter accumulation [*van Breemen*, 1995]. The switch from mineral soil to peat soil can occur within decades [*Hulme*, 1994], and the rate of lateral expansion can reach 8 m a^{-1} [*Rydin and Jeglum*, 2006]. The rapidity and abruptness with which peatland initiation and expansion can occur suggests a threshold response from one regime (mineral soil) to another (peat). At a continental scale, postglacial peatland expansion is described very well by a simple sigmoidal model [*Gorham et al.*, 2007]. The phase of "explosive" peatland expansion (i.e., the near-vertical part of the sigmoidal curve) occurred much earlier in Siberia than in North America [*Smith et al.*, 2004] and was accompanied by similarly abrupt increases in storage of organic carbon and emission of CH_4 to the atmosphere [*Gajewski et al.*, 2001; *Smith et al.*, 2004]. The difference in timing of the explosive phase may indicate a disproportionate response to changes in climate, although this hypothesis should be tested explicitly.

2.2. Successional Dynamics

Once initiated, peatlands undergo a wide variety of vegetation changes at scales ranging from individual microforms ("microsuccession," occurring at length scales of 10^0–10^1 m, e.g., hollow to hummock transition) to the whole peatland ("macrosuccession," occurring at length scales of 10^2–10^3 m, e.g., fen to bog transition). Regional analyses, which compile paleoecological records of sediment cores from many sites, show a complicated network of developmental pathways inferred from transitions in plant macrofossil composition [e.g., *Walker*, 1970; *Aaviksoo et al.*, 1993; *Bunting and Warner*, 1998]. In these records, self-replacement of vegetation types is very common; that is, the peatland tends to persist in one vegetation state over many sampling intervals. When transitions do occur, they can follow one of a number of alternative pathways rather than a single dominant pathway. More detailed analyses of single cores suggest that long periods of little or no change are punctuated by occasional, brief episodes of abrupt change, both in terms of vegetation and carbon storage [*Belyea and Malmer*, 2004; *Yu*, 2006a]. These step-like transitions can occur even through periods where environmental forcing is gradual or weak [*Belyea and Malmer*, 2004; *Yu*, 2006a]. These results support the idea that successional dynamics are nonlinear: a peatland will remain static in one regime for a long period of time despite external forcing and then occasionally undergo rapid transition to one of a number of alternative regimes, with the change in state disproportionately large compared with any change in the strength of environmental forcing.

2.3. Ombrotrophication (Fen-Bog Transition)

Ombrotrophication is a particular type of macrosuccession that occurs when the surface of a fen becomes isolated from minerotrophic water and undergoes transition to poor fen or bog, triggering a vegetation switch to dominance by *Sphagnum* mosses and marked changes in soil water chemistry. In large part, isolation from minerotrophic water occurs because of height growth of the peat deposit. The timing of fen-bog transition, therefore, is highly dependent on local factors such as time of peat initiation, local rate of peat growth, and local topography of the mineral substrate [*Bunting and Warner*, 1998; *Anderson et al.*, 2003]. The process is driven largely by *Sphagnum* "engineering" the physical and chemical environment [*van Breemen*, 1995], and the new, ombrotrophic state (i.e., bog or poor fen) will persist if *Sphagnum* achieves a threshold abundance. Detailed paleoecological analyses of peat cores show that ombrotrophication occurs rapidly on a time scale of decades [*Kuhry et al.*, 1993]. In some cases, the records show one or two previous, failed attempts before the successful transition occurs (P. Kuhry, personal communication, 2005). Once the peatland has reached a critical stage of development, a period of drier climate may help to trigger the transition [*Hughes*, 2000; *Anderson et al.*, 2003]. These observations suggest a step-like shift from one state (fen) to another (bog or poor fen). Although regional climate change may help push the system up and over the step, the transition is driven mainly by internal, positive feedbacks involving the effects of *Sphagnum* on the physicochemical environment. Given that fens and bogs differ markedly in rates of CH_4 emission [*Bubier et al.*, 2005] and carbon storage [*Yu*, 2006a], ombrotrophication is almost certain to result in a step-like change in carbon cycling.

2.4. Permafrost Formation and Thaw

In the discontinuous permafrost zone, peat plateau landforms undergo a cyclical succession, switching between permafrost and nonpermafrost regimes [*Zoltai*, 1993; *Camill and Clark*, 2000]. In the nonpermafrost state, peat accumulates differentially, leading to formation of hummocks, which allow colonization by trees [*Camill and Clark*, 2000]. In winter, the (evergreen) canopy reduces snow depth [*Camill and Clark*, 2000], and the frozen peat conducts heat rapidly to the atmosphere [*Zoltai*, 1993]; in summer, the tree canopy shades the peat surface [*Camill and Clark*, 2000] and the dry surface peat acts to insulate deeper layers [*Williams*, 1968; *Zoltai*, 1993]. As a result, temperature within the hummocks is reduced and permafrost begins to form. The permafrost features gradually expand and coalesce, forming a densely forested permafrost plateau. Disturbance of forest cover on the permafrost plateau by fire or windthrow initiates permafrost thaw in isolated "collapse scars" [*Camill and Clark*, 2000]. Without the sheltering effect of the trees, the peat within the scar thaws, and thermokarst features such as wet lawns or small lakes form as the surface collapses. Over time, differential peat accumulation can lead to the formation of hummocks within the collapse scar. The unfrozen area expands in size until the thawing edge of the collapse scar is stabilized, either by shade of fringing trees or by the insulating effects of *Sphagnum* hummocks [*Camill and Clark*, 2000]. The dominance of local factors (tree size and density and differential peat microtopography) in this cycle suggests that responses to climate change will be nonlinear. In continental Canada, the southern limit of the discontinuous permafrost zone has shifted northward since the Little Ice Age, but this shift has lagged increases in temperature, leaving behind relict permafrost in regions where it could not form today [*Turetsky et al.*, 2007]. Over the past 50 years, the rate of permafrost thaw has increased, with the rate increasing more quickly in southern regions than in those farther north, possibly linked to greater increases in winter and spring temperatures [*Camill*, 2005]. In Siberia over the past 30 years, thaw lake area has increased by about 12% in the continuous permafrost zone and decreased by about the same percentage in the discontinuous permafrost zone [*Smith et al.*, 2005], suggesting that the link of climate change with permafrost degradation and lake drainage is mediated by regional factors. These temporal lags and differential responses in permafrost distribution to climate forcing are suggestive of nonlinear dynamics. Carbon dynamics are also likely to be nonlinear and transient, because collapse scars have higher rates of CH_4 emission but also higher rates of peat accumulation than do permafrost landforms [*Camill et al.*, 2001; *Turetsky et al.*, 2007].

2.5. Water Table and CO_2 Flux Dynamics

Long-term instrumental records of climate variables, water table height, and atmospheric carbon flux have been collected at a number of peatlands, and these high-resolution records can be tested for concordance of climate and ecosystem variables. Yu [2006b] analyzed frequency variation in such records for three continental fens and one maritime bog, using power density spectra. His analyses showed that water table behavior is highly dependent on history and past events, suggesting a nonlinear response to climate and dominance of rare (low frequency) events. Precipitation records were not included in the analysis, but temperature (and relative humidity for the maritime bog site) showed a spectrum very different from that of water table height. These results support the idea that the peatland is self-regulating with respect to water table height [Ivanov, 1981; Ingram, 1983] and that it responds to external forcing at unexpected frequencies. Yu's [2006b] analyses of CO_2 flux (for the three fen sites only) showed a more complicated spectrum, showing concordance with water table height at time scales greater than 1 month and less than 1 day but concordance with temperature at intermediate time scales. In its role as a control on CO_2 flux, water table height is likely to be an indirect indicator of soil moisture at subdaily time scales and of microform type at time scales greater than 1 day, whereas temperature may be a direct control as well as an indirect indicator of vascular leaf area at intermediate times scales [cf. Riutta et al., 2007; Laine et al., 2007]. In any case, the analyses suggest that peatland CO_2 flux behaves in a nonlinear way, with dominant controls switching across time scales of observation.

3. STABILIZING AND DESTABILIZING FORCES

The empirical examples presented in section 2 provide qualitative evidence suggestive of nonlinear behavior: peatland dynamics seem to be characterized by long periods of stasis punctuated by occasional, abrupt shifts to alternative regimes. Further investigation of time series data using appropriate quantitative techniques is required to verify the existence of regime shifts in peatlands and to explore how they are related to possible drivers [see Andersen et al., 2009].

What forces and mechanisms might underlie threshold dynamics and regime shifts in peatlands? As in other complex systems, negative feedback mechanisms that act to damp fluctuations in the system compete with positive feedback mechanisms that act to amplify initially small fluctuations [Holling and Gunderson, 2002; Rial et al., 2004]. Stabilizing forces dominate most of the time, but occasionally, the balance will shift in favor of destabilizing forces. The system then undergoes abrupt change until stabilizing forces once again dominate. External forcing may sometimes tip the balance, but perturbations are either damped or amplified primarily by feedback mechanisms internal to the system.

As in other complex systems [Werner, 1999; Holling and Gunderson, 2002; Rial et al., 2004], forces that act to damp or amplify fluctuations in peatlands are linked to processes interacting across spatial and temporal scales [Belyea and Baird, 2006]. A fundamental concept borrowed from physics is the "enslaving principle" [Haken, 2004]: when stabilizing forces dominate, the dynamics of components that have the potential to change rapidly ("slaves") are entrained by the dynamics of a few components that always change more slowly ("masters"). In a dune system, for example, the long-term movement of sand grains (slaves) is determined by migration of the sand dune (master), which itself is a dynamic pattern emerging from the collective behavior of many individual sand grains [Werner, 1999]. The number of degrees of freedom of the system (i.e., its entropy) is greatly reduced as the fast variables (slaves) almost instantaneously come into a slowly varying "quasi steady state" dictated by the slow variables (masters) [Rinaldi and Scheffer, 2000]. In some systems, destabilizing forces occasionally dominate when the fast components "revolt" [Holling and Gunderson, 2002] and temporarily escape the constraints imposed by the slow components. At certain values of the slow variable ("bifurcation points"), the system can undergo a qualitative change of behavior in which the fast variables leave the quasi steady state and begin to vary much more rapidly [Rinaldi and Scheffer, 2000]. Depending on the system, the resulting bifurcation can take one of several different forms, including catastrophic shift to a new state. The classic ecological example is the "driver-state hysteresis" type of regime shift between clear and turbid conditions in shallow lakes, which occurs as external nutrient loading slowly increases or decreases across two distinct thresholds [Scheffer et al., 1993].

In studies of ecosystems, theoretical investigation of threshold dynamics and regime shifts focuses on concepts of resilience and alternative stable states [Holling and Gunderson, 2002]. "Ecosystem resilience" refers to the capacity of an ecosystem to absorb perturbation through changes in function rather than fundamental structure: the more resilient the ecosystem, the larger the disturbance it can absorb without change in fundamental structure or loss of key processes [Holling, 1973]. When the resilience of an ecosystem is exceeded, the system undergoes rapid transformation to another (alternative) state. In the case of peatlands, these transformations may occur across biomes (e.g., switch from forest to peatland or from permafrost to nonpermafrost landform) or within the ecosystem (e.g., switch from fen to bog or from homogeneous vegetation to a two-phase mosaic of hummocks and hollows). *Holling and Gunderson* [2002] point out that resilience is not fixed in time or space but operates at a range of scales and may change as the ecosystem develops. For example, a switch from homogeneous vegetation to a two-phase mosaic of hummocks and hollows (transformation at within-ecosystem scale) may allow the peatland to persist under a changing climate (persistence at biome scale).

The most pressing challenges to understanding nonlinear dynamics in ecosystems are to identify, first, the slow (master) variables, which stabilize the system, and, second, the subset of fast (slave) variables, which have the potential to destabilize the system at critical points. In section 3.1, I examine some of the stabilizing and destabilizing forces that arise in peatland ecosystems through hydrological processes, organic matter dynamics, and surface energy exchanges and attempt to identify some of the key slow and fast variables involved. This analysis is purely qualitative, somewhat speculative, and certainly biased by my experience and understanding of northern peatlands.

3.1. Hydrological Processes

Depth-integrated rates of water flow through peat are highly dependent on fluctuations in the water table, and this dependence provides a potential mechanism for stabilization of peatland hydrology on daily and seasonal time scales. Saturated hydraulic conductivity (K_{sat}) and specific yield generally decrease nonlinearly with depth, because pore spaces become smaller as peat decays and compresses [*Boelter*, 1969; *Ingram*, 1983]. As the water table rises closer to the peatland surface, transmissivity (i.e., K_{sat} integrated through the entire thickness of saturated peat) increases and water is discharged more rapidly. At the same time, more water can be stored per unit rise in water table because the pore volume is greater near the surface and because the total peat volume can expand because of the elastic nature of near-surface peat [*Kellner and Halldin*, 2002]. Consequently, excess water from a rainfall event is either stored within the peat or rapidly discharged through near-surface peat, minimizing the occurrence of saturation-excess overland flow. Conversely, as the water table falls during a period without rain, transmissivity decreases and water is discharged more slowly. Total peat volume may also contract as water is lost [*Kellner and Halldin*, 2002]. The net effect is that the peatland is self-regulating with respect to its water table, which tends to be maintained most of the time within a narrow range of elevations [*Ingram*, 1983].

The influence of evaporation and transpiration on peatland dynamics is complicated, because the two processes have different controls and their rates depend on a multitude of factors, including the relative abundances of mosses and vascular plants, water table depth, atmospheric conditions, and local advection [*Kellner*, 2001]. Rates of evaporation from mosses are high when the water table is close to the peat surface [*Nichols and Brown*, 1980] but may (or may not) decline as water table depth increases [*Ingram*, 1983; *Lafleur et al.*, 2005]. During prolonged drought, the mosses may lose their pigments and turn white [*Gerdol et al.*, 1996], reducing net radiation by the increase in albedo. A surface crust may also form as the moss capitula dry out, "sealing" the surface and increasing resistance to evaporation. In contrast to moss evaporation, transpiration by vascular plants is unaffected by drought as long as the water table remains within the rooting zone [*Lafleur et al.*, 2005]. When the water table is close to the surface, however, transpiration may decline [*Lafleur et al.*, 2005], presumably because of plant stress caused by oxygen deficiency. Other than during extreme drought when the water table drops below the rooting zone causing decreases in both evaporation and transpiration, losses of water to the atmosphere may be largely decoupled from peatland hydrological conditions and therefore offer no direct mechanism for stabilization or destabilization of water table dynamics. Spatial differences in rate, however, may contribute to an indirect mechanism that involves local redistribution of water.

Spatial heterogeneity in hydraulic properties and water flux rates can promote local flow networks that redistribute water (and possibly carbon and nutrients) among microforms. Limited data suggest that transmissivity and storativity are higher in hollows than in hummocks because the water table in a hollow is situated in more porous, less decomposed peat [*Ivanov*, 1981; *Kellner and Halldin*, 2002]. After a rainfall event, the absolute elevation of the water table may be slightly higher in hummocks than in hollows, such that hummocks or ridges can act as local watersheds, with a tendency for water to flow down a hydraulic gradient to adjacent hollows or pools [*Price and Maloney*, 1994]. During wet periods, flooded pools may expand laterally onto adjacent microforms and coalesce, storing excess water and rapidly discharging it by overland flow [*Quinton and Roulet*, 1998]. During drier periods, the pools may shrink and become isolated from one another, so that no overland flow occurs and water is lost by evaporation over a smaller area of open water [*Quinton and Roulet*, 1998]. These mechanisms are stabilizing, because pool expansion and contraction modulate water losses in response to fluctuating inputs. In some situations, however, densely vegetated ridges may act as water pumps, with high rates of transpiration drawing water (and nutrients) from surrounding hollows [*Eppinga et al.*, 2008]. The flux of nutrients may promote further growth of vascular plants on the ridges, which, in turn, may promote higher rates of advection of water and nutrients from surrounding hollows [*Rietkerk et al.*, 2004; *Eppinga et al.*, 2008]. This mechanism is destabilizing in the long term, because it amplifies initially small differences in water loss. Whether the net redistribution of water is from ridge to pool (i.e., precipitation-driven flow, stabilizing) or from pool to ridge (i.e., transpiration-driven flow, destabilizing) may change seasonally with weather conditions [*Eppinga et al.*, 2008] and may also vary regionally with climate wetness (M. B. Eppinga et al., Resource contrast in patterned peatlands increases along an evapotranspiration gradient, manuscript in preparation, 2009).

Local flow networks provide a mechanism for shifting water losses between evapotranspiration and runoff, and the way in which these networks function may depend at least partially on regional climate, specifically the excess of precipitation, P, over evapotranspiration, E. In wetter climates (high $P - E$), the amount of water lost by runoff must be greater than in drier climates (low $P - E$). Since hydraulic gradient is controlled mainly by topography of the peatland surface, the increased discharge in wetter climates must be accommodated by an area-averaged increase in transmissivity. This increase could be accomplished through either a general decrease in thickness of the unsaturated zone (i.e., water table positioned within peat of high K_{sat}) or an expansion of high-transmissivity microforms, such as hollows and pools [*Ivanov*, 1981]. Evidence from stratigraphic analyses of large peat exposures provide support for the idea that peatlands respond to decadal changes in climate wetness through expansion and contraction of microforms of contrasting transmissivity [*Barber*, 1981]. In most cases, these responses are likely to damp fluctuations in climate wetness: when the climate is wetter, hollows expand and allow greater runoff losses; when the climate is drier, hummocks expand and reduce runoff losses. This stabilizing effect may be weakened or reversed in continental climates, because the expansion of hummocks would tend to increase losses by the mechanism of "transpiration-driven flow" mentioned above (Eppinga et al., manuscript in preparation, 2009).

The formation of surface water bodies on peatlands (ponds, pools, and lakes) may become destabilizing in the long term. Once pool initiation begins, it seems to proceed in one direction: individual pools expand laterally and coalesce to cover a greater proportion of the peatland, and the pool complex as a whole extends outward, with new pools initiating on the periphery of the complex [*Foster et al.*, 1988; *Belyea and Lancaster*, 2002]. Catastrophic drainage may occur once the pools have expanded to such an extent that the intervening ridges are structurally compromised [*Foster et al.*, 1988]. Pool initiation and expansion, therefore, may be stabilizing at the within-ecosystem scale (by regulating water losses) but destabilizing at the biome scale (by eventually leading to peatland degradation). Pool formation is also likely to have a strong effect of decreasing CO_2 sequestration and increasing CH_4 emission [*Hamilton et al.*, 1994].

3.2. Organic Matter Dynamics

At the scale of individual microforms (1–10 m), height of the peat surface relative to the upper boundary of the saturated zone (hereinafter referred to as "thickness of the unsaturated zone") is a key variable controlling plant and microbial community structure, peat formation, and carbon flux [*Sjörs*, 1990; *Waddington and Roulet*, 1996; *Alm et al.*, 1997; *Belyea and Clymo*, 2001; *Laine et al.*, 2007]. Mechanistically, this control is related to differences in production of litter by vegetation (i.e., species composition, especially the relative abundance of vascular plants) and differences in the mass of litter/peat exposed to aerobic decomposition (i.e., thickness of the "acrotelm") [*Belyea and Clymo*, 2001]. Although the upper boundary of the saturated zone is relatively homogeneous across a scale of meters, the peat surface is highly heterogeneous across this scale, with thickness of the unsaturated zone varying tremendously, from high and dry hummocks to low and wet hollows or open water pools. In defining the upper boundary of the saturated zone, daily and seasonal fluctuations in the water table (and the fact that some of the pore space in the "saturated" zone may be filled by biogenic gas bubbles rather than water [e.g., *Strack et al.*, 2005]) are ignored. Water storage change, therefore, refers to long-term rise or fall of the lowest water table over years or decades due, for example, to growth of the groundwater mound. As such, storage change is a slow variable, which, most of the time, controls the fast variables of peat formation and carbon flux (i.e., CH_4 emission and CO_2 exchange).

Changes in thickness of the unsaturated zone depend on the relative rates of two processes: formation of new litter/peat and water storage change in the saturated zone [*Belyea and Clymo*, 2001]. If new litter/peat is added at the peatland surface at the same rate as water storage increases, then thickness of the unsaturated zone remains in a steady state (Figure 1a). If litter/peat forms more quickly or if storage increases more slowly, then the microform increases in height, e.g., from lawn to hummock (Figure 1b). Conversely, if litter/peat forms more slowly or if storage increases more quickly, then the microform decreases in height (Figure 1c), e.g., from lawn to hollow. Surface wetness, as indicated by thickness of the unsaturated zone, is not simply a function of the climatic water balance, as is implicitly assumed in many climate reconstructions based on proxies such as plant macrofossils, testate amoebae, or peat humification [e.g., *Blackford*, 2000]. If storage change is held constant, then change in thickness of the unsaturated zone depends entirely

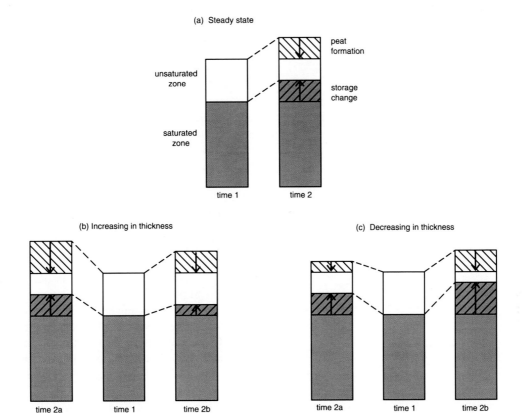

Figure 1. Changes in thickness of the unsaturated zone, resulting from the relative rates of new litter/peat formation and water storage change. (a) Steady state. Peat formation equals storage change. (b) Thickening of the unsaturated zone. Peat formation increases relative to storage change (time 2a), or storage change decreases relative to peat formation (time 2b). The alternative scenarios result in the same thickness of the unsaturated zone but different absolute elevations of the peat surface. (c) Thinning of the unsaturated zone. Peat formation decreases relative to storage change (time 2a), or storage change increases relative to peat formation (time 2b). The alternative scenarios result in the same thickness of the unsaturated zone but different absolute elevations of the peat surface.

on the rate of peat formation. Nonlinear responses arising from negative and positive feedback mechanisms must be considered.

The negative and positive feedbacks become evident if we impose small perturbations on an otherwise constant slow variable, storage change, and examine the dynamics of the system when the fast variable, peat formation, is allowed to vary in response to these small perturbations [*Rinaldi and Scheffer*, 2000]. Empirical studies show that there is a nonlinear relationship between rate of peat formation and thickness of the unsaturated zone, with peat forming more quickly in intermediate microforms (lawns and low hummocks) than in either wetter (hollows and pools) or drier (high hummocks) microforms [*Alm et al.*, 1997; *Belyea and Clymo*, 2001; *Laine et al.*, 2007]. Since the relationship between rate of peat formation and thickness of the unsaturated zone is humpbacked, there are potentially two points where peat

formation equals storage change, one on the rising limb and one on the falling limb of the "hump" [*Belyea and Clymo*, 2001]. The point on the rising limb is unstable, because small changes in thickness of the unsaturated zone (e.g., due to a run of dry or wet years) will be amplified by positive feedback. The point on the falling limb is stable, because small changes in thickness of the unsaturated zone will be damped by negative feedback. Hence, stabilizing forces operate on the falling limb of the hump, and destabilizing forces operate on the rising limb [*Belyea and Clymo*, 2001].

The dynamics of the system can be illustrated by a thought experiment (Figure 2). In a run of wet years, storage change will increase slightly, leading to a slight decrease in thickness of the unsaturated zone as the upper boundary of the saturated zone rises. If the system is on the rising limb, the rate of peat formation will decrease, and, in turn, thickness of the unsaturated zone will decrease even further, amplifying

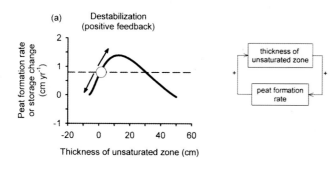

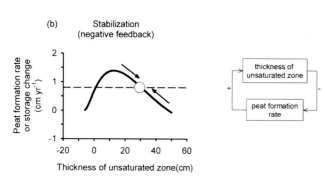

Figure 2. Stabilization and destabilization of organic matter dynamics. The solid curve is rate of litter/peat formation as a function of thickness of the unsaturated zone (TUZ), whereas the dashed line is storage change (constant). TUZ is in a steady state when peat formation equals storage change (intersection of curve and line). The equation for the curve is from *Belyea and Clymo* [2001]. (a) Destabilization (positive feedback) at the unstable point, where rate of peat formation increases with increasing TUZ. In a dry year, the unsaturated zone thickens (see Figure 1b). In a wet year, the unsaturated zone thins (see Figure 1c). (b) Stabilization (negative feedback) at the stable point, where rate of peat formation decreases with increasing TUZ. In both dry and wet years, TUZ is maintained close to a steady state (see Figure 1a).

the initial perturbation (Figure 2a). Conversely, if the system is on the falling limb, the rate of peat formation will increase, and, in turn, thickness of the unsaturated zone will increase, counteracting the initial perturbation (Figure 2b). Complementary responses would be observed with the opposite sort of perturbation. In a run of dry years, storage change will decrease slightly, leading to a slight increase in thickness of the unsaturated zone as the upper boundary of the saturated layer falls. If the system is on the rising limb, the rate of peat formation will increase, and, in turn, thickness of the unsaturated zone will increase even further, amplifying the initial perturbation (Figure 2a). Conversely, if the system is on the falling limb, the rate of peat formation will decrease, and, in turn, thickness of the unsaturated zone will decrease, counteracting the initial perturbation (Figure

2b). Hence, a high hummock (falling limb, stable point) is resistant to perturbations and will tend to persist, whereas a hollow (rising limb, unstable point) is sensitive to perturbations and may undergo rapid transition to either a hummock or a pool. At the unstable point, the trajectory of change may vary stochastically across space; the system may undergo a bifurcation or "divergent succession" [*Sjörs*, 1990], such that some hollows become hummocks, whereas others become pools, depending on small differences in initial conditions.

Microform dynamics are contingent on the shape of the curve relating peat formation and thickness of the unsaturated zone, as well as the point(s) at which it is intersected by storage change. The magnitude and position of the hump may depend on macrovariables, such as regional climate (e.g., maritime versus continental) and trophic state (e.g., fen versus bog) [*Sjörs*, 1990; *Laine et al.*, 2007; *Riutta et al.*, 2007]. At present, there are few data on which to make generalizations, but one might surmise that the hump will be smaller and shifted toward a thinner unsaturated zone in systems where aerobic decomposition proceeds more rapidly (e.g., maritime climates and fens) [*Sjörs*, 1990]. The positions of the stable and unstable points depend on where this curve is intersected by storage change, which is inextricably linked to growth of the peat deposit [*Belyea and Baird*, 2006]. One might surmise that as the peatland grows, leading to slower rates of peat accumulation and slower rates of storage change, the stable and unstable points will drop down the *y* axis. These macroscale differences in storage change and shape of the humpbacked curve have not yet been explored theoretically but may have profound implications for the dynamics of peatland systems.

An empirical example of how these stabilizing and destabilizing forces might operate is illustrated in Figure 3, which interprets microform dynamics at Store Mosse Mire, Sweden [see *Belyea and Malmer*, 2004]. This bog has undergone two abrupt shifts in vegetation since transition from fen about 5000 years ago. In the first vegetation shift (about 2500 years ago), a homogeneous cover of *Sphagnum fuscum* was replaced by a homogeneous cover of codominating *S. rubellum* and *S. fuscum*. In the most recent vegetation shift (about 1000 years ago), the surface diverged into a two-phase mosaic of hummocks and hollows, with *S. magellanicum* dominant. The three stages were characterized by long periods of little change in vegetation but a gradual decrease in rates of vertical growth and carbon sequestration ("within stage" persistence). The two major shifts in vegetation were associated with abrupt increases in rates of vertical growth and carbon sequestration ("between-stage" transformation). Within stages, stabilizing forces (i.e., negative feedbacks) dominated, even though the rate of storage increase gradu-

(a) Within stage (persistence)

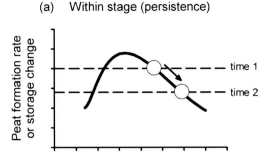

(b) Between stages (transformation)

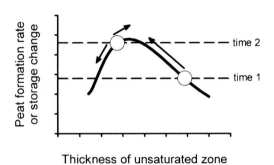

Figure 3. Interpretation of microform dynamics at Store Mosse Mire, Sweden. See *Belyea and Malmer* [2004] for details. The solid curve is rate of litter/peat formation as a function of thickness of the unsaturated zone (TUZ), whereas the dashed line is storage change. (a) Persistence within vegetation stages. Storage change gradually decreases from time 1 to time 2, owing to growth of the groundwater mound. The unsaturated zone gradually thickens, and the system is resilient to small changes in climate wetness. (b) Transformation from homogeneous vegetation to a two-phase mosaic of hummocks and hollows. At the initial steady state (time 1), increased climate wetness causes storage change to increase, and so the unsaturated zone thins. As a result, the steady state is pushed over the hump from a stable point on the falling limb to an unstable point on the rising limb of the curve. At the new steady state (time 2), random spatial variation causes the system to diverge into hummocks and hollows.

ally slowed with growth of the groundwater mound (Figure 3a). The unsaturated zone gradually thickened, and as a result the rates of vertical growth and carbon sequestration gradually decreased. At the transition between the two most recent vegetation stages (about 1000 years ago), dominance temporarily shifted from stabilizing to destabilizing forces (i.e., from negative to positive feedbacks) (Figure 3b). The system was initially at a stable point (point 1). Storage

change increased in response to climate forcing, and as a consequence the unsaturated zone thinned to such an extent that the system moved to an unstable point (point 2). Owing to random spatial variations, some areas of the unsaturated zone thickened to form hummocks, whereas other areas thinned to form hollows. An initially large increase in rates of vertical growth and carbon sequestration (point 2) rapidly dissipated as the surface diverged into hummocks and hollows (points not shown). An important lesson learned from this example is that resilience to climate change decreased within stages, with the result that the regime shift from homogeneous vegetation to a two-phase mosaic of hummocks and hollows was triggered late in development by relatively weak climate forcing.

3.3. Energy Exchanges

Land surface characteristics such as albedo, leaf area, and canopy complexity determine how much incoming solar radiation is absorbed [*Thompson et al.*, 2004] and show substantial spatial variation both within peatlands and between peatlands and surrounding lands. Vegetation types with less reflective leaves, larger area of absorbing surface, and greater vertical complexity (e.g., coniferous forest) receive higher net radiation compared with more reflective and simpler vegetation types (e.g., tundra and peatland) [*Thompson et al.*, 2004; *Arneth et al.*, 2006; *Vygodskaya et al.*, 2007]. These differences in vegetation structure are particularly important during the snow season, when vegetation types dominated by shrubs and ground-layer cryptogams have higher snow cover, and much higher albedo, than coniferous forests [*Arneth et al.*, 2006; *Bonan*, 2008]. As a result, boreal forest expansion has a net warming effect, whereas peatland expansion by paludification has a net cooling effect in terms of radiation balance [*Vygodskaya et al.*, 2007; *Bonan*, 2008].

Land surface characteristics also affect the partitioning of net radiation among sensible, latent, and ground heat fluxes. Stomatal control has only a weak influence on surface exchange processes on peatlands owing to the dominance of nonvascular plants and the high coverage of surface water bodies and saturated peat surface [*Arneth et al.*, 2006]. During the snow-free period, energy exchanges on peatlands [*Kurbatova et al.*, 2002] and small water bodies [*Nagarajan et al.*, 2004] are dominated by latent heat flux, whereas those in forests are dominated by sensible heat flux and turbulent transfer [*Baldocchi et al.*, 2000; *Thompson et al.*, 2004]. As a result, most energy input to peatlands is used to evaporate water, and so summer air temperatures are cooler than those over surrounding forests [*Williams*, 1968; *Baldocchi et al.*, 2000; *Kurbatova et al.*, 2002]. During winter, sensible heat fluxes from lakes can be up to an order of magnitude larger than from surrounding land surfaces owing to the high thermal conductivity of the lake ice [*Jeffries et al.*, 1999]; frozen, saturated peat has a similarly high thermal conductivity [*Williams*, 1968]. Consequently, the winter heat loss from land surfaces with extensive lake coverage may be comparable to that from oceans and may have a similar effect of increasing winter air temperatures [*Jeffries et al.*, 1999]. Viewed on an annual cycle, transfers of latent heat in summer and sensible heat in winter would have a stabilizing effect on peatlands by damping seasonal fluctuations in air temperature.

During spring snowmelt, peatland albedo decreases drastically from about 0.8 to about 0.2, with concomitant increases in net radiation, whereas forest albedo and net radiation undergo much smaller changes [*Arneth et al.*, 2006]. Snowmelt occurs later in peatlands than in forests and seems to be sustained by negative sensible heat fluxes, i.e., energy drawn from the overlying air [*Arneth et al.*, 2006]. In order to understand whether these differences in energy partitioning will tend to stabilize or destabilize conditions on peatlands, it is necessary to consider horizontal fluxes across land surface types.

Energy exchanges with the atmosphere differ markedly among land surface types, and the contrasts within peatlands (and between peatlands and surrounding lands) may affect atmospheric circulation at local or regional scales. Topography may also be important, especially in relation to katabatic flow of cold air into low-lying peatlands during clear and dry periods ("frost hollow effect" [*Williams*, 1968]). When surface heterogeneity in land surface types is fine-grained (length scale less than 5 km), effects on atmospheric circulation are limited to the lower part of the atmospheric boundary layer and extend over short ranges [*Pielke et al.*, 1998]. On peatlands with a two-phase mosaic, local advection of warm air from hummocks and ridges can enhance evaporation rates from small hollows and pools ("oasis effect" [*Price and Maloney*, 1994]). Combined with "precipitation-driven flow" of water from ridges to hollows, this mechanism will tend to stabilize hydrological conditions during wet periods, because evaporation from hollows and pools will be enhanced. Conversely, if these features retain surface water during dry periods, the oasis effect might combine with "transpiration-driven flow" to amplify evaporation and transpiration losses from both pools and ridges, destabilizing hydrological conditions. Where peatlands extend over large areas as in continental Canada and Siberia, effects may propagate to upper parts of the atmospheric boundary layer and over longer ranges. Modeling studies suggest that peatlands increase summer precipitation through regional recycling of evaporated water [*Krinner*, 2003]. This mechanism will tend to stabilize hydrological conditions during dry periods, as recycled precipitation will at least partially compensate for

water lost by increased summer evapotranspiration. Other studies suggest that peatlands can have transient effects on atmospheric circulation by amplifying small disturbances that occur for other reasons. In simulations for central Siberia, warm air advection enhanced snowmelt during a 2-week period in spring, but the system rapidly switched to cold air advection as meltwater evaporated [*Gutowski et al.*, 2007]. Eventually, the perturbation propagated outside the peatland-dominated area, and evaporation rates were amplified in nonpeatland areas "downwind" [*Gutowski et al.*, 2007]. These transient effects are destabilizing in the sense that they promote rapid switch between snow-covered and snow-free conditions. The weather/climate effects of spatial variations in peatland energy exchange are just starting to be explored, and studies that integrate local and regional feedbacks [cf. *Scheffer et al.*, 2005; *Janssen et al.*, 2008] should be encouraged.

4. POTENTIAL FEEDBACKS ON THE CLIMATE SYSTEM

The stabilizing and destabilizing forces discussed above (and probably others) control the nonlinear dynamics of peatlands, i.e., whether they maintain steady biogeochemical and biogeophysical function in the face of environmental forcing or whether they undergo rapid transformation to new states. The nature of this ecological response, persistence or transformation, will largely determine whether peatlands have a net warming or cooling effect on future global climate. Much research effort has been directed at understanding and quantifying peatland carbon cycles, but predictions on whether peatland response to climate change will amplify or damp recent increases in atmospheric CO_2 and CH_4 concentration remain uncertain [*Moore et al.*, 1998]. The effects of peatlands on the surface energy balance and patterns of atmospheric circulation have received much less attention, but they have the potential for significant impacts at regional scales and on time scales of decades [*Schuur et al.*, 2008; *Bonan*, 2008]. The magnitude and direction of these biogeochemical and biogeophysical feedbacks are likely to depend on the resilience of peatlands to climatic perturbation and their capacity to adapt to new climatic boundary conditions.

The Intergovernmental Panel on Climate Change (IPCC) predicts that increases in evapotranspiration will lead to drier summers in continental regions [*Meehl et al.*, 2007]. For peatlands located in areas where there is still an excess of precipitation over evapotranspiration, the stabilizing forces outlined above are likely to lead to deepening of hollows or pools, thickening of the unsaturated zone in lawns and hummocks, expansion of microforms with low transmissivity, and increased abundance of vascular plants, especially shrubs and coniferous trees. The drier climate may also trigger widespread ombrotrophication, with similar effects on peatland surface structure. Owing to deepening of the water column in pools and thickening of the unsaturated zone elsewhere, more CH_4 may be oxidized before it reaches the atmosphere, and so there may be a negative feedback on atmospheric CH_4 concentration. Conversely, increased vascular plant cover may lead to more CH_4 production (because of the use of plant exudates by methanogens) and greater bypassing of CH_4 oxidation (by plant-mediated CH_4 flux), and so there may be a positive feedback on atmospheric CH_4 concentration. The structural changes may have a positive feedback on atmospheric CO_2 by accelerating aerobic microbial decomposition, although afforestation may at least partially offset this effect. Changes in albedo and energy partitioning associated with changes in land surface characteristics may reduce regional recycling of precipitation and therefore accelerate drying of peatlands.

The IPCC also predicts that lands at high latitude will warm by up to 7°C to 8°C by 2100 [*Meehl et al.*, 2007]. The potential effects on permafrost carbon have been reviewed thoroughly by *Schuur et al.* [2008]. Decreases in albedo resulting from expansion of shrub cover on tundra and northward displacement of the tree line will promote local summer warming, a positive feedback on climate change [*Schuur et al.*, 2008]. Increased fire frequency may have a negative feedback if albedo changes associated with postfire succession (i.e., replacement of evergreen trees by deciduous trees and increased snow cover in winter) have a cooling effect strong enough to offset the warming effect of peat combustion and release of CO_2 and CH_4 [*Schuur et al.*, 2008]. Some authors argue that transfer of permafrost carbon to aerobic environments (i.e., by thickening of the active layer or by talik formation) is likely to have a much stronger climate forcing than will transfer to anaerobic environments (i.e., by formation of thermokarst lakes) [*Schuur et al.*, 2008]. Others estimate that northern lakes currently account for about 6% of global CH_4 emissions and that these emissions will accelerate with permafrost thaw [*Walter et al.*, 2007]. Widespread formation of thermokarst lakes, as is occurring in the continuous permafrost zone of Siberia [*Smith et al.*, 2005], may also alter fluxes of latent and sensible heat and hence affect regional patterns of atmospheric circulation [*Schuur et al.*, 2008].

5. RESEARCH NEEDS

One of the most pressing questions we should ask when trying to understand nonlinear dynamics in peatlands is how hydrological processes, organic matter dynamics, and energy exchanges interact to promote persistence or transforma-

tion. Although we have some understanding of a few specific mechanisms, we do not yet understand how feedbacks operating at different spatial and temporal scales might interact. Most importantly, we have not yet identified all of the slow variables that promote persistence or the key fast variables that may trigger regime shift once a threshold is crossed.

Improving our understanding of nonlinear behavior in peatlands could be approached in three complementary ways. First, long records of instrumental data and paleoecological records could be analyzed quantitatively to detect past regime shifts and to identify candidate driving variables [*Andersen et al.*, 2009]. For example, eddy covariance towers at a few peatland sites have been recording hourly CO_2 and CH_4 flux for close to 10 years (FLUXNET Synthesis Data Set, available at http://www.fluxdata.org, accessed 23 January 2009), and these records could be analyzed in relation to meteorological variables. Second, postulated feedback mechanisms (at least at smaller spatial scales) could be tested experimentally. For example, the role of precipitation- and transpiration-driven flow in stabilizing or destabilizing hydrological conditions could be explored in plot-scale rainfall-addition or rainfall-exclusion experiments. Third, interactions between feedback mechanisms operating at different scales could be explored in simulation models to ascertain whether microscale feedbacks can amplify landscape-scale feedbacks [e.g., *Janssen et al.*, 2008]. For example, the interactions between local redistribution of water between ridges and pools, regional recycling of evaporated water, and transfer of sensible heat from surrounding forests to peatlands could be simulated to determine their combined effect on peatland hydrological conditions and carbon gas fluxes. In each of these three approaches, the focus needs to be on identifying the key variables and interactions which control nonlinear behavior.

In order to answer the question of whether peatlands will have a net cooling or warming effect on future global climate, it will be necessary to represent their nonlinear dynamics in Earth system models that couple processes operating in the atmosphere, biosphere, hydrosphere, and geosphere. The unique characteristics of peatlands present real practical challenges to development of these coupled models, and some of these issues are considered in two other chapters of this monograph [*Frolking et al.*, this volume; *Baird et al.*, this volume]. Reducing the complexity of peatlands by identifying key variables and interactions that control nonlinear behavior is essential to predicting feedbacks on the climate system.

Acknowledgment. Two referees provided comments that helped to clarify the manuscript.

REFERENCES

Aaviksoo, K., M. Ilomets, and M. Zobel (1993), Dynamics of mire communities: A Markovian approach (Estonia), in *Wetlands and Shallow Continental Water Bodies*, vol. 2, *Case Studies*, edited by B. C. Patten, S. E. Jørgensen, and H. Dumont, pp. 23–43, SPB Acad. Publ., The Hague, Netherlands.

Alm, J., A. Talanov, S. Saarnio, J. Silvola, E. Ikkonen, H. Aaltonen, H. Nykänen, and P. J. Martikainen (1997), Reconstruction of the carbon balance for microsites in a boreal oligotrophic pine fen, Finland, *Oecologia*, *110*, 423–431.

Andersen, T., J. Carstensen, E. Hernández-García, and C. M. Duarte (2009), Ecological thresholds and regime shifts: Approaches to identification, *Trends Ecol. Evol.*, *24*(1), 49–57.

Anderson, R. L., D. R. Foster, and G. Motzin (2003), Integrating lateral expansion into models of peatland development in temperate New England, *J. Ecol.*, *91*(1), 68–76.

Arneth, A., J. Lloyd, O. Shibistova, A. Sogachev, and O. Kolle (2006), Spring in the boreal environment: Observations on pre- and post-melt energy and CO_2 fluxes in two central Siberian ecosystems, *Boreal Environ. Res.*, *11*, 311–328.

Baird, A. J., L. R. Belyea, and P. J. Morris (2009), Upscaling of peatland-atmosphere fluxes of methane: Small-scale heterogeneity in process rates and the pitfalls of "bucket-and-slab" models, *Geophys. Monogr. Ser.*, doi:10.1029/2008GM000826, this volume.

Baldocchi, D., F. M. Kelliher, T. A. Black, and P. Jarvis (2000), Climate and vegetation controls on boreal zone energy exchange, *Global Change Biol.*, *6*, 69–83.

Barber, K. E. (1981), *Peat Stratigraphy and Climatic Change: A Palaeoecological Test of the Theory of Cyclic Bog Regeneration*, Balkema, Rotterdam, Netherlands.

Belyea, L. R., and A. J. Baird (2006), Beyond "the limits to peat bog growth": Cross-scale feedback in peatland development, *Ecol. Monogr.*, *76*(3), 299–322.

Belyea, L. R., and R. S. Clymo (2001), Feedback control of the rate of peat formation, *Proc. R. Soc. London, Ser. B*, *268*, 1315–1321.

Belyea, L. R., and J. Lancaster (2002), Inferring landscape dynamics of bog pools from scaling relationships and spatial patterns, *J. Ecol.*, *90*(2), 223–234.

Belyea, L. R., and N. Malmer (2004), Carbon sequestration in peatland: Patterns and mechanisms of response to climate change, *Global Change Biol.*, *10*, 1043–1052.

Blackford, J. (2000), Palaeoclimatic records from peat bogs, *Trends Ecol. Evol.*, *15*(5), 193–198.

Boelter, D. H. (1969), Physical properties of peats as related to degree of decomposition, *Soil Sci. Soc. Am. J.*, *33*, 606–609.

Bonan, G. B. (2008), Forests and climate change: Forcings, feedbacks, and the climate benefits of forests, *Science*, *320*, 1444–1449.

Bubier, J., T. Moore, K. Savage, and P. Crill (2005), A comparison of methane flux in a boreal landscape between a dry and a wet year, *Global Biogeochem. Cycles*, *19*, GB1023, doi:10.1029/2004GB002351.

Bunting, M. J., and B. G. Warner (1998), Hydroseral development in southern Ontario: Patterns and controls, *J. Biogeogr.*, *25*, 3–18.

Camill, P. (2005), Permafrost thaw accelerates in boreal peatlands during late-20th century climate warming, *Clim. Change*, *68*, 135–152.

Camill, P., and J. S. Clark (2000), Long-term perspectives on lagged ecosystem responses to climate change: Permafrost in boreal peatlands and the grassland/woodland boundary, *Ecosystems*, *3*, 534–544.

Camill, P., J. A. Lynch, J. S. Clark, J. B. Adams, and B. Jordan (2001), Changes in biomass, aboveground net primary production, and peat accumulation following permafrost thaw in the boreal peatlands of Manitoba, Canada, *Ecosystems*, *4*, 461–478.

Eppinga, M. B., M. Rietkerk, W. Borren, E. D. Lapshina, W. Bleuten, and M. J. Wassen (2008), Regular surface patterning of peatlands: Confronting theory with field data, *Ecosystems*, *11*, 520–536.

Foster, D. R., H. E. Wright Jr., M. Thelaus, and G. A. King (1988), Bog development and landform dynamics in central Sweden and south-eastern Labrador, Canada, *J. Ecol.*, *76*, 1164–1185.

Frolking, S., N. Roulet, and J. Fuglestvedt (2006), How northern peatlands influence the Earth's radiative budget: Sustained methane emission versus sustained carbon sequestration, *J. Geophys. Res.*, *111*, G01008, doi:10.1029/2005JG000091.

Frolking, S., N. Roulet, and D. Lawrence (2009), Issues related to incorporating northern peatlands into global climate models, *Geophys. Monogr. Ser.*, doi:10.1029/2008GM000809, this volume.

Gajewski, K., A. Viau, M. Sawada, D. Atkinson, and S. Wilson (2001), *Sphagnum* peatland distribution in North America and Eurasia during the past 21,000 years, *Global Biogeochem. Cycles*, *15*(2), 297–310.

Gerdol, R., A. Bonora, R. Gualandri, and S. Pancaldi (1996), CO_2 exchange, photosynthetic pigment composition, and cell ultrastructure of *Sphagnum* mosses during dehydration and subsequent rehydration, *Can. J. Bot.*, *74*, 726–734.

Gorham, E., C. Lehman, A. Dyke, J. Janssens, and L. Dyke (2007), Temporal and spatial aspects of peatland initiation following deglaciation in North America, *Quat. Sci. Rev.*, *26*, 300–311.

Gutowski, W. J., Jr., H. Wei, C. J. Vörösmarty, and B. M. Fekete (2007), Influence of Arctic wetlands on Arctic atmospheric circulation, *J. Clim.*, *20*, 4243–4254.

Haken, H. (2004), *Synergetics: Introduction and Advanced Topics*, 3rd ed., Springer, Berlin.

Hamilton, J. D., C. A. Kelly, J. W. M. Rudd, R. H. Hesslein, and N. T. Roulet (1994), Flux to the atmosphere of CH_4 and CO_2 from wetland ponds on the Hudson Bay lowlands (HBLs), *J. Geophys. Res.*, *99*(D1), 1495–1510.

Holling, C. S. (1973), Resilience and stability of ecological systems, *Annu. Rev. Ecol. Syst.*, *4*, 1–23.

Holling, C. S., and L. H. Gunderson (2002), Resilience and adaptive cycles, in *Panarchy: Understanding Transformations in Human and Natural Systems*, edited by L. H. Gunderson and C. S. Holling, pp. 25–62, Island Press, Washington, D. C.

Hughes, P. D. M. (2000), A reappraisal of the mechanisms leading to ombrotrophy in British raised mires, *Ecol. Lett.*, *3*, 7–9.

Hulme, P. D. (1994), A palaeobotanical study of paludifying pine forest on the island of Hailuoto, northern Finland, *New Phytol.*, *126*, 153–162.

Ingram, H. A. P. (1983), Hydrology, in *Mires—Swamp, Bog, Fen and Moor, Ecosyst. World*, vol. 4, edited by A. J. P. Gore, pp. 67–158, Elsevier, Oxford, U. K.

Ivanov, K. E. (1981), *Water Movement in Mirelands*, 276 pp., Academic, London.

Janssen, R. H. H., M. B. J. Meinders, E. H. van Nes, and M. Scheffer (2008), Microscale vegetation-soil feedback boosts hysteresis in a regional vegetation-climate system, *Global Change Biol.*, *14*, 1104–1112.

Jeffries, M. O., T. Zhang, K. Frey, and N. Kozlenko (1999), Estimating late-winter heat flow to the atmosphere from the lake-dominated Alaskan North Slope, *J. Glaciol.*, *45*, 315–324.

Kellner, E. (2001), Surface energy fluxes and control of evapotranspiration from a Swedish *Sphagnum* mire, *Agric. For. Meteorol.*, *110*, 101–123.

Kellner, E., and S. Halldin (2002), Water budget and surface-layer water storage in a *Sphagnum* bog in central Sweden, *Hydrol. Processes*, *16*, 87–103.

Korhola, A., J. Alm, K. Tolonen, J. Turunen, and H. Jungner (1996), Three-dimensional reconstruction of carbon accumulation and CH_4 emission during nine millennia in a raised mire, *J. Quat. Sci.*, *11*(2), 161–165.

Krinner, G. (2003), Impact of lakes and wetlands on boreal climate, *J. Geophys. Res.*, *108*(D16), 4520, doi:10.1029/2002JD002597.

Kuhry, P., B. J. Nicholson, L. D. Gignac, D. H. Vitt, and S. E. Bayley (1993), Development of *Sphagnum*-dominated peatlands in boreal continental Canada, *Can. J. Bot.*, *71*, 10–22.

Kurbatova, J., A. Arneth, N. N. Vygodskaya, O. Kolle, A. V. Varlargin, I. M. Milyukova, N. M. Tchebakova, E.-D. Schulze, and J. Lloyd (2002), Comparative ecosystem-atmosphere exchange of energy and mass in a European Russian and a central Siberian bog I. Interseasonal and interannual variability of energy and latent heat fluxes during the snowfree period, *Tellus, Ser. B*, *54*, 497–513.

Lafleur, P. M., R. A. Hember, S. W. Admiral, and N. T. Roulet (2005), Annual and seasonal variability in evapotranspiration and water table at a shrub-covered bog in southern Ontario, Canada, *Hydrol. Processes*, *19*, 3533–3550.

Laine, A., K. A. Byrne, G. Kiely, and E.-S. Tuittila (2007), Patterns in vegetation and CO_2 dynamics along a water level gradient in a lowland blanket bog, *Ecosystems*, *10*, 890–905.

Meehl, G. A., et al. (2007), Global climate projections, in *Climate Change 2007: The Physical Science Basis: Contribution of Working Group I to the Fourth Assessment Report of the Intergovernmental Panel on Climate Change*, edited by S. Solomon et al., pp. 747–846, Cambridge Univ. Press, Cambridge, U. K.

Moore, T. R., N. T. Roulet, and J. M. Waddington (1998), Uncertainty in predicting the effect of climatic change on the carbon cycling of Canadian peatlands, *Clim. Change*, *40*, 229–245.

Nagarajan, B., M. K. Yau, and P. H. Schuepp (2004), The effects of small water bodies on the atmospheric heat and water budgets over the Mackenzie River Basin, *Hydrol. Processes*, *18*, 913–938.

Nichols, D. S., and J. M. Brown (1980), Evaporation from a *Sphagnum* moss surface, *J. Hydrol.*, *48*, 289–302.

Pielke, R. A., Sr., R. Avissar, M. Raupach, A. J. Dolman, X. Zeng, and A. S. Denning (1998), Interactions between the atmosphere and terrestrial ecosystems: Influence on weather and climate, *Global Change Biol.*, *4*, 461–475.

Price, J. S., and D. A. Maloney (1994), Hydrology of a patterned bog-fen complex in southeastern Labrador, Canada, *Nord. Hydrol.*, *25*(5), 313–330.

Quinton, W. L., and N. T. Roulet (1998), Spring and summer runoff hydrology of a subarctic patterned wetland, *Arct. Alp. Res.*, *30*(3), 285–294.

Rial, J. A., et al. (2004), Nonlinearities, feedbacks and critical thresholds within the Earth's climate system, *Clim. Change*, *65*, 11–38.

Rietkerk, M., S. C. Dekker, M. J. Wassen, A. W. M. Verkroost, and M. F. P. Bierkens (2004), A putative mechanism for bog patterning, *Am. Nat.*, *163*(5), 699–708.

Rinaldi, S., and M. Scheffer (2000), Geometric analysis of ecological models with slow and fast processes, *Ecosystems*, *3*, 507–521.

Riutta, T., J. Laine, M. Aurela, J. Rinne, T. Vesala, T. Laurila, S. Haapanala, M. Pihlatie, and E.-S. Tuittila (2007), Spatial variation in plant community functions regulates carbon gas dynamics in a boreal fen ecosystem, *Tellus, Ser. B*, *59*, 838–852.

Rydin, H., and J. Jeglum (2006), *The Biology of Peatlands*, 343 pp., Oxford Univ. Press, Oxford, U. K.

Scheffer, M., S. H. Hosper, M.-L. Meijer, B. Moss, and E. Jeppesen (1993), Alternative equilibria in shallow lakes, *Trends Ecol. Evol.*, *8*(8), 275–279.

Scheffer, M., M. Holmgren, V. Brovkin, and M. Claussen (2005), Synergy between small- and large-scale feedbacks of vegetation on the water cycle, *Global Change Biol.*, *11*, 1003–1012.

Schuur, E. A. G., et al. (2008), Vulnerability of permafrost carbon to climate change: Implications for the global carbon cycle, *BioScience*, *58*, 701–714.

Sjörs, H. (1990), Divergent successions in mires, a comparative study, *Aquilo Ser. Bot.*, *28*, 67–77.

Smith, L. C., G. M. MacDonald, A. A. Velichko, D. W. Beilman, O. K. Borisova, K. E. Frey, K. V. Kremenetski, and Y. Sheng (2004), Siberian peatlands a net carbon sink and global methane source since the early Holocene, *Science*, *303*, 353–356.

Smith, L. C., Y. Sheng, G. M. MacDonald, and L. D. Hinzman (2005), Disappearing Arctic lakes, *Science*, *308*, 1429.

Strack, M., E. Kellner, and J. M. Waddington (2005), Dynamics of biogenic gas bubbles in peat and their effects on peatland biogeochemistry, *Global Biogeochem. Cycles*, *19*, GB1003, doi:10.1029/2004GB002330.

Thompson, C., J. Beringer, F. S. Chapin III, and A. D. McGuire (2004), Structural complexity and land-surface energy exchange along a gradient from arctic tundra to boreal forest, *J. Veg. Sci.*, *15*, 397–406.

Turetsky, M. R., R. K. Wieder, D. H. Vitt, R. J. Evans, and K. D. Scott (2007), The disappearance of relict permafrost in boreal North America: Effects on peatland carbon storage and fluxes, *Global Change Biol.*, *13*, 1922–1934.

van Breemen, N. (1995), How *Sphagnum* bogs down other plants, *Trends Ecol. Evol.*, *10*(7), 270–275.

Vygodskaya, N. N., P. Y. Groisman, N. M. Tchebakova, J. A. Kurbatova, O. Panfyorov, E. I. Parfenova, and A. F. Sogachev (2007), Ecosystems and climate interactions in the boreal zone of northern Eurasia, *Environ. Res. Lett.*, *2*, 045033, doi:10.1088/1748-9326/2/4/045033.

Waddington, J. M., and N. T. Roulet (1996), Atmosphere-wetland carbon exchanges: Scale dependency of CO_2 and CH_4 exchange on the developmental topography of a peatland, *Global Biogeochem. Cycles*, *10*(2), 233–245.

Walker, D. (1970), Direction and rate in some British postglacial hydroseres, in *Studies in the Vegetational History of the British Isles*, edited by D. Walker and R. G. West, pp. 117–139, Cambridge Univ. Press, Cambridge, U. K.

Walter, K. M., L. C. Smith, and F. S. Chapin III (2007), Methane bubbling from northern lakes: Present and future contributions to the global methane budget, *Philos. Trans. R. Soc., Ser. A, 365*, 1657–1676.

Werner, B. T. (1999), Complexity in natural landform patterns, *Science, 284*, 102–104.

Williams, G. P. (1968), The thermal regime of a *Sphagnum* peat bog, in *Proceedings of the Third International Peat Congress, Quebec*, pp. 195–200, Int. Peat Soc., Jyväskylä, Finland.

Yu, Z. (2006a), Holocene carbon accumulation of fen peatlands in boreal western Canada: A complex ecosystem response to climate variation and disturbance, *Ecosystems, 9*, 1278–1288.

Yu, Z. (2006b), Power laws governing hydrology and carbon dynamics in northern peatlands, *Global Planet. Change, 53*, 169–175.

Zoltai, S. C. (1993), Cyclic development of permafrost in the peatlands of northwestern Alberta, Canada, *Arct. Alp. Res., 25*(3), 240–246.

L. R. Belyea, Department of Geography, Queen Mary University of London, Mile End Road, London E1 4NS, UK (l.belyea@qmul.ac.uk)

Issues Related to Incorporating Northern Peatlands
Into Global Climate Models

Steve Frolking

Institute for the Study of Earth, Oceans, and Space, University of New Hampshire, Durham, New Hampshire, USA

Nigel Roulet

Department of Geography and McGill School of the Environment, McGill University, Montreal, Quebec, Canada

David Lawrence

Climate and Global Dynamics Division, National Center for Atmospheric Research, Boulder, Colorado, USA

Northern peatlands cover ~3–4 million km^2 (~10% of the land north of 45°N) and contain ~200–400 Pg carbon (~10–20% of total global soil carbon), almost entirely as peat (organic soil). Recent developments in global climate models have included incorporation of the terrestrial carbon cycle and representation of several terrestrial ecosystem types and processes in their land surface modules. Peatlands share many general properties with upland, mineral-soil ecosystems, and general ecosystem carbon, water, and energy cycle functions (productivity, decomposition, water infiltration, evapotranspiration, runoff, latent, sensible, and ground heat fluxes). However, northern peatlands also have several unique characteristics that will require some rethinking or revising of land surface algorithms in global climate models. Here we review some of these characteristics, deep organic soils, a significant fraction of bryophyte vegetation, shallow water tables, spatial heterogeneity, anaerobic biogeochemistry, and disturbance regimes, in the context of incorporating them into global climate models. With the incorporation of peatlands, global climate models will be able to simulate the fate of northern peatland carbon under climate change, and estimate the magnitude and strength of any climate system feedbacks associated with the dynamics of this large carbon pool.

1. INTRODUCTION

A substantial amount of carbon has accumulated as peat (partially decomposed organic matter) in northern peatlands or mires through the Holocene [*Gorham*, 1991]. This carbon is situated on what we can think of as two thermodynamic state boundaries that are strongly controlled by the both the climate system and the peatlands themselves. Both of these

Carbon Cycling in Northern Peatlands
Geophysical Monograph Series 184
Copyright 2009 by the American Geophysical Union.
10.1029/2008GM000809

state boundaries have a very strong influence on the fate of peatland carbon; will it remain as peat or be transformed into dissolved or particulate organic matter or into gaseous CO_2 or CH_4, and if transformed, how rapidly will this occur?

For carbon in many northern peatlands, one of these state boundaries is the solid/liquid phase boundary of water at 0°C. A significant fraction of northern peatlands are under-lain or embedded in permafrost (perennially frozen ground which lies below a surface active layer that seasonally thaws and is generally less than 1 m thick). In Canada, more than one third of peatlands have permafrost [*Tarnocai*, 2006]. *Smith et al.* [2007] estimated that about one third of northern peatlands are in zones of continuous permafrost, with an-other 40% of northern peatlands in discontinuous, sporadic, and isolated permafrost zones. Organic carbon in permafrost is relatively inert both physically and biogeochemically while frozen, although laboratory incubations have shown that microbial metabolism and methane production can occur, albeit at very low rates, at temperatures well below 0°C [*Brouchkov and Fukuda*, 2002; *Rivkina et al.*, 2004]. However, any gas produced by this slow metabolic activity will remain within the permafrost because diffusive gas loss from permafrost is negligible [*Rivkina et al.*, 2004, 2007]. A number of studies have established that the old organic mat-ter frozen into permafrost readily decomposes if thawed and that microbial populations that can decompose the organic matter are present and viable in the permafrost [*Rivkina et al.*, 1998, 2004, 2007; *Zimov et al.*, 2006].

Warming in recent decades has been stronger at high northern latitudes than in the rest of the world [*Serreze and Francis*, 2006], a trend that is projected to continue [*Meehl et al.*, 2007], and this will affect permafrost. *Zhang et al.* [2006] used a soil physics model to estimate that the area underlain by permafrost in Canada decreased by ~5% from 1850 to 1990. *Yi et al.* [2006, 2007] used the land module of a general circulation model (GCM) to simulate permafrost dynamics under warming for discontinuous and continuous permafrost sites and a range of soil properties. Their results were sensitive to surface cover and soil properties, with sur-face peat substantially slowing the rate of thaw. *Lawrence et al.* [2008] also included organic soils in their land sur-face model; they found that organic soils slowed the rate of permafrost thawing but, nonetheless, projected a significant decline in near-surface permafrost during the 21st century, using a GCM forced by a strong warming scenario (+7.5° over Arctic land during 1900–2100). If frozen peat thaws, it will become more readily decomposable, and both it and any decomposition products will become much more susceptible to loss to the atmosphere, leaching or thermokarst erosion. On the other hand, if permafrost develops or expands in a northern peatland (perhaps due to a drying-induced change

in peat thermal properties), the peat that freezes will become less susceptible to decomposition or transport.

The second state boundary is biogeochemical, the bound-ary between oxia and anoxia. Peatland water table depth is the first-order control of the partitioning of the peat profile into aerobic and anaerobic zones. A peatland's water table is generally within 0.5 m of the peat surface, and this rela-tively stable, high water table is a result of both the climate and topographic setting and the hydrological properties of the peat itself. Above the water table, the peat is generally oxic, while below the water table, it is generally anoxic. This anoxia affects the decomposition pathways of organic mat-ter, both by slowing its overall rate relative to aerobic de-composition and generating reduced carbon compounds as intermediate- and end-products, including methane (CH_4), a strong greenhouse gas. The relative proportion of CO_2 and CH_4 in carbon gas losses from peatlands has important climate consequences due to their different radiative im-pacts [e.g., *Laine et al.*, 1996; *Whiting and Chanton*, 2001; *Minkkinen et al.*, 2002; *Frolking et al.*, 2006]. Water table depth is a direct expression of peatland hydrology and is strongly influenced by precipitation, peat hydraulic prop-erties, and a peatland's hydrologic setting within a larger watershed.

A third factor affecting the fate of peat in northern peat-lands is locational; almost all peatland carbon is within sev-eral meters of the atmosphere, some peat (i.e., that in fens) is also well-integrated into regional hydrological flow paths, and little of the peat is physically isolated in mineral soil aggregates or adsorbed onto mineral surfaces, which can shelter the organic matter from decomposing organisms and reduce its sensitivity to climate change [*Davidson and Jans-sens*, 2006; *Trumbore and Czimczik*, 2008]. In this way peat, though technically soil carbon, is more similar to a vegeta-tion carbon pool. If the peat carbon is mobilized through decomposition or erosion/dissolution, gaseous forms will likely enter the atmosphere, and in some peatland systems, dissolved or particulate organic matter or dissolved inor-ganic carbon will likely flow out of the peatland and further down the drainage network [e.g., *Moore*, this volume].

All biogeochemical cycling in vegetation/soil systems is sensitive to climate change through temperature, soil moisture, and other climatic controls on cycling rates and metabolic activity. These direct sensitivities are generally considered to be nonlinear but smoothly varying responses that are relatively small for small changes in climate. The nature of the physical and biogeochemical state boundaries on which much northern peat is poised means that the fate of the large northern peatland carbon pool may be very sensi-tive to relatively small changes in climate. Northern peatland geographic location ensures that it will experience climate change earlier and more rapidly than many other biomes [*Christensen et al.*, 2007], and peat's position at the soil surface means that any response in terms of carbon mobilization and greenhouse gas emissions will rapidly influence the climate system.

Global climate models are needed to provide the best available representations of future climate for assessing the fate of the large pool of carbon in northern peatlands, and those representations will improve if climate feedback effects that can be generated by the dynamics of northern peatlands are included explicitly and if the local climate temperature and moisture conditions of northern peatlands are modeled directly. In other words, climate change projections should be more accurate if the next generation of coupled climate-carbon Earth system models [e.g., *Friedlingstein et al.*, 2006] include northern peatlands as a specific terrestrial biome with some unique properties.

2. DEVELOPMENTS OF COUPLED CARBON-CLIMATE MODELS, WITH REPRESENTATION OF ECOSYSTEMS

Over the past several decades, as atmosphere-ocean GCMs have developed in complexity, and as computational power has increased, the land surface representation in these models has gone from a simple bulk surface representation of albedo, aerodynamic roughness, and soil moisture availability to more explicit modeling of the hydrological cycle and to partitioning energy, and water fluxes between the ground and vegetation [*Sellers et al.*, 1997]. Further developments have included layered soils and plant physiological control over canopy stomatal conductance [*Sellers et al.*, 1997; *Le Treut et al.*, 2007; *Bonan*, 2008].

In the last several years, explicit treatment of the carbon cycle and vegetation dynamics has been incorporated into some GCMs; in these models, the biosphere and atmosphere operate as a coupled system [*Cox et al.*, 2000; *Friedlingstein et al.*, 2006; *Le Treut et al.*, 2007; *Randall et al.*, 2007; *Bonan*, 2008]. This coupling of the carbon and climate cycles into a single dynamic model has demonstrated the importance of modeling the inherent feedbacks between the climate system and the carbon cycle because they can substantially change the climate response to anthropogenic forcing of greenhouse gas concentrations [e.g., *Cox et al.*, 2000]. The carbon cycle component of these coupled models typically considers a few to about 20 different plant functional types and several plant and soil carbon pools. The plant functional types can be dynamic (i.e., redistribute geographically due to quasi-competitive responses to climate change) or static. Soil carbon pools are spun up to be in approximate equilibrium with the climate forcing (without explicit consideration of peat-lands), and their dynamics during the simulation are controlled by inputs (vegetation productivity and litterfall) and output (decomposition losses), both responding to changing temperature and moisture conditions.

In an early work, *Cox et al.* [2000] found that carbon cycle feedbacks on the climate system had a positive feedback on warming because climate warming/drying led to the collapse of wet neotropical forest ecosystems and a large net flux of carbon from the land surface to the atmosphere. More recently, an intercomparison was conducted with 11 coupled carbon-climate models (both GCMs and Earth system models of intermediate complexity (EMICs)) using historical anthropogenic greenhouse gas emissions and a future emissions scenario (A2) developed by the Intergovernmental Panel on Climate Change (IPCC) Special Report on Emissions Scenarios (SRES). All model results indicated that increasing CO_2 concentrations alone would enhance the rate at which CO_2 was taken up by both the land and ocean (a negative feedback), but that the climate change reduced the rate at which CO_2 is removed from the atmosphere [*Friedlingstein et al.*, 2006]. However, the model results exhibited substantial variability in their quantification of the strengths of these feedbacks, and in the relative importance of the land and ocean [*Friedlingstein et al.*, 2006]. The models had a range of representations of the terrestrial carbon cycle, some with dynamic vegetation models and some without, but none included peatlands as a possible land cover type. In the recent coupled carbon-climate cycle modeling study of *Yoshikawa et al.* [2008], the two northern regions identified as having strong system feedbacks, Siberia and western boreal North America, are also regions where a substantial fraction of the landscape is peatlands [e.g., *Wieder et al.*, 2006]. The terrestrial ecosystem component of this coupled carbon-climate system model [*Ito and Oikawa*, 2002] has general representations of plant and soil functioning, but no specific representation of unique characteristics of peatlands (see section 4 below).

Incorporating peatlands and their carbon cycling into coupled carbon-climate models poses a number of challenges (see section 4 below). Any representation of peatland carbon cycling will have to comply with strict water, energy, and carbon conservation constraints that are imposed by global climate models for climate change integrations. Another requirement is global applicability (e.g., regionally specific solutions should be avoided). There is an additional, more philosophical modeling goal; limit the amount of information, such as surface data sets (e.g., a wetland map) that are prespecified and not permitted to evolve with the rest of the Earth system.

GCMs are computationally demanding, and typical simulations are for periods of hundreds of years or less. This is long enough to simulate many issues relevant to peatlands as part of the coupled carbon-climate system (e.g., weather-driven interannual variability in C balances; impacts on the peatland C cycle of drought, fire, pollution, harvest or climate change). However, radiocarbon dating of peat cores shows that most sites have been accumulating peat (carbon) persistently for millennia [e.g., *Turunen et al.*, 2002; *Yu et al.*, 2003; *Smith et al.*, 2004]. During this time, the peatlands have not been static; for any peatland, there may have been variations or changes in vegetation cover, hydrological status, peat depth, and peat (C) accumulation rate [e.g., *Yu et al.*, 2003]. At this time, GCM groups are not running continuous simulations for several millennia, though they may be within another decade or less, though probably not as a regular practice. However, several EMICS have been developed that include many GCM processes in simplified or parameterized forms that substantially reduce computation time [*Claussen et al.*, 2002]. These models have been designed for a number of applications, including paleoclimatic reconstructions of the Holocene. Both the CLIMBER-2 model [*Brovkin et al.*, 2002, 2008] and the McGill paleoclimate model [*Wang et al.*, 2005] have done Holocene climate-carbon cycle simulations. Neither explicitly included peatlands.

Spatial resolution of GCM simulations has decreased toward ~100-km grid cells [*Le Treut et al.*, 2007], and representation of subgrid heterogeneity, characterized as a mosaic of tiles, each with a different land cover, is now common though not universal [*Pitman*, 2003]. Surface energy, water, and carbon fluxes are calculated on each tile before being aggregated and passed to an atmospheric submodel. EMICS have variable spatial resolutions for their representations of the atmosphere, oceans, and land, generally coarser than GCMs. The land representation in CLIMBER-2 was ~10° latitude × 50° longitude, or ~1000 km × 5000 km [*Brovkin et al.*, 2002], and for the McGill paleoclimate model, it was ~5° × 5° [*Wang et al.*, 2005]. Both GCM and EMIC spatial resolutions present a challenge for representing land surface heterogeneity in vegetation cover, soils, topography, biogeochemical processes, and human management, which occur at scales from <1 m to >10^6 m.

Although the steady growth in computational resources has permitted global models to keep advancing to finer spatial resolution, the resolution is still not fine enough to address fine-scale variability evident in northern peatlands (i.e., variability over scales of 1–10,000 m) [*Baird et al.*, this volume]. For even the highest resolution GCMs, the degree of subgrid surface heterogeneity remains large, especially in the northern high latitudes. To a certain degree, heterogeneity in surface cover has been accounted for by grid cell tiling of vegetation cover. By comparison, soils are treated much more homogeneously. Typically, all vegetation types within a grid cell [for example, up to four plant functional types in a standard configuration for the National Center for Atmospheric Research Community Land Model (CLM)] share the same nonheterogeneous soil column. Wetland distribution is typically either prescribed, based on satellite or other global wetland distribution estimates, or is defined as the fraction of the water table that intersects the surface which is a function of mean grid cell water table depth and surface topography [*Gedney et al.*, 2004; *Niu et al.*, 2005]. Peatlands are typically ignored, or as in *Lawrence and Slater* [2008] represented without regard to spatial heterogeneity across a grid cell. In global carbon cycle models used in GCMs, there is sometimes no relationship between soil carbon, which is a grid cell level quantity, and wetlands, which is a diagnostic quantity that is a function of the grid cell water balance and surface topography.

3. NORTHERN PEATLANDS IN THE COUPLED CLIMATE-CARBON SYSTEM

Northern peatlands, like other terrestrial ecosystems, influence the Earth's climate system through their impact on the land-surface energy balance. Land-surface albedo and roughness are direct functions of vegetation community composition and landscape heterogeneity (e.g., fraction that is open water). The surface energy balance partitioning of the net radiation energy inputs into sensible and latent heat fluxes also depends on the nature of the surface and vegetation cover and on the availability of evaporable water on the vegetation and in the soil.

Northern peatlands, again like other terrestrial systems, also influence the Earth's climate through their impact on the composition of the atmosphere, particularly the greenhouse gases CO_2 and CH_4. Northern peatlands have been a persistent atmospheric CO_2 sink for millennia (0.02–0.03 kg C m^{-2} a^{-1} over the long term [*Gorham*, 1995; *Tolonen et al.*, 1992; *Smith et al.*, 2004]). About 250–400 Pg C is sequestered in ~3–4 million km^2 of northern peatlands [*Gorham*, 1991; *Turunen et al.*, 2002]. It is not known if, overall, northern peatlands still sequester C at that rate; multiyear site measurements show a variable annual C balance and generally a net uptake [*Lafleur et al.*, 2003; *Aurela et al.*, 2002; *Roulet et al.*, 2007; *Nilsson et al.*, 2008].

Northern peatlands are currently also a source of ~10–40 Tg CH_4 a^{-1} [*Prather et al.*, 2001] and, along with tropical wetlands, likely emitted a large fraction of global total methane flux through the Holocene, when the anthropogenic sources that dominate current budgets were small to negligible. Peatland methane emissions are strongly related to hydrology [*Bubier et al.*, 1995; *Waddington et al.*, 1996;

MacDonald et al., 1998], net primary productivity [*Whiting and Chanton*, 1993; *Waddington et al.*, 1996], and vegetation composition [*Bubier*, 1995; *King et al.*, 1998; *Joabsson et al.*, 1999]. All of these factors are interrelated, and they interact to control methane fluxes [e.g., *Treat et al.*, 2007], so predictions based on any one factor inevitably have a limited range of application.

Frolking and Roulet [2007] have shown that the net fluxes of CO_2 and CH_4 from northern peatlands through the Holocene were large enough to influence the global climate system. They estimated a contemporary radiative forcing impact of about -0.4 W m^{-2} (a net cooling) as a result of the effect that peatland development through the Holocene has on the current atmospheric burdens of CO_2 and CH_4. Current peatland carbon content, accumulated over the past ~10,000 years, is roughly equivalent to 100–200 ppmv CO_2 in the atmosphere (~25–50%), so simulations of Holocene climate dynamics should include a representation of peatlands as a significant component of the global carbon cycle.

Peatlands are generally viewed as sluggish, slowly evolving, self-stabilizing ecosystems [e.g., *Charman*, 2002] and, under relatively stable climatic conditions, their large carbon pool as relatively inert [e.g., *Clymo*, 1984]. However, relatively rapid changes in peatland vegetation and net carbon fluxes are possible; these changes include fire burning for hours to months [*Turetsky et al.*, 2004], industrial harvest occurring over weeks to years [e.g., *Tuittila et al.*, 2003; *Petrone et al.*, 2001], permafrost thaw/collapse occurring over years to decades [e.g., *Camill et al.*, 2001; *Malmer et al.*, 2005; *Johansson et al.*, 2006; *Wickland et al.*, 2006], decadal changes in vegetation composition—including tree encroachment attributed to gradual drying and N-deposition as well as internal or autogenic processes [e.g., *Gunnarsson et al.*, 2000], drainage/drought impacts occurring over years to decades [e.g., *Laine et al.*, 1995; *Minkkinen et al.*, 1999, 2002], and pollution inputs and related vegetation changes over years to decades [e.g., *Bobbink et al.*, 1998]. All of these factors are likely to change over the coming century with changes in climate, atmospheric chemistry, and human activity. Through destabilization, disturbance, or other changes in ecosystem structure or physiology, the carbon in peatlands can be released to the atmosphere as CO_2, CO, and/or CH_4 or can transfer as dissolved organic carbon (DOC) and/or particulate organic carbon downstream. Even without disturbance/destabilization, the net carbon balance of northern peatlands is expected to change with climate change, but the nature, magnitude, and even the sign of that change is uncertain [*Moore et al.*, 1998; *Gorham*, 1991]. Since changing fluxes of CO_2 and CH_4 from northern peatlands will affect the climate system, the best way to model these feed-

back loops is to incorporate peatlands directly into a climate model.

4. NORTHERN PEATLAND ECOSYSTEM PROPERTIES THAT WILL REQUIRE NEW CLIMATE MODEL DEVELOPMENTS

Wetlands differ from other terrestrial landscapes due to the presence of water at or near the soil surface for most or all of the year, soils that frequently have limited oxygen content, and specialized plants that are able to grow in these conditions. Peatlands (or mires, in Europe) are a subclass of wetlands that have substantial accumulations of partially decomposed plant detritus at the soil surface [*Charman*, 2002; *Rydin and Jeglum*, 2006; *Wieder and Vitt*, 2006; *Mitsch and Gosselink*, 2007]. Modeling peatland carbon cycling, as a stand-alone model or within a regional ecosystem/biogeochemistry model or GCM, requires special attention to several unique peatland properties related to soil physics and hydrology, landscape spatial heterogeneity, vegetation physiology, and ecosystem biogeochemistry (Table 1).

4.1. Soil

To be classified as a peatland, there must be a surface layer of organic soil or peat that is at least 0.3 m (United States) or 0.4 m (Canada) thick; typically, peat depths are one to several meters but may exceed 10 m. This peat is predominantly organic matter, with a small mineral component (<30%, and often only a few percent), and thus peat physical properties, e.g., pore size distribution, bulk density, thermal and hydraulic conductivities, differ significantly from those of mineral soils [e.g., *Boelter*, 1964, 1969; *Walmsey*, 1977; *Hillel*, 1980]. Peat ash-free bulk densities are typically 0.02 to 0.35 g cm^{-3} [e.g., *Walmsey*, 1977]; based on extensive sampling in western Canada [*Zoltai et al.*, 2000], shrubby and treed fens generally have a median of bulk density of 0.1–0.15 g cm^{-3}, while bogs and open fens have a median of bulk density of 0.06–0.1 g cm^{-3} [Zicheng Yu, personal communication]. Peat porosities are >0.8 cm^3 cm^{-3} [*Verry and Boelter*, 1978]. Mineral soil bulk densities are typically 1.1–1.6 g cm^{-3}, and porosities are typically 0.3–0.6 cm^3 cm^{-3} [e.g., *Hillel*, 1980].

The low bulk density and high porosity of peat give it significantly different thermal properties than mineral soils [*Hillel*, 1980], a factor that can be important in permafrost development and decay [*Zoltai*, 1993]. Peat heat capacity and thermal conductivity are highly dependent on moisture content [*Farouki*, 1981]. *Kettridge and Baird* [2007] developed peat-specific predictive relationships of vertical variations in heat capacity through the unsaturated zone of poorly

Table 1. Peatland Characteristics That Will Require Model Development for Inclusion in GCMs[a]

Peatland Characteristic	Modeling Issues
Thick organic soils (section 4.1)	Thermal and hydraulic properties differ from mineral soils, and can be more variable both vertically and horizontally, e.g., saturated hydraulic conductivity can vary by orders of magnitude in a single vertical peat profile.
	Appropriate soil depth and layering may differ from current formulations appropriate for mineral soils.
	Soil profile is inherently dynamic over moderate timescales (decades or longer), without disturbance or erosion, including changes in soil thickness and soil hydraulic properties as a function of net peat accumulation, changes in vegetation composition, and peat decomposition
	Peat properties are partially determined by overlying vegetation, creating stronger link between vegetation and soils than is typical for GCMs.
Fine-scale spatial heterogeneity (section 4.2)	Northern peatlands have significant variability in microtopography, vegetation, and water table depth over scales of meters to kilometers.
Abundant nonvascular plant cover (section 4.3)	Nonvascular plants have different physiology and phenology than vascular plants currently modeled.
Anaerobic biogeochemistry (section 4.4)	Carbon and nitrogen cycle in ways and at rates not characteristic of drained mineral soils.
Unique disturbance characteristics (section 4.5)	Peatlands burn, but little is known about peatland recovery after fire.
	Thermokarst dynamics and erosion in permafrost/peat soils generates major changes in surface characteristics; this is not currently modeled in GCMs.

[a]See discussion in section 4.

decomposed *Sphagnum* peat and between peat thermal conductivity and heat capacity.

Most land surface models in GCMs use some variant of Richards' equation for modeling soil water dynamics [e.g., *Cox et al.*, 1999], and the required hydraulic parameters (saturated hydraulic conductivity, porosity, specific yield) and functions relating soil water content to matric potential and unsaturated hydraulic conductivity come from parameterizations developed as functions of soil texture [e.g., *Clapp and Hornberger*, 1978]. Peatland soils require new parameterizations, and some work has been done on this [e.g., *Letts et al.*, 2000]. One particular challenge is that peats often have a very steep decline in hydraulic conductivity (often more than two orders of magnitude) in the top tens of centimeters of peat [e.g., *Paavilainen and Päivänen*, 1995]; this can pose numerical problems in solving Richard's equation for the relatively coarse vertical representation of soils common in GCMs. *Pauwels and Wood* [1999a, 1999b] incorporated a moss (organic) soil layer into a land-surface energy balance model, the type of model that would be a land-surface modeling scheme in a GCM. The addition of a moss layer improved model simulations of soil temperature and moisture in boreal forest stands with thick organic horizons; the model was not tested against peatland data. *Beringer et al.* [2001] incorporated a moss and lichen layer into the soil representation of a GCM land surface model. This surface layer enhanced soil infiltration and insulated the soil, making it cooler in summer and warmer in winter. Moss and lichen

metabolism and carbon cycling were not modeled. In their work on incorporating the thermal and hydrologic influences of organic soils into a global climate model, *Lawrence and Slater* [2008] used data on soil C content of the upper 1.5 m of soil from the *Global Soil Data Task* [2000], available at $1° \times 1°$ resolution, to derive a gridded soil carbon data set. The soil C was distributed over seven soil layers representing the top 1.38 m of the soil, with a prescribed soil carbon density profile. The original field data for this database come from ~21,000 soil profiles [*Tempel et al.*, 1996], or about 1 profile per 6000 km^2 of the earth's ice-free land surface. The approach of *Lawrence et al.* [2008] is acknowledged to be a first attempt at representing the physical influence of organic-rich soil (of which peatlands are a particular class) in the climate system. The coarse resolution of the source, and regridding of the data, meant that the organic matter was effectively spread over the grid cell and that the depth of the "peatlands," which typically would occupy only a fraction of a grid cell, was shallower than many observed peat profiles. More importantly, this approach did not fully address peatland carbon cycling because bryophytes were not represented (see section 4.3 below), and the soil carbon pools were fixed in time and space.

GCMs that include ecosystems and a carbon cycle require initialization of the vegetation cover and vegetation and soil carbon stocks; this is typically done by "spinning up" the model, that is, by running the model with a fixed or regularly repeating climate pattern until the soil and vegetation carbon pools reach quasi-steady state [e.g., *Thornton and Rosenbloom*, 2005]. This steady state is determined over the longest time period of the initialization climate data, so seasonal and interannual variability might still occur with a 20-year climate file, but variability on time-scales longer than 20 years would be minimal [*Thornton and Rosenbloom*, 2005]. However, it seems that northern peatlands can be still accumulating carbon as peat ~5000 to 10,000 years after initial formation [e.g., *Roulet et al.*, 2007; *Nilsson et al.*, 2008], and many may not have reached steady state. This ongoing accumulation is slow, 0.03 kg C m^{-2} a^{-1} corresponds to 0.3 mm a^{-1} of peat with a bulk density of 100 kg m^{-3}, assuming 0.5 kg C per kg peat. This timescale is longer, but not unreasonably longer, than the approximately 3000-year spin up timescale of the Biome-BGC carbon-nitrogen cycle model [*Thornton and Rosenbloom*, 2005]. *Thornton and Rosenbloom* [2005] show that the approximately 3000-year spin-up to equilibrium can be reduced by up to 73% by implementing an accelerated spin-up algorithm. If such an algorithm can be adapted so that it is applicable to peat accumulation, and also considering the ongoing increases in computing capacity, the long peat accumulation timescale would no longer be so daunting.

In the meantime, this slow approach to steady state may not compromise a GCM initialization algorithm. In a simple peat accumulation model, the surface or acrotelm peat (the top 0.3–0.5 m), which is the portion of the peat that has the most dynamic seasonal water and carbon cycling, reaches equilibrium much more rapidly, and the slow long-term accumulation happens in the deeper, less dynamic anaerobic zone or catotelm [*Clymo*, 1984]. *Belyea and Baird* [2006] argue that peatlands are complex adaptive systems and that the acrotelm is probably never really in steady state; however, from the practical point of view of initializing a peatland for a GCM, it may approach steady state if forced for millennia by a steady climate. The catotelm has only minimal direct interaction with the atmosphere and short-term climate system. This would change only in the case of a major disturbance (e.g., fire, or anthropogenic activity like harvest or draining) that exposed the catotelm to the atmosphere. Long-term peat accumulation will also be relevant in applications of EMICS to Holocene climate dynamics, but is not yet considered in those models [e.g., *Brovkin et al.*, 2002, 2008; *Wang et al.*, 2005].

4.2. Hydrology and Landscape Spatial Heterogeneity

As with all other ecosystems, soil temperature and moisture play an important role in peatland C cycling. However, in peatlands, the role of water is a dominant one. The excess amount of water stored in peatlands controls the predominantly anoxic conditions that reduce decomposition, so that net ecosystem production is persistently positive (i.e., a CO$_2$ sink). Methane, an end-product of anoxic decomposition, is an important greenhouse gas. Hydrology plays a key role in the relative strength of peatlands as a CO$_2$ sink and as a CH$_4$ source and thus on peatland net climate impact. Hence, it is necessary to understand and simulate the hydrology of peatlands to be able to explain and simulate their carbon exchanges.

Peatlands are unique ecosystems in the degree to which they influence their own hydrology. Because the accumulation of meters of peat occurs over millennia, it becomes the substrate that controls the position of the water table and the moisture condition for plants. Due to the near-surface decomposition and collapse of the original plant material, a peat profile develops a stratification with less decomposed fibric peat, with large pores and low density, near the surface, and more decomposed, relatively high-density peat with finer pores deeper in the profile. This transition from less to more decomposed peat, with accompanying large changes in hydraulic properties, occurs over several tens of centimeters around the long-term average water table position. This characteristic, along with the balance and source of inputs and outputs of water, ultimately control structure and function of peatlands. The tight coupling of peat structure and function and peatland hydrology has led *Belyea and Baird* [2006] to suggest peatlands be considered complex adaptive systems, with important internal dynamic feedbacks governing their development and behavior. The significance of the very steep changes in properties and the existence of a near surface, hydrologically "active" layer, or acrotelm, and a deeper, hydrologically much less active layer, the catotelm [*Ingram*, 1978] has been recognized for a long time [*Clymo*, 1984; *Charman*, 2002; *Rydin and Jeglum*, 2006]. In ombrotrophic peatlands the peat surface accumulates to an elevation above the local topography, and the water supply is only by atmospheric inputs, while minerotrophic peatlands receive small to large quantities of water that has been in contact with the mineral sediments either beneath or adjacent to a peatland. Peatlands span a gradient along this water/nutrient supply axis [*Vitt*, 2006], but for the purposes of simulation, the functional structure of peatlands and the biogeochemistry of carbon cycling the division between the ombrotrophic "bog"-like systems and minerotrophic "fen"-like systems may be sufficient as a first approximation [e.g., *Frolking et al.*, 2001].

Simulating the hydrology of a peatland to determine the position of the water table and the distribution of moisture above the water table represents a challenge even in the case of individual peatlands. There have been models of peatland hydrology, but these have been primarily for estimating the discharge from peatlands and have been based on relatively simple empirical functions [e.g., *Guertin et al.*, 1987; *Verry et al.*, 1988]. There has also been a very long tradition of modeling peatland drainage for forestry and agricultural practices [e.g., *Konyha et al.*, 1988]. However, there have been few modeling attempts to simulate peatland hydrology that are appropriate for climate and carbon simulations. One example is the recent work of *Borren and Bleuten* [2006], who combined the MODFLOW groundwater model with a dynamic digital elevation model (driven by peat accumulation) and a paleoclimate time series to simulate the coupled carbon and water cycles through the Holocene of an 800-km^2 peatland complex in the West Siberian Lowlands. Peat hydraulic properties were constant unless the simulated peatland switched from fen to bog, which depended on water fluxes and assumed nutrient availability, so the primary feedback between peatland development and hydraulic properties was through the digital elevation model. They found that long-term peat accumulation and lateral expansion were limited by hydrology and that model sensitivity to hydrological parameters was high.

Comer et al. [2000] and *Letts et al.* [2000] modified the Canadian Land Surface Scheme (CLASS) for the inclusion of peat and the simulation of the position of the water table. The revised CLASS was quite successful at simulating the moisture dynamics in fens and was able to reproduce the evapotranspiration losses for fens and bogs reasonably well [*Comer et al.*, 2000], but failed to reproduce the runoff from bogs. CLASS does not include subsurface lateral flow, and this is the most important runoff pathway in bogs [e.g., *Verry et al.*, 1988; *Evans et al.*, 1999]. *Yurova et al.* [2007] and *Yurova and Lankreijera* [2007] have combined a soil organic matter model (ROMUL) and a surface climate-ecosystem model (GUESS) to simulate the coupling of hydrology and carbon dynamics in a Swedish northern fen. Others have developed one-dimensional surface hydrological models that would be suitable for climate simulations but few have been evaluated for multiple years [e.g., *Weiss et al.*, 2006].

However, the problem becomes significantly more difficult when a peatland has either significant internal redistribution of water (i.e., spatial heterogeneity within a peatland) or when it receives a meaningful proportion of the water required to maintain a high water table from beyond its boundaries (e.g., groundwater inputs). Spatial heterogeneity within northern peatlands is extremely common [e.g., *Couwenberg and Joosten*, 2005] and may play an important role in the peatland water and carbon balance. *Sonnentag et al.* [2008] modified the Boreal Ecosystem Productivity Simulator for peatland hydrology, and ran it in a version of the Terrain-Lab model to simulate the effect of the mesoscale spatial variability (on the order of 1 km) on the hydrology and net ecosystem productivity of a peatland. They found a strong

correlation between the spatial variability of simulated evapotranspiration (ET) and gross primary productivity (GPP), and that ignoring the effect of spatial variability systematically underestimated ET and GPP by ~10%. If external water inputs to a peatland are significant, the peatland needs to be simulated in the context of its hydrological setting within the surrounding watershed. If the input of water from outside the peatland is a function of the surface topography, then lateral inputs could potentially be simulated by some topography-based flow modeling [e.g., *Gedney and Cox*, 2003]. However, in many regions where annual potential evapotranspiration minus precipitation is very small or even negative, e.g., boreal western Canada, it has been shown that external inputs of water are essential for peatlands and that the external input is not related to surface topography but rather to the complex structure of the underlying geology and surface deposits [*Devito et al.*, 2005]. This situation is going to present a serious challenge for climate simulations given the current hydrology in climate models and the state of global data sets of underlying geology and surface deposits. *Baird et al.* [this volume] present a detailed discussion of spatial heterogeneity issues related to northern peatlands and outline a multiscale scheme for addressing them.

At the global scale, there have been a number of attempts to estimate and map the distribution of peatlands and wetlands [e.g., *Matthews and Fung*, 1987; *Lehner and Döll*, 2004]. These distributions are based on a combination of large-scale topographic modeling and globally mapped surface characteristics such as indices of inundation, soil properties, and vegetation. There have been attempts to simulate the distribution of wetlands using topographic wetness indexes. *Kirkby et al.* [1995] mapped the distribution of northern European wetlands using the topographic wetness index first proposed by *Beven and Kirkby* [1979]. Recently, *Gedney and Cox* [2003] used the same topographic index to simulate grid-scale runoff in a GCM and found improved estimates of global runoff (bias reduction from 37 to 25%), improved estimates of precipitation (though not statistically significant improvements), and resulting patterns of saturated areas given by the topographic index (a new model result) that were consistent with major wetlands areas identified in the wetland distribution maps developed by *Aselmann and Crutzen* [1989]. *Gedney and Cox* [2003] concluded that any improvement in subgrid scale representation of soil moisture heterogeneity is an important step toward improving GCM projections of climate and hydrological changes.

The final aspect of hydrology in peatlands necessary to simulating the carbon balance is the transport/export of DOC. Multiyear observations of carbon balances on several peatlands have shown that the export of DOC is between

10 and 20 g C m^{-2} a^{-1}, and this is the same order of magnitude as the long-term accumulation of organic matter in the peatlands [*Roulet et al.*, 2007; *Nilsson et al.*, 2008]. The production of DOC is a function of decomposition, but only a small fraction of the DOC produced is exported; the controls on DOC export are both hydrological and biogeochemical [*Kalbitz et al.*, 2000]. Simulation of DOC export in peatland models has used a fixed DOC concentration and water export [*Frolking et al.*, 2002], but this simply means DOC export tracks runoff, which is clearly not always the case [*Fraser et al.*, 2001]. *Yurova et al.* [2007] developed a model based on convection and dispersion equations to simulate DOC concentration within a peatland and in peatland discharge. A simplified parameterization of export, based on this approach, might be found for coarse scale climate, carbon simulations. *Moore* [this volume] reviews DOC export from northern peatlands.

4.3. Nonvascular Vegetation

Plant functional types in peatlands are generally similar to those found in other terrestrial ecosystems, e.g., woody plants (deciduous and evergreen trees and shrubs), graminoids (sedges, rushes, grasses), forbs (other herbaceous plants), and bryophytes (nonvascular plants such as mosses and lichens). The primary differences between peatland and nonpeatland systems are that in peatlands (1) in some cases, bryophytes can account for a majority of total vegetation biomass and productivity [*Moore et al.*, 2002], and (2) a fraction of the typical soil 0.5 m "root zone" [*Jackson et al.*, 1996] is often saturated, so plant rooting strategies and vertical root distributions may be different. The physiology of nonvascular plants is typically not represented in global carbon-climate models.

Bryophytes have no roots nor vascular system [*Proctor*, 2000], so land surface model developments related to soil moisture, root distributions, and leaf stomatal control on water and carbon exchanges [e.g., *Sellers et al.*, 1997] are not directly relevant. Bryophyte metabolic rates are strongly related to their leaf water content [*Proctor*, 1982, 2000]; this presents a significant challenge to the vertical resolving power of climate model soil hydrology because bryophyte metabolism is sensitive to the water content of only the top few centimeters of the soil (peat and moss), rather than a thicker root zone. Bryophytes are also able to respond very quickly to changing environmental conditions, e.g., the seasonal temperature cycle and the subseasonal wetting and drying associated with weather patterns [*Proctor*, 1982, 2000], requiring new functions or algorithms for vegetation seasonal phenology. Bryophytes may be responsible for a significant fraction of the net ecosystem productivity

in the "shoulder seasons," in early spring before vascular plants have emerged from their winter dormancy [*Moore et al.*, 2006] and perhaps also in the autumn when vascular plants, particularly deciduous plants, have senesced. Lacking roots, algorithms for carbon allocation in bryophytes can be simpler than for vascular plants [e.g., *Frolking et al.*, 2002].

Many peatlands are ombrotrophic, receiving the bulk of their nutrient inputs from wet and dry deposition. Mosses, which often develop a fairly complete ground cover under any emergent vascular vegetation, intercept and efficiently absorb much of wet nutrient deposition before it can percolate to the vascular root zone [e.g., *Aerts et al.*, 1992; *Malmer et al.*, 1994; *Nordbakken et al.*, 2003]. Thus, the nutrients only become available to the vascular plants after they have cycled through the mosses and are re-mineralized during decomposition of moss litter. *Limpens et al.* [2003] found that nitrogen additions favored vascular plants and suppressed *Sphagnum* growth in greenhouse mesocosm studies. In the competition for light, vascular plants have the advantage because they can grow above the moss layer, although rapidly growing moss can engulf vascular seedlings [*Limpens et al.*, 2003]. In a nutrient addition manipulation study on an ombrotrophic bog, *Bubier et al.* [2007] found that enhanced shrub growth shaded the underlying *Sphagnum* through increased leaf area and increased leaf litter fall; *Sphagnum* cover diminished substantially, but *Polytrichum* cover increased, and overall moss biomass diminished by about 50% over a 5-year treatment period. They could not determine whether *Polytrichum* growth was due to nutrient enrichment or diminished competition from declining *Sphagnum* species. Shading and cooler moss temperature may also have had an impact [*Bubier et al.*, 2007]. *Pastor et al.* [2002] presented a simple model of vegetation dynamics, simulating competition between a single vascular plant type and a single moss type. The model centers around access to different nutrient sources, wet deposition for mosses and mineralized nutrients for vascular plants. Moss and vascular plants also compete for other resources (e.g., light), simulated as a reduction in growth rate proportional to the biomass of the other plant type. This simple model generated nonlinear, dynamic behavior, and several different stable states can emerge that influence the capacity of the system to store and release carbon and nutrients. *Frolking et al.* [2001] used a simple model of peat accumulation to show that accumulation rates were sensitive to the fraction of total productivity generated by mosses and by vascular plants. Finally, bryophytes, and particularly *Sphagnum* species, play an important role in peat accumulation and peatland development, through their effects on the chemistry of peatland waters, their interactions with vascular plants, and their production of decay-resistant litter [*Rydin et al.*, 2006].

Published data on above- and belowground vascular plant biomass and productivity for wetlands (peatlands, wet tundra, freshwater marsh, and salt marsh) show that wetlands, overall, have root biomass values and above- to belowground biomass ratios similar to those of nonforested systems (e.g., grasslands) [*Jackson et al.*, 1996; *Mokany et al.*, 2006; T. R. Moore, personal communication, 2008]. Wetlands generally have shallower root distributions than other terrestrial systems, and the below- to aboveground vascular plant biomass ratio for peatlands of ~0.8 (T. R. Moore, personal communication, 2008) is substantially higher than the root:shoot biomass ratios reported in *Mokany et al.* [2006] for boreal forests, i.e., ~0.4 for shoot biomass >75 Mg ha^{-1} and ~0.25 for shoot biomass <75 Mg ha^{-1}). There is no indication whether or not the boreal forest data summarized in *Mokany et al.* [2006] included nonvascular biomass.

4.4. Biogeochemistry

Complete modeling of the peatland carbon cycle, and its impact on the global atmospheric burdens of CO_2 and CH_4, cannot be done without explicit or implicit representation of anaerobic decomposition of organic matter, and its impact on organic carbon accumulation rates, nutrient mineralization rates, nutrient availability, and methane emissions. A key challenge is to develop a comprehensive representation of the impacts of anaerobic conditions in the soil on all of these processes, along with a robust algorithm for the soil physics and hydrology that reliably simulates both the position of the water table, the peat water content above the water table, and the moss water content. This is not done at this time in any global coupled carbon-climate model.

Peatland methane emissions are the net of methane production and oxidation, and are influenced by transport mechanisms from peat to the atmosphere (diffusion, bubbling, plant-mediated transport). Peatland methane emissions have been modeled for the past decade or two, and insights from this work should be useful as climate models address these biogeochemical issues. *Frolking and Crill* [1994] modeled peat temperature and moisture profiles and correlated this to net methane flux. *Granberg et al.* [2001] and *Kettunen* [2003] extended this to modeling of plant productivity, methane production, oxidation, transport, and net flux. There are similarly constructed global-scale models of methane emissions from wetlands [e.g., *Walter et al.*, 2001; *Cao et al.*, 1996]. There have been studies with GCMS of climate change impacts on wetland methane emissions [e.g., *Gedney and Cox*, 2003; *Shindell et al.*, 2004]; however, these studies have not directly incorporated the wetlands into the

climate model, but have instead used GCM climate change projections to drive a model of wetland extent and methane emissions, again without a full simulation of the carbon cycle. There are peatland carbon cycle models that include anaerobic suppression of decomposition and vegetation productivity, but do not simulate methane production, oxidation, and transport [e.g., *Frolking et al.*, 2002; *Yurova et al.*, 2007; *St-Hilaire et al.*, 2009]. There are several terrestrial biogeochemical models that simulate wetland biogeochemistry, including explicit or implicit representation of aerobic and anaerobic processes in the soil at varying levels of detail [e.g., DNDC, *Zhang et al.*, 2002; *Li et al.*, 2004; ecosys, *Grant and Roulet*, 2002; Biome-BGC, *Bond-Lamberty et al.*, 2007; NASA-CASA, *Potter et al.*, 2001; TEM, *Zhuang et al.*, 2004]. These models can also simulate upland systems, but to date, they have not been applied to wetlands/peatlands at global scales, and only the TEM model has been applied at regional scales for a landscape mosaic of peatlands and uplands [*Zhuang et al.*, 2004].

Gedney et al. [2004] used a GCM-driven topographic index estimation of saturated area (discussed in section 4.2 above) to simulate climate change impacts on CH_4 emissions from wetlands. In this exercise, they calibrated the CH_4 production rate and temperature sensitivity to match current global methane emissions and then estimated changes in wetland area and CH_4 emissions over the period up to 2100. A limitation to this approach is that it is essentially an estimate of inundated area, while many wetlands have water tables slightly below the surface, and the slight variations in the location of this water table has a significant impact on methane fluxes. *Bubier et al.* [2005] estimated a 60% increase in methane flux at the landscape scale due to small changes in water table depth (2–5 cm) and slight warming (0.5°) of small wetlands in a wet and a dry year. This variability, at large scales, has been suggested as a cause for observed interannual variability in the atmospheric methane burden over the past decade or two [*Dlugokencky et al.*, 2001; *Bousquet et al.*, 2006]. A difference of water table elevation from the surface to only 0.3 m depth is the difference between a peatland emitting a large amount of CH_4 and very little or none (i.e., all the CH_4 the peatland produces being oxidized) [*Granberg et al.*, 1997]. Topographic index hydrological modeling [e.g., *Beven and Kirkby*, 1978] may not be sufficient to the task of simulating the subtle differences in water table depth in the relatively flat landscapes that are common in much of the domain of northern peatlands. However, the shallow groundwater modeling of *Borren and Bleuten* [2006] did simulate water table depth variability at relevant vertical resolution over a large peatland complex. This will be further complicated by the effect of melting permafrost and the subsequent changes in landscape topography, peatland hydrology, and ecosystem structure and function. At a site in northern Sweden, the melting of the permafrost has led to a several order of magnitude increase in CH_4 emissions and significant changes in CO_2 exchange and DOC export [*Christensen et al.*, 2004; *Malmer et al.*, 2005; *Johansson et al.*, 2006], though the latter change was not as clear as that of CH_4. How long this impact on carbon fluxes will persist is not known.

Another challenge for modeling peatlands is the range of soil pH that occurs across peatland types; peatland pH interacts with vegetation species composition, nutrient availability, and productivity [e.g., *Bubier*, 1995]. Perhaps the first order effects can be captured in a simple peatland classification (e.g., bog and fen) with model parameterizations specific to the vegetation types that dominate these broad classes, in a similar way to how GCMs would disaggregate forests into a few classes (evergreen or deciduous, needle-leaved, or broad-leaved).

4.5. Disturbance

For many natural landscapes, disturbance by wind or fire plays an important role in determining landscape characteristics that are of central importance to the coupled climate-carbon system [e.g., *Foster et al.*, 1998]. Northern peatland disturbances include fire (both natural and human-caused); flooding/inundation due to beaver activity, reservoir construction, or thermokarst activity associated with permafrost degradation; water table drawdown as peatlands are drained for forestry, agriculture, or peat extraction; pollution/nutrient deposition (e.g., nitrogen and sulfur deposition); and linear disturbances such as roads and seismic lines that can fragment peatlands and alter their hydrology [*Turetsky and St. Louis*, 2006]. Only a few of these disturbances are unique to peatlands. Many that are common to all ecosystems (fire, forestry and agriculture, atmospheric deposition) are beginning to be incorporated into global climate models. For example, GCM simulations planned for the fifth IPCC Assessment will include, to a degree that will vary from model to model, the dominant human land use activities of agriculture and forestry [e.g., *Washington et al.*, 2008], along with radiative forcing from anthropogenic greenhouse gases. Simulation of their impacts on peatlands may follow after inclusion of peatlands in the models. Linear disturbances, beaver activity, peatland draining, and peat extraction are not likely to be included in climate models in the near future.

Fire is a dominant form of disturbance in boreal forest ecosystems [e.g., *Stocks et al.*, 2002], and boreal fire recurrence intervals range from <100 to ~1000 years [*Balshi et al.*, 2007]. In forests, severe fires have a significant impact on forest structure and age distribution [e.g., *Oliver and Larson*, 1996] and thus on the forest's direct interactions with the climate system through surface albedo and roughness. Peatlands also can burn, and their fire regimes may be similar to upland boreal forests, at least in Central Canada [*Turetsky et al.*, 2004]. This has not been well documented across the pan-boreal domain of northern peatlands, however, and little is known about the statistics of peatland burn severity [*Turetsky et al.*, 2004]. Fire algorithms are dependent on fuel availability and near-surface soil moisture with different combustion efficiencies assumed for vegetation, litter, and root biomass [*Thonicke et al.*, 2001; *Thornton et al.*, 2007]. Parameterizations will need to be developed for peatland fires.

Thermokarst landscapes arise in regions where melting permafrost and draining water cause the ground to settle unevenly. There is evidence that the surface water conditions can change quickly as permafrost thaws and hydrological flow paths are modified [e.g., *Jorgenson et al.*, 2006]. There is more evidence of thermokarst impacts on lakes because they are more easily detected in spaceborne remote sensing than peatlands, but the hydrological impacts on lakes will also apply to wetlands. Thermokarst (or thaw) lakes can form in permafrost regions when massive ground-ice wedges melt, causing the ground surface to subside and lakes to form. Further ground melting can eventually lead to drainage and disappearance of thermokarst lakes [*Smith et al.*, 2005]. *Riordan et al.* [2006] found that the area and number of small water bodies in non-arctic Alaska decreased from 1950 to 2002, and attributed this change to increased drainage and or an increase in evapotranspiration. Soil thermal regimes and ground thaw are very dependent on soil thermal properties and water content, so a net drying (or wetting) of soils, particularly organic soils [e.g., *Lawrence et al.*, 2008] will have an impact on permafrost formation and degradation rates. Modeling this dynamic nature of the land surface [e.g., *West and Plug*, 2008; *Plug and West*, 2009] may prove to be difficult in a GCM. In principal, a model could be structured to permit the fractional area of wetland and upland zones to evolve over time, although in practice, this may be technically challenging when it comes to maintaining water, energy, and carbon conservation as soil characteristics evolve.

5. CONCLUDING COMMENTS

Peatland carbon cycling is affected by weather and climate, and the Earth's climate system is affected by peatland carbon cycling. This inherent feedback suggests that peatlands should be incorporated into global climate models. However, northern peatlands have several unique characteristics that will make it difficult to represent their behavior in the Earth system within the vegetation classes of current global climate model land surface schemes. These characteristics include deep organic soils [a topic being addressed in current GCM model development; *Lawrence and Slater*, 2008], bryophyte vegetation, shallow water tables, and anoxic soil profiles, a high degree of spatial heterogeneity in vegetation and hydrology, and some unique disturbance/recovery characteristics. A primary requirement for a successful representation is for peatland distribution to correspond to wetland distribution, i.e., the peatlands form and persist where the land is wet. GCMs will need to be able to interrelate soil moisture heterogeneity with soil carbon heterogeneity. One potential solution is to split each grid cell into static wetland (peatland) and upland zones as is done in the TEM model [*Zhuang et al.*, 2004]. Under this configuration, in wetland zones, carbon will accumulate, and peatlands can accumulate due to anoxic conditions that limit decomposition. A limitation to this approach is the inherent assumption of stationarity, e.g., that wetland distribution is fixed in time and does not respond to changes in the surface water balance (though changes in the surface water balance would alter water table depth within wetland zones). A more dynamic representation of peatlands would include peatland initiation (probably as paludification) and growth and development, where hydrology and peatland vegetation would interact, rather than being prescribed as coincident.

Global climate models have a global domain, and peatlands do not only occur in the north, though a large majority of peatland research to date has been in the boreal and temperate zones. Incorporating peatlands into global climate models will need to account for all peatlands, not just northern peatlands. Tropical peatlands occupy ~0.3–0.6 million km^2, about 10% of the total global peatland area, and may contain ~20% of global peat carbon [*Charman*, 2002; *Page et al.*, 2002, 2004]. These tropical peatlands also have accumulated peat over millennia, often to depths ~10 m [*Page et al.*, 2004]. Many of the same issues will arise in representing tropical peatlands in global climate models (deep organic soils, shallow water tables and anoxia, potentially unique disturbance regimes, spatial heterogeneity); bryophytes probably play a less important role in most tropical peatlands than they do in northern peatlands [*Page et al.*, 2006]. Before we can know how well a peatland land surface scheme successfully developed for northern peatlands will work for tropical peatlands, there is a need for more basic observational data from tropical peatlands, vegetation ecology, hydrology, biogeochemical cycling, and palynology.

Acknowledgments. We thank C.C. Treat, A.J. Baird, and one anonymous reviewer for comments on the manuscript. S.F. was supported by NSF grant ATM-0628399 and NASA IDS Program grant NNX07AH32G. Funding for N.T.R. climate research was provided by the Natural Sciences and Engineering Research Council of Canada and the Canadian Foundation for Climate and Atmospheric Sciences. D.L. was supported by the U.S. DOE, Office of Science (BER), Cooperative Agreement DE-FC02-97ER62402.

REFERENCES

Aerts, R., B. Wallén, and N. Malmer (1992), Growth limiting nutrients in *Sphagnum*-dominated bogs subject to low and high atmospheric nitrogen supply, *J. Ecol.*, *80*, 131–140.

Aselmann, I., and P. J. Crutzen (1989), Freshwater wetlands: Global distribution of natural wetlands and rice paddies, their net primary productivity, seasonality and possible methane emissions, *J. Atmos. Chem.*, *8*, 307–358.

Aurela, M., T. Laurila, and J.-P. Tuovinen (2002), Annual CO_2 balance of a subarctic fen in northern Europe: Importance of the wintertime efflux, *J. Geophys. Res.*, *107*(D21), 4607, doi:10.1029/2002JD002055.

Baird, A. J., L. R. Belyea, and P. J. Morris (2009), Upscaling peatland-atmosphere fluxes of methane: Small-scale heterogeneity in process rates and the pitfalls of "bucket-and-slab" models, *Geophys. Monogr. Ser.*, doi:10.1029/2008GM000826, this volume.

Balshi, M. S., et al. (2007), The role of historical fire disturbance in the carbon dynamics of the pan-boreal region: A process-based analysis, *J. Geophys. Res.*, *112*, G02029, doi:10.1029/2006JG000380.

Belyea, L. R., and A. J. Baird (2006), Beyond "The limits to peat bog growth": Cross-scale feedback in peatland development, *Ecol Monogr.*, *76*, 299–322.

Beringer, J., A. H. Lynch, F. S. Chapin, M. Mack, and G. B. Bonan (2001), The representation of arctic soils in the land surface model: The importance of mosses, *J. Climate*, *14*, 3324–3335.

Beven, K. J., and M. J. Kirkby (1979), A physically-based, variable contributing area model of basin hydrology, *Hydrol. Sci. Bull.*, *24*, 43–69.

Bobbink, R., M. Hornung, and J. G. M. Roelofs (1998), The effects of air-borne nitrogen pollutants on species diversity in natural and semi-natural European vegetation, *J. Ecol.*, *86*, 717–738.

Boelter, D. (1964), Water storage properties of several peats in situ, *Soil Sci. Soc. Am. Proc.*, *28*, 433–435.

Boelter, D. (1969), Physical properties of peats related to degree of decomposition, *Soil Sci. Soc. Am. Proc.*, *33*, 606–609.

Bonan, G. B. (2008), Forests and climate change: Forcings, feedbacks, and the climate benefits of forests, *Science*, *320*, 1444–1449.

Bond-Lamberty, B., S. T. Gower, and D. E. Ahl (2007), Improved simulation of poorly drained forests using Biome-BGC, *Tree Physiol.*, *27*, 703–715.

Borren, W., and W. Bleuten (2006), Simulating Holocene carbon accumulation in a western Siberian watershed mire using a three-dimensional dynamic modeling approach, *Water Resour. Res.*, *42*, W12413, doi:10.1029/2006WR004885.

Bousquet, P., et al. (2006), Contribution of anthropogenic and natural sources to atmospheric methane variability, *Nature*, *443*, 439–443.

Brouchkov, A., and M. Fukuda (2002), Preliminary measurements on methane content in permafrost, central Yakutia, and some experimental data, *Permafrost Periglacial Processes*, 3, 187–197.

Brovkin, V., J. Bendtsen, M. Claussen, A. Ganopolski, C. Kubatzki, V. K. Petoukhov, and A. Andreev (2002), Carbon cycle, vegetation and climate dynamics in the Holocene: Experiments with the CLIMBER-2 model, *Global Biogeochem. Cycles*, 16(4), 1139, doi:10.1029/2001GB001662.

Brovkin, V., J.-H. Kim, M. Hofmann, and R. Schneider (2008), A lowering effect of reconstructed Holocene changes in sea surface temperatures on the atmospheric CO_2 concentration, *Global Biogeochem. Cycles*, 22, GB1016, doi:10.1029/2006GB002885.

Bubier, J. L. (1995), The relationship of vegetation to methane emission and hydrochemical gradients in northern peatlands, *J. Ecol.*, 83, 403–420.

Bubier, J. L., T. R. Moore, L. Bellisario, N. T. Comer, and P. M. Crill (1995), Ecological controls on methane emissions from a Northern peatland complex in the zone of discontinuous permafrost, Manitoba, Canada, *Global Biogeochem. Cycles*, 9, 455–470.

Bubier, J., T. Moore, K. Savage, and P. Crill (2005), A comparison of methane flux in a boreal landscape between a dry and a wet year, *Global Biogeochem. Cycles*, 19, GB1023, doi:10.1029/2004GB002351.

Bubier, J. L., T. R. Moore, and L. A. Bledzki (2007), Effects of nutrient addition on vegetation and carbon cycling in an ombrotrophic bog, *Global Change Biol.*, 13, 1168–1186.

Camill, P., J. A. Lynch, J. S. Clark, J. B. Adams, and B. Jordan (2001), Changes in biomass, aboveground net primary production, and peat accumulation following permafrost thaw in the boreal peatlands of Manitoba, Canada, *Ecosystems*, 4, 461–478.

Cao, M., S. Marshall, and K. Gregson (1996), Global carbon exchange and methane emission from natural wetlands: Application of a process-based model, *J. Geophys. Res.*, 101, 14,399–14,414.

Charman, D. (2002), *Peatlands and Environmental Change*, 301 pp., John Wiley, Chichester U. K.

Christensen, J. H., et al. (2007), Regional climate projections, in *Climate Change 2007: The Physical Science Basis: Contribution of Working Group I to the Fourth Assessment Report of the Intergovernmental Panel on Climate Change*, edited by S. Solomon et al., pp. 847–940, Cambridge Univ. Press, Cambridge, U. K.

Christensen, T. R., T. Johansson, H. J. Åkerman, M. Mastepanov, N. Malmer, T. Friborg, P. Crill, and B. H. Svensson (2004), Thawing sub-arctic permafrost: Effects on vegetation and methane emissions, *Geophys. Res. Lett.*, 31, L04501, doi:10.1029/2003GL018680.

Clapp, R. B., and G. M. Hornberger (1978), Empirical equations for some soil hydraulic properties, *Water Resour. Res.*, 14, 601–604.

Claussen, M., et al. (2002), Earth system models of intermediate complexity: Closing the gap in the spectrum of climate system models, *Clim. Dyn.*, 18, 579–586.

Clymo, R. (1984), The limits of peat bog growth, *Philos. Trans. R. Soc. London B*, 303, 605–654.

Comer, N. T., P. Lafleur, N. T. Roulet, M. G. Letts, M. R. Skarupa, and D. L. Verseghy (2000), A test of the Canadian Land Surface Scheme (CLASS) for a variety of wetland types, *Atmos. Oceans*, 38, 161–179.

Couwenberg, J., and H. Joosten (2005), Self-organization in raised bog patterning: The origin of microtope zonation and mesotope diversity, *J. Ecol.*, 93, 1238–1248.

Cox, P. M., R. A. Betts, C. B. Bunton, R. L. H. Essery, P. R. Rowntree, and J. Smith (1999), The impact of new land surface physics on the GCM simulation of climate and climate sensitivity, *Clim. Dyn.*, 15, 183–203.

Cox, P. M., R. A. Betts, C. D. Jones, S. A. Spall, and I .J. Totterdell (2000), Acceleration of global warming due to carbon-cycle feedbacks in a coupled climate model, *Nature*, 408, 184–187.

Davidson, E. A., and I. A. Janssens (2006), Temperature sensitivity of soil carbon decomposition and feedbacks to climate change, *Nature*, 440, 165–173.

Devito, K. J., I. F. Creed, and C. J. D. Fraser (2005), Controls on runoff from a partially harvested aspen-forested headwater catchment, Boreal Plain, Canada, *Hydrol. Process.*, 19, 3–25.

Dlugokencky, E. J., B. P. Walter, K. A. Masarie, P. M. Lang, and E. S. Kasischke (2001), Measurements of an anomalous global methane increase during 1998, *Geophys. Res. Lett.*, 28, 499–502.

Evans, M. G., T. P. Burt, J. Holden, and J. K. Adamson (1999), Runoff generation and water table fluctuations in blanket peat: Evidence from UK data spanning the dry summer of 1995, *J. Hydrol.*, 221, 141–160.

Farouki, O. T. (1981), *Thermal Properties of Soils*, CRREL Monograph 81-1, 136 pp., U.S. Army Corps of Engineers, Cold Regions Research and Engineering Laboratory, Hanover, NH.

Foster, D. R., D. H. Knight, and J. F. Franklin (1998), Landscape patterns and legacies resulting from large, infrequent forest disturbances, *Ecosystems*, 1, 497–510.

Fraser, C. J. D., N. T. Roulet, and P. M. Lafleur (2001), Groundwater flow patterns in a large peatland, *J. Hydrol.*, 246, 142–154.

Friedlingstein, P., et al. (2006), Climate-carbon cycle feedback analysis: results from the C4MIP model intercomparison, *J. Clim.*, 19, 3337–3353.

Frolking, S., and P. Crill (1994), Climate controls on temporal variability of methane flux from a poor fen in southeastern New Hampshire: Measurement and modeling, *Global Biogeochem. Cycles*, 8, 385–397.

Frolking, S., and N. T. Roulet (2007), Holocene radiative forcing impact of northern peatland carbon accumulation and methane emissions, *Global Change Biol.*, 13, 1079–1088.

Frolking, S., N. T. Roulet, T. R. Moore, P. J. H. Richard, M. Lavoie, and S. D. Muller (2001), Modeling northern peatland decomposition and peat accumulation, *Ecosystems*, 4, 479–498.

Frolking, S., N. T. Roulet, T. R. Moore, P. Lafleur, J. L. Bubier, and P. M. Crill (2002), Modeling the seasonal to annual carbon balance of Mer Bleue Bog, Ontario, Canada, *Global Biogeochem. Cycles*, 16(3), 1030, doi:10.1029/2001GB001457.

Frolking, S., N. Roulet, and J. Fuglestvedt (2006), How northern peatlands influence the Earth's radiative budget: Sustained methane emission versus sustained carbon sequestration, *J. Geophys. Res.*, 111, G01008, doi:10.1029/2005JG000091.

Gedney, N., and P. M. Cox (2003), The sensitivity of global climate model simulations to the representation of soil moisture heterogeneity, *J. Hydrometeorol.*, *4*, 1265–1275.

Gedney, N., P. M. Cox, and C. Huntingford (2004), Climate feedback from wetland methane emissions, *Geophys. Res. Lett.*, *31*, L20503, doi:10.1029/2004GL020919.

Global Soil Data Task (2000), Global gridded surfaces of selected soil characteristics (IGBPDIS), http://www.daac.ornl.gov/, Distrib. Active Arch. Cent., Oak Ridge Natl. Lab., Oak Ridge, Tenn.

Gorham, E. (1991), Northern peatlands: role in the carbon cycle and probable responses to climatic warming, *Ecol. Appl.*, *1*, 182–195.

Gorham, E. (1995), The biogeochemistry of northern peatlands and its possible response to global warming, in *Biotic Feedbacks in the Global Climatic System: will the warming speed the Warming?*, edited by G. M. Woodwell and F. T. MacKenzie, pp. 169–187, Oxford Univ. Press, New York.

Granberg, G., C. Mikkelä, I. Sundh, B. H. Svensson, and M. Nilsson (1997), Sources of spatial variation in methane emission from mires in northern Sweden—A mechanistic approach in statistical modeling, *Global Biogeochem. Cycles*, *11*, 135–150.

Granberg, G., M. Ottosson-Löfvenius, H. Grip, I. Sundh, and M. Nilsson (2001), Effect of climatic variability from 1980 to 1997 on simulated methane emission from a boreal mixed mire in northern Sweden, *Global Biogeochem. Cycles*, *15*, 977–991.

Grant, R. F., and N. T. Roulet (2002), Methane efflux from boreal wetlands: Theory and testing of the ecosystem model Ecosys with chamber and tower flux measurements, *Global Biogeochem. Cycles*, *16*(4), 1054, doi:10.1029/2001GB001702.

Guertin, D. P., Barten. P. K.. and K. N. Brooks (1987), The peatland hydrologic impact model: Development and testing, *Nordic Hydrol.*, *18*, 79–100.

Gunnarsson, U., H. Rydin, and H. Sjörs (2000), Diversity and pH changes after 50 years on the boreal mire Skattlosbergs Stormosse, Central Sweden, *J. Veg. Sci.*, *11*, 277–286.

Hillel, D. (1980), *Fundamentals of Soil Physics*, Elsevier, New York.

Ingram, H. A. P. (1978), Soil layers in mires: Function and terminology, *J. Soil Sci.*, *29*, 224–227.

Ito, A., and T. Oikawa (2002), A simulation model of the carbon cycle in land ecosystems Sim-CYCLE: A description based on dry-matter production theory and plot-scale validation, *Ecol. Modell.*, *151*, 147–179.

Jackson, R. B., J. Canadell, J. R. Ehleringer, H. A. Mooney, O. E. Sala, and E. D. Schulze (1996), A global analysis of root distributions for terrestrial biomes, *Oecologia*, *108*, 389–411.

Joabsson, A., T. R. Christensen, and B. Wallen (1999), Vascular plant controls on methane emissions from northern peatforming wetlands, *Trends Ecol. Evol.*, *14*, 385–388.

Johansson, T., N. Malmer, P. M. Crill, T. Friborg, J. H. Åkerman, M. Mastepanov, and T. R. Christensen (2006), Decadal vegetation change in a northern peatland, greenhouse gas fluxes and net radiative forcing, *Global Change Biol.*, *12*, 2352–2369.

Jorgenson, M. T., Y. L. Shur, and E. R. Pullman (2006), Abrupt increase in permafrost degradation in Arctic Alaska, *Geophys. Res. Lett.*, *33*, L02503, doi:10:1029/2005GL024960.

Kalbitz, K., S. Solinger, J.-H. Park, B. Michalzik, and E. Matzner (2000), Controls on the dynamics of dissolved organic matter in soils: A review, *Soil Sci.*, *165*, 277–304.

Kettridge, N., and A. Baird (2007), In situ measurements of the thermal properties of a northern peatland: Implications for peatland temperature models, *J. Geophys. Res.*, *112*, F02019, doi:10.1029/2006JF000655.

Kettunen, A. (2003), Connection methane fluxes to vegetation cover and water table fluctuations at microsite level: A modeling study, *Global Biogeochem. Cycles*, *17*(2), 1051, doi:10.1029/2002GB001958.

King, J. Y., W. S. Reeburgh, and S. K. Regli (1998), Methane emission and transport by arctic sedges in Alaska: Results of a vegetation removal experiment, *J. Geophys. Res.*, *103*, 29083–29092.

Kirkby, M. J., P. E. Kneale, S. L. Lewis, and R. T. Smith (1995), Modelling the form and distribution of peat mires, in *Hydrology and Hydrochemistry of British Wetlands*, edited by J. M. R. Hughes and A. L. Heathwaite, pp. 83–94, John Wiley, Chichester, U. K.

Konyha, K. D., K. D. Robbins, and R. W. Skaggs (1988), Evaluating peat mining hydrology using DRAINMOD, *J. Irrig. Drain.*, *114*, 490–504.

Lafleur, P. M., N. T. Roulet, J. L. Bubier, S. Frolking, and T. R. Moore (2003), Interannual variability in the peatland-atmosphere carbon dioxide exchange at an ombrotrophic bog, *Global Biogeochem. Cycles*, *17*(2), 1036, doi:10.1029/2002GB001983.

Laine, J., H. Vasander, and R. Laiho (1995), Long-term effects of water level drawdown on the vegetation of drained pine mires in southern Finland, *J. Appl. Ecol.*, *32*, 785–802.

Laine, J., J. Silvola, K. Tolenen, J. Alm, H. Nykänen, H. Vasander, T. Sallantaus, I. Savolainen, J. Sinisalo, and P. Marikainen (1996), Effect of water-level drawdown on global climatic warming: Northern peatlands, *Ambio*, *25*, 179–184.

Lawrence, D. M., and A. G. Slater (2008), Incorporating organic soil into a global climate model, *Clim. Dyn.*, *30*, 145–160.

Lawrence, D. M., A. G. Slater, V. E. Romanovsky, and D. J. Nicolsky (2008), Sensitivity of a model projection of near-surface permafrost degradation to soil column depth and representation of soil organic matter, *J. Geophys. Res.*, *113*, F02011, doi:10.1029/2007JF000883.

Lehner, B., and P. Döll (2004), Development and validation of a global database of lakes, reservoirs and wetlands, *J. Hydrol.*, *296*, 1–22.

Le Treut, H., R. Somerville, U. Cubasch, Y. Ding, C. Mauritzen, A. Mokssit, T. Peterson, and M. Prather (2007), Historical overview of climate change, in *Climate Change 2007: The Physical Science Basis: Contribution of Working Group I to the Fourth Assessment Report of the Intergovernmental Panel on Climate Change*, edited by S. Solomon et al., pp. 93–127, Cambridge Univ. Press, Cambridge, U. K.

Letts, M. G., N. T. Roulet, N. T. Comer, M. R. Skarupa, and D. L. Verseghy (2000), Parameterization of peatland hydraulic properties for the Canadian Land Surface Scheme, *Atmos. Ocean*, *38*, 141–160.

Li, C., J. Cui, G. Sun, and C. Trettin (2004), Modeling impacts of management on carbon sequestration and trace gas emissions in forested wetland ecosystems, *Environ. Manag.*, *33*, 176–186.

Limpens, J., F. Berendse, and H. Klees (2003), N deposition affects N availability in interstitial water, growth of Sphagnum and invasion of vascular plants in bog vegetation, *New Phytol.*, *157*, 339–347.

MacDonald, J. A., D. Fowler, K. J. Hargreaves, U. Skiba, I. D. Leith, and M. B. Murray (1998), Methane emission rates from a northern wetland; response to temperature, water table and transport, *Atmos. Environ.*, *32*, 3219–3227.

Malmer, N., B. M. Svensson, and B. Wallén (1994), Interactions between Sphagnum mosses and field layer vascular plants in the development of peat-forming systems, *Folia Goebot. Phytotax., Praha*, *29*, 483–496.

Malmer, N., T. Johansson, and T. R. Christensen (2005), Vegetation, climate changes, and net carbon sequestration in a Northern-Scandinavian subarctic mire over 30 years, *Global Change Biol.*, *11*, 1895–1109.

Matthews, E., and I. Fung (1987), Methane emission from natural wetlands: global distribution, area, and environmental characteristics of sources, *Global Biogeochem. Cycles*, *1*, 61–86.

Meehl, G. A., et al. (2007), Global climate projections, in *Climate Change 2007: The Physical Science Basis: Contribution of Working Group I to the Fourth Assessment Report of the Intergovernmental Panel on Climate Change*, edited by S. Solomon et al., pp. 747–845, Cambridge Univ. Press, Cambridge, U. K.

Minkkinen, K., H. Vasander, S. Jauhiainen, S. Karsisto, and J. Laine (1999), Post-drainage changes in vegetation composition and carbon balance in Lakkasuo mire, Central Finland, *Plant Soil*, *207*, 107–120.

Minkkinen, K., R. Korhonen, I. Savolainen, and J. Laine (2002), Carbon balance and radiative forcing of Finnish peatlands 1900–2100—The impact of forestry drainage, *Global Change Biol.*, *8*, 785–799.

Mitsch, W. J., and J. G. Gosselink (2007), *Wetlands*, 4th ed., 600 pp., John Wiley, New York.

Mokany, K., R. J. Raison, and A. S. Prokushkini (2006), Critical analysis of root:shoot ratios in terrestrial ecosystems, *Global Change Biol.*, *12*, 84–96.

Moore, T. R. (2009), Dissolved organic carbon production and transport in Canadian peatlands, *Geophys. Monogr. Ser.*, doi: 10.1029/2008GM000816, this volume.

Moore, T. R., N. T. Roulet, and J. M. Waddington (1998), Uncertainty in predicting the effect of climatic change on the carbon cycling of Canadian peatlands, *Clim. Change*, *40*, 229–245.

Moore, T. R., J. L. Bubier, S. Frolking, P. Lafleur, and N. T. Roulet (2002), Plant biomass and production and CO_2 exchange in an ombrotrophic bog, *J. Ecol.*, *90*, 25–36.

Moore, T. R., P. M. Lafleur, D. M. I. Poon, B. W. Heumann, J. W. Seaquist, and N. T. Roulet (2006), Spring photosynthesis in a cool temperate bog, *Global Change Biol.*, *12*, 2323–2335.

Nilsson, M., J. Sagerfors, I. Buffam, H. Laudon, T. Eriksson, A. Grelle, L. Klemedtsson, P. Weslien, and A. Lindroth (2008), Contemporary carbon accumulation in a boreal oligotrophic minerogenic mire—A significant sink after accounting for all C-fluxes, *Global Change Biol.*, *14*, 2317–2332.

Niu, G.-Y., Z.-L. Yang, R. E. Dickinson, and L. E. Gulden (2005), A simple TOPMODEL-based runoff parameterization (SIMTOP) for use in global climate models, *J. Geophys. Res.*, *110*, D21106, doi:10.1029/2005JD006111.

Nordbakken, J. F., M. Ohlson, and P. Hogberg (2003), Boreal bog plants: Nitrogen sources and uptake of recently deposited nitrogen, *Environ. Pollut.*, *126*, 191–200.

Oliver, C. D., and B. C. Larson (1996), *Forest Stand Dynamics*, updated ed., 520 pp., John Wiley, New York.

Paavilainen, E., and J. Päivänen (1995), *Peatland Forestry: Ecology and Principles*, 248 pp., Springer, Berlin.

Page, S. E., F. Siegert, J. O. Rieley, H.-D. V. Boehm, A. Jayak, and S. Limink (2002), The amount of carbon released from peat and forest fires in Indonesia during 1997, *Nature*, *420*, 61–65.

Page, S. E., R. A. J. Wüst, D. Weiss, J. O. Rieley, W. Shotyk, and S. H. Limin (2004), A record of Late Pleistocene and Holocene carbon accumulation and climate change from an equatorial peat bog (Kalimantan, Indonesia): Implications for past, present and future carbon dynamics, *J. Quat. Sci.*, *19*, 625–635.

Page, S. E., J. O. Rieley, and R. Wüst (2006), Lowland tropical peatlands of Southeast Asia, in *Peatlands: Evolution and Records of Environmental and Climate Change*, edited by I. P. Martini, A. Martinez Cortizas, and W. Chesworth, pp. 145–171, Elsevier, Amsterdam, Netherlands.

Pastor, J., B. Peckham, S. Bridgham, J. Weltzin, and J. Q. Chen (2002), Plant community dynamics, nutrient cycling, and alternative stable equilibria in peatlands, *Am. Nat.*, *160*, 553–568.

Pauwels, V. R. N., and E. F. Wood (1999a), A soil-vegetation-atmosphere transfer scheme for the modeling of water and energy balance processes in high latitudes: 1. Model improvements, *J. Geophys. Res.*, *104*, 27,811–27,822.

Pauwels, V. R. N., and E. F. Wood (1999b), A soil-vegetation-atmosphere transfer scheme for the modeling of water and energy balance processes in high latitudes: 2. Application and validation, *J. Geophys. Res.*, *104*, 27,823–27,839.

Petrone, R. M., J. M. Waddington, and J. S. Price (2001), Ecosystem scale evapotranspiration and net CO_2 exchange from a restored peatland, *Hydrol. Process.*, *15*, 2839–2845.

Pitman, A. J. (2003), The evolution of, and revolution in, land surface schemes designed for climate models, *Int. J. Climatol.*, *23*, 479–510.

Plug, L. J., and J. J. West (2009), Thaw lake expansion in a two-dimensional coupled model of heat transfer, thaw subsidence, and mass movement, *J. Geophys. Res.*, *114*, F01002, doi:10.1029/2006JF000740.

Potter, C., J. Bubier, P. Crill, and P. Lafleur (2001), Ecosystem modeling of methane and carbon dioxide fluxes for boreal forest sites, *Can. J. For. Res.*, *31*, 208–223.

Prather, M., et al. (2001), Atmospheric chemistry and greenhouse gases, in *Climate Change 2001: The Scientific Basis: Contribution of Working Group I to the Third Assessment Report of the Intergovernmental Panel on Climate Change*, edited by J. T. Houghton et al., pp. 239–287, Cambridge Univ. Press, Cambridge, U. K.

Proctor, M. C. F. (1982), Physiological ecology: water relations, light and temperature responses, carbon balance, in *Bryophyte Ecology*, edited by A. J. E. Smith, pp. 333–382, CRC Press, London.

Proctor, M. C. F. (2000), Physiological ecology, in *Bryophyte Biology*, edited by A. W. Shaw and B. Goffinet, pp. 225–247, Cambridge Univ. Press, Cambridge, U. K.

Randall, D. A., et al. (2007), Climate Models and Their Evaluation, in *Climate Change 2007: The Physical Science Basis: Contribution of Working Group I to the Fourth Assessment Report of the Intergovernmental Panel on Climate Change*, edited by S. Solomon et al., pp. 589–662, Cambridge Univ. Press, Cambridge, U. K.

Riordan, B., D. Verbyla, and A. D. McGuire (2006), Shrinking ponds in subarctic Alaska based on 1950–2002 remotely sensed images, *J. Geophys. Res.*, *111*, G04002, doi:10.1029/2005JG000150.

Rivkina, E., D. Gilichinsky, S. Wagener, J. Tiedje, and J. McGrath (1998), Biogeochemical activity of anaerobic microorganisms from buried permafrost sediments, *Geomicrobiology*, *15*, 187–193.

Rivkina, E., K. Laurinavichius, J. McGrath, J. Tiedje, V. Shcherbakova, and D. Gilichinsky (2004), Microbial life in permafrost, *Adv. Space Res.*, *33*, 1215–1221.

Rivkina, E., V. Shcherbakova, K. Laurinavichius, L. Petrovskaya, K. Krivushin, G. Kraev, S. Pecheritsina, and D. Gilichinsky (2007), Biogeochemistry of methane and methanogenic archaea in permafrost, *FEMS Microbiol. Ecol.*, *61*, 1–15.

Roulet, N. T., P. M. Lafleur, P. J. H. Richard, T. R. Moore, E. R. Humphreys, and J. L. Bubier (2007), Contemporary carbon balance and late Holocene carbon accumulation in a northern peatland, *Global Change Biol.*, *13*, 397–411.

Rydin, H., and J. Jeglum (2006), *The Biology of Peatlands*, 343 pp., Oxford Univ. Press, Oxford, U. K.

Rydin, H., U. Gunnarsson, and S. Sundberg (2006), The role of Sphagnum in peatland development and persistence, in *Boreal Peatland Ecosystems*, edited by R. K. Wieder and D. H. Vitt, pp. 47–65, Springer, New York.

Sellers, P. J., et al. (1997), Modeling the exchanges of energy, water, and carbon between continents and the atmosphere, *Science*, *275*, 502–509.

Serreze, M. C., and J. A. Francis (2006), The arctic amplification debate, *Clim. Change*, *76*, 241–264.

Shindell, D. T., B. P. Walter, and G. Faluvegi (2004), Impacts of climate change on methane emissions from wetlands, *Geophys. Res. Lett.*, *31*, L21202, doi:10.1029/2004GL021009.

Smith, L. C., G. M. MacDonald, A. A. Velichko, D. W. Beilman, O. K. Borisova, K. E. Frey, K. V. Kremenetski, and Y. Sheng (2004), Siberian peatlands a net carbon sink and global methane source since the early Holocene, *Science*, *303*, 353–356.

Smith, L. C., Y. Sheng, G. M. MacDonald, and L. D. Hinzman (2005), Disappearing Arctic lakes, *Science*, *308*, 1429.

Smith, L. C., Y. Sheng, and G. M. Macdonald (2007), A first pan-Arctic assessment of the influence of glaciation, permafrost, topography and peatlands on northern hemisphere lake distribution, *Permafrost Periglacial Proces*ses, *18*, 201–208.

Sonnentag, O., J. M. Chen, N. T. Roulet, W. Ju, and A. Govind (2008), Spatially explicit simulation of peatland hydrology and

carbon dioxide exchange: The influence of mesoscale topography, *J. Geophys. Res.*, *113*, G02005, doi:10.1029/2007JG000605.

St-Hilaire, F., J. H. Wu, N. T. Roulet, S. Frolking, P. M. Lafleur, E. R. Humphreys, and V. Arora (2009), McGill Wetland Model: Evaluation of a peatland carbon simulator developed for global assessments, *Biogeosciences*, in press.

Stocks, B. J., et al. (2002), Large forest fires in Canada, 1959–1997, *J. Geophys. Res.*, *108*, 8149, doi:10.1029/2001JD000484, [printed *108*(D1), 2003].

Tarnocai, C. (2006), The effect of climate change on carbon in Canadian peatlands, *Global Planet. Change*, *53*, 222–232.

Tempel, P., N. H. Batjes, and V. W. P. van Engelen (1996), IGBP-DIS soil data set for pedotransfer function development, *Working Pap. Preprint 96/06*, Int. Soil Ref. and Inf. Cent., Wageningen, Netherlands. (Available at http://www.isric.org/UK/About+Soils/Soil+data/)

Thonicke, K., S. Venevsky, S. Sitch, and W. Cramer (2001), The role of fire disturbance for global vegetation dynamics: Coupling fire into a Dynamic Global Vegetation Model, *Global Ecol. Biogeogr.*, *10*, 661–667.

Thornton, P. E., and N. A. Rosenbloom (2005), Ecosystem model spin-up: Estimating steady state conditions in a coupled terrestrial carbon and nitrogen cycle model, *Ecol. Modell.*, *189*, 25–48.

Thornton, P. E., J.-F. Lamarque, N. A. Rosenbloom, and N. M. Mahowald (2007), Influence of carbon-nitrogen cycle coupling on land model response to CO_2 fertilization and climate variability, *Global Biogeochem. Cycles*, *21*, GB4018, doi:10.1029/2006GB002868.

Tolonen, K., H. Vasander, A. W. H. Damman, and R. S. Clymo (1992), Preliminary estimate of long-term carbon accumulation and loss in 25 boreal peatlands, *Suo*, *43*, 277–280.

Treat, C. C., J. L. Bubier, R. K. Varner, and P. M. Crill (2007), Timescale dependence of environmental and plant-mediated controls on CH_4 flux in a temperate fen, *J. Geophys. Res.*, *112*, G01014, doi:10.1029/2006JG000210.

Trumbore, S. E., and C. I. Czimczik (2008), An uncertain future for soil carbon, *Science*, *321*, 1455–1456.

Tuittila, E.-S., H. Vasander, and J. Laine (2003), Success of reintroduced *Sphagnum* in a cut-away peatland, *Boreal Environ. Res.*, *8*, 245–250.

Turetsky, M. R., and V. L. St. Louis (2006), Disturbance in boreal peatlands, in *Boreal Peatland Ecosystems*, edited by R. K. Wieder and D. H. Vitt, pp. 359–380, Springer, Berlin.

Turetsky, M. R., B. D. Amiro, E. Bosch, and J. S. Bhatti (2004), Historical burn area in western Canadian peatlands and its relationship to fire weather indices, *Global Biogeochem. Cycles*, *18*, GB4014, doi:10.1029/2004GB002222.

Turunen, J., E. Tomppo, K. Tolonen, and A. Reinikainen (2002), Estimating carbon accumulation rates of undrained mires in Finland—Application to boreal and subarctic mires, *Holocene*, *12*, 79–90.

Verry, E. S., and D. H. Boelter (1978), Peatland hydrology, in *Wetland Functions and Values: The State of Our Understanding*, *Proceedings of the National Symposium on Wetlands*, pp. 389–402, Am. Water Resour. Assoc., Minneapolis, Minn.

Verry, E. S., K. N. Brooks, and P. K. Barten (1988), Streamflow response from an ombrotrophic mire, in *Proceedings of the Inter-*

national Symposium on the Hydrology of Wetlands in Temperate and Cold Regions, pp. 51–59, Acad. of Finland, Helsinki.

Vitt, D. H. (2006), Functional characteristics and indicators of boreal peatlands, in Boreal Peatland Ecosystems, edited by R. K. Wieder and D. H. Vitt, pp. 9–24, Springer, New York.

Waddington, J. M., N. T. Roulet, and R. V. Swanson (1996), Water table control of CH_4 emission enhancement by vascular plants in boreal peatlands, J. Geophys. Res., 101, 22,775–22,785.

Walmsley, M. E. (1977), Physical and chemical properties of peat, in Muskeg and the Northern Environment in Canada, edited by N. W. Radforth and C. O. Brawner, pp. 82–129, Univ. of Toronto Press, Toronto, Ont., Canada.

Walter, B. P., M. Heimann, and E. Matthews (2001), Modeling modern methane emissions from natural wetlands: 1. Model description and results, J. Geophys. Res., 106, 34,189–34,206.

Wang, Y., L. A. Mysak, and N. T. Roulet (2005), Holocene climate and carbon cycle dynamics: Experiments with the 'green' McGill Paleoclimate Model, Global Biogeochem. Cycles, 19, GB3022, doi:10.1029/2005GB002484.

Washington, W. M., J. Drake, L. Buja, D. Anderson, D. Bader, R. Dickinson, D. Erickson, P. Gent, S. Ghan, P. Jones, et al. (2008), The use of the Climate-Science Computational End Station (CCES) development and grand challenge team for the next IPCC assessment: An operational plan, J. Phys. Conf. Ser., 125, doi:10.1088/1742-6596/125/1/012024.

Weiss, R., N. J. Shurpali, T. Sallantaus, R. Laiho, J. Laine, and J. Alm (2006), Simulation of water table level and peat temperatures in boreal peatlands, Ecol. Modell., 192, 441–456.

West, J. J., and L. J. Plug (2008), Time-dependent morphology of thaw lakes and taliks in deep and shallow ground ice, J. Geophys. Res., 113, F01009, doi:10.1029/2006JF000696.

Whiting, G. J., and J. P. Chanton (1993), Primary production control of methane emissions from wetlands, Nature, 364, 794–795.

Whiting, G. J., and J. P. Chanton (2001), Greenhouse carbon balance of wetlands: Methane emission versus carbon sequestration, Tellus, Ser. B, 53, 521–528.

Wickland, K. P., R. G. Striegl, J. C. Neff, and T. Sachs (2006), Effects of permafrost melting on CO_2 and CH_4 exchange of a poorly drained black spruce lowland, J. Geophys. Res., 111, G02011, doi:10.1029/2005JG000099.

Wieder, R. K., and D. H. Vitt (Eds.) (2006), Boreal Peatland Ecosystems, 435 pp., Springer, Berlin.

Wieder, R. K., D. H. Vitt, and B. W. Benscoter (2006), Peatlands and the boreal forest, in Boreal Peatland Ecosystems, edited by R. K. Wieder and D. H. Vitt, pp. 1–8, Springer, New York.

Yi, S., M. A. Arain, and M.-K. Woo (2006), Modifications of a land surface scheme for improved simulation of ground freeze-thaw in northern environments, Geophys. Res. Lett., 33, L13501, doi:10.1029/2006GL026340.

Yi, S., M. Woo, and M. A. Arain (2007), Impacts of peat and vegetation on permafrost degradation under climate warming, Geophys. Res. Lett., 34, L16504, doi:10.1029/2007GL030550.

Yoshikawa, C., M. Kawamiya, T. Kato, Y. Yamanaka, and T. Matsuno (2008), Geographical distribution of the feedback between future climate change and the carbon cycle, J. Geophys. Res., 113, G03002, doi:10.1029/2007JG000570.

Yu, Z., D. H. Vitt, I. D. Campbell, and M. J. Apps (2003), Understanding Holocene peat accumulation pattern of continental fens in western Canada, Can. J. Bot., 81, 267–282.

Yurova, A. Y., and H. Lankreijera (2007), Carbon storage in the organic layers of boreal forest soils under various moisture conditions: A model study for Northern Sweden sites, Ecol. Modell., 204, 475–484.

Yurova, A., A. Wolf, J. Sagerfors, and M. Nilsson (2007), Variations in net ecosystem exchange of carbon dioxide in a boreal mire: Modeling mechanisms linked to water table position, J. Geophys. Res., 112, G02025, doi:10.1029/2006JG000342.

Zhang, Y., C. Li, C. C. Trettin, H. Li, and G. Sun (2002), An integrated model of soil, hydrology, and vegetation for carbon dynamics in wetland ecosystems, Global Biogeochem. Cycles, 16(4), 1061, doi:10.1029/2001GB001838.

Zhang, Y., W. Chen, and D. W. Riseborough (2006), Temporal and spatial changes of permafrost in Canada since the end of the Little Ice Age, J. Geophys. Res., 111, D22103, doi:10.1029/2006JD007284.

Zhuang, Q., J. M. Melillo, D. W. Kicklighter, R. G. Prinn, A. D. McGuire, P. A. Steudler, B. S. Felzer, and S. Hu (2004), Methane fluxes between terrestrial ecosystems and the atmosphere at northern high latitudes during the past century: A retrospective analysis with a process-based biogeochemistry model, Global Biogeochem. Cycles, 18, GB3010, doi:10.1029/2004GB002239.

Zimov, S. A., S. P. Davydov, G. M. Zimova, A. I. Davydova. E. A. G. Schuur, K. Durra, and F. S. Chapin III (2006), Permafrost carbon: Stock and decomposability of a globally significant carbon pool, Geophys. Res. Lett., 33, L20502, doi:10.1029/2006GL027484.

Zoltai, S. C. (1993), Cyclic development of permafrost in the peatlands of northwestern Alberta, Canada, Arct. Alp. Res., 25, 240–246.

Zoltai, S. C., R. M. Siltanen, and J. D. Johnson (2000), A wetland database for the western boreal, subarctic, and arctic regions of Canada, Inf. Rep. NOR-X-368, North. For. Cent., Can. For. Serv., Edmonton, Alberta, Canada.

S. Frolking, Complex Systems Research Center, Institute for the Study of Earth Oceans and Space, 8 College Road, University of New Hampshire, Durham, NH 03824, USA. (steve.frolking@ unh.edu)

D. Lawrence, Climate and Global Dynamics Division, National Center for Atmospheric Research, Boulder, CO 80305, USA.

N. Roulet, Department of Geography and McGill School of the Environment, McGill University, Montreal, QC H3A 2K6, Canada.

Upscaling of Peatland-Atmosphere Fluxes of Methane: Small-Scale Heterogeneity in Process Rates and the Pitfalls of "Bucket-and-Slab" Models

A. J. Baird

School of Geography, University of Leeds, Leeds, UK

L. R. Belyea and P. J. Morris

Department of Geography, Queen Mary University of London, London, UK

Most models of carbon balance processes in northern peatlands were developed to apply at the scale of small peatland features such as hummocks and hollows, which may have linear extents of $10^0–10^1$ m, or at the scale of individual peatlands (up to 10^4 m). However, many have been applied at the much larger scale of cells in land surface schemes (typically, 1° latitude × 1° longitude or larger), which are linked to global circulation models. In these larger representations, as with the scale for which the models were originally developed, they take some account of vertical variability through the peat profile but ignore horizontal variability in vegetation and peat properties. That is, they contain a single vegetation type, a horizontally uniform slab of peat, and predict a single water table position: they represent the peatland as a "bucket and slab." In this chapter, we examine the accuracy of assuming that carbon balance processes in spatially variable peatlands can be represented as a bucket and slab. We focus on methane (CH_4) fluxes from peatlands and, using scaling theory, suggest there is a need to consider small-scale patterning ($10^1–10^2$ m) when estimating whole-peatland behavior. In other words, we propose that the bucket-and-slab approach may not be suitable for upscaling peatland carbon cycling processes. We present an alternative, but untested, approach as well as an agenda for future work on how interactions across scales affect peatland functioning.

1. THE NEED FOR MODELS OF PEATLAND-ATMOSPHERE CARBON EXCHANGES

Through the Holocene, northern peatlands have been a persistent net sink of atmospheric carbon dioxide (CO_2) and

Carbon Cycling in Northern Peatlands
Geophysical Monograph Series 184
Copyright 2009 by the American Geophysical Union.
10.1029/2008GM000826

a persistent source of methane (CH_4) [e.g., *Frolking et al.*, 2006; *Smith et al.*, 2004]. Although modeling work [e.g., *Frolking et al.*, 2006] indicates that the net effect of peatland growth through the Holocene has been one of global cooling, there is considerable uncertainty about whether peatlands will have a global warming or cooling effect in the next 20–200 years as the climate changes [e.g., *Belyea and Baird*, 2006; *Belyea and Clymo*, 2001; *Rydin and Jeglum*, 2006; *Smith et al.*, 2004]. This uncertainty is coupled with concern, because northern peatlands constitute a huge store of carbon; it has been estimated that northern peatlands contain between 270 and 455 Pg (10^9 t) of carbon (C) [cf. *Gorham*, 1991; *Smith et al.*, 2004; *Turunen et al.*, 2002], which is equivalent to 35–60% of the total amount of carbon stored in the atmosphere (which is approximately 750 Pg). The functioning of this carbon store could change dramatically over the coming decades as the climate changes [*Rydin and Jeglum*, 2006]. For example, in Arctic and subarctic regions, melting of permafrost may be accompanied by the "explosive" development [cf. *Smith et al.*, 2004; *Yu et al.*, this volume] of new peatlands on former tundra soils which could lead to a rapid and substantial drawdown of atmospheric CO_2. On the other hand, permafrost melting may lead to widespread formation of thermokarst pools and lakes and enhanced rates of CH_4 production [e.g., *Turetsky et al.*, 2007], CH_4 being a much more potent greenhouse gas than CO_2. Peatland pools are hot spots of CH_4 loss to the atmosphere [*Hamilton et al.*, 1994; *Waddington et al.*, 1996], and thermokarst lakes (including those in nonpeatland areas) may have made a substantial contribution to the >30% increase in atmospheric CH_4 concentration that occurred at the Pleistocene-Holocene transition [*Walter et al.*, 2007]. In the southern part of their range, peatlands may shrink as higher temperatures lead to higher rates of peat decay or desiccation, cracking, and erosion. It is also likely that many peatlands will experience changes in the composition of their vegetation, which, in turn, could dramatically alter carbon balance processes [cf. *Christensen et al.*, 2003; *Strack et al.*, 2006; *Ström et al.*, 2003, 2005].

Because the response of peatlands to climate change is likely to be complicated, a predictive modeling approach is required in which the key processes involved in peatland carbon cycling are explicitly represented. However, if models are to be of use, they need to be coupled with global circulation models (GCMs) (also called global climate models) so that peatland-atmosphere feedbacks are properly accounted for. How such coupling might work forms the focus of *Frolking et al.* [this volume]. Although we briefly consider such coupling, our main aim is to consider a related issue: how understanding peatland carbon balance at the small scale (10^0–10^1 m) can be used to derive estimates of peatland-atmosphere carbon exchanges at the regional scales used in GCMs.

2. VARIABILITY OF CARBON BALANCE PROCESSES WITHIN PEATLANDS

2.1. GCM Scales

GCMs work at large scales. A typical GCM is the United Kingdom Hadley Centre's coupled ocean-atmosphere model HadCM3 [cf. *Cox et al.*, 1999]. In plan, the atmospheric model has a resolution of 2.5° × 3.75° (approximately 300 × 300 km at midlatitudes). GCMs can be coupled with land surface models that treat the land surface as a dynamic system in which vegetation and near-surface soil properties such as carbon storage can change over time. Example applications of coupled GCM–land surface models are given by *Cox et al.* [2000] and *Notaro et al.* [2007]. A recently developed land surface scheme is the Joint UK Land Environment Simulator (JULES) (http://www.jchmr.org/jules/ (accessed on 27 October 2008) [*Blyth et al.*, 2006; *Sitch et al.*, 2008]), which combines two previous models: one which describes surface-atmosphere energy and water exchanges (Met Office Surface Exchange Scheme (MOSES) [cf. *Cox et al.*, 1999]) and one which describes surface vegetation dynamics (Top-down Representation of Interactive Foliage and Flora Including Dynamics (TRIFFID) [cf. *Cox*, 2001]). Models such as JULES operate at a similar scale to individual computational cells used in GCMs. Sometimes, different vegetation types may be represented as tiles within a GCM cell, but these tiles still represent large areas: MOSES, for example, recognizes nine land surface (or cover) types, so the maximum number of tiles representing a GCM cell would be nine. In addition, soil variability at the GCM subgrid scale is still often ignored. Effectively, land surface schemes adopt an approach similar to the "mean field" approach described in section 3 and make the implicit assumption that the different types of soil within a GCM land surface cell can be described satisfactorily by an equivalent single soil type. Additionally, variability in vegetation is only rather crudely represented, if at all. How reasonable is the "mean field" assumption?

2.2. Scales of Carbon Cycling in Peatlands

Before we attempt to answer this question, we consider the fundamental scales at which dynamics in the peatland carbon store may be conceptualized. The largest individual peatlands are usually of the order of 10^4 m in length or at least an order of magnitude smaller than the cells in GCMs. There is clear evidence that carbon balance processes within peatlands show wide variability; for example, within-peatland variability of CH_4 emissions may be an order of magnitude greater than variability between individual peatlands of the same generic type (such as continental raised bog) [*Waddington et al.*, 1996].

Variability in vegetation and soil properties in peatlands may be described with a range of terms that are not always used consistently by different peatland scientists. Particular difficulty attends to the use of prefixes such as micro, meso, and macro, which can mean vastly different scales depending on context and scientific discipline. Even within peatland

Figure 1. Examples of fundamental SL1 units at Cors Fochno, an oceanic raised bog in west Wales (photograph credit: A. Baird). The hummock vegetation is dominated by *Calluna vulgaris* (L.) Hull, *Erica tretralix* L., *Myrica gale* L., *Eriophorum vaginatum* L., and *Sphagnum capillifolium* (Ehrh.) Hedw. The lawn vegetation is dominated by *Sphagnum pulchrum* (Lindb. ex Braithw.) Warnst. and *Rhynchospora alba* (L.) Vahl. The hollow/pool contains algal communities and *Menyanthes trifoliata* L. The horizontal distance in the foreground of the image is approximately 250 cm.

can be separated from each other by their distinctive plant communities and other attributes like water table depth (see Figure 1, Table 1, and below). We can think of them as fundamental in the sense that they represent the smallest scale at which we can identify plant communities in peatlands; at smaller scales we may find individual species of plant but not groupings of different species. However, more importantly, they are fundamental in terms of their ecohydrological functioning and are the smallest units that have separable and distinctive biogeochemical and water table behavior and regimes of net peat accumulation [*Belyea and Clymo*, 2001]. At length scales of 10^1–10^2 m, which we call scale level 2 (SL2), the peatland may be characterized by nonrandom arrangements of the fundamental units. For example, aggregations of hummocks, called ridges or "strings," may form (principally) along the contours of the peatland and may be interspersed with "flarks" of lawns, hollows, or pools (see Figure 2). At length scales of 10^2–10^4 m (SL3), such as the dome of peat comprising a raised bog, the peatland surface may display variation in the different types of SL2 patterns. Thus, the central area of a raised bog may be dominated by large, irregularly shaped hollows or pools, the margin may be dominated by a high cover of hummocks, and the intervening annulus may be dominated by a striped pattern of contour-parallel strings and flarks [*Ingram*, 1983; *Ivanov*, 1981].

On bogs, or ombrotrophic peatlands, hummocks (Figure 1) are SL1 mounds or raised areas of peat (their surface may be 20–60 cm above the local water table) with a plant assemblage often consisting of small-leaved *Sphagnum* mosses such as *S. fuscum* (Schimp.) Klinggr., and *S. capillifolium* (Ehrh.) Hedw., ericaceous shrubs such as *Calluna vulgaris* (L.) Hull, and stunted trees of species such as *Pinus sylvestris* L. Lawns, hollows, and pools may be found between hummocks. The water table rarely rises to the surface of lawns, whereas hollows may be inundated during wet periods in

science, different workers use terms such as "mesotope" differently, which inevitably creates problems when we wish to understand process interactions across scales. What is a problem for peatland scientists may be a bigger source of confusion to other groups of scientists such as climate modelers. Therefore, we propose here a simple classification of scales. Peatlands may be considered as consisting of arrays of fundamental units, scale level 1 (SL1), called hummocks, lawns, hollows, and pools (see below for descriptions), which typically vary in length from 10^0 to 10^1 m and which

Table 1. Vegetation Patterns Within Individual Northern Peatlands and Their Characteristic Spatial Scales[a]

Scale Level	Example	Characteristic Length Scale (m)
1	Fundamental functional unit: hummocks, lawns, hollows, and pools. See Figure 1.	10^0–10^1
2	Aggregations or collections of SL1 units. Hummocks may coalesce to form ridges or strings, and these may be separated by linear pools, hollows, or lawns. See Figure 2.	10^1–10^2
3	The pattern of vegetation across an entire peatland unit like the dome of a raised bog. SL2 patterning may vary between the central dome and the margin of the dome. Thus, the center may be characterized by a random arrangement of large pools, the margin may be characterized by a high cover of hummocks, and the intervening annulus may be characterized by largely contour-parallel strings and flarks.	10^2–10^4

[a]Individual means a single peatland, like an individual raised bog.

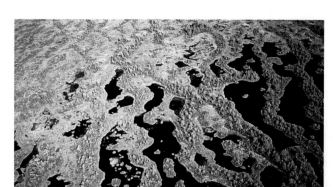

Figure 2. Aggregations of fundamental units in plan. High oblique view over Männikjärve bog, Estonia, showing linear aggregations of hummocks, pools, and hollows. Scale varies between foreground and background. The ridges separating the large pools in the upper right of the photograph are approximately 20 m wide. For further information see *Aber and Aber* [2001] and *Aber et al.* [2002]. Image was acquired by kite aerial photography in September 2001. Copyright J. S. Aber and S. W. Aber, used here with permission.

autumn, winter (in areas without long-lasting snow cover), and spring. The vegetation of lawns and hollows usually comprises large-leaved sphagna such as *S. papillosum* Lindb. (lawns) and *S. cuspidatum* Ehrh. ex Hoffm. (hollows), and sedges such as *Eriophorum angustifolium* Honck. and *Rhynchospora alba* (L.) Vahl. (Figure 1); some hollows may show little moss or sedge cover and, instead, may be mud-bottomed with algal communities (Figure 1). Pools are areas of the peatland that are perennially inundated. They sometimes contain aquatic sphagna and emergent plants such as *Menyanthes trifoliata* L.

How do carbon balance processes vary between SL1 units on peatlands? This question has been addressed both practically via field measurements and more theoretically via proposing simple models of SLI unit growth over time. In terms of the former, *Waddington and Roulet* [1996], for example, found that CH_4 fluxes from pools at the margin of an ombrotrophic bog in central Sweden exceeded fluxes from adjacent ridges by between a factor of 8 and one of 50. Mean integrated summer fluxes, as measured using static chambers, were 12.90 ± 0.72 g CH_4 m^{-2} in 1992 and 12.55 ± 0.35 g CH_4 m^{-2} in 1993 for pools. For adjacent ridges, they were 0.26 ± 0.06 g CH_4 m^{-2} in 1992 and 1.61 ± 0.35 g CH_4 m^{-2} in 1993. Several studies have found that rates of peat formation and net ecosystem exchange (NEE) of CO_2 (here we use the convention that NEE is positive when the flux is from atmosphere to peatland) vary nonlinearly with water table depth, with higher rates occurring in lawns and low hummocks than in either wetter (hollows and pools) or drier (high hummocks) SL1 units [e.g., *Alm et al.*, 1997; *Belyea and Clymo*, 2001; *Laine et al.*, 2007]. On an Irish blanket bog, for example, annual NEE varied from a net uptake of ~650 g CO_2 m^{-2} for sites at the ecotone between high lawn and hummock to no net uptake or small losses to the atmosphere for permanently inundated hollows [*Laine et al.*, 2007]. *Belyea and Clymo* [2001] concluded that the past and present role of peatlands in the global carbon cycle could be understood only if feedbacks involving SL1 units were taken into account [see also *Belyea*, this volume]. This idea was developed by *Belyea and Baird* [2006], who suggested that peatlands behave as complex adaptive systems and that models of peatland carbon dynamics have to consider explicitly the cross-scale linkages between SL1 units and the entire peatland. These ideas will be elaborated on in sections 3 and 4; we now consider the scales at which carbon balance processes are considered in current peatland models. We do so by focusing on models that predict fluxes of CH_4 between peatlands and the atmosphere, partly because most existing peatland models also focus on CH_4. A chapter that looked at CO_2 fluxes as well would offer a wider perspective but would be too long to include in this volume. Nevertheless, much of what we say applies to peatland-atmosphere CO_2 exchanges and not just to CH_4.

2.3. Models of Carbon Cycling in Peatlands

There are many models that simulate peatland-atmosphere carbon exchanges. Examples include the wetland/peatland CH_4 models presented by *Cao et al.* [1996], *Frolking et al.* [2002], *Grant and Roulet* [2002], *Kettunen* [2003], *Walter et al.* [1996] (and later *Walter et al.* [2001a, 2001b]), and *Zhang et al.* [2002]. Models have also been developed to simulate peatland development over time scales of 10^2–10^3 years, and examples include those of *Clymo* [1984], *Clymo et al.* [1998], *Frolking et al.* [2001], *Gilmer et al.* [2000], *Hilbert et al.* [2000], and *Yu et al.* [2003].

A feature of all but one of these models (Ecosys, see section 4) is that they ignore horizontal variability in vegetation and peat properties; that is, they treat the peatland as a slab or single column of peat. The models vary somewhat in terms of the scale for which they were originally developed, but all were designed initially to apply to scales much smaller than the scale of cells or tiles in land surface schemes linked with GCMs. For example, the models of *Kettunen* [2003] and *Walter et al.* [1996] were initially developed to apply at SL1. Both consider depth variability in CH_4 production and consumption (oxidation), and both use measured values of the water table as inputs. A later version of the *Wal-*

ter et al. [1996] model presented by *Walter et al.* [2001a, 2001b] simulates water table position using a simple store and flow model and was applied by the authors to scales of $1° \times 1°$, which is similar to the scales used in coupled land surface models and GCMs. The Wetland Methane Emission Model of *Cao et al.* [1996], although simpler in terms of its mechanistic descriptions of CH_4 production and oxidation, is very similar in its basic conceptual structure to the model of *Walter et al.* [2001a, 2001b], contains a "bucket" hydrology (in part) (see below), and has been applied at the $1° \times 1°$ scale. The Peatland Carbon Simulator model of *Frolking et al.* [2002] was developed to apply to individual peatlands (i.e., SL3), but the authors discuss how it might be applied at larger scales, in particular how it might be linked with GCMs. The current version of the model requires soil climate variables (water table, soil moisture, and temperature) as inputs, although the authors note that these variables can be simulated using other models.

With most of the wetland/peatland CH_4 models that have been developed, it is clear that the authors assume that the models can be applied successfully at large (SL3 and greater) scales, even though the models do not take account of lateral variability in peat physical and chemical properties or in vegetation. When applied at large scales, these models, in essence, consider peatlands as horizontally homogenous slabs of peat in which the gross hydrological functioning of the slab is treated as a simple store and on which there is a single community of plants. Some hydrological submodels account for vertical variability of hydrological variables such as soil wetness in the zone above the water table, but even then they may be considered akin to a leaky bucket that receives water as precipitation and loses water via one evaporation/transpiration outlet (representing the single vegetation type) and one or two liquid water outflow(s) (overland flow and subsurface seepage). Thus, most existing peatland CH_4 models may be considered to be "bucket-and-slab" models and therefore a variant of land surface models such as JU-LES (see section 2.1).

What does it mean to scale up processes in the peatland CH_4 models using a bucket-and-slab approach? Clearly, there is a need to allocate properties to each land surface scheme tile or cell. Decisions are required on whether or not a cell contains peatland, and, if so, how many tiles to use to represent the peatland. Difficulties may arise if, at the scale of the tile, there is still variation in peatland types with differences in vegetation and hydrological behavior between the different types. It is then necessary to choose a single peat profile and vegetation type that captures this variability and produces peatland-atmosphere CH_4 fluxes that are an accurate representation of real-world fluxes integrated across the heterogeneous landscape.

In section 3, we consider the theoretical basis of scaling from SL1 to larger scales. In particular, we address the question of whether it is possible to ignore SL1 variability in larger-scale models. In other words, we address the question of whether a heterogeneous peatland landscape can be accurately represented as an equivalent homogenous one. As noted in section 2.2, we do so from the perspective of peatland-atmosphere CH_4 fluxes, but much of what we say also applies to CO_2 exchange. We restrict our attention to the relationship between peatland-atmosphere CH_4 fluxes at SL1 and at SL3 (whole peatland). Our analysis is deliberately simple; we use an empirical relationship between CH_4 emissions and water table position and consider how variability of water table levels across a peatland affects whole-peatland CH_4 fluxes. Our CH_4–water table model is one of the empirical equations of *Bubier et al.* [1993] in which CH_4 emissions from boreal peatlands in northern Ontario, Canada, decline exponentially with increasing water table depth below the peatland surface. Similar relationships have been presented by *Bubier* [1995] and *Bubier et al.* [1995]. We could have used one of the aforementioned peatland CH_4 models, but our conclusions would not have changed qualitatively, provided there is a nonlinear relationship between CH_4 emissions from a peatland and water table level.

3. UPSCALING CARBON BALANCE PROCESSES IN PEATLANDS: PROBLEMS WITH THE BUCKET AND SLAB

Problems of treating a peatland as a bucket and slab arise mainly because of spatial and temporal variability [cf. *Strayer et al.*, 2003]. First, peatland-atmosphere exchanges of CH_4 and CO_2 respond in a nonlinear way to spatial variations and temporal fluctuations in driving variables. CH_4 flux decreases in a monotonically and strongly nonlinear way as water table depth increases. As a result, those places on the peatland surface that are wet (i.e., low water table depth), and the times when the peatland surface is wet (the wet season and periods of rainfall more generally), make a disproportionately large contribution to the total annual flux. In contrast, peatland uptake of CO_2 shows a humped-back relationship with water table depth, such that uptake is greatest in places and at times where the water table is at intermediate levels [e.g., *Alm et al.*, 1997; *Laine et al.*, 2007]. Second, water flow is directional. As a result, the water balance of the peatland (and hence the distribution of water table depths within it) depends on the spatial arrangement of SL1 units with contrasting hydraulic properties and contrasting rates of evapotranspiration or net rainfall (see section 3.2). Third, the CH_4 flux at any point in time and space is dependent on past events owing to temporal lags in response and structural "legacies" (or ecological "memory") that affect present-day behavior. For example, rates of decomposition and CH_4 production in the litter and peat below a hummock may be affected by the history of SL1 units at that position in the peatland. Thus, a part of a peatland that has been a hummock for hundreds of years may have quite different rates of CH_4 loss or CO_2 uptake from a part which has only recently undergone "microsuccession" to a hummock and which was previously a lawn. We examine each of the problems in more detail below. We do so initially from the perspective of upscaling from scale level 1 measurements. This allows us to assess the accuracy with which models that do not consider SL1 or SL2 variability predict CH_4 fluxes from whole peatlands. We focus on spatial variability, although temporal variability, especially diurnal and seasonal variability, also deserves greater attention when considering peatland carbon balance processes.

3.1. Interactions Between Spatial Variability and Nonlinearity

The first problem associated with spatial variability can be illustrated by examining biases that arise when different methods are used to upscale fluxes from SL1 units to larger areas. Such upscaling corresponds to situations where measurements of CH_4 flux have been made at the scale of the flux chamber (typically smaller than a SL1 unit and described as a "point" below) and where we wish to estimate whole-peatland CH_4 fluxes [e.g., *Pelletier et al.*, 2007]. However, we can also think of the problem as being one of how well bucket-and-slab models reflect fluxes in a peatland that displays pronounced spatial variability in CH_4 fluxes at sub-peatland scales (SL1 and SL2). Broadly speaking, there are three common approaches to upscaling: distributed, mean field, and lumped methods. Each is described below.

3.1.1. Distributed method. The most direct way of obtaining the spatially averaged flux per unit area, F_{XY}, over a peatland is to take the average of n point fluxes, $F(x, y)$:

$$F_{XY} = \frac{1}{n} \sum_1^n F(x, y).$$ (1)

In most applications, $F(x, y)$ will be measured only at a small number of points (x, y) because of the cost and time required for data collection. If the point flux can be predicted from a more easily measured (or modeled) point variable, $a(x, y)$, then F_{XY} can be estimated by computing the point flux as a function of the independent variable, $G[a(x,y)]$, weighted by the probability density function of the independent variable, $P[a(x,y)]$, and integrated over the full range of a and the spatial domain XY:

$$F_{XY} = \iiint_{YXa} G[a(x,y)] P[a(x,y)] da \, dx \, dy.$$ (2)

If the flux or predictor variable is measured for discrete classes (e.g., land surface types) rather than continuously, then the estimate is obtained by summing the class-specific estimate, F_k or $G(a_k)$, weighted by the relative area covered by that class, w_k, over all classes, K:

$$F_{XY} = \sum_K^0 w_k F_k \text{ or } F_{XY} = \sum_0^K w_k G(a_k).$$ (3)

The main limitation of the distributed method is that it requires a large amount of information on the distribution of land surface classes (equation (3)) or of a continuous predictor variable (equation (2)). Nevertheless, this method of estimating fluxes from distinct land surface classes is probably the most commonly used for upscaling chamber-based estimates of carbon gas flux from peatlands [e.g., *Waddington et al.*, 1996; *Pelletier et al.*, 2007], although most applications of the method have used a small number of measurements (relative to peatland area). Some studies [e.g., *Laine et al.*, 2006] have taken a hybrid approach, in which fluxes for a particular class of SL1 unit are integrated through the seasonal cycle using independent driving variables such as water table depth (e.g., a time-equivalent version of equation (2)) and integrated across space by weighting the seasonal, class-specific fluxes by the relative abundances of the different classes (e.g., equation (3)).

3.1.2. Mean field method. A simple method for computing the scale level 3 (SL3) (and greater) flux is to use the spatial average of the predictor variable, μ_a, as follows:

$$F_{XY} \cong G(\mu_a).$$ (4)

This method is known as the mean field method, and its main advantage is that we need to know only the mean value of the predictor variable. A serious disadvantage is that the method may give highly biased estimates if the predictor variable shows spatial variance and the function that relates predictor and response variables is nonlinear.

In the case of peatland-atmosphere exchanges of CO_2 and CH_4, large bias errors are almost inevitable. Predictor variables such as water table depth and vascular plant biomass or leaf area show high spatial variance at spatial scales from SL2 to SL3 (see sections 2 and 3.1.1). Moreover, empirical and process-based functions relating fluxes to these predictor variables are highly nonlinear for both CH_4 [e.g., *Bubier et al.*, 1993; *Bubier*, 1995; *Bubier et al.*, 1995] and CO_2 [cf. *Alm et al.*, 1997; *Laine et al.*, 2007]. Despite these potential biases, the mean field method is used implicitly in bucket-and-slab models of peatland-atmosphere carbon exchanges. In other words, it is assumed that these models provide some sort of representative average of the predictor variable.

3.1.3. Lumped method. Scale transition theory [*Hu and Islam*, 1997; *Baldocchi et al.*, 2005; *Melbourne and Chesson*, 2006] provides a theoretical framework for correcting mean field estimates by taking into account the interaction between the nonlinearity of the function $G(a)$ and the spatial variance, σ_a^2, of the predictor variable a. The magnitude of the bias error is the difference between the true whole-system (SL3) flux, F_{XY}, and the mean field estimate. The correction term (i.e., the second term on the right-hand side of equation (5)) is estimated by a second-order approximation of a Taylor's series expansion:

$$F_{XY} \approx G(\mu_a) + \frac{1}{2}G''(\mu_a)\sigma_a^2, \qquad (5)$$

where $G''(\mu_a)$ is the second derivative of $G(a)$ evaluated at μ_a. If necessary, this approximation can be extended to include additional functions and predictor variables [*Hu and Islam*, 1997; *Melbourne and Chesson*, 2006].

The obvious advantages of this method over the distributed method are that it is much more efficient computationally and that it has much more manageable data requirements. For retrospective analyses, the need for intensive ground-based estimates of the predictor variable a might be obviated by high spatial or spectral resolution remotely sensed data [e.g., *Harris et al.*, 2005, 2006; *Harris*, 2008; *Harris and Bryant*, this volume], which potentially could provide estimates of the mean and spatial variance. For prognostic analyses, the mean value would be provided as the output of a bucket-and-slab model, whereas spatial variance could be estimated for present-day conditions and then assumed invariant into the future. We are not aware of this method being applied to peatland carbon flux estimates, but it is commonly applied to other ecosystem types in studies of surface energy balance [cf. *Baldocchi et al.*, 2005].

Using the lumped method on some hypothetical peatlands helps to illustrate the biases associated with the mean field method. Here we apply the "combined" model of *Bubier et al.* [1993] to several normal distributions of peatland water table depth, described by mean, μ_a, and standard deviation, σ_a. By substituting the function and parameter values of *Bubier et al.* [1993] into equation (5) and simplifying, we can compute a lumped estimate of the spatial average of peatland CH_4 flux, F_{XY}, for any combination of μ_a and σ_a:

$$F_{XY} \approx G(\mu_a) + \frac{1}{2}[0.00649483\,G(\mu_a)]\sigma_a^2, \qquad (6)$$

where $G(\mu_a) = 10^{1.815-0.035a}$. The mean field estimate is represented by $\sigma_a = 0$ cm, i.e., the lowest curve in Figure 3a and the y intercepts in Figure 3b. Owing to the interaction in the second term on the right-hand side of equation (6), the mean field method grossly underestimates CH_4 flux when spatial variance is large, especially in places where the water table is close to the peatland surface (i.e., μ_a close to zero) (Fig-

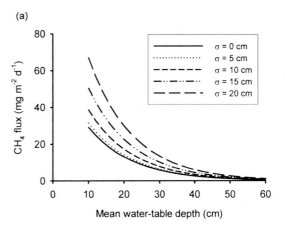

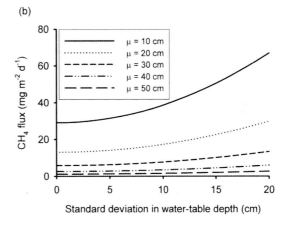

Figure 3. Effects of spatial heterogeneity in water table depth on CH_4 flux (as mg CH_4 m^{-2} d^{-1}). The equation used to compute CH_4 flux is from *Bubier et al.* [1993]. Fluxes have been computed using the "lumped method" of upscaling (see text). The "mean field method" is represented by $\sigma = 0$ cm (no spatial variance). The maximum σ shown (20 cm) is reasonable for heterogeneous peatlands (L. R. Belyea, unpublished data, 2004). (a) Effect of increasing mean water table depth, μ, for different degrees of spatial variance, expressed by standard deviation, σ. (b) Effect of increasing spatial variance of water table depth, expressed by standard deviation, σ, for different mean values, μ. Note that the mean field method gives highly biased estimates when spatial variance is large, especially when mean water table depth is close to the peatland surface (i.e., μ approaches 0 cm).

ure 3). As discussed in section 2.2 (see also Figures 1 and 2), many real peatlands show large spatial variance in water table depth, so this issue raises considerable doubt over the validity of the bucket-and-slab approach.

Although the lumped method might appear to provide a simple statistical solution to the bias errors of the mean field method, there is a further complication. On peatlands with two types of SL1 feature, such as hummocks and hollows or hummocks and pools (see Figure 2), the distribution of water table depths is often bimodal [*Eppinga et al.*, 2008], and hence the predictor variable (e.g., water table depth) is not normally distributed (see also section 3.2). In such cases, the lumped model can be extended by representing the distribution of water table depths, P, as the union of two normal distributions with different means, μ_1 and μ_2, and spatial

variances, expressed by standard deviations σ_1 and σ_2 [*Eppinga et al.*, 2008]:

$$P = qN(a,\mu_1,\sigma_1) + (q-1)N(a,\mu_2,\sigma_2), \quad (7)$$

where q is a proportion between 0 and 1 (Figure 4a) and N denotes normal. By applying equation (6) separately to each mode of water table depth and then adding the results (weighted by q and $q-1$) together, we see that CH_4 flux increases with increasing bimodality, especially when spatial variance is high (Figure 4b). As is the case with the unimodal distribution (Figure 3), the mean field method ($\sigma = 0$ cm; $\mu_2 - \mu_1 = 0$ cm) severely underestimates CH_4 flux when the distribution is bimodal. Codominance (in terms of cover) by two types of SL1 feature such as those shown in Figure 2

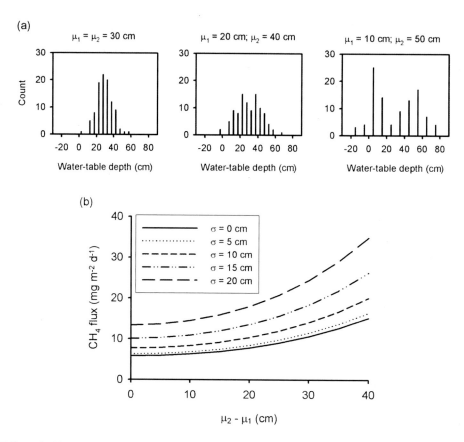

Figure 4. Effect of a bimodal distribution of water table depth on CH_4 flux. The equation used to calculate CH_4 flux is from *Bubier et al.* [1993]. Fluxes have been computed using the "lumped method" of upscaling, taking into account the bimodal distribution (see text). The maximum degree of bimodality shown (40 cm) is conservative for strongly patterned peatlands [cf. *Eppinga et al.*, 2008]. (a) Histograms of water table depth for three hypothetical sites with increasing bimodality, $\mu_2 - \mu_1$. In all cases, $q = 0.5$ and $\sigma = 10$ cm. (b) Effect of increasing bimodality for different degrees of spatial variance expressed by standard deviation, σ. The "mean field method" is represented by $\sigma = 0$ cm (no spatial variance) and $\mu_2 - \mu_1 = 0$ cm (unimodal distribution). In all cases, the mean water table depth is 30 cm. Note that CH_4 flux increases with increasing bimodality, especially when spatial variance is high.

occurs on many peatlands, and the effect of this bimodality on lumped estimates of CH_4 flux is a further reason why bucket-and-slab models cannot work for many peatlands.

Can the lumped method and its bimodal extension be used to improve bucket-and-slab models? It is important to note that spatial variance σ_a^2, and hence the magnitude of the correction term [Baldocchi et al., 2005], decreases as the window size over which it is estimated increases [Levin, 1992]. If the window size is too coarse, the magnitude of the correction term will be underestimated. This is a problem for peatlands because driving variables such as water table depth and vegetation biomass are heterogeneous on spatial scales approximating the pixel size of even the highest-resolution satellite sensors (e.g., 0.8 m for IKONOS panchromatic band). Because several pixels must be included in the sampling window used to calculate spatial statistics, estimates of σ_a^2 derived directly from satellite-based remote sensing data may lead to underestimation of the correction term. A potential solution to this problem involves finding a predictive relationship between spatial variance and window size [Baldocchi et al., 2005] and then inflating the correction term accordingly. For peatlands with bimodal aggregations of SL1 units, however, the challenges are greater still. First, the bucket-and-slab model must give not one but two values for the mean water table depth (μ_1 and μ_2). Second, spatial statistics are required for two landscape elements (q, σ_1, and σ_2). For peatlands that show weak bimodality (e.g., the middle plot in Figure 4a), it may be difficult to classify the peatland into just two distinct landscape elements. Moreover, it may be difficult to obtain reliable scaling relationships if spatial variances of the landscape elements are unequal ($\sigma_1^2 \neq \sigma_2^2$). A solution that uses multiple classes, i.e., one that considers SL1 units, is perhaps the most logical choice [cf. Strayer et al., 2003].

We now turn to the second and third problems identified in section 3. Our consideration of these reinforces the idea that accurate simulation of whole-peatland carbon exchanges will require explicit consideration of SL1 units and their spatial arrangement.

3.2. Interactions Between Spatial Arrangement and Directionality

To examine the effect of the spatial arrangement of SL1 features on whole-peatland water balance (and, therefore, carbon balance), we considered a theoretical raised bog. To quantify the effect of spatial arrangement on overall peatland water balance, we used the frequency distribution of the peatland water table. We conducted three simulations using the newly developed model DigiBog, details of which will be reported in another paper (A. J. Baird et al., Growing bogs in computers: A new conceptual model and a new computer model of bog hydrology—DigiBog, manuscript in preparation, 2009). The gross shape of the theoretical bog was generated using the groundwater mound equation used by Ingram [1982] and by assuming a drought-year net rainfall (i.e., precipitation minus evapotranspiration) of 15 cm a^{-1} and a catotelm hydraulic conductivity (K) of 0.00125 cm s^{-1}. Plan model dimensions of 200 m (along slope) × 70 m (across slope) gave a maximum catotelm thickness of 397 cm, thinning elliptically to 28.2 cm. The model aquifer was divided into a horizontal (x, y) grid of square-sectioned columns (see below) with plan dimensions of 1 m × 1 m. Groundwater flow within the bog aquifer was simulated using a solution to the following version of the Boussinesq equation [cf. McWhorter and Sunada, 1977]:

$$\frac{\partial h}{\partial t} = \frac{\partial}{\partial x}\left(\frac{K(h)}{s(h)}h\frac{\partial h}{\partial x}\right) + \frac{\partial}{\partial y}\left(\frac{K(h)}{s(h)}h\frac{\partial h}{\partial y}\right), \quad (8)$$
$$+ P(t) - E(h,t),$$

where h is water table elevation (L) above an impermeable layer of mineral soil or rock, t is time (T), x and y are the plan coordinates, K is hydraulic conductivity (L T^{-1}), s is drainable porosity (dimensionless), P is the rate of rainfall addition to the water table (L T^{-1}), and E is the rate of evaporative (or transpirative) loss of water from the water table (L T^{-1}). Groundwater flow in the two plan directions is simulated; however, because vertical variability in K and s is also accounted for in the model, it can be thought of as 2½-dimensional. DigiBog ignores vertical flow, which may be important in some peatlands and hydrogeological settings [e.g., Reeve et al., 2000, this volume].

In DigiBog, a peatland is conceptualized as a grid of square-sectioned columns between which water can flow according to a simple finite difference solution to equation (8). The variation in K and s with depth through the profile of each column is represented by layers which make up the column. These layers can have any thickness, and a column's height is defined as the accumulated thickness of the layers. To simulate water movement between columns, it is necessary to specify a depth-averaged K for that part of a column below the water table. This is done for each time step by taking a weighted average of the K values of all of the layers in a column below the water table. Account is also taken of situations where the water table resides within a layer rather than at its boundary. The intercolumn K is calculated as the harmonic mean of the two depth-averaged K values in the neighboring columns. The mean water table elevation between columns is given by the arithmetic mean of the two water tables in the neighboring columns. The model has been tested against various analytical models and has been shown to provide very accurate solutions to equation (8).

For the simulations reported here, each model column consisted of two vertically stacked layers of peat, the lower layer representing the less permeable catotelm, the thickness of which was determined for each column using the groundwater mound equation. Above the catotelm was a more permeable layer, 20 cm thick in all columns, representing the acrotelm. K in the acrotelm for each column was set to one of two values, a "low" value of 0.005 cm s^{-1} or a "high" value of 0.05 cm s^{-1}, with the two values representing peat that had been laid down by different peatland plant communities. Notionally, the lower value might be thought of as representing peat formed in a hummock, and the higher value may represent peat formed under a lawn; however, microtopographical variations between such features were not accounted for in the simulations. The hydraulic conductivity values assumed for catotelm peat are within plausible bounds for central areas of a raised bog, although they may be somewhat high for marginal zones [cf. *Baird et al.*, 2008]. The K values for the acrotelm sit within the ranges reported by *Baird et al.* [2006] for a raised bog in west central Wales. The drainable porosity of all of the peat types was set to 0.3. The peat deposit was assumed to sit atop an impermeable mineral substrate, and the model's upslope and lateral boundaries were assumed to be impermeable (reflective). At the downslope boundary, to which water in the model drained, the water table was held constant.

In the first model simulation, run 1, we assumed that the two different types of column containing low- and high-K acrotelm were randomly arranged (Plate 1a). In the second simulation, run 2, we set a nonrandom arrangement of low- and high-K acrotelm values, which formed closely spaced, cross-slope stripes (Plate 1b). In the third simulation, run 3, the high- and low-K columns were again arranged as cross-slope stripes (Plate 1c) but at a broader scale than in the second simulation. In each simulation, the proportion of columns containing high- and low-K acrotelm was approximately the same: between 45 and 50%. In each case, initial conditions were such that the water table was at the peatland surface. Each simulation represented 10 days of drainage in which no rain fell and no water was lost from the peatland via evaporation and transpiration. We could have included evapotranspiration, and that would have led to different spatial patterns of water tables, depending on the functional relationship between evapotranspiration and water table depth and the functional relationship between evapotranspiration and other independent variables such as peat and vegetation type. Such functional relationships are still being investigated [see *Belyea*, this volume], and more field data are needed. However, such detail is not necessary to explore how spatial arrangement of peat hydraulic properties may affect a peatland's water balance, and our results should be viewed in this context.

The results from the simulations are shown in Plates 1d–1f (spatial pattern) and Figure 5 (frequency diagrams). In Figure 5, it can be seen that the frequency distribution of water tables varies considerably between the simulations despite the proportions of columns with high and low K being similar. In the random arrangement (run 1, Figure 5a), the distribution is unimodal but skewed with a modal water table depth of 2.5–3.0 cm and a range of values from 2.0 to 14.0 cm. In the arrangement with multiple stripes (run 2, Figure 5b), the distribution is again unimodal but symmetrical with a modal water table of 3.0–3.5 cm and a range extending from −3.0 (i.e., surface ponding) to 14.0 cm. Finally, in run 3, the distribution is bimodal, with modes at 2.0–2.5 cm and 5.0–5.5 cm and a range from −7.0 to 12.5 cm.

Given the importance of water table depth in controlling fluxes of CO_2 and CH_4 (see sections 2.3 and 3), the spatial arrangements of SL1 units, through their effect on the frequency distributions of water table depth, are likely to have a large impact on upscaled carbon balance. Of course, our simulations are for a theoretical bog and for just one short period; they cannot, therefore, be compared with work that considers the combined effect of rainfall and evapotranspiration on the water table distribution over longer periods of time [e.g., *Eppinga et al.*, 2008]. However, our general conclusion, that spatial arrangement matters, has also been reached by *Swanson* [2007] in a simple mathematical exploration of how different arrangements of SL1 units affect flow rates through peat and rates of peat/carbon accumulation. Our results and those of *Swanson* [2007] show the importance of small-scale behavior on the larger peatland system and suggest that we need to know more than the average properties of a peatland in order to simulate its behavior.

3.3. Response Lags and Structural Legacies and SL3 and SL1 Succession

It was noted earlier that rates of decomposition and CH_4 production in the litter and peat below a hummock may be affected by the history of SL1 units at that position in the peatland. Although there is evidence that some surface features of a peatland such as hummocks are long-lived during the development of the peatland [e.g., *Barber*, 1981], it is also known that plant communities on peatlands change in response to internal (structural) changes as the peatland develops over time and in response to external drivers such as climate [*Belyea and Malmer*, 2004; *Kettridge et al.*, 2008]. Changes in vegetation feed back into the peatland carbon balance in a complex way. The type of peat that builds up

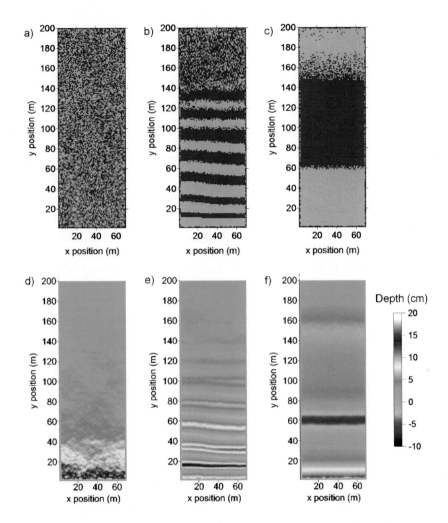

Plate 1. Details of the DigiBog runs. (top) Planimetric distribution of high- and low-K acrotelm: runs (a) 1, (b) 2, and (c) 3. Dark gray shows the higher-K acrotelm, and the lighter gray indicates the lower-K acrotelm. The lowest part of the peatland is at the bottom of each diagram, so the general flow of water is down the page (from high to low y position). (bottom) The 10-day water table depths for each simulation: runs (d) 1, (e) 2, and (f) 3.

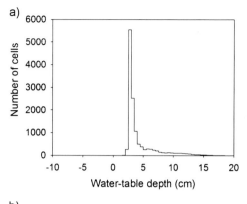

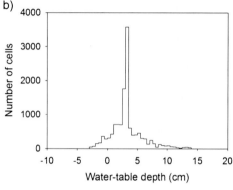

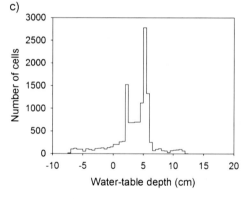

Figure 5. Frequency distributions of water table depth for each of the DigiBog runs: (a) 1, (b) 2, and (c) 3. Number of cells refers to the number of computational cells in DigiBog. Note the difference in the *y* axis scales.

in different places within a peatland depends on its botanical composition (the proportion of woody, sedge, and moss material), the conditions under which the peat formed (water table regime and water chemistry), and the temperature regime within the peat soil. These attributes and processes are linked, sometimes tightly. For example, water table regime at any point in a peatland affects the vegetation type, which, in turn, affects evaporative and transpirative losses

of water from the peatland and, therefore, the water table regime, which, as shown earlier, affects peatland-atmosphere CH_4 fluxes and CO_2 exchange. Water table position is also affected by the distribution of peat types (e.g., whether the peat has a high or low *K*) across the peatland, as shown in section 3.2. As part of the feedback, peat properties depend on the mix of plants making up the peat, which, as we have noted, depends on water table regime. The overall effect of these feedbacks is that the water table regime and carbon balance processes at any point across a peatland depend on the history of the peatland's development and the structural legacies of previous patterns of vegetation on the peatland, which are manifested as the subsurface pattern of different peat types [*Belyea and Baird*, 2006; *Kettridge et al.*, 2008].

It was also noted in section 1 that new peatlands may develop on thawed tundra soils and that existing peatlands may undergo dramatic change, including in their vegetation composition. Thus, any model that seeks to predict change in peatland carbon balance over the next 20–200 years will have to consider successional processes. The available evidence suggests that gross changes in whole peatland vegetation composition is driven, in part at least, by smaller-scale (both in time and space) successional processes; that is, it is necessary to consider how SL1 units undergo change from one vegetation type to another. The need to consider cross-scale linkages for modeling peatland vegetation composition and carbon balance has been articulated by *Belyea and Baird* [2006] and is evident in the approaches used by *Swanson and Grigal* [1988], *Couwenberg and Joosten* [2005], and *Rietkerk et al.* [2004], where vegetational changes across whole peatlands (SL3) are predicted on the basis of what happens at SL1 and how different patterns of SL1 units, through their effect on peat hydrochemical and physical properties, affect whole-peatland water movements and nutrient transfers. Again, the suggestion is that we need to consider what happens at SL1 and the linkages between SL1, SL2 and SL3.

The unavoidable outcome of section 3 is that if accurate simulations of peatland-atmosphere carbon exchanges are to be attained, we need to model SL1 units explicitly in some way. This is an uncomfortable finding because it suggests that existing approaches for upscaling carbon fluxes to and from peatlands are unsuitable, at least for peatlands that show distinct patterning in vegetation and water table depths. In section 4, we consider how some of the problems we have articulated may be addressed.

4. REPLACING THE BUCKET AND SLAB: DEALING WITH HETEROGENEITY

If it is not always appropriate to adopt a bucket-and-slab upscaling approach when looking at peatland-atmosphere carbon exchanges, how should we deal with the problem of heterogeneity? This question has been dealt with in part by *Gedney and Cox* [2002]. *Gedney and Cox* [2002] still proposed using a simple bucket model at the scale of a land surface scheme cell to simulate the "mean" water table in the cell. However, they then "decomposed" this single water table depth into a series of water table depths across the landscape represented by the cell using the TOPMODEL approach [*Beven and Kirkby*, 1979]. In essence, the approach assumes that the most important control on local soil wetness within a landscape is topography. By assuming an exponential decline in soil hydraulic conductivity (K) with depth below the surface and by assuming a catchment at any one time is in steady state (so that downslope subsurface water flow at any point in the landscape is balanced by recharge from upslope), it is possible to establish a relationship between the mean water table depth across the catchment (or land surface scheme's cell) and the local water table depth. Therefore, using this approach allows one to generate a series of local water table depths across a catchment or land surface scheme's cell and to use these to predict CH_4 emissions or CO_2 exchange for each local water table. This approach clearly requires digital elevation data for areas to which the land surface scheme is to be applied, which is a constraint for some areas [cf. *Gedney and Cox*, 2002]. The assumption of an exponential decline in K with depth is not necessarily a constraint because other functions may be used. The relationship between local and the mean water table depth across the catchment (land surface scheme (LSS) cell) can also be used to derive parameters for the simple store and flow model that represents the entire catchment or LSS cell, therefore giving confidence in the predictions from the model. However, although such a relationship may be derived in theory for some well-researched and relatively small catchments (up to, say, 10^8–10^9 m^2), it is unlikely that a single bucket-and-slab model representing a whole LSS cell would give a sensible mean water table for the various subcatchments within the cell because of inter-catchment variability. So pronounced differences in relief, soil type, and vegetation between catchments would make it impossible for a single bucket-and-slab model to give a water table depth that represents each catchment properly. Perhaps most critically, it may not be reasonable to expect water tables within peatlands to show the relationship with topography as suggested by *Gedney and Cox* [2002]. Although many fens are found in areas of flow convergence and at the base of slopes, including river floodplains, raised bogs and blanket peats may not be found in such areas. Furthermore, with the latter two, water tables are often nearest the surface in central areas close to the flow divide (e.g., crest of a raised bog's dome) where contributing ar-

eas are small in size; i.e., the opposite of the TOPMODEL assumption.

Other possibilities for dealing with heterogeneity include the massively parallel approach advocated by the authors of Ecosys (see www.ecosys.rr.ualberta.ca/, accessed on 27 October 2008), which was used by *Grant and Roulet* [2002] to model peatland-atmosphere CH_4 exchanges. The model can be used to simulate soils in three dimensions, and it is suggested on the model's Web site that different groups of cells in the model (which may be of the order of 10^1 m) could be simulated on separate processors, making it possible to scale up from local (SL1 and SL2) processes to the regional scale (10^5 m), i.e., to the scale of a cell within a land surface scheme. Although on the face of it such an approach would seem to be impossible because of the unrealistic parameterization demands of the model, it is important to remember that, increasingly, it will be possible to gain high-resolution information on some model parameters from remote sensing data, whether derived from airborne or spaceborne sensors [cf. *Harris and Bryant*, this volume]. Nevertheless, data such as the hydraulic properties of peat soils cannot be obtained remotely (at least not operationally), and it is difficult to envisage a situation where a model such as Ecosys could be parameterized successfully for massively parallel applications. It is also difficult to conceive of there being enough processing power to run such a model for all northern peatlands, although Moore's law [*Moore*, 1965] teaches us to be careful when making assumptions about what size of task a computer can run.

A third possibility to replacing the bucket and slab is that of treating SL1 units in peatlands as noninteracting entities and running separate simulations for each entity. Thus, a particular region may have a 70% cover of raised bogs and a 30% cover of floodplain fens. The bogs themselves may contain, on average, a 30% cover of hummocks and a 70% cover of lawns. Such data, in theory at least, could be obtained via airborne or (possibly) spaceborne remote sensing [cf. *Harris and Bryant*, this volume]. To simulate regional-scale carbon fluxes for this system, three models would be needed using this approach: one for the fens, one for the hummocks, and one for the lawns. Each model would assume a homogenous vegetation cover and a spatially uniform water table. To arrive at regional carbon fluxes, the outputs of these models would be scaled by actual area and factors of 0.3, 0.21, and 0.49 for the fen, bog hummocks, and bog lawns respectively. Although an improvement on schemes that use a single bucket and slab, this approach ignores interactions between different SL1 units and the directionality of water flow in peatlands. In effect, it is replacing one bucket and slab with several buckets and slabs, but it still suffers from problems noted earlier that arise from ignoring SL1 units.

5. AN AGENDA FOR FUTURE RESEARCH

Thus far our argument may be taken as having been somewhat negative. We have highlighted the importance of being able to model peatland-atmosphere carbon exchanges over the coming decades but have suggested that existing approaches to such modeling are unlikely to succeed, at least for peatlands that are spatially heterogeneous. We have also shown that there is no easy solution to this problem. What then is to be done? Figure 6 provides one possible route to a solution in the case of peatland-atmosphere CH$_4$ fluxes, although it should be emphasized that, at this stage, the route is not operational. The route relies initially on a top-down classification of peatlands. The starting point is the scale of a land surface scheme cell. Climate data and low-resolution digital elevation models are used to define broad climotopographic zones in which northern peatlands occur. Thus, we might expect groupings and mosaics of peatland types in, say, Scandinavia, where there is a strong maritime influence on climate, to differ from the groupings in western Siberia, where the climate is continental. For each climotopographic zone, higher-resolution remote sensing and DEM data are used to identify different peatland landscapes, each comprising distinctive assemblages of peatland types

(i.e., individual peatlands such as treed raised bog and ladder fen). For each peatland type, a representative sample of individual peatlands (e.g., $n = 5$) is selected for detailed simulation using a model capable of representing SL1 variation and capable of allowing peat to build up and decay and surface vegetation composition to change over time. A number of research groups appear to be developing models with such capability [cf. *Eppinga et al.*, 2008; N. T. Roulet, personal communication, 2008]. The simulations for each SL3 peatland type are scaled upward for use in a GCM using information on total coverage of that peatland type within each peatland landscape and each peatland zone. Changes in peatland form can also be fed back into the classification of peatland landscapes. For example, a particular landscape may be reclassified at some point if fens start to develop into bogs (a fen-bog transition).

Our suggested approach is more efficient than the massively parallel approach described earlier for Ecosys because it only considers a selection of peatland types; not every peatland is simulated. However, even modeling a selection of each peatland type presents considerable challenges. Models such as Ecosys are "parameter hungry" and provide very detailed descriptions of, for example, microbial breakdown of peat. Such detail will not be possible with our approach be-

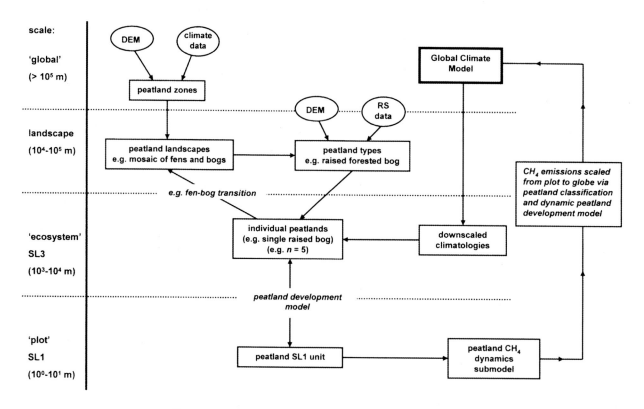

Figure 6. Proposed new upscaling approach. See text for details.

cause of the constraints imposed by parameterization. Where possible, a reduced-complexity approach should be adopted, although the exact form such an approach should take has not been fully researched. To decide on what level of complexity reduction is acceptable will require model intercomparison studies and testing different models with the same data sets. Even with simplified models, there will still be challenges with parameterization. Given the resolution of spaceborne remote sensing data, airborne data will be needed to identify detailed peatland shape, including variation in elevation of the peatland surface at SL1 (i.e., lidar data will be needed) and patterns of vegetation at SL2 [cf. *Harris and Bryant*, this volume]. There will also be a need to parameterize the subsurface properties of the peatlands being modeled, and we are still a long way from such parameterization for the full suite of peatland types that exist. For example, work has only started on revealing SL2-SL3 hydraulic structures within peatlands [e.g., *Lapen et al.*, 2005; *Baird et al.*, 2008], although the use of geophysical methods such a ground-penetrating radar [*Kettridge et al.*, 2008] offers promise for the rapid characterization of such structures.

It is perhaps astonishing that little has been done to date by climate modelers on representing peatland-climate feedbacks in the latest generation of GCMs linked with land surface schemes. However, the problem lies in the fact that peatlands are understudied compared to other ecosystems. Rightfully, attention has been given to simulating feedbacks between climate and, for example, the Amazon rain forest [e.g., *Cox et al.*, 2000], but northern peatlands contain between 3 and 5 times as much carbon as the Amazon [cf. *Fearnside*, 1997] and are also a dynamic and potentially fragile carbon store. Therefore, it is imperative that we invest heavily in researching peatland function both in the field and through the development of improved models; the importance of such work cannot be overstated.

Acknowledgments. James Aber is thanked for his kind permission to use Figure 2. Two anonymous reviewers provided formal reviews of an earlier version of the paper, while Steve Frolking, Nick Kettridge, and Nigel Roulet kindly provided informal reviews: all sets of review comments helped us improve the manuscript.

REFERENCES

Aber, J. S., and S. W. Aber (2001), Potential of kite aerial photography for peatland investigations with examples from Estonia, *Suo, 52*, 45–56.

Aber, J. S., K. Aaviksoo, E. Karofeld, and S. W. Aber (2002), Patterns in Estonian bogs as depicted in color kite aerial photographs, *Suo, 53*, 1–15.

Alm, J., A. Talanov, S. Saarnio, J. Silvola, E. Ikkonen, H. Aaltonen, H. Nykänen, and P. J. Martikainen (1997), Reconstruction of the carbon balance for microsites in a boreal oligotrophic pine fen, Finland, *Oecologia, 110*, 423–431.

Baird, A. J., P. A. Eades, B. W. J. Surridge, and A. Harris (2006), Cors Fochno Hydrological Research and Management Study: Final Report of Theme 1, *CCW Contract Sci. Rep. 718*, 74 pp., Countryside Counc. for Wales Bangor, U. K.

Baird, A. J., P. A. Eades, and B. W. J. Surridge (2008), The hydraulic structure of a raised bog and its implications for ecohydrological modelling of bog development, *Ecohydrology, 1*, 289–298, doi:10.1003/eco33.

Baldocchi, D. D., T. Krebs, and M. Y. Leclerc (2005), "Wet/dry daisyworld": A conceptual tool for quantifying the spatial scaling of heterogeneous landscapes and its impact on the subgrid variability of energy fluxes, *Tellus, Ser. B, 57*, 175–188.

Barber, K. E. (1981), *Peat Stratigraphy and Climatic Change: A Palaeoecological Test of the Theory of Cyclic Bog Regeneration*, 219 pp., Balkema, Rotterdam, Netherlands.

Belyea, L. R. (2009), Nonlinear dynamics of peatlands and potential feedbacks on the climate system, *Geophys. Monogr. Ser.*, doi:10.1029/2008GM000829, this volume.

Belyea, L. R., and A. J. Baird (2006), Beyond the "limits to peat bog growth": Cross-scale feedback in peatland development, *Ecol. Monogr., 76*(3), 299–322.

Belyea, L. R., and R. S. Clymo (2001), Feedback control of the rate of peat formation, *Proc. R. Soc., Ser. B, 268*, 1315–1321.

Belyea, L. R., and N. Malmer (2004), Carbon sequestration in peatland: Patterns and mechanisms of response to climate change, *Global Change Biol., 10*, 1043–1052.

Beven, K., and M. Kirkby (1979), A physically based, variable contributing area model of basin hydrology, *Hydrol. Sci. Bull., 24*, 43–69.

Blyth, E., M. Best, P. Cox, R. Essery, O. Boucher, R. Harding, C. Prentice, P. L. Vidale, and I. Woodward (2006), JULES: A new community land surface model, *Global Change Newsl., 66*, 9–11.

Bubier, J. L. (1995), The relationship of vegetation to methane emission and hydrochemical gradients in northern peatlands, *J. Ecol., 83*, 403–420.

Bubier, J., A. Costello, T. R. Moore, N. T. Roulet, and K. Savage (1993), Microtopography and methane flux in boreal peatlands, northern Ontario, Canada, *Can. J. Bot., 71*, 1056–1063.

Bubier, J. L., T. R. Moore, L. Bellisario, N. T. Comer, and P. M. Grill (1995), Ecological controls on methane emissions from a northern peatland complex in the zone of discontinuous permafrost, Manitoba, Canada, *Global Biogeochem. Cycles, 9*(4), 455–470.

Cao, M., S. Marshall, and K. Gregson (1996), Global carbon exchange and methane emissions from natural wetlands: Application of a process-based model, *J. Geophys. Res., 101*(D9), 14,399–14,414.

Christensen, T. R., N. Panikov, M. Mastepanov, A. Joabsson, A. Stewart, M. Öquist, M. Sommerkorn, S. Reynaud, and B. Svensson (2003), Biotic controls on CO$_2$ and CH$_4$ exchange in wetlands—A closed environment study, *Biogeochemistry, 64*, 337–354.

Clymo, R. S. (1984), The limits to peat bog growth, *Philos. Trans. R. Soc. London, Ser. B, 303*, 605–654.

Clymo, R. S., J. Turunen, and K. Tolonen (1998), Carbon accumulation in peatland, *Oikos*, *81*, 368–388.

Couwenberg, J., and H. Joosten (2005), Self-organisation in raised bog patterning: The origin of microtope zonation and mesotope diversity, *J. Ecol.*, *93*, 1238–1248.

Cox, P. M. (2001), Description of the "TRIFFID" dynamic global vegetation model, *Hadley Cent. Tech. Rep. 24*, 16 pp., Met Off., Bracknell, U. K.

Cox, P. M., R. A. Betts, C. B. Bunton, R. L. H. Essery, P. R. Rowntree, and J. Smith (1999), The impact of new land surface physics on the GCM simulation of climate and climate sensitivity, *Clim. Dyn.*, *15*, 183–203.

Cox, P. M., R. A. Betts, C. D. Jones, S. A. Spall, and I. J. Totterdell (2000), Acceleration of global warming due to carbon-cycle feedbacks in a coupled climate model, *Nature*, *408*, 184–187.

Eppinga, M. B., M. Rietkerk, W. Borren, E. D. Lapshina, W. Bleuten, and M. Wassen (2008), Regular surface patterning of peatlands: Confronting theory with field data, *Ecosystems*, *11*, 520–536.

Fearnside, P. M. (1997), Greenhouse gases from deforestation in Brazilian Amazonia: Net committed emissions, *Clim. Change*, *35*, 321–360.

Frolking, S., N. T. Roulet, T. R. Moore, P. J. H. Richard, M. Lavoie, and S. D. Muller (2001), Modeling northern peatland decomposition and peat accumulation, *Ecosystems*, *4*, 479–498.

Frolking, S., N. T. Roulet, T. R. Moore, P. M. Lafleur, J. L. Bubier, and P. M. Crill (2002), Modeling seasonal to annual carbon balance of Mer Bleue Bog, Ontario, Canada, *Global Biogeochem. Cycles*, *16*(3), 1030, doi:10.1029/2001GB001457.

Frolking, S., N. Roulet, and J. Fuglestvedt (2006), How northern peatlands influence the Earth's radiative budget: Sustained methane emission versus sustained carbon sequestration, *J. Geophys. Res.*, *111*, G01008, doi:10.1029/2005JG000091.

Frolking, S., N. Roulet, and D. Lawrence (2009), Issues related to incorporating northern peatlands into global climate models, *Geophys. Monogr. Ser.*, doi:10.1029/2008GM000809, this volume.

Gedney, N., and P. M. Cox (2002), The sensitivity of global climate model simulations to the representation of soil moisture heterogeneity, *Hadley Cent. Tech. Note 41*, 18 pp., Met Off., Bracknell, U. K.

Gilmer, A. J., N. M. Holden, S. M. Ward, A. Brereton, and E. P. Farrell (2000), A model of organic matter accumulation in a developing fen/raised bog complex, *Suo*, *51*, 155–167.

Gorham, E. (1991), Northern peatlands: Role in the carbon cycle and probable responses to climate warming, *Ecol. Appl.*, *1*(2), 182–195.

Grant, R. F., and N. T. Roulet (2002), Methane efflux from boreal wetlands: Theory and testing of the ecosystem model Ecosys with chamber and tower flux measurements, *Global Biogeochem. Cycles*, *16*(4), 1054, doi:10.1029/2001GB001702.

Hamilton, J. D., C. A. Kelly, J. W. M. Rudd, R. H. Hesslein, and N. T. Roulet (1994), Flux to the atmosphere of CH_4 and CO_2 from wetland ponds on the Hudson Bay lowlands (HBLs), *J. Geophys. Res.*, *99*(D1), 1495–1510.

Harris, A. (2008), Spectral reflectance and photosynthetic properties of *Sphagnum* mosses exposed to progressive drought, *Ecohydrology*, *1*, 35–42, doi:10.1002/eco.5.

Harris, A., and R. G. Bryant (2009), Northern peatland vegetation and the carbon cycle: A remote sensing approach, *Geophys. Monogr. Ser.*, doi:10.1029/2008GM000818, this volume.

Harris, A., R. G. Bryant, and A. J. Baird (2005), Detecting moisture stress in *Sphagnum* spp., *Remote Sens. Environ.*, *97*(3), 371–381.

Harris, A., R. G. Bryant, and A. J. Baird (2006), Mapping the effects of water stress on *Sphagnum*: Preliminary observations using airborne remote sensing, *Remote Sens. Environ.*, *100*(3), 363–378.

Hilbert, D. W., N. Roulet, and T. Moore (2000), Modelling and analysis of peatlands as dynamical systems, *J. Ecol.*, *88*, 230–242.

Hu, Z., and S. Islam (1997), A framework for analyzing and designing scale invariant remote sensing algorithms, *IEEE Trans. Geosci. Remote Sens.*, *35*, 747–755.

Ingram, H. A. P. (1982), Size and shape in raised mire ecosystems: A geophysical model, *Nature*, *297*, 300–303.

Ingram, H. A. P. (1983), Hydrology, in *Mires: Swamp, Bog, Fen, and Moor*, edited by A. J. P. Gore, pp. 67–158, Elsevier, Amsterdam.

Ivanov, K. E. (1981), *Water Movement in Mirelands*, translated from Russian by A. Thompson and H. A. P. Ingram, Academic, London.

Kettridge, N., X. Comas, A. Baird, L. Slater, M. Strack, D. Thompson, H. Jol, and A. Binley (2008), Ecohydrologically important subsurface structures in peatlands revealed by ground-penetrating radar and complex conductivity surveys, *J. Geophys. Res.*, *113*, G04030, doi:10.1029/2008JG000787.

Kettunen, A. (2003), Connecting methane fluxes to vegetation cover and water table fluctuations at microsite level: A modeling study, *Global Biogeochem. Cycles*, *17*(2), 1051, doi:10.1029/2002GB001958.

Laine, A., M. Sottocornola, G. Kiely, K. A. Byrne, D. Wilson, and E.-S. Tuittila (2006), Estimating net ecosystem exchange in a patterned ecosystem: Example from blanket bog, *Agric. For. Meteorol.*, *138*, 231–243.

Laine, A., K. A. Byrne, G. Kiely, and E.-S. Tuittila (2007), Patterns in vegetation and CO_2 dynamics along a water level gradient in a lowland blanket bog, *Ecosystems*, *10*, 890–905.

Lapen, D. R., J. S. Price, and R. Gilbert (2005), Modelling two-dimensional, steady-state groundwater flow and flow sensitivity to boundary conditions in blanket peat complexes, *Hydrol. Processes*, *19*, 371–386, doi:10.1002/hyp.1507.

Levin, S. A. (1992), The problem of pattern and scale in ecology, *Ecology*, *73*, 1943–1967.

McWhorter, D. B., and D. K. Sunada (1977), *Ground-water Hydrology and Hydraulics*, Water Resour. Publ., Fort Collins, Colo.

Melbourne, B. A., and P. Chesson (2006), The scale transition: Scaling up population dynamics with field data, *Ecology*, *87*, 1478–1488.

Moore, G. E. (1965), Cramming more components onto integrated circuits, *Electron. Mag.*, *38*(8).

Notaro, M., S. Vavrus, and Z. Liu (2007), Global vegetation and climate change due to future increases in CO_2 as projected by a fully coupled model with dynamic vegetation, *J. Clim.*, *20*, 70–90, doi:10.1175/JCLI3989.1.

Pelletier, L., T. R. Moore, N. T. Roulet, M. Garneau, and V. Beaulieu-Audy (2007), Methane fluxes from three peatlands in the La Grande Rivière watershed, James Bay lowland, Canada, *J. Geophys. Res.*, *112*, G01018, doi:10.1029/2006JG000216.

Reeve, A. S., D. I. Siegel, and P. H. Glaser (2000), Simulating vertical flow in large peatlands, *J. Hydrol.*, *227*, 207–217.

Reeve, A. S., Z. D. Tyczka, X. Comas, and L. D. Slater (2009), The influence of permeable mineral lenses on peatland hydrology, *Geophys. Monogr. Ser.*, doi:10.1029/2008GM000825, this volume.

Rietkerk, M., S. C. Dekker, M. J. Wassen, A. W. M. Verkroost, and M. F. P. Bierkens (2004), A putative mechanism for bog patterning, *Am. Nat.*, *163*, 699–708.

Rydin, H., and J. K. Jeglum (2006), *The Biology of Peatlands*, 343 pp., Oxford Univ. Press, New York.

Sitch, S., et al. (2008), Evaluation of the terrestrial carbon cycle, future plant geography and climate-carbon cycle feedbacks using five Dynamic Global Vegetation Models (DGVMs), *Global Change Biol.*, *14*, 1–25, doi:10.1111/j.1365-2486.2008.01626.x.

Smith, L. C., G. M. MacDonald, A. A. Velichko, D. W. Beilman, O. K. Borisova, K. E. Frey, K. V. Kremenetski, and Y. Sheng (2004), Siberian peatlands a net carbon sink and global methane source since the early Holocene, *Science*, *303*, 353–356.

Strack, M., M. F. Waller, and J. M. Waddington (2006), Sedge succession and peatland methane dynamics: A potential feedback to climate change, *Ecosystems*, *9*, 278–287, doi:10.1007/s100021-005-0070-1.

Strayer, D. L., H. A. Ewing, and S. Bigelow (2003), What kind of spatial and temporal details are required in models of heterogeneous systems?, *Oikos*, *102*, 654–662.

Ström, L., A. Ekberg, M. Mastepanov, and T. R. Christensen (2003), The effect of vascular plants on carbon turnover and methane emissions from a tundra wetland, *Global Change Biol.*, *9*, 1185–1192.

Ström, L., M. Mastepanov, and T. R. Christensen (2005), Species-specific effects of vascular plants on carbon turnover and methane emissions from wetlands, *Biogeochemistry*, *75*, 65–82, doi:10.1007/s10533-004-6124-1.

Swanson, D. K. (2007), Interaction of mire microtopography, water supply, and peat accumulation in boreal mires, *Suo*, *58*, 37–47.

Swanson, D. K., and D. F. Grigal (1988), A simulation model of mire patterning, *Oikos*, *53*, 309–314.

Turetsky, M. R., R. K. Wieder, D. H. Vitt, R. J. Evans, and K. D. Scott (2007), The disappearance of relict permafrost in boreal North America: Effects on peatland carbon storage and fluxes, *Global Change Biol.*, *13*, 1922–1934.

Turunen, J., E. Tomppo, K. Tolonen, and A. Reinikainen (2002), Estimating carbon accumulation rates of undrained mires in Finland—Application to boreal and subarctic regions, *Holocene*, *12*, 69–80.

Waddington, J. M., and N. T. Roulet (1996), Atmosphere-wetland carbon exchanges: Scale dependency of CO_2 and CH_4 exchange on the developmental topography of a peatland, *Global Biogeochem. Cycles*, *10*(2), 233–245.

Waddington, J. M., N. T. Roulet, and R. V. Swanson (1996), Water table control of CH_4 emission enhancement by vascular plants in boreal peatlands, *J. Geophys. Res.*, *101*(D17), 22,775–22,785.

Walter, B. P., M. Heimann, R. D. Shannon, and J. R. White (1996), A process-based model to derive methane emissions from natural wetlands, *Geophys. Res. Lett.*, *23*(25), 3731–3734.

Walter, B. P., M. Heimann, and E. Matthews (2001a), Modeling modern methane emissions from natural wetlands: 1. Model description and results, *J. Geophys. Res.*, *106*(D24), 34,189–34,206.

Walter, B. P., M. Heimann, and E. Matthews (2001b), Modeling modern methane emissions from natural wetlands: 2. Interannual variations 1982–1993, *J. Geophys. Res.*, *106*(D24), 34,207–34,219.

Walter, K. M., M. E. Edwards, G. Grosse, S. A. Zimov and F. S. Chapin III (2007), Thermokarst lakes as a source of atmospheric CH_4 during the last deglaciation, *Science*, *318*, 633–636.

Yu, Z., D. H. Vitt, I. D. Campbell, and M. J. Apps (2003), Understanding Holocene peat accumulation pattern of continental fens in western Canada, *Can. J. Bot.*, *81*, 267–282.

Yu, Z., D. W. Beilman, and M. C. Jones (2009), Sensitivity of northern peatland carbon dynamics to Holocene climate change, *Geophys. Monogr. Ser.*, doi:10.1029/2008GM000822, this volume.

Zhang, Y., C. Li, C. C. Trettin, H. Li, and G. Sun (2002), An integrated model of soil, hydrology, and vegetation for carbon dynamics in wetland ecosystems, *Global Biogeochem. Cycles*, *16*(4), 1061, doi:10.1029/2001GB001838.

A. J. Baird, School of Geography, University of Leeds, Leeds LS2 9JT, UK. (a.j.baird@leeds.ac.uk)

L. R. Belyea and P. J. Morris, Department of Geography, Queen Mary University of London, Mile End Road, London E1 4NS, UK.

Sensitivity of Northern Peatland Carbon Dynamics to Holocene Climate Change

Zicheng Yu,[1] David W. Beilman,[2,3] and Miriam C. Jones[1]

In this paper, we evaluate the long-term climate sensitivity and global carbon (C) cycle implications of northern peatland C dynamics by synthesizing available data and providing a conceptual framework for understanding the dominant controls, processes, and interactions of peatland initiation and C accumulation. Northern peatlands are distributed throughout the climate domain of the boreal forest/taiga biome, but important differences between peatland regions are evident in annual temperature vs. precipitation (*T-P*) space, suggesting complex hydroclimatic controls through various seasonal thermal-moisture associations. Of 2380 available basal peat dates from northern peatlands, nearly half show initiation before 8000 calendar years (cal years) B.P. Peat-core data from sites spanning peatland *T-P* space show large variations in apparent C accumulation rates during the Holocene, ranging from 8.4 in the Arctic to 38.0 g C m^{-2} a^{-1} in west Siberia, with an overall time-weighted average rate of 18.6 g C m^{-2} a^{-1}. Sites with multiple age determinations show millennial-scale variations, with the highest C accumulation generally at 11,000–8000 cal years B.P. The early Holocene was likely a period of rapid peatland expansion and C accumulation. For example, maximum peat expansion and accumulation in Alaska occurred at this time when climate was warmest and possibly driest, suggesting the dominant role of productivity over decomposition processes or a difference in precipitation seasonality. Northern peatland C dynamics contributed to the peak in atmospheric CH_4 and the decrease in CO_2 concentrations in the early Holocene. This synthesis of data, processes, and ideas provides baselines for understanding the sensitivity of these C-rich ecosystems in a changing climate.

[1]Department of Earth and Environmental Sciences, Lehigh University, Bethlehem, Pennsylvania, USA.

[2]CHRONO Centre for Climate, the Environment and Chronology, School of Geography, Archaeology and Palaeoecology, Queen's University Belfast, Belfast, UK.

[3]Now at Department of Geography, University of Hawai'i at Manoa, Honolulu, Hawaii, USA.

Carbon Cycling in Northern Peatlands
Geophysical Monograph Series 184
Copyright 2009 by the American Geophysical Union.
10.1029/2008GM000822

1. INTRODUCTION

Northern peatland ecosystems have cycled and stored substantial amounts of global land carbon (C) over the Holocene (the last 11,700 years). Today, peatlands are one of the largest terrestrial biosphere C pools and are the largest natural source of methane (CH_4) in the northern hemisphere. Owing to their large accumulated C mass and dynamic greenhouse gas fluxes, these ecosystems have been an important component of the high-latitude C cycle for thousands of years.

Peatland ecosystems and their C-rich peat archives have been studied for several decades, mostly for reconstructing past climate [*Charman*, 2002], and have been central to early ideas about recurrent climate changes [*Sernander*, 1908]. Over recent decades, attention has also turned to carbon cycling and the implications of long-term peatland ecosystem dynamics and climate sensitivity [*Clymo*, 1984; *Gorham*, 1991; *Rydin and Jeglum*, 2006]. Peat C accumulation is determined by the balance of biological inputs (plant growth and litter production) and outputs (organic matter decomposition); both of these processes are sensitive to climate change and climate variability or are indirectly affected by climate through related processes.

In this chapter, we provide a conceptual framework for understanding the dominant controls, processes, and interactions of northern peatland initiation and long-term peat C accumulation and dynamics using climate data and peat-core data. We use modern instrumental climate data to explore the climate envelope of today's northern peatland distribution. We synthesize spatial and temporal patterns of peat C accumulation rates during the Holocene in different regions and discuss climatic and autogenic influences. We also discuss the implications of peatland dynamics for the global carbon cycle. Understanding the causal connection between peat C dynamics and past climate would provide insight into the possible future response of these C-rich ecosystems to climate change in different regions and over different timescales.

2. CLIMATE CONTROLS OVER DISTRIBUTION, INITIATION, AND EXPANSION OF NORTHERN PEATLANDS

Northern peatlands occur mostly in boreal and subarctic regions in the northern hemisphere. A cool climate, low evaporation rates, and high effective moisture (precipitation minus evaporation) are essential for the formation and development of northern peatlands on suitable substrates and in suitable topographic settings. Despite the generally short summer seasons at high latitudes and the moderate net primary production (NPP) of peatland vegetation, peatlands accumulate excess organic matter as peat owing to depressed decomposition in waterlogged and anoxic conditions and the chemical recalcitrance of some peatland plant tissues. Extensive development of northern peatland ecosystems has occurred in west Siberia, central Canada, northwest Europe, and Alaska (Plate 1). Due to different regional climate and deglaciation histories, the timing of peatland initiation varies greatly from region to region [*Kuhry and Turunen*, 2006], but the majority of today's peatlands first expanded in the early and mid-Holocene [*MacDonald et al.*, 2006; *Gorham et al.*, 2007] (Plate 1).

To explore the climate domain of northern peatlands, particularly in relation to the boreal forest/taiga ecoregion [*Olson et al.*, 2001], we compared the distribution of peatlands [*MacDonald et al.*, 2006] to gridded instrumental climate data (0.5° × 0.5° grids) for land north of 45°N (1960–1990 [*Rawlins and Willmott*, 1999]). Northern peatlands typically occur where mean annual temperatures are between −12° and 5°C and mean annual precipitation is between 200 and 1000 mm. This distribution spans most of the climate domain of the boreal ecoregion (Plate 2). Peatlands are often abundant in regions that receive <500 mm of mean annual precipitation, which is a broader climate range than suggested by previous analyses [*Gignac and Vitt*, 1994; *Wieder et al.*, 2006]. This wide range in annual temperature versus precipitation (*T-P*) space contrasts with the uneven geographic distribution of northern peatlands (Plate 1), including broad regions of cold East Siberia taiga with relatively few peatlands. This contrast indicates that climate seasonality and local factors, such as topography and geologic substrate (parent material), have probably exerted critical secondary controls on Holocene peatland expansion and C accumulation.

Although peatland regions display considerable overlap in annual *T-P* space, many important regional differences are evident (Plate 2). For example, and considering the largest wetland regions, peatlands of the West Siberia Lowland (WSL) and Mackenzie River Basin (MRB) include peatland areas with mean annual temperatures about 7°C colder than those of the Hudson Bay Lowland (HBL). In contrast, the warmer peatlands in these three regions (between −2° and 2°C) receive highly variable annual precipitation, and many HBL peatlands receive twice as much precipitation as those in the WSL or MRB (900 versus 400 mm), owing to the climatic influence of Hudson Bay. Peatlands in Alaska span a similar range in mean annual temperature as those in the HBL, but span a very broad precipitation range from 150 to >1500 mm, owing to coastal and orographic influences. The distinct character of the peatland regions in modern *T-P* space suggests that the maintenance of peatland hydrology suitable for long-term peat C accumulation is the result of various thermal-moisture associations and precipitation seasonality. In the same way, climate histories and temperature and precipitation associations in the past were likely also very different between regions. As a result, a regional perspective would be most informative in understanding and projecting C cycling responses to climate change. In particular, peatlands located in regions near the limits of peatland climate space may be the first to experience expansion and shrinkage under changing regional climates.

Preliminary results from similar analysis of relative humidity (RH) show that northern peatlands have high annual RH values ranging from 65 to 95%. Peatlands with the highest RH occur in regions with a mean annual temperature around −10°C. A surprising pattern is that peatlands with the highest RH (and also a wide range of RH) tend to oc-

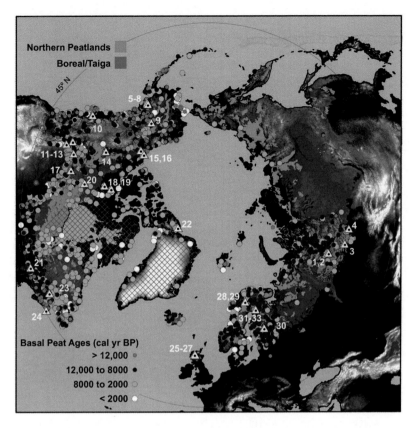

Plate 1. The distribution of northern peatland regions (light blue [*MacDonald et al.*, 2006]), the boreal/taiga biome (green [*Olson et al.*, 2001]), basal peat ages north of 45°N latitude (circles with ages in legends [*MacDonald et al.*, 2006; *Gorham et al.*, 2007]; *n* = 2380), and 33 sites with detailed peat C accumulation data (yellow triangles; site numbers as in Table 1). Terrain is from the ETOPO2 data set. The extent of the Laurentide Ice Sheet at 8000 calendar years B.P. [*Dyke et al.*, 2004] is shown by the crosshatched pattern.

cur at low annual precipitation of <550 mm. Further analysis of seasonal patterns of climate parameters, including temperature, precipitation, and relative humidity, would provide additional insights into understanding climate controls of peatland distribution.

At the hemispheric scale, northern peatlands expanded rapidly following the last glacial termination, in response to changes in large-scale boundary conditions and climate controls. These include ice retreats and availability of new land surface [*Dyke et al.*, 2004], large increases in summer insolation [*Berger and Loutre*, 1991], increases in greenhouse gases [*Brook et al.*, 2000; *Monnin et al.*, 2004], deglacial warming [*Kaufman et al.*, 2004], and possibly increasing moisture conditions [*Wolf et al.*, 2000]. At the regional scale, peatland initiation and expansion, either by means of paludification (formation or expansion onto non-wetland terrestrial ecosystems) or terrestrialization (lake-infilling) processes [*Kuhry and Turunen*, 2006], followed regional climate changes. For example, Holocene thermal maximum (HTM)

conditions in the early Holocene might have promoted peatland initiation and expansion in south-central Alaska (see sections 3 and 5 below). In eastern Canada, including the HBL and Labrador, the deglaciation and subsequent climate warming occurred much later during the mid-Holocene, resulting in peatland expansion later in time (Plate 1). At the basin scale, regional and local factors interact to control peatland dynamics. Local expansion is highly nonlinear as a function of both local terrain and regional climate [*Korhola et al.*, 1996; *Bauer et al.*, 2003]. Therefore, climate sensitivity of peatland expansion to Holocene climate change needs to be evaluated on different spatial scales.

3. SPATIAL AND TEMPORAL PATTERNS OF CARBON ACCUMULATION DURING THE HOLOCENE

Overall peat accumulation patterns, i.e., convex versus concave peat mass versus age curves, show long-term trajectories of peatland changes that do not necessarily re-

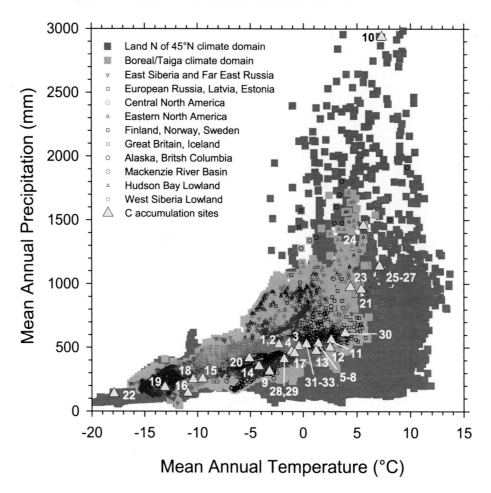

Plate 2. The climate space of mean annual temperature and precipitation (*T-P* space) of total land area north of 45°N latitude (dark gray), the boreal/taiga biome (light gray), and northern peatland regions based on 0.5° × 0.5°-gridded instrumental climate data for the period 1960–1990 [*Rawlins and Willmott*, 1999]. The location in climate space of C accumulation sites is shown by yellow triangles (site numbers as in Table 1).

veal peatland sensitivity to climate changes at millennial or Holocene scales. However, different trajectories imply different underlying fundamental processes [*Yu*, 2006a], which may relate to persistent climatic or ecological controls. For example, a concave accumulation pattern, with increasing long-term apparent accumulation rates over time, indicates that cumulative decomposition is the dominant control on net peat accumulation, i.e., more recent peat yields high apparent accumulation rates as less time has elapsed for decomposition. This concave pattern is often observed in oceanic raised bogs [*Clymo*, 1984]. In contrast, a convex pattern, with decreasing accumulation rates over time, suggests that either allogenic (e.g., directional change in climate) or autogenic (e.g., peatland height growth) factors have played a major role. This convex pattern is often observed in continental climates, especially in fens [*Yu*, 2006b]. Several conceptual

models show that the trajectory of peatland development is controlled by initial conditions, external influences, and internal processes [*Belyea and Baird*, 2006]. Any deviations from an overall trajectory are at least partly the result of the direct or indirect influence of climate variability.

Numerous studies have presented C accumulation data from northern peatlands at the site or regional scale. We compiled available data to examine the variability of apparent C accumulation rates within and between sites during the Holocene (Plate 3). Peatland development is affected not only by climate but also by local edaphic and autogenic factors [*Kuhry and Turunen*, 2006] (Plate 4), which may explain the spatial heterogeneity of peatland C accumulation rates across the boreal biome and within a region (Plate 3b). We calculated time-weighted average C accumulation rates for each site using multiple calibrated ^{14}C ages and bulk

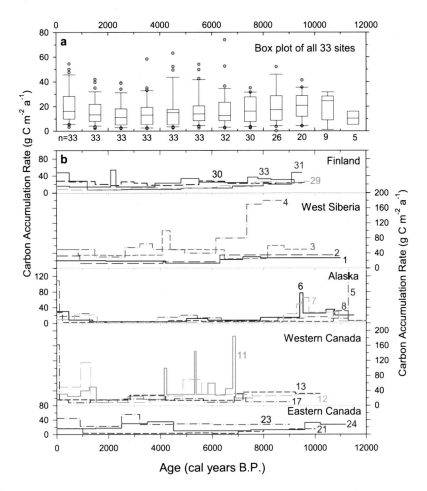

Plate 3. Variation in long-term apparent C accumulation rates from 33 northern peatland sites with bulk density and C measurements and multiple radiocarbon or tephra dates (Table 1). (a) Box plot of C rates in 1000-year bins from 33 sites. The horizontal lines within boxes indicate the medians. Numbers below the panel indicate the number of sites used in each bin. (b) Variations in C accumulation rates from selected sites with the highest number of dates across five peatland regions (site numbers as in Table 1).

density measurements. We then averaged the rates from all sites within each region, also considering the time lengths of peat accumulation at individual sites. These reconstructed rates of peat C accumulation are apparent rates in that they typically underestimate the true rate of past carbon uptake (net ecosystem production), since often many thousands of years of deep C decomposition have occurred. Thus, reconstructed C accumulation rates necessitate a careful interpretation relative to modern C flux measurements from eddy flux tower and chamber techniques [e.g., *Roulet et al.*, 2007] despite being reported in the same measurement units (g C m^{-2} a^{-1}). The highest apparent C accumulation rates over the Holocene are observed in west Siberia (38.0 g C m^{-2} a^{-1}; $n = 4$) followed by western Canada (20.3 g C m^{-2} a^{-1}; $n = 7$), and the lowest rates are found in the High Arctic

(8.4 g C m^{-2} a^{-1}; $n = 5$). The overall time-weighted average rate is 18.6 g C m^{-2} a^{-1} during the Holocene based on 33 sites (Table 1). Peatland regions showing high C accumulation rates appear to occur in the intermediate ranges of annual *T-P* space (Plates 2 and 3), such as in west Siberia and western Canada, in particular around 0° to 2.5°C of mean annual temperature and 450–550 mm of annual precipitation. In other words, these climate conditions may be optimal in producing a balance between primary production and decomposition that strongly promotes C accumulation. On the other hand, the regions showing low C accumulation rates appear to be located either at the extreme cold end of climate space, e.g., sites in the Arctic (Plate 2, Table 1) or at high temperature and high precipitation locations, e.g., sites in eastern Canada and perhaps also in Finland (Plates 2 and 3). This

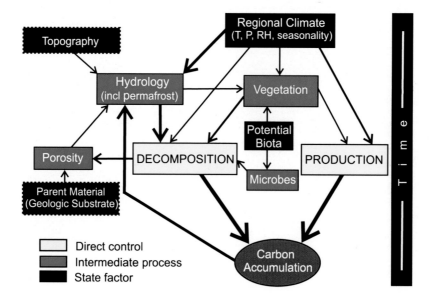

Plate 4. Conceptual model of controls over growth and carbon accumulation in northern peatland ecosystems. Carbon accumulation is directly controlled by plant production and organic matter decomposition (yellow boxes), which are affected by intermediate processes (blue boxes) and ultimately by state factors (black boxes). The hierarchy of controls and processes are applicable to the initiation, expansion, and accumulation of peatlands. Two state factors, geologic substrate (parent material) and topography (dashed black boxes), are most relevant to early peatland establishment by initiation and expansion. The line thickness of the arrows indicates relative importance of controls. *T*, temperature; *P*, precipitation; RH, relative humidity.

observation suggests that high total precipitation does not necessarily lead to high C accumulation, as high precipitation may be offset by increased evaporation under high temperatures, or the seasonal precipitation regime may be such that the site is quite dry during the growing season, thereby reducing NPP and increasing decomposition.

The temporal pattern of peatland initiation across the boreal region is not uniform. For the most part, peatlands older than 12,000 cal years B.P. are found in Alaska, southern parts of central and eastern North America, and the Pacific coast that remained unglaciated during the last glacial maximum (Plate 1) [*MacDonald et al.*, 2006; *Gorham et al.*, 2007]. The highest C accumulation rates in Alaska occurred at 11,000–9000 cal years B.P. (Plates 3b and 5d) during the period of Holocene thermal maximum (HTM) conditions in Alaska (Plate 5a) [*Kaufman et al.*, 2004]. The HTM, with warm summers and strong climate seasonality, was caused by increased summer insolation in the early Holocene [*Berger and Loutre*, 1991]. The C accumulation peaks occurred later in other regions, due to either delayed deglaciation, different timing of the HTM, or other climate factors. For example, warm summer conditions were delayed in eastern North America owing to the presence and

cooling effect of the remnant Laurentide ice sheet, so high C accumulation rates in eastern Canada at 5000–3000 cal years B.P. (Plate 3b) were likely caused by a warm and humid climate as the result of shifting atmospheric circulation [*Kaufman et al.*, 2004]. In western Canada, peak C accumulation occurred in the mid-Holocene, which could be due to a warm climate at that time [*Schweger and Hickman*, 1989] or the dominance of fen peatlands that were nutrient-rich at young ages. Across northern Siberia, warm Holocene conditions persisted until 5000 cal years B.P. as reflected by the northern expansion of the tree line [*MacDonald et al.*, 2008]. An overall slowdown of carbon accumulation seems to have occurred after ~4000 cal years B.P. across many sites, especially in west Siberia and in western Canada (Plate 3b). The decline in C accumulation has been attributed to neoglacial climate cooling and permafrost development [*Zoltai*, 1993; *Peteet et al.*, 1998; *Vitt et al.*, 2000; *Beilman et al.*, 2009]. Relatively high C accumulation rates in the late Holocene are partly attributable to younger peat that has experienced less decomposition. Averaging across all 33 sites, apparent C accumulation rates are highest in the early Holocene, a pattern that holds when only averaging the 20 oldest sites with basal ages of 10,000 cal years B.P.

Table 1. Peat Carbon Accumulation Sites From Northern Peatlands

Site	Site Name	Location	Peatland Type[a]	Latitude	Longitude	Dating Method[b]	No. of Dates	Basal Age (cal years B.P.)	Time-Weighted Holocene Accumulation Rates (g C m^{-2} a^{-1})	Reference
1	Salym-Gyugan Mire, site 3	West Siberia, Russia	bog	60°10'N	72°50'E	conventional	6	10,500	21.9	Turunen et al. [2001]
2	Salym-Gyugan Mire, site 4	West Siberia, Russia	bog	60°10'N	72°50'E	conventional	4	11,000	24.4	Turunen et al. [2001]
3	Vasyugan V21	West Siberia, Russia	bog	56°50'N	78°25'E	conventional	11	9710	42.6	Borren et al. [2004]
4	86-Kvartal Zh0	West Siberia, Russia	fen	56°20'N	84°35'E	conventional	9	8700	70.6	Borren et al.[2004]
5	Kenai Gasfield	Alaska, USA	fen	60°27'N	151°14'W	AMS	12	11,408	13.1	Z. C. Yu (unpublished data, 2008)
6	No Name Creek	Alaska, USA	fen	60°38'N	151°04'W	AMS	11	11,526	12.3	Z. C. Yu (unpublished data, 2008)
7	Horsetrail fen	Alaska, USA	rich fen	60°25'N	150°54'W	AMS	10	13,614	10.7	Jones [2008]
8	Swanson fen	Alaska, USA	poor fen	60°47'N	150°49'W	AMS	9	14,225	5.7	Jones [2008]
9	Fairbanks	Alaska, USA	taiga bog	64°52'N	147°46'W	conventional	4	5509	24.1	Billings [1987]
10	Diana Lake bog	British Columbia, Canada	slope bog	54°09'N	130°15'W	AMS	5	8500	8.6	Turunen and Turunen [2003]
11	Upper Pinto fen	Alberta, Canada	rich fen	53°35'N	118°01'W	AMS	20	7600	30.6	Yu et al. [2003a]
12	Goldeye Lake fen	Alberta, Canada	fen	52°27'N	116°12'W	AMS	6	10,000	31.7	Yu [2006b]
13	Slave Lake bog	Alberta, Canada	bog	55°01'N	114°09'W	conventional	6	10,200	22.4	Kuhry and Vitt [1996]
14	Martin River peatland	NWT, Canada	permafrost	61°48'N	121°24'W	conventional/ AMS/tephra	7	8010	18.3	Robinson [2006]
15	CC-P, Campbell Ck peatland	Nunavut, Canada	fen	68°17'N	133°15'W	conventional/ AMS/tephra	4	10,050	6.1	Vardy et al. [2000]
16	KJ-B, KukJuk peatland	Nunavut, Canada	fen	69°29'N	132°40'W	AMS	4	8000	3.4	Vardy et al. [2000]
17	Patuanak	Saskatchewan, Canada	permafrost	55°51'N	107°41'W	AMS	11	9050	15.6	D. W. Beilman and Z. C. Yu (unpublished data, 2008)
18	BB1	Nunavut, Canada	fen	64°43'N	105°34'W	AMS	2	7720	8.4	Vardy et al. [2000]
19	TK1P2	Nunavut, Canada	fen	66°27'N	104°50'W	AMS	2	6700	12.5	Vardy et al. [2000]
20	Selwyn Lake (SL1)	Saskatchewan, Canada	permafrost bog	59°53'N	104°12'W	AMS	13	6690	12.1	Sannel and Kuhry [2009]

Table 1. (continued)

Site	Site Name	Location	Peatland Type[a]	Latitude	Longitude	Dating Method[b]	No. of Dates	Basal Age (cal year B.P.)	Time-Weighted Holocene Accumulation Rates (g C m^{-2} a^{-1})	Reference
21	Mirabel Bog (average of 7 cores)	Québec, Canada	bog	45°41'N	74°02'W	conventional/AMS	2–7	10,000	7.0	*Muller et al.* [2003]
22	Ellesmere Island (average of 4 cores)	Canada	rich fen	82°N	68°W	conventional/AMS	3	7507	12.9	*LaFarge-England et al.* [1991]
23	Miscou	New Brunswick, Canada	N/A	47°56'N	64°30'W	AMS	7	9000	30.6	*Gorham et al.* [2003]
24	Fourchou	Nova Scotia, Canada	N/A	45°56'N	60°16'W	AMS	8	11,200	18.7	*Gorham et al.* [2003]
25	The Glen Carron bog	Scotland	bog	57°31'N	5°09'W	conventional	6	10,140	10.5	*Anderson* [2002]
26	The Glen Torridon bog	Scotland	bog	57°34'N	5°22'W	conventional	7	9490	20.5	*Anderson* [2002]
27	The Eilean Subbainn bog	Scotland	bog	57°41'N	5°41'W	conventional	4	8550	17.7	*Anderson* [2002]
28	Hanhijänkä	Finland	palsa mire	68°24'N	23°33'E	conventional	7	9800	12.4	*Mäkilä and Moisanen* [2007]
29	Luovuoma (average of 3 center cores)	Finland	fen	68°24'N	23°33'E	conventional	5–6	9800	13.7	*Mäkilä and Moisanen* [2007]
30	S. Haukkasuo, (average of 3 cores)	Finland	raised bog	60°49'N	26°57'E	conventional	5–13	9500	22.5	*Mäkilä* [1997]
31	Ruosuo P8	Finland	aapa mire	65°39'N	27°19'E	conventional	7	9500	12.9	*Mäkilä et al.* [2001]
32	Ruosuo P20	Finland	aapa mire	65°39'N	27°19'E	conventional	9	9500	16.2	*Mäkilä et al.* [2001]
33	Saarisuo B8	Finland	fen	65°39'N	27°19'E	conventional	11	9600	22.4	*Mäkilä et al.* [2001]

[a]N/A, not available.
[b]AMS, accelerator mass spectrometry.

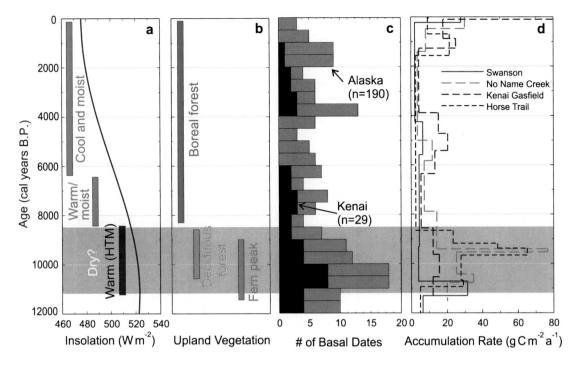

Plate 5. Case study of peatland carbon-climate connections during the Holocene from the Kenai Peninsula, Alaska. (a) Summer insolation [*Berger and Loutre*, 1991] and general Holocene climate history [*Edwards et al.*, 2001; *Kaufman et al.*, 2004]; (b) Non-wetland vegetation history [*Anderson et al.*, 2006; *Jones*, 2008]; (c) Basal peat ages from the Kenai (*n* = 29; black bars [*Reger et al.*, 2007], with additions) and across Alaska (*n* = 190; gray bars; Plate 1); (d) Carbon accumulation rates at four peatland sites on the Kenai Peninsula, Alaska (Swanson Fen and Horse Trail [*Jones*, 2008]; No Name Creek and Kenai Gas Field [Z. C. Yu, unpublished, 2008]). HTM, Holocene Thermal Maximum.

4. CONCEPTUAL MODEL OF LONG-TERM CARBON ACCUMULATION IN NORTHERN PEATLANDS

Peat initiation, persistence, and accumulation are controlled directly by the balance of production and decomposition (Plate 4). NPP of peatland vegetation and the decomposition of plant litter and soil organic matter are two key ecosystem processes, and are affected by numerous physical and biological factors. In common with all other soil types, the formation and development of peat, as organic soils (histosols), can be conceptually organized around five state factors [*Jenny*, 1941], i.e., parent material (geologic mineral substrate), topography, climate, potential biota, and time [*Gorham*, 1957; *Amundson and Jenny*, 1997; *Chapin et al.*, 2002]. These factors affect the two key processes (production and decomposition) either directly or through additional intermediate processes. Some state factors are more important during peatland initiation and the onset of C accumulation processes and less important after the peatland ecosystem is established. For example, topography and geologic substrate (parent material) can promote or prevent peatland initiation

through their effects on porosity and permeability, and the supply and flow of nutrients [*Gorham*, 1957].

In a number of fundamental respects, peatland soils are different from mineral soils. These important differences include the presence of a stable water table and the resulting two-layer structure of peat [*Ivanov*, 1981], the lack, or secondary role, of organo-mineral molecular interactions and the physical aggregate structure of mineral soils [*Sollins et al.*, 1996], and the strong feedbacks between biological and physical processes [*Belyea and Clymo*, 2001], including interactions between fast biological and slow peat-forming processes [*Belyea and Baird*, 2006]. The upper layer of peat above the water table (called the "acrotelm") is often aerated and contains many plant roots, while the lower layer (the "catotelm") is permanently waterlogged [*Ivanov*, 1981]. The acrotelm contains recent plant litter inputs and less-decomposed organic matter, which promotes a higher hydraulic conductivity and the lateral movement of water in near-surface peat layers. Also, most microbial activity occurs in the acrotelm, where aerobic decomposition rates can be orders of magnitude higher than the anaerobic decomposition in the catotelm [*Clymo*, 1984]. The proportion of peat entering the catotelm depends on the burial rate and residence time of litter and organic matter in the acrotelm, which is controlled by plant productivity and water table dynamics. Slow processes in deep peat can be as important as fast processes near the peatland surface for long-term peatland development [*Belyea and Baird*, 2006]. Compared to dry upland mineral soils, peatlands experience stronger internal interactions and feedbacks between hydrology and plant production and C accumulation. For example, as a peatland grows and the groundwater mound within it develops, the peatland surface can become progressively isolated from the surrounding mineral-rich water, slowing down plant production and peatland growth due to reduced nutrient input (Plate 4) [*Damman*, 1979; *Yu et al.*, 2003a; *Belyea and Baird*, 2006]. Also, the peatland may experience drying conditions, if the growth of the groundwater mound does not keep pace with the increase in height of the peatland surface. On the other hand, highly humified deep catotelm peat with low hydraulic conductivity can buffer short-term climate variations by maintaining a relatively high and stable water table in an otherwise dry climate period.

Climate affects both primary production and decomposition processes, either directly or indirectly through vegetation and hydrology. Both the annual character and the seasonality of temperature and precipitation are important in determining peatland water balances. Increases in summer temperatures may directly stimulate photosynthesis and NPP. For example, in a metadata analysis, temperature has been identified as the single most important factor controlling *Sphagnum* production [*Gunnarsson*, 2005]. Climate also affects plant productivity through its influence on hydrology and resultant vegetation and peatland types (e.g., ombrotrophic *Sphagnum*-dominated bogs versus minerotrophic fens dominated by brown mosses or sedges). Even within a peatland type, climate can affect the relative dominance of plant functional types (e.g., vascular plants versus mosses) that have different NPP [*Campbell et al.*, 2000a]. Less studied and discussed is the indirect influence of temperature on productivity through its influences on decomposition and peatland nutrient availability. Temperature has direct impact on organic matter decomposition, owing to the inherent temperature mediation of microbial activity and biochemical reactions, though the temperature sensitivity of soil decomposition is debatable [*Trumbore et al.*, 1996; *Davidson and Janssens*, 2006]. The characteristic waterlogging of peat soils, which makes belowground processes distinct from non-wetland terrestrial ecosystems, is affected by state factors of climate, topography, and parent material (geologic substrate; Plate 4). Low oxygen content under waterlogged conditions, together with acidic waters partly

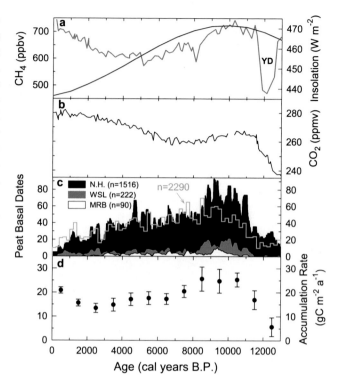

Plate 6. Northern peatland carbon dynamics and its implications for the Holocene global carbon cycle. (a) Summer insolation (red smooth curve [*Berger and Loutre*, 1991]) and atmospheric CH_4 concentration from Greenland ice core [*Brook et al.*, 2000]. YD, Younger Dryas. (b) Atmospheric CO_2 concentration from Dome C in Antarctica [*Monnin et al.*, 2004]. (c) Frequency distribution of basal peat ages as calibrated age-ranges per year from northern peatlands ($n = 1516$; N.H., northern hemisphere [*MacDonald et al.*, 2006]) and subsets showing basal ages from the West Siberia Lowland (WSL; $n = 222$) and Mackenzie River Basin (MRB; $n = 90$). Step lines shows the number of basal dates in each 200-year bin from combined data sets in *MacDonald et al.* [2006] and *Gorham et al.* [2007] ($n = 2290$). (d) Means of carbon accumulation rates from 33 northern peatland sites in each 1000-year bin, with error bars as standard errors (Table 1; Plate 3).

related to cation exchange by the *Sphagnum* plants, limit microbial activity and decomposition [*van Breemen*, 1995]. Different peatland plant litter types and litter chemistry are important factors that affect the inherent decomposability of organic matter [*Moore and Basiliko*, 2006; *Turetsky et al.*, 2008]. Climate change may also shift interactions between vegetation and plant parasites (fungi) in peatland ecosystems [*Wiedermann et al.*, 2007], which may affect plant communities, NPP, and C balances. Changes in seasonality, especially the seasonal association of thermal and moisture conditions, would affect growing season lengths and water budgets, thereby impacting productivity and decomposition [*Cecil and Dulong*, 2003]. Thus, climate variation has direct and indirect influences on these key processes and, depending on their relative importance, on peat carbon accumulation. Peatlands near the edges of the climate envelope may be more sensitive to a shift in the balance between productivity and decomposition. This may be especially true for peatlands near the high temperature and low precipitation ends of climate space (Plate 2).

5. PEATLAND RESPONSE TO CLIMATE CHANGE AND IMPLICATIONS FOR THE GLOBAL CARBON CYCLE

Postglacial initiation, expansion, and subsequent variation in C accumulation of northern peatlands have responded to global climate change. At hemispheric and regional scales, northern peatlands show rapid expansion during the initial warming of the last deglaciation, especially in the early Holocene (Plate 6c) [*MacDonald et al.*, 2006]. Our synthesis of peatland initiation and C accumulation data in northern peatlands consistently shows the highest initiation and C accumulation rates at 11,000–8000 cal years B.P. (Plates 6c and 6d). In western continental Canada, one of the best-studied peatland regions [*Vitt et al.*, 2000], 71 basal dates from paludified peatlands show regular millennial-scale variation in peatland initiation [*Campbell et al.*, 2000b], which is also corroborated by a detailed peat C accumulation analysis during the mid- and late Holocene [*Yu et al.*, 2003b]. These oscillations in peatland expansion and C accumulation appear to correlate with millennial-scale climate variations in the Holocene [*Bond et al.*, 2001].

As a case study, four fen sites from the Kenai Peninsula, Alaska show a remarkably similar pattern in C accumulation over the Holocene, with a peak C accumulation at 11,000–9000 cal years B.P. (Plate 5d). There is good evidence to indicate that a warm climate in south-central Alaska prevailed during the early Holocene, but the moisture conditions are less clear (Plate 5a) [*Kaufman et al.*, 2004]. Widespread *Populus*-dominated deciduous forests (Plate 5b) and low levels of some lakes at that time [*Anderson et al.*, 2006] suggest dry non-wetland soils where potential evaporation was greater than precipitation. On the basis of the evidence for a warm climate, dry forests, and rapid peat C accumulation in the early Holocene, we suggest that moisture conditions could have been different for non-wetland ecosystems than for peatlands. Enhanced climate seasonality in the early Holocene, especially different seasonal associations between temperature and precipitation, could have resulted in contrasting moisture conditions on peatlands and non-wetland terrestrial ecosystems. For example, droughts in late summer

are major constraints for growing season lengths and vegetation production on peatlands [*Aurela et al.*, 2007], while upland soil moisture and lake hydrology are mostly controlled by winter snowfall and snow melt recharge in springs [*Shuman and Donnelly*, 2006]. Here, we speculate that high summer temperatures in the early Holocene promoted greater vapor transport from surrounding warmer oceans, resulting in significant increases in summer precipitation in Alaska. High summer precipitation would prevent summer droughts often experienced by peatlands, facilitating a longer growing season and greater vegetation production, but the decrease in winter and spring precipitation under cold winters would cause lower lake levels and dry upland soils, even during the summer season. Therefore, greater NPP has more likely controlled high C accumulation than low decomposition in these peatlands during the early Holocene.

Similar to the overall pattern in northern peatlands (Plate 6c), peatland initiation was widespread across Alaska in the early Holocene, including the Kenai Peninsula (Plate 5c). A sustained and widespread peak in fern (Polypodiaceae) spores in Kenai lakes and peats during this time (Plate 5b) [*Ager*, 2000; *Anderson et al.*, 2006; *Jones*, 2008] suggests extensive peatland expansion during the first two millennia of the Holocene because we have observed abundant fern growth around expanding peatland margins today in south-central Alaska. A climate shift associated with the establishment of boreal forest on the Kenai in the early mid-Holocene [*Anderson et al.*, 2006] is associated with a decrease in C accumulation in all four sites (Plate 5d) and a decline in peatland initiation (Plate 5c). An increase in C accumulation around 6000–4000 cal years B.P. is a robust signal across sites (Plate 5d) that might have been in response to a cool and moist climate beginning ~6000 cal years B.P. [*Mann and Hamilton*, 1995]. In contrast, peatland expansion appears to have slowed at this time (Plate 5c), suggesting that optimal climate conditions for peatland initiation are not exactly the same as those conditions that promote C accumulation in extant peatlands, which may be related to state factor interactions.

Peatland C dynamics have, in turn, had an influence on the global carbon cycle during the Holocene. Ice core records from Greenland and Antarctica show a dramatic increase and sustained peak in atmospheric CH_4 concentration between 11,500 and 8000 cal years B.P. (Plate 6a) [*Chappellaz et al.*, 1997; *Brook et al.*, 2000]. Ice core $^{13}CH_4$ values and trends show that this early Holocene CH_4 increase was strongly influenced by ^{13}C-enriched CH_4 emissions, suggesting a number of biosphere sources rather than catastrophic or sustained marine clathrate dissociation [*Schaefer et al.*, 2006]. Changes in the interpolar CH_4 gradient suggest a dominant tropical source, but also an increase in northern sources at this time [*Brook et al.*, 2000]. This sharp rise is coincident with the rapid expansion of many northern peatlands (Plate 6c) [*MacDonald et al.*, 2006], particularly across western North America and Eurasia (Plate 1). Also, the prevalence of northern peatlands and their sequestered terrestrial C would favor methanogenic microbial activity in warm summers during the HTM (see section 3). Thus, northern peatlands were a likely and substantial CH_4 source that contributed to the early Holocene atmospheric CH_4 rise.

The large C pools in northern peatlands may have also affected past changes in atmospheric CO_2 concentration. While CH_4 concentration rose sharply and remained high in the early Holocene, CO_2 concentration decreased by several parts per million between 11,000 and 8000 cal years B.P., equivalent to an uptake of about 100 Pg C [*Indermühle et al.*, 1999]. Simple terrestrial ecosystem models suggest that a similar mass of new C was sequestered as biomass and soil C during the re-growth of northern forests [*Joos et al.*, 2004; *Kohler et al.*, 2005]. In addition to non-wetland ecosystems, the C sequestered by northern peatlands was a further, possibly substantial, contribution to this postglacial land C buildup. An estimate of the northern peatland C pool at 8000 cal years B.P. is possible on the basis of our data synthesis. Of the ^{14}C basal peat ages north of 45°N from available land area at that time (not covered by the Laurentide ice sheet; Plate 1), 975 (out of a total of 2173) are older than 8000 cal years B.P. (Plate 1). Assuming that the Holocene expansion of peatlands to their modern extent (about 4 million km^2) was roughly proportional to the cumulative frequency of peat basal ages, northern peatlands may have covered 1.8 million km^2 of land area by that time. The mean of apparent C accumulation rates for the 12,000–8000 cal years B.P. period from the available sites is 23.1 ± 3.4 g C m^{-2} a^{-1} ($n = 5 - 26$; Plate 3; Table 1), a value that likely underestimates the true early Holocene C uptake rate (see section 3). The combination of these conservative approximations suggests that the northern peatland C pool at 8000 cal years B.P. was 73–98 Pg C, an estimate that is similar to or greater than previous estimates [29–58 Pg C by *MacDonald et al.* [2006] and 92 Pg C by *Adams and Faure* [1998]).

6. CONCLUDING REMARKS: OUTSTANDING ISSUES AND FUTURE DIRECTIONS

Carbon dynamics of northern peatlands have shown sensitive responses to changes in boundary conditions and climate and have had noticeable influence on the global carbon cycle during the Holocene. Our synthesis and analysis of climate-peatland distributions and peat-core data provide a framework for understanding the dominant controls of peat-land C dynamics and climate sensitivity. We emphasize that primary productivity and decomposition are two ecosystem processes that have direct controls over carbon accumulation in peatlands. These key controls are, in turn, affected by intermediate processes (i.e., hydrology and vegetation) and ultimately by broader state factors. Our spatial analysis of climate data and northern peatland distributions shows that most peatlands occur within a mean annual air temperature range of −12° to 5°C and a mean annual precipitation range of 200 to 1000 mm, which spans the climate space of the boreal/taiga ecoregion. Also, peatlands in different regions show a distinct regional character in annual *T-P* space, suggesting the complex control of regional effective moisture and resultant peatland water budgets.

About a half of northern peatlands initiated before 8000 cal years B.P. in the early Holocene on the basis of >2000 basal peat dates from the northern hemisphere. Also, peat C accumulation appears to have been highest in the first few millennia of the Holocene, especially in regions that experienced Holocene thermal maximum conditions at that time. We observe that peatlands having high C accumulation rates tend to occur in regions with intermediate temperature and precipitation. These regions have the largest peatland areas in the world, including west Siberia and western Canada. On the other hand, high precipitation may not necessarily result in high C accumulation, e.g., in eastern Canada and British Columbia, suggesting that water budgets and carbon balance between production and decomposition are key to net C accumulation.

The early Holocene peatland expansion and C accumulation contributed to the peak in global CH_4 concentration and the decline in CO_2 concentration at this time. Also, the estimates on the basis of our synthesis of the largest available data sets show that, in the early Holocene before 8000 cal years B.P., northern peatlands alone may have sequestered about 100 Pg of atmospheric carbon.

This synthesis of data and ideas has identified some major outstanding issues and key future research directions.

1. Our data compilation and synthesis show major data gaps for peatland initiation and carbon accumulation histories in the Russian Far East, East Siberia, and the Hudson Bay Lowland (Plate 1). These regions represent geographic locations of intermediate temperatures and high precipitation in modern climate space (Plate 2), where peat-core data are lacking. Therefore, filling these gaps will further inform our understanding of the climate sensitivity of peatland C dynamics.

2. Further refined analysis of the large data sets of available basal peat ages could provide useful information for understanding climate control and sensitivity of peatland expansion, including separation of paludified and terrestrial-

ized peatlands since these peatland formation pathways have very different climate controls.

3. There is a need to develop and integrate process-based peatland dynamic models that take into account interactions and feedbacks of local and regional factors as well as fast and slow processes affecting production and decomposition to determine net peat C accumulation.

4. Developing novel peat-based proxies will facilitate our understanding of climate sensitivity of specific ecosystem processes, including independent proxies for productivity and decomposition. Also, new proxies to indicate hydrological and permafrost dynamics would improve our understanding of these important processes.

Acknowledgments. We thank Andy Baird and an anonymous reviewer for their constructive comments, and Daniel Brosseau and Julie Loisel for assistance with data analysis. Yu and Jones were supported by the United States National Science Foundation (NSF)—Biocomplexity in the Environment: Carbon and Water in the Earth System Program (ATM 0628455) for their peatland research in Alaska. Beilman was supported by the Marie Curie Incoming International Fellow Program of the European Commission (MC IIF 40974). We acknowledge the stimulating discussions and ideas presented at the NSF-supported PeatNet workshop in March 2008, which were part of the impetus for this paper.

REFERENCES

Adams, J. M., and H. Faure (1998), A new estimate of changing carbon storage on land since the last glacial maximum, based on global land ecosystem reconstruction, *Global Planet. Change, 16/17*, 3–24.

Ager, T. A. (2000), Holocene vegetation history of the Kachemak Bay area, Cook Inlet, South-central Alaska, *Geologic Studies in Alaska by the U.S. Geological Survey, 1998, U.S. Geol. Surv. Prof. Pap. 1615*, pp. 147–165.

Amundson, R., and H. Jenny (1997), On a state factor model of ecosystems, *BioScience, 47*, 536–543.

Anderson, D. E. (2002), Carbon accumulation and C/N ratios of peat bogs in north-west Scotland, *Scott. Geogr. J., 118*, 323–341.

Anderson, R. S., D. J. Hallett, E. Berg, R. B. Jass, J. L. Toney, C. S. de Fontaine, and A. DeVolder (2006), Holocene development of boreal forest and fire regimes on the Kenai Lowlands of Alaska, *Holocene, 16*, 791–803.

Aurela, M., T. Riutta, T. Laurila, J.-P. Tuovinen, T. Vesala, E.-S. Tuittila, J. Rinne, S. Haapanala, and J. Laine (2007), CO₂ exchange of a sedge fen in southern Finland—The impact of a drought period, *Tellus, 59B*, 826–837.

Bauer, I. E., L. D. Gignac, and D. H. Vitt (2003), Development of a peatland complex in boreal western Canada: Lateral site expansion and local variability in vegetation succession and long-term peat accumulation, *Can. J. Bot., 81*, 833–847.

Beilman, D. W., G. M. MacDonald, L. C. Smith, and P. J. Reimer (2009), Carbon accumulation in peatlands of West Siberia over the last 2000 years, *Global Biogeochem. Cycles, 23*, GB1012, doi:10.1029/2007GB003112.

Belyea, L. R., and A. J. Baird (2006), Beyond 'the limits to peat bog growth': Cross-scale feedback in peatland development, *Ecol. Monogr., 76*, 299–322.

Belyea, L. R., and R. S. Clymo (2001), Feedback control of the rate of peat formation, *Proc. R. Soc. London, Ser. B, 268*, 1315–1321.

Berger, A., and M. F. Loutre (1991), Insolation values for the climate of the last 10 million years, *Quat. Sci. Rev., 10*, 297–317.

Billings, W. D. (1987), Carbon balance of Alaskan tundra and taiga ecosystems: Past, present, and future, *Quat. Sci. Rev., 6*, 165–177.

Bond, G., B. Kromer, J. Beer, R. Muscheler, M. N. Evans, W. Showers, S. Hoffmann, R. Lotti-Bond, I. Hajdas, and G. Bonani (2001), Persistent solar influence on North Atlantic climate during the Holocene, *Science, 294*, 2130–2136.

Borren, W., W. Bleuten, and E. D. Lapshina (2004), Holocene peat and carbon accumulation rates in the southern taiga of western Siberia, *Quat. Res., 61*, 42–51.

Brook, E. J., S. Harder, J. Severinghaus, E. J. Steig, and C. M. Sucher (2000), On the origin and timing of rapid changes in atmospheric methane during the last glacial period, *Global Biogeochem. Cycles, 14*, 559–572.

Campbell, C., D. H. Vitt, L. A. Halsey, I. D. Campbell, M. N. Thormann, and S. E. Bayley (2000a), Net primary production and standing biomass in northern continental wetlands, Northern Forestry Centre Information Report NOR-X-369, 57 pp., Can. For. Service, Edmonton, AB.

Campbell, I. D., C. Campbell, Z. C. Yu, D. H. Vitt, and M. J. Apps (2000b), Millennial-scale rhythms in peatlands in the western interior of Canada and in the global carbon cycle, *Quat. Res., 54*, 155–158.

Cecil, C. B., and F. T. Dulong (2003), Precipitation models for sediment supply in warm climates, in *Climate Controls on Stratigraphy, SEPM Special Publication, No. 77*, edited by C. B. Cecil, and N. T. Edgar, pp. 21–27, SEPM.

Chapin, F. S., P. A. Matson, and H. A. Mooney (2002), *Principles of Terrestrial Ecosystem Ecology*, 472 pp., Springer, New York.

Chappellaz, J., T. Blunier, S. Kints, A. Dallenbach, J. M. Barnola, J. Schwander, D. Raynaud, and B. Stauffer (1997), Changes in the atmospheric CH₄ gradient between Greenland and Antarctica during the Holocene, *J. Geophys. Res., 102*, 15,987–15,997.

Charman, D. (2002), *Peatlands and Environmental Change*, 301 pp., John Wiley, New York.

Clymo, R. S. (1984), The limits to peat bog growth, *Philos. Trans. R. Soc. London, Ser. B, 303*, 605–654.

Damman, A. W. H. (1979), Geographic patterns in peatland development in eastern North America, *Proceedings of the International Symposium on Classification of Peat and Peatlands, International Peat Society*, pp. 42–57, Hyytiälä, Finland, Sept. 17–21.

Davidson, E. A., and I. A. Janssens (2006), Temperature sensitivity of soil carbon decomposition and feedbacks to climate change, *Nature*, *440*, 165–173.

Dyke, A. S., D. Giroux, and L. Robertson (2004), *Paleovegetation Maps, Northern North America, 18000 to 1000 BP, Geol. Surv. Can. Open File 4682*, Ottawa, Canada.

Edwards, M. E., C. J. Mock, B. P. Finney, V. A. Barber, and P. J. Bartlein (2001), Potential analogues for paleoclimatic variations in eastern interior Alaska during the past 14,000 yr: Atmospheric-circulation controls of regional temperature and moisture responses, *Quat. Sci. Rev.*, *20*, 189–202.

Gignac, L. D., and D. H. Vitt (1994), Responses of northern peatlands to climate change: Effects on bryophytes, *J. Hattori Bot. Lab.*, *75*, 119–132.

Gorham, E. (1957), The development of peat lands, *Q. Rev. Biol.*, *32*, 145–166.

Gorham, E. (1991), Northern peatlands: Role in the carbon cycle and probable responses to climatic warming, *Ecol. Appl.*, *1*, 182–195.

Gorham, E., J. A. Janssens, and P. H. Glaser (2003), Rates of peat accumulation during the postglacial period in 32 sites from Alaska to Newfoundland, with special emphasis on northern Minnesota, *Can. J. Bot.*, *81*, 429–438.

Gorham, E., C. Lehmn, A. Dyke, J. Janssens, and L. Dyke (2007), Temporal and spatial aspects of peatland initiation following deglaciation in North America, *Quat. Sci. Rev.*, *26*, 300–311.

Gunnarsson, U. (2005), Global patterns of *Sphagnum* productivity, *J. Bryol.*, *27*, 269–279.

Indermühle, A., et al. (1999), Holocene carbon-cycle dynamics based on CO_2 trapped in ice at Taylor Dome, Antarctica, *Nature*, *398*, 121–126.

Ivanov, K. E. (1981), *Water Movement in Mirelands* (translated from Russian by A. Thomson, and H. A. P. Ingram), 276 pp., Academic Press, Toronto.

Jenny, H. (1941), *Factors of Soil Formation*, 281 pp., McGraw-Hill, New York.

Jones, M. C. (2008), Climate and vegetation change in a 14,200 cal yr BP peatland from the Kenai Peninsula, Alaska: A record of pollen, macrofossils, stable isotopes, and carbon storage, Ph.D. dissertation, 164 pp., Columbia Univ., New York.

Joos, F., S. Gerber, I. C. Prentice, B. L. Otto-Bleisner, and P. J. Valdes (2004), Transient simulations of Holocene atmospheric carbon dioxide and terrestrial carbon since the Last Glacial Maximum, *Global Biogeochem. Cycles*, *18*, GB2002, doi:10.1029/2003GB002156.

Kaufman, D. S., et al. (2004), Holocene thermal maximum in the western Arctic (0–180°W), *Quat. Sci. Rev.*, *23*, 529–560.

Kohler, P., F. Joos, S. Gerber, and R. Knutti (2005), Simulated changes in vegetation distribution, land carbon storage, and atmospheric CO_2 in response to a collapse of the North Atlantic thermohaline circulation, *Clim. Dyn.*, *25*, 689–708.

Korhola, A., J. Alm, K. Tolonen, J. Turunen, and H. Junger (1996), Three-dimensional reconstruction of carbon accumulation and CH_4 emission during nine millennia in a raised mire, *J. Quat. Sci.*, *11*, 161–165.

Kuhry, P., and J. Turunen (2006), The postglacial development of boreal and subarctic peatlands, in *Boreal Peatland Ecosystems*, Ecological Studies Series, vol. 188, edited by R. K. Wieder and D. H. Vitt, pp. 25–46, Springer, New York.

Kuhry, P., and D. H. Vitt (1996), Fossil carbon/nitrogen ratios as a measure of peat decomposition, *Ecology*, *77*, 271–275.

LaFarge-England, C., D. H. Vitt, and J. England (1991), Holocene soligenous fens on a high Arctic fault block, Northern Ellesmere Island (82°N), N.W.T., Canada, *Arct. Alp. Res.*, *23*, 80–98.

MacDonald, G. M., D. W. Beilman, K. V. Kremenetski, Y. W. Sheng, L. C. Smith, and A. A. Velichko (2006), Rapid early development of circumarctic peatlands and atmospheric CH_4 and CO_2 variations, *Science*, *314*, 285–288.

MacDonald, G. M., K. V. Kremenetski, and D. W. Beilman (2008), Climate change and the northern Russian treeline zone, *Philos. Trans. R. Soc. Ser. B*, *363*, 2285–2299.

Mäkilä, M. (1997), Holocene lateral expansion, peat growth and carbon accumulation on Haukkasuo, a raised bog in southeastern Finland, *Boreas*, *26*, 1–14.

Mäkilä, M., and M. Moisanen (2007), Holocene lateral expansion and carbon accumulation of Luovuoma, a northern fen in Finnish Lapland, *Boreas*, *36*, 198–210.

Mäkilä, M, M. Saarnisto, and T. Kankainen (2001), Aapa mires as a carbon sink and source during the Holocene, *J. Ecol.*, *89*, 589–599.

Mann, D. H., and T. D. Hamilton (1995), Late Pleistocene and Holocene paleoenvironments of the North Pacific coast, *Quat. Sci. Rev.*, *14*, 449–471.

Monnin, E., et al. (2004), Evidence for substantial accumulation rate variability in Antarctica during the Holocene, through synchronization of CO_2 in the Taylor Dome, Dome C and DML ice cores, *Earth Planet. Sci. Lett.*, *224*, 45–54.

Moore, T., and N. Basiliko (2006), Decomposition in boreal peatlands, in *Boreal Peatland Ecosystems*, *Ecol. Stud. Ser.*, vol. 188, edited by R. K. Wieder and D. H. Vitt, pp. 125–143, Springer, New York.

Muller, S. D., P. H. Richard, and A. C. Larouche (2003), Holocene development of a peatland (southern Québec): A spatio-temporal reconstruction based on pachymetry, sedimentology, microfossils and macrofossils, *Holocene*, *13*, 649–664.

Olson, D. M. et al. (2001), Terrestrial ecoregions of the world: A new map of life on earth, *BioScience*, *51*, 933–938.

Peteet, D., A. Andreev, W. Bardeen, and F. Mistretta (1998), Long-term Arctic peatland dynamics, vegetation and climate history of the Pur-Taz region, Western Siberia, *Boreas*, *27*, 115–127.

Rawlins, M. A., and C. J. Willmott (1999), Arctic land-surface air temperature and precipitation: 1960–1990 gridded monthly time series, version 1.01, *Center for Clim. Res.*, Univ. Delaware, Newark, DE.

Reger, R. D., A. G. Sturmann, E. E. Berg, and P. A. C. Burns (2007), *A Guide to the Late Quaternary History of Northern and Western Kenai Peninsula, Alaska*, Division of Geological and Geophysical Surveys, Alaska, Guidebook 8.

Robinson, S. D. (2006), Carbon accumulation in peatlands, south-western Northwest Territories, Canada, *Can. J. Soil Sci.*, *86*, 305–319.

Roulet, N. T., P. F. Lafleur, P. J. H. Richard, T. R. Moore, E. R. Humphreys, and J. Bubier (2007), Contemporary carbon balance and late Holocene carbon accumulation in a northern peatland, *Global Change Biol.*, *13*, 397–411.

Rydin, H., and J. Jeglum (2006), *The Biology of Peatlands*, 343 pp., Oxford Univ. Press, New York.

Sannel, A. B. K., and P. Kuhry (2009), Holocene peat growth and decay dynamics in sub-arctic peat plateaus, west-central Canada, *Boreas*, *38*, 13–24.

Schaefer, H., M. J. Whiticar, E. J. Brook, V. V. Petrenko, D. F. Ferretti, and J. P. Severinghaus (2006), Ice record of $\delta^{13}C$ for atmospheric CH_4 across the Younger Dryas-Preboreal transition, *Science*, *313*, 1109–1112.

Schweger, C. E., and M. Hickman (1989), Holocene paleohydrology of central Alberta—Testing the general-circulation-model climate simulations, *Can. J. Earth Sci.*, *26*, 1826–1833.

Sernander, R. (1908), On the evidence of Postglacial changes of climate furnished by the peat-mosses of northern Europe, *Geol. Foren. Forh.*, *30*, 465–478.

Shuman, B., and J. P. Donnelly (2006), The influence of seasonal precipitation and temperature regimes on lake levels in the northeastern United States during the Holocene, *Quat. Res.*, *65*, 44–56.

Sollins, P., P. Homman, and B. A. Caldwell (1996), Stabilization and destabilization of soil organic matter: Mechanisms and controls, *Geoderma*, *74*, 65–105.

Trumbore, S. E., O. A. Chadwick, and R. Amundson (1996), Rapid exchange between soil carbon and atmospheric carbon dioxide driven by temperature change, *Science*, *272*, 393–396.

Turetsky, M. R., S. E. Crow, R. J. Evans, D. H. Vitt, and R. K. Weider (2008), Trade-offs in resource allocation among moss species control decomposition in boreal peatlands, *J. Ecol.*, *96*, 1297–1305.

Turunen, C., and J. Turunen (2003), Development history and carbon accumulation of a slope bog in oceanic British Columbia, Canada, *Holocene*, *13*, 225–238.

Turunen, J., T. Tahvanainen, and K. Tolonen (2001), Carbon accumulation in west Siberian mires, Russia, *Global Biogeochem. Cycles*, *15*, 285–296.

van Breemen, N. (1995), How *Sphagnum* bogs down other plants, *Trends Ecol. Evol.*, *10*, 270–275.

Vardy, S. R, B. G. Warner, J. Turunen, and R. Aravena (2000), Carbon accumulation in permafrost peatlands in the Northwest Territories and Nunavut, Canada, *Holocene*, *10*, 273–280.

Vitt, D. H., L. A. Halsey, I. E. Bauer, and C. Campbell (2000), Spatial and temporal trends in carbon storage of peatlands of continental western Canada through the Holocene, *Can. J. Earth Sci.*, *37*, 683–693.

Wieder, R. K., D. H. Vitt, and B. W. Benscoter (2006), Peatlands and the boreal forest, in *Boreal Peatland Ecosystems*, *Ecol. Stud. Ser.*, vol. 188, edited by R. K. Wieder and D. H. Vitt, pp. 1–8, Springer, New York.

Wiedermann, M. M., A. Nordin, U. Gunnarsson, M. B. Nilsson, and L. Ericson (2007), Global change shifts vegetation and plant-parasite interactions in a boreal mire, *Ecology*, *88*, 454–464.

Wolfe, B. B., T. W. D. Edwards, R. Aravena, S. L. Forman, B. G. Warner, A. A. Velichko, and G. M. MacDonald (2000), Holocene paleohydrology and paleoclimate at treeline, north-central Russia, inferred from oxygen isotope records in lake sediment cellulose, *Quat. Res.*, *53*, 319–329.

Yu, Z. C. (2006a), Modeling ecosystem processes and peat accumulation in boreal peatlands, in *Boreal Peatland Ecosystems*, *Ecol. Stud. Ser.*, vol. 188, edited by R. K. Wieder and D. H. Vitt, pp. 313–329, Springer, New York.

Yu, Z. C. (2006b), Holocene carbon accumulation of fen peatlands in boreal western Canada: Complex ecosystem response to climate variation and disturbance, *Ecosystems*, *9*, 1278–1288.

Yu, Z. C., D. H. Vitt, I. D. Campbell, and M. J. Apps (2003a), Understanding Holocene peat accumulation pattern of continental fens in western Canada, *Can. J. Bot.*, *81*, 267–282.

Yu, Z. C., I. D. Campbell, C. Campbell, D. H. Vitt, G. C. Bond, and M. J. Apps (2003b), Carbon sequestration in peat highly sensitive to Holocene wet-dry climate cycles at millennial time scales, *Holocene*, *13*, 801–808.

Zoltai, S. C. (1993), Cyclic development of permafrost in the peatlands of Northwestern Alberta, Canada, *Arct. Alp. Res.*, *25*, 240–246.

D. W. Beilman, Department of Geography, University of Hawai'i at Manoa, 2424 Maile Way, Honolulu, HI 96822, USA. (beilman@hawaii.edu)

M. C. Jones and Z. Yu, Department of Earth and Environmental Sciences, Lehigh University, 31 Williams Drive, Bethlehem, PA 18015, USA. (mcj208@lehigh.edu; ziy2@lehigh.edu)

Direct Human Impacts on the Peatland Carbon Sink

Jukka Laine

Finnish Forest Research Institute, Parkano, Finland

Kari Minkkinen

Department of Forest Ecology, University of Helsinki, Helsinki, Finland

Carl Trettin

Center for Forested Wetlands Research, U.S. Forest Service, Cordesville, South Carolina, USA

Northern peatlands occupy over 3 million km^2 globally and contain the largest carbon (C) pool (typically >100 kg C m^{-2}) among terrestrial ecosystems. Agriculture, forestry, and peat harvesting are the principal human-induced activities that alter the peatland and hence the distribution and flux of carbon. As a prerequisite to those uses, the peatland is usually drained, which has long-term effects on the site hydrology and corresponding direct linkages to changes in C dynamics in the vegetation and soils. Soil organic matter decomposition is stimulated following drainage, typically reported as increased CO_2 emissions and peat subsidence. The vegetation also changes following drainage, regardless of cropping or harvesting practices, and this change influences the net ecosystem C exchange rate. Peatland drainage tends to reduce CH_4 emissions, as a result of a greater aerated surface soil volume. The net global effect of active management of northern peatlands has been to reduce the C pool, because agricultural use is dominant. There remain considerable uncertainties in estimating the C pools, fluxes, and responses to management in peatlands; the issues span inadequate inventories to sampling and measurements in complex settings.

1. INTRODUCTION

Peatlands are widely recognized as an important terrestrial carbon (C) pool. Their role in the global C cycle and the potential interactions with changing climatic conditions have been the subject of several recent reviews [*Bridgham et al.*, 2006; *Roulet*, 2000; *Trettin et al.*, 2006] and volumes [*Strack*, 2008]. The following is a synthesis of the effects of land use change on the C cycle of northern peatlands intended to provide context for future work toward a better understanding of these ecologically important and sociologically valuable ecosystems.

2. DIRECT HUMAN INFLUENCE: LAND USE CHANGE

2.1. Extent and Importance

Agriculture, forestry, and peat extraction for fuel and horticultural use are the major causes of change in peatland

Carbon Cycling in Northern Peatlands
Geophysical Monograph Series 184

Table 1. Area of Northern Peatlands Used in Agriculture and Forestry[a]

	Total Peatland Area	Peatland Area Used for Agriculture		Peatland Area Used for Forestry	
	km^2	km^2	%	km^2	%
Europe (without Russia)	325,000	50,600	15	97,540	30
Russia	1,390,000	70,400	5	40,000	3
North America	1,307,000[b]	231,000	18	4,250	<1
China	10,440	2,610	25	700	7

[a]Adapted from *Oleszczuk et al.* [2008], *Minkkinen et al.* [2008], *Bridgham et al.* [2006], and *Rydin and Jeglum* [2006].
[b]Includes 513×10^3 km^2 permafrost peatlands.

condition. Conversion to agriculture has been the most extensive form of land use change, with approximately 13% of European peatlands currently used for this purpose, and somewhat more in North America and China (Table 1). In some European countries, more than half of the original peatland area is presently used for agriculture [*Oleszczuk et al.*, 2008]. The majority of the agricultural use is for grazing as meadows and pastures, but lands are also cultivated for cereal crops. Agricultural use significantly alters the peatland C balance as a result of changes to vegetation, hydrology, and biogeochemical processes.

The utilization of peatlands for forestry is concentrated in Nordic and Baltic countries where approximately 30% of peatland area has been drained for production forestry (Table 1). The use varies from extensive management regimes that may consist of harvesting and natural regeneration to intensive management prescriptions that may involve drainage, water management, and fertilization. Drainage is a common management prescription due to the constraints of saturated soil conditions on site productivity and operability; as a result, over 10 million ha of peatlands have been drained in the Nordic countries and Russia for silvicultural purposes [*Paavilainen and Päivänen*, 1995]. In Finland alone, more than 5 million ha is in production forestry. Production forestry in peatlands has important benefits to the local and national economy; in Finland, more than 20% of annual forest growth takes place in peatland forests [*Päivänen*, 2008].

The use of peat for domestic energy purposes by local communities has been common in many parts of the world for centuries. Large-scale use of peat in energy production started in the 20th century, and today, Finland, Ireland, Russia, Belarus, and Sweden account for almost 90% of the world's use [*Holmgren et al.*, 2008], with an average annual consumption of 3.3 Mt [*Paappanen et al.*, 2006]. Peat is also excavated and sold as a growing medium for horticultural use, but the volume used annually is approximately 50% that of fuel peat. Germany and Canada account for over half of horticultural peat extraction. The area used for peat produc-

tion is very small in comparison to agricultural or forestry use: approximately 0.5 million ha. Although the amount of peat used for energy and horticultural use is relatively small, these represent important economic uses for communities and industries.

2.2. Impacts on the Ecosystem

Conversion of peatlands to other land uses typically involves alterations of the hydrology and plant community, both of which have consequences to the carbon cycle, carbon storage, and greenhouse gas emissions. The effects of land use conversion of peatlands are to change the rates of carbon sequestration, organic matter decomposition, and fluxes from the ecosystem. Assessing the impact on the C cycle necessarily involves the consideration of functional linkages among the vegetative community, soil biogeochemistry, the water balance, and the specific land use practices. The most evident effect of peatland drainage and utilization is the acceleration of oxidative processes in the surface peat.

The impacts of converting peatlands to agriculture on ecosystem properties and carbon storage were recently summarized by *Oleszczuk et al.* [2008]. For agricultural use, intensive drainage is used to achieve suitable hydrological conditions with sufficient aeration in the soil for the cultivated crops and pasture. The agricultural conversion process also involves the removal of the original peatland vegetation which is replaced with crop or pastoral species. The combination of altered hydrology and vegetation, and cultivation measures (e.g., tillage, fertilization), drastically change the carbon sequestration and sink functions of the land [*Kasimir-Klemedtsson et al.*, 1997; *Maljanen et al.*, 2004].

Similarly, drainage and water management systems are commonly associated with peatland forestry operations, but the intensity of drainage is much less than under agricultural systems. As a result of silvicultural drainage, the plant community changes to one dominated by tree stands and forest flora, but some of the original peatland species remain. As a result, the use of northern peatlands for forestry has smaller impacts on the ecosystem functions than those used for agriculture, but they tend to be more complex.

Utilization of a peatland for energy and horticultural products totally changes the structure and functions of the original ecosystem, as all the vegetation is removed at the onset of operations, and eventually, the majority of the peat is removed. During the production phase, CH_4 and N_2O emissions of the site are avoided, but the intensive drainage of these sites maintains high oxidative microbial activity in the peat layer resulting in large CO_2 emissions [*Alm et al.*, 2007; *Holmgren et al.*, 2008].

3. LAND USE IMPACTS ON THE CARBON SINK FUNCTIONS

3.1. Agriculture

Under agricultural use, the peatland is drained and cultivated, resulting in increased aeration of the surface soil which enhances microbial activity and results in increased CO_2 emissions. Despite the relatively large area of converted peatlands in North America, most of the available information comes from European studies, which show a wide range in emission estimates (Table 2). These estimates seem to depend more on the measurement methodology and agricultural system and less on the geographical location of the study sites [*Oleszczuk et al.*, 2008]. High emissions have been reported from cereal crop cultivation using intensive drainage and annual tillage and also from fallow fields with bare soils [*Maljanen et al.*, 2007]. Cultivated peat soils reflect the agricultural legacy after the use ends, as evidenced by rather high C losses from sites abandoned from cultivation [*Maljanen et al.*, 2007].

Increased aeration in the surface soil generally enhances CH_4 oxidation, resulting in decreased emissions; this effect often converts the soil into a CH_4 sink [*Kasimir-Klemedtsson et al.*, 1997; *Flessa et al.*, 1998]. The activity of CH_4-oxidizing bacteria may, however, be reduced by nitrogen fertilizer applications [*Van den Pol-van Dasselaar et al.*, 1999; *Reay and Nedwell*, 2004]. The CH_4 balance is highly dependent on the soil hydrology, and poorly drained soils may continue to emit methane during cultivation [*Maljanen et al.*, 2003a]. In addition to increasing soil oxidation, agricultural use of

Table 2. Greenhouse Gas Emissions From European Peatlands Converted to Agriculture[a]

	CO_2	CH_4	N_2O
Emission range	7000–22,000	−5–4	2–56

[a]Direct flux measurements modified from *Oleszczuk et al.* [2008]. Emissions are given in kg ha^{-1} a^{-1}.

peatlands reduces the net organic matter input to the soil, as a result of the crop utilization. The effect is to ensure that the peatland is a net source of C to the atmosphere, as a result of the increased soil oxidation and reduced input. In many European countries, greenhouse gas (GHG) emissions from agricultural peatlands dominate national emissions of GHGs from peat utilization in total [*Oleszczuk et al.*, 2008].

3.2. Forestry

Impacts of silvicultural drainage on soil CO_2 emissions are similar to agricultural soils whereby improved aeration increases the organic matter decomposition rate and the CO_2 efflux from soil [e.g., *von Arnold et al.*, 2005a, 2005b; *Byrne and Farrel*, 2005; *Minkkinen et al.*, 2007; *Silvola et al.*, 1996]. The increase in efflux is, however, smaller than in agricultural soils because resultant soil conditions are different under silviculture. Since peatland forestry is mainly based on the use of natural tree stands, tillage is not needed during stand development, which may take 50–100 years. Forest fertilization may take place in some sites with unbalanced nutrient status, but it is far less common than in agricultural uses. Perennial plant cover remains, which means that CO_2 fixation and organic C input to the soil continues despite the increased oxidation that is stimulated by the drainage. For these reasons, soils under forestry use are much less disturbed, and functionally, they resemble natural peatlands as opposed to intensively cultivated lands associated with agriculture.

The purpose of forestry drainage is to improve site conditions such that they are suitable for growing trees commercially and to improve the operability of the site. The net effect of peatland silviculture is that trees function as an effective CO_2 sink [*Laiho and Laine*, 1997]. This sink includes the C bound in aboveground and belowground tree biomass and in the C deposited as litter on and in the soil [*Laiho et al.*, 2003], where it decomposes at variable rates, depending on litter quality and conditions of the environment. Ground vegetation also contributes to the litter input to the soil; although the amount is highly variable, it is a significant proportion of the total (25–95% [*Laiho et al.*, 2003]). If this increase of biomass and litterfall is greater than the efflux from soil and litter decomposition, the drained peatland ecosystem remains a C sink. In the boreal zone, especially in the Nordic countries where peatland forestry is actively practiced, this is often the case, although differences in soil conditions, tree species, and climate, affect the ecosystem C balance. There are few eddy covariance measurement sites in drained peatland forests, but several of these (Sweden [*Lindroth et al.*, 2007], Scotland [*Hargreaves et al.*, 2003], Finland [*Laurila et al.*, 2007]) support this view. Examples showing that the balance may in some cases be slightly [*Lohila et al.*, 2007] or highly negative [*Lindroth et al.*, 1998], however, also exist. National greenhouse gas assessments [e.g., *Statistics Finland*, 2009] indicate that generally, integrated over large areas, the tree stand C sink in forested peatlands exceeds the C efflux from peat soil.

The CO_2 sink function of the peatland usually remains as long as the tree stand grows. After final fellings, primary production dramatically decreases, as the productive tree stand is harvested, and remaining ground vegetation suffers from the sudden change in microclimatic conditions. The fine roots of the harvested trees comprise a large stock of rather easily decomposable fresh organic matter, whereas the C pools bound in the stumps and coarse roots may persist for longer periods (e.g., decades) before they are decomposed [*Laiho and Penttilä*, 2006]. Most of the C bound in the harvested trees is lost to the atmosphere, at first from the slash (needles, branches, etc.) left at the felling site, then from short-lived wood products like paper, but a small proportion (<10%) remains in wood products for longer periods (tens of years) [*Seppälä and Siekkinen*, 1993; *Minkkinen et al.*, 2002].

Following forestry drainage, CH_4 emissions decrease [e.g., *Nykänen et al.*, 1998; *von Arnold et al.*, 2005a]. The decrease is caused by (1) increase in soil aeration and (2) decrease in easily decomposable organic matter input to waterlogged soil layers through aerenchymatous mire plants. These environmental changes in soil matrix simultaneously decrease microbial CH_4 production and increase its oxidation. The changes are faster and more pronounced in the nutrient-rich sites, which become CH_4 sinks, and act in a similar manner to mineral forest soils. However, even in the most nutrient-poor sites, CH_4 emissions clearly decrease from the natural state, if the water table level is at all decreased.

An opposite impact can sometimes be created by the drainage ditches themselves. Ditches may sometimes act as CH_4 hotspots, if they remain continuously wet [*Roulet and Moore*, 1995]. In some cases, CH_4 emissions from ditches may totally compensate for the decreased emissions from the drained land between the ditches [*Minkkinen and Laine*, 2006]. It has not been established if this phenomenon is prevalent in drained forested peatlands, but clearly, a perspective that integrates the varying conditions across the entire peatland site merits consideration when attempting to derive unit-area estimates of emissions.

Carbon is also released from peatlands in the drainage water. The leaching of organic C increases during and immediately after construction of the drainage network, but because the groundwater flow through the peatland is decreased by ditches that trap the inflowing water, the long-term increase in organic C leaching is small (approximately 10%, i.e., 1 g C m^{-2} a^{-1} [*Sallantaus*, 1994]), or it may even decrease [*Lundin and Bergquist*, 1990].

Based on available information from C studies in forested peatlands, we can state that the C sink function of a peatland is altered when it is drained for forestry. The direction of the change is, however, not unidirectional nor unambiguous. Some peatlands become C sources, some remain sinks, some may even increase their C sequestration rates. If the peat soil becomes a C source, ecosystem C balance will also, at some point in time, turn negative, when the previously bound tree C pool is released to the atmosphere through decomposition of the woody biomass.

4. HOW TO REDUCE UNCERTAINTIES IN EMISSION ESTIMATES

There are considerable uncertainties associated with estimating the C stocks and greenhouse gas emissions from managed and unmanaged peatlands alike. The issues involve sampling, measurement, and analyses, particularly when trying to assess change over time.

Measuring changes in the peat C stock requires accurate estimates of peat volume, bulk density, and C content. Unfortunately, most studies do not consider the total peat depth nor do they have a reference depth to support periodic remeasurements. Without a basal reference, it is very difficult to estimate changes in peat volume given the inevitable subsidence that is associated with most uses. Accurate bulk density measurements are also required, and seldom reported, in order to assess changes in peat C pools; for example, small variations in bulk density (<0.01 g cm^{-3}) can translate into large variations (>10 t ha^{-1}) in C pool estimates.

The surface of a peatland, whether natural or managed, is a mosaic of microtopography and different plant assemblages; these have been traditionally termed hummocks, hollows, and lawns. These microforms differ in terms of vegetation, hydrologic conditions, and soil properties; hence, the emission rates also vary considerably among these features [e.g., *Bubier et al.*, 2003]. Consideration of the microtopography is particularly important when using chambers to estimate efflux rates. Failure to consider the spatial heterogeneity in the peatland will inevitably result in biased estimates when point measures are extrapolated to a unit area basis. Accordingly, detailed assessment of the surface topography is needed when integrating results from chamber measurements to the site scale. In contrast to chambers, eddy covariance measurements functionally integrate flux rates over the tower fetch area, thereby effectively integrating fluxes among surface microtopographic positions. The problem with the eddy covariance measurements is that it does not distinguish among the different flux sources, but gives the total ecosystem flux. Accordingly, other methods (e.g., chamber measurements) must be used to derive the soil fluxes from heterotrophic processes, from the total ecosystem flux.

The flux of carbon and dissolved gases in the moving groundwater is seldom accounted for when estimating the C balance of a peatland, and it is a major uncertainty that should be addressed. For example, *Nilsson et al.* [2008] showed that 15–20 g C m^{-2} is lost from the peatland annually in the out-flow waters. Unfortunately, there are few recent or current study sites that have the capabilities for estimating gaseous and hydrologic fluxes in peatlands. Watershed-scale experimental sites, where the atmospheric and hydrologic fluxes can be considered simultaneously, are needed to provide estimates of the hydrologic fluxes.

There are also issues with the seasonality of GHG measurements. Most of the reported literature focuses on sampling during the "growing season" or snow-free period, suggesting that this is the period of major fluxes. However, recent chamber [e.g., *Mäkiranta et al.*, 2007] and eddy covariance [e.g., *Lohila et al.*, 2007] measurements demonstrate the importance of winter fluxes, even through the snowpack. Accordingly, sampling regimes that address the full seasonal cycle are needed, especially in light of predicted changes in climatic conditions.

5. CLIMATIC IMPACTS OF THE LAND USE CHANGE IN PEATLANDS

The climatic impacts of land use change can be described by radiative forcing, a term that indicates the effect that greenhouse gas emissions or sequestration have on the Earth's radiative balance. The global warming potential (GWP) of a CH$_4$ molecule is approximately 24 times greater than that of CO$_2$, but since CH$_4$ decomposes faster in the atmosphere (circa 12 years), the effect is shorter, while that of CO$_2$ continues much longer. The changes in the Earth's radiative balance leads to alterations in the surface temperatures.

Generally, pristine peatlands are a net sink of CO$_2$ and a source of CH$_4$ [*Frolking et al.*, 2006]. Over the course of thousands of years of mire development, the net radiative forcing has been for a cooling climatic effect, even considering the higher GWP of CH$_4$ emissions [*Frolking and Roulet*, 2007]. However, conversion to agricultural or forestry use changes the net radiative forcing independently of the natural history of the peatland. Impacts on the radiative forcing as a result of land use change are conveyed through alterations in the C and N cycles. The change may be a warming or a cooling effect depending on the altered gas emissions and the time horizon under consideration. The net radiative forcing from a managed peatland is strongly influenced by the difference between CO$_2$ sequestration, as a result of managing plant productivity, and releases of CO$_2$ and CH$_4$ from the soil, which are largely regulated by alterations to soil drainage and tillage practices.

5.1. Agriculture

Besides releasing large amounts of CO$_2$, drained organic soils can also contribute to the atmospheric N$_2$O load [e.g., *Kasimir-Klemedtsson et al.*, 1997; *Flessa et al.*, 1998; *Maljanen et al.*, 2003b]. Decomposition of the N-containing compounds in organic matter, linked to denitrification processes, is the primary source of the N$_2$O with fertilizer being a minor contributing factor [*Regina et al.*, 2004]. Despite a decrease in CH$_4$ emissions, combined with large areas, these make organic soil croplands an important land use, which increases radiative forcing. In Finland, for instance, agricultural use is responsible for some 4–6 Tg CO$_2$ eq reported for the United Nations Framework Convention for Climate Change (UNFCCC), 25% of the reported peat-based emissions [*Lapveteläinen et al.*, 2007].

5.2. Forestry

The decrease or total cessation of CH$_4$ emissions, following forestry drainage, causes negative radiative forcing, which means cooling, compared to the undrained conditions. If C sequestration increases as a whole, because of greater increases in tree stand and litter C stocks over soil C losses, the change in radiative forcing is, again, negative.

However, when climate impact is estimated, changes in non-carbon gases must also be accounted for. Drainage for forestry has been shown to increase N$_2$O emissions significantly only at nutrient-rich peatland sites [*Martikainen et al.*, 1993], where the pH is high enough for nitrification. At those sites, however, N$_2$O emissions may become large enough to exceed the impacts of CH$_4$ and CO$_2$ [*Klemedtsson et al.*, 2005], turning such sites to net greenhouse gas sources soon after drainage.

The development of a forest canopy on a previously open or sparsely treed mire causes the albedo of the site to decrease. A decreasing albedo causes a warming impact on climate, since less radiation is reflected back to the sky. Quantitative estimates of the impacts of forestry drainage on albedo and climatic warming do not exist, however.

Greenhouse gas emissions and especially radiative forcing impacts are dynamic phenomena. The aforementioned predictions are valid as long as the first post-drainage tree stand is growing and binding C. After fellings, C starts to liberate back to the atmosphere, first from decomposing slash, then from biomass burning and finally from the wood products. Even if a new tree stand is established subsequently on the same site, it cannot decrease the atmospheric C pool more than the first generation did (unless the production capacity of the site for some reason increases). If the soil acts as a C sink or C neutral, the cooling impact of drainage will prevail.

However, if a permanent C loss from soil is created, the climatic impact will eventually be warming.

5.3. Peat in Energy Production

Some 90% of the climate impact of energy production using peat fuel comes from the combustion process: 100–108 g CO_2 per MJ. However, a life cycle analysis of the climate impact considers also the other emission sources, such as emissions from the initial stage of the fuel peat production, those during the production phase, and potential sequestration of CO_2 during after-use of the abandoned production area, the cutaway [*Kirkinen et al.*, 2007].

Holmgren et al. [2008] have recently summarized results from life cycle analyses with comparisons to fossil fuels, such as coal and natural gas. The climate impact, expressed as modeled radiative forcing appears to be highly dependent on the emissions of the site prior to start of the production phase. Clearly, sites with high greenhouse gas emissions during the initial phase give much lower radiative forcing values for the production chain, as the high initial emissions are avoided with the start of the production [*Kirkinen et al.*, 2007; *Holmgren et al.*, 2008].

These analyses clearly direct peat fuel production to sites where anthropogenic impacts on the peatland carbon exchange are most drastic, namely, peatlands under agriculture and forestry. In many cases, these production chains lead to lower radiative forcings than coal over a 100-year time span and often lower than natural gas over a 150-year time span [*Holmgren et al.*, 2008]. The problem in the interpretation of life cycle analyses is the choice of appropriate time spans.

6. SUMMARY AND PERSPECTIVES

Northern peatlands are a terrestrial resource of global importance, providing necessary ecological functions and valued goods and services to many communities and economies. Many of the ecosystem services provided by peatlands are derived either directly or indirectly from the soil C accumulation processes. The C-sink function of a peatland involves intricate interactions among plant communities, hydrology, soils, climate, and biogeochemical processes, which yield, over time, a positive accumulation of organic matter. As a result, peatlands have the largest carbon pools (typically >100 kg C m^{-2}) among terrestrial ecosystem types. Carbon accumulation in peatlands is regulated by the hydrologic regime, the factor that is almost always altered as a result of land use change.

Agriculture, forestry, and peat harvesting (e.g., energy and horticultural products) are the principal human-induced activities that alter the peatland. As a prerequisite to those uses, the peatland is usually drained, with the drainage being most intense under agriculture and peat harvesting. Forestry drainage is not as intensive because of the longer rotations and the adaptability of the forest overstory and understory species. Soil organic matter decomposition is stimulated following drainage, resulting in peat oxidation, typically reported as increased CO_2 emissions and peat subsidence. The vegetation also changes following drainage, regardless of cropping or harvesting practices, and this change influences the net ecosystem C exchange rate. Where net sequestration has been enhanced, there is growing evidence that some managed peatlands may continue to function as a net C sink, despite the presence of drainage and active management. Peatland drainage also tends to reduce CH_4 emissions, as a result of a greater aerated surface soil volume. Accordingly, active management of peatlands results in changes in the above- and belowground C pools and fluxes. The net effect of active management of northern peatlands has been to reduce the soil C pool because agricultural use is dominant.

There are considerable uncertainties in estimating the peatland C pools, fluxes, and responses to management. The issues span inadequate inventory procedures to sampling complex settings. Fundamentally, assessing changes in the soil C pool in peatlands is difficult because the pool size is large, and the annual changes are usually small. Accordingly, the reporting objective must be clearly stated at the onset in order to devise the most appropriate assessment. Among our recommendations are the improved precision in peat pool assessments and consideration of spatial variability when using chambers. Another critical need is sites that are monitored over long periods; it will be the long-term observatories that provide the basis for confirming our understanding of the effects of management practices on C dynamics and ecosystem functions of peatlands. This information is also needed as a basis for predicting how managed and natural peatlands will respond to climate change.

REFERENCES

Alm, J., et al. (2007), Emission factors and their uncertainty for the exchange of CO_2, CH_4 and N_2O in Finnish managed peatlands, *Boreal Environ. Res.*, *12*, 191–209.

Bridgham, S. D., J. P. Megonigal, J. K. Keller, N. B. Bliss, and C. Trettin (2006), The carbon balance of North American wetlands, *Wetlands*, *26*(4), 889–916.

Bubier, J. L., G. Bhatia, T. R. Moore, N. T. Roulet, and P. M. Lafleur (2003), Spatial and temporal variability in growing-season net ecosystem carbon dioxide exchange at a large peatland in Ontario, Canada, *Ecosystems*, *6*, 353–367, doi:10.1007/s10021-003-0125-0.

Byrne, K. A., and E. P. Farrell (2005), The effect of afforestation on soil carbon dioxide emissions in blanket peatland in Ireland, *Forestry*, *78*, 217–227.

Flessa, H., U. Wild, M. Klemisch, and J. Pfadenhauer (1998), Nitrous oxide and methane fluxes from organic soils under agriculture, *Eur. J. Soil Sci.*, *49*, 327–335.

Frolking, S., and N. T. Roulet (2007), Holocene radiative forcing impact of northern peatland carbon accumulation and methane emissions, *Global Change Biol.*, *13*, 1079–1088.

Frolking, S., N. Roulet, and J. Fuglestvedt (2006), How northern peatlands influence the Earth's radiative budget: Sustained methane emission versus sustained carbon sequestration, *J. Geophys. Res.*, *111*, G01008, doi:10.1029/2005JG000091.

Hargreaves, K. J., R. Milne, and M. G. R. Cannell (2003), Carbon balance of afforested peatland in Scotland, *Forestry*, *76*, 299–317.

Holmgren, K., J. Kirkinen, and I. Savolainen (2008), Climate impact of peat fuel utilisation, in *Peatlands and Climate Change*, edited by M. Strack, pp. 70–97, Int. Peat Soc., Jyväskylä, Finland.

Kasimir-Klemedtsson, Å., L. Klemedtsson, K. Berglund, P. J. Martikainen, J. Silvola, and O. Oenema (1997), Greenhouse gas emissions from farmed organic soils: A review, *Soil Use Manage.*, *13*, 245–250.

Kirkinen, J., K. Minkkinen, T. Penttilä, S. Kojola, R. Sievänen, J. Alm, S. Saarnio, N. Silvan, J. Laine, and I. Savolainen (2007), Greenhouse impact due to different peat fuel utilisation chains in Finland—A life-cycle approach, *Boreal Environ. Res.*, *12*, 211–223.

Klemedtsson, L., K. von Arnold, P. Weslien, and P. Gundersen (2005), Soil CN ratio as a scalar parameter to predict nitrous oxide emissions, *Global Change Biol.*, *11*, 1142–1147.

Laiho, R., and J. Laine (1997), Tree stand biomass and carbon content in an age sequence of drained pine mires in southern Finland, *For. Ecol. Manage.*, *93*, 161–169.

Laiho, R., and T. Penttilä (2006), Root system carbon pools in northern peatland forests: Effects of stand density manipulations, Abstract, in *Forest Ecosystem Carbon and Its Economic Implications—Seminar Presenting Results of the Research Programme "Pools and Fluxes of Carbon in Finnish Forests and Their Socio-Economic Implications," 16.3.2006, Helsinki, Finland*.

Laiho, R., H. Vasander, T. Penttilä, and J. Laine (2003), Dynamics of plant-mediated organic matter and nutrient cycling following water-level drawdown in boreal peatlands, *Global Biogeochem. Cycles*, *17*(2), 1053, doi:10.1029/2002GB002015.

Lapveteläinen, T., K. Regina, and P. Perälä (2007), Peat based emissions in Finland's national greenhouse gas inventory, *Boreal Environ. Res.*, *12*, 225–236.

Laurila, T., et al. (2008), Ecosystem-level carbon sink measurements on forested peatlands, in *Greenhouse Impacts of the Use of Peat and Peatlands in Finland Research Programme Final Report*, edited by S. Sarkkola, pp. 38–40, Ministry of Agriculture and Forestry 11a/2007.

Lindroth, A., A. Grelle, and A. S. Moren (1998), Long-term measurements of boreal forest carbon balance reveal large temperature sensitivity, *Global Change Biol.*, *4*, 443–450.

Lindroth, A., L. Klemedtsson, A. Grelle, P. Weslien, and O. Langvall (2007), Measurement of net ecosystem exchange, productivity and respiration in three spruce forests in Sweden shows unexpectedly large soil carbon losses, *Biogeochemistry*, *89*, 43–60, doi:10.1007/s10533-007-9137-8.

Lohila, A., T. Laurila, L. Aro, M. Aurela, J.-P. Tuovinen, J. Laine, P. Kolari, and K. Minkkinen (2007), Carbon dioxide exchange above a 30-year-old Scots pine plantation established on organic soil cropland, *Boreal Environ. Res.*, *12*, 141–157.

Lundin, L., and B. Bergquist (1990), Effects on water chemistry after drainage of a bog for forestry, *Hydrobiologia*, *196*, 167–181.

Mäkiranta, P., J., et al. (2007), Soil greenhouse gas emissions from afforested organic soil croplands and cutaway peatlands, *Boreal Environ. Res.*, *12*, 159–175.

Maljanen, M., A. Liikanen, J. Silvola, and P. J. Martikainen (2003a), Methane fluxes on agricultural and forested boreal organic soils, *Soil Use Manage.*, *19*, 73–79.

Maljanen, M., A. Liikanen, J. Silvola, and P. J. Martikainen (2003b), Nitrous oxide emissions from boreal organic soil under different land-use, *Soil Biol. Biochem.*, *35*, 689–700.

Maljanen, M., V.-M. Komulainen, J. Hytönen, P. J. Martikainen, and J. Laine (2004), Carbon dioxide, nitrous oxide and methane dynamics in boreal organic agricultural soils with different soil characteristics, *Soil Biol. Biochem.*, *36*, 1801–1808.

Maljanen, M., J. Hytönen, P. Mäkiranta, J. Alm, K. Minkkinen, J. Laine, and P. J. Martikainen (2007), Greenhouse gas emissions from cultivated and abandoned organic croplands in Finland, *Boreal Environ. Res.*, *12*, 133–140.

Martikainen, P. J., H. Nykänen, P. Crill, and J. Silvola (1993), Effect of a lowered water-table on nitrous-oxide fluxes from northern peatlands, *Nature*, *366*, 51–53.

Minkkinen, K., and J. Laine (2006), Vegetation heterogeneity and ditches create spatial variability in methane fluxes from peatlands drained for forestry, *Plant Soil*, *285*, 289–304.

Minkkinen, K., R. Korhonen, I. Savolainen, and J. Laine (2002), Carbon balance and radiative forcing of Finnish peatlands 1900–2100—The impact of forestry drainage, *Global Change Biol.*, *8*, 785–799.

Minkkinen, K., J. Laine, N. J. Shurpali, P. Mäkiranta, J. Alm, and T. Penttilä (2007), Heterotrophic soil respiration in forestry-drained peatlands, *Boreal Environ. Res.*, *12*, 115–126.

Minkkinen, K., K. Byrne, and C. Trettin (2008), Climate impacts of peatland forestry, in *Peatlands and Climate Change*, edited by M. Strack, pp. 98–122, Int. Peat Soc., Jyväskylä, Finland.

Nilsson, M., J. Sagerfors, I. Buffam, H. Laudon, T. Eriksson, A. Grelle, L. Klemedtsson, P. Weslien, and A. Lindroth (2008), Contemporary carbon accumulation in a boreal oligotrophic minerogenic mire—A significant sink after accounting for all C-fluxes, *Global Change Biol.*, *14*, 2317–2332.

Nykänen, H., J. Alm, J. Silvola, K. Tolonen, and P. J. Martikainen (1998), Methane fluxes on boreal peatlands of different fertility and the effect of long-term experimental lowering of the water table on flux rates, *Global Biogeochem. Cycles*, *12*, 53–69.

Oleszczuk, R., K. Regina, H. Szajdak, H. Höper, and V. Maryganova (2008), Impacts of agricultural utilization of peat soils on the greenhouse gas balance, in *Peatlands and Climate Change*, edited by M. Strack, pp. 70–97, Int. Peat Soc., Jyväskylä, Finland.

Paappanen, T., A. Leinonen, and K. Hillebrand (2006), Fuel peat industry in EU—Summary report, *Res. Rep. VTT-R-00545-06*, VTT, Jyväskylä, Finland.

Paavilainen, E. and J. Päivänen (1995), *Peatland Forestry—Ecology and Principles*, 248 pp., Springer, Berlin.

Päivänen, J. (2008), Peatland forestry—The Finnish case, in *After Wise Use—The Future of Peatlands*, Proceedings of the 13th International Peat Congress, vol. 1, Oral Presentations, edited by C. Farrell and J. Feehan, pp. 499–501.

Reay, D. S., and D. B. Nedwell (2004), Methane oxidation in temperate soils: Effects of inorganic N, *Soil Biol. Biochem.*, *35*, 2059–2065.

Regina, K., E. Syväsalo, A. Hannukkala, and M. Esala (2004), Fluxes of N_2O from farmed peat soils in Finland, *Eur. J. Soil Sci.*, *55*, 591–599.

Roulet, N. T. (2000), Peatlands, carbon storage, greenhouse gases and the Kyoto Protocol: Prospects and significance for Canada, *Wetlands*, *20*, 605–615.

Roulet, N. T., and T. R. Moore (1995), The effect of forestry drainage practices on the emission of methane from northern peatlands, *Can. J. For. Res.*, *25*, 491–499.

Rydin, H., and J. K Jeglum (2006), *The Biology of Peatlands*, 343 pp., Oxford Univ. Press, New York.

Sallantaus, T. (1994), Response of leaching from mire ecosystems to changing climate, in *The Finnish Research Programme on Climate Change, Second Progress Report*, edited by M. Kanninen, pp. 291–296, Acad. of Finland, Helsinki.

Seppälä, H., and V. Siekkinen (1993), Puun käyttö ja hiilitasapaino, Tutkimus puun käytön vaikutuksesta hiilenkiertokulkuun Suomessa 1990, *Metsäntutkimuslaitoksen Tied.*, *473*, 1–51.

Silvola, J., J. Alm, U. Ahlholm, H. Nykänen, and P. J. Martikainen (1996), CO_2 fluxes from peat in boreal mires under varying temperature and moisture conditions, *J. Ecol.*, *84*, 219–228.

Statistics Finland (2009), Greenhouse gas emissions in Finland 1990-2007: National Inventory Report under the UNFCCC and the Kyoto Protocol, report, 412 pp., Helsinki. (Available at http://www.stat.fi/tup/khkinv/fi_nir_030409.pdf).

Strack, M. (Ed.) (2008), *Peatlands and Climate Change*, Int. Peat Soc., Jyväskylä, Finland.

Trettin, C., R. Laiho, K. Minkkinen, and J. Laine (2006), Influence of climate change factors on carbon dynamics in northern forested peatlands, *Can. J. Soil Sci.*, *86*, 269–280.

Van den Pol-van Dasselaar, A., M. L. van Beusichem, and O. Oenema (1999), Effects of nitrogen input and grazing on methane fluxes of extensively and intensively managed grasslands in the Netherlands, *Biol. Fert. Soils*, *29*, 24–30.

von Arnold, K., P. Weslien, M. Nilsson, B. H. Svensson, and L. Klemedtsson (2005a), Fluxes of CO_2, CH_4 and N_2O from drained coniferous forests on organic soils, *For. Ecol. Manage.*, *210*, 239–254.

von Arnold, K., B. Hånell, J. Stendahl, and L. Klemedtsson (2005b), Greenhouse gas fluxes from drained organic forestland in Sweden, *Scand. J. For. Res.*, *20*, 400–411.

J. Laine, Finnish Forest Research Institute, Kaironiementie 54, FI-39700 Parkano, Finland. (jukka.laine@metla.fi)

K. Minkkinen, Department of Forest Ecology, University of Helsinki, Latokartanonkaari, FI-00014 Helsinki, Finland.

C. Trettin, Center for Forested Wetlands Research, U.S. Forest Service, 3734 Highway 402, Cordesville, SC 29434, USA.

Northern Peatland Vegetation and the Carbon Cycle: A Remote Sensing Approach

A. Harris

School of Geography, University of Southampton, Southampton, UK

R. G. Bryant

Department of Geography, University of Sheffield, Sheffield, UK

Rates of carbon exchange in northern peatlands are dependent on the composition, structure, and spatial arrangement of vegetation. While in situ observations can provide detailed information for a given location, remote sensing is the only viable means of collecting land-surface data in a spatially continuous manner across a range of spatial scales. In this paper, we review and evaluate many existing and emerging remote sensing approaches used to retrieve peatland land-surface data of relevance to the carbon cycle. We review studies documented in the scientific literature that use remotely sensed data to (1) generate vegetation maps, which may be used to extrapolate field observations, calibrate and extrapolate carbon models, and inform peatland management efforts; and (2) retrieve vegetation biophysical properties, which can be used to parameterize process-based models [e.g., leaf area index (LAI)]. There has been considerable progress in the development and implementation of remote sensing approaches that provide data relating to peatland carbon processes. However, there remain a number of methodological challenges, which limit the effectiveness of remote sensing data in some instances. Consequently, we propose that future research approaches focus on (1) continued development, testing, and validation of approaches to overcome difficulties caused by the heterogeneous nature of peatland vegetation surfaces (e.g., mixture modeling); (2) assessment of spatial errors and uncertainty in image classifications, (3) synergistic use of multiple data sets, (4) development of scaling algorithms, and (5) continued development of radiative transfer models that can be applied to heterogeneous peatland plant assemblages.

1. INTRODUCTION

Peatlands represent a diverse array of wetlands that accumulate partially decomposed organic material. Bogs and fens are the most common types of northern peatland. Bogs are ombrotrophic environments that receive water and solutes solely from rainfall and other atmospheric inputs and are therefore acid and low in plant nutrients, whereas fens are minerotrophic peatlands, which receive inputs from groundwater or surface runoff and tend to be more alkaline and nutrient rich [*Rydin and Jeglum*, 2006]. Despite low rates of productivity and decomposition, northern peatlands play an important role in the global carbon cycle through the

Carbon Cycling in Northern Peatlands
Geophysical Monograph Series 184
10.1029/2008GM000818

sequestration of carbon into peat via photosynthesis and the release of greenhouse gases (CO_2 and CH_4), through the decomposition of organic material [*Moore and Knowles*, 1989; *Gorham*, 1991; *Smith et al.*, 2004]. In their natural state, most peatlands act as net atmospheric carbon sinks. However, small changes in water balance can cause these ecosystems to shift from carbon sinks to carbon sources over short timescales [*Aurela et al.*, 2001; *Oechel et al.*, 1993]. The balance between carbon sequestration and release is also spatially variable across a single peatland because of the presence of niche species, which form distinct patterns in relation to subtle gradients in water movement and chemistry [*Rydin and Jeglum*, 2006]. Thus, variations in carbon fluxes often reflect associated differences in the rates of productivity and decomposition among these species [*Dorrepaal et al.*, 2005, 2007]. Alterations to moisture, temperature, or nutrient regimes of northern peatlands can trigger changes in vegetation composition [*Foster et al.*, 1993; *Minkkinen et al.*, 2002], rates of primary production [*Belyea and Malmer*, 2004; *Bubier et al.*, 2007; *Waddington et al.*, 1998; *Wiedermann et al.*, 2007] and plant litter decomposition [*Dorrepaal et al.*, 2005], thereby affecting the overall carbon dynamics of these ecosystems [*Minkkinen et al.*, 2002; *Strack and Waddington*, 2007]. Given that climate change scenarios predict that greatest global temperature increases will occur at higher latitudes where many peatlands are located [*IPCC*, 2007], it is important to understand the response of these ecosystems to past and current climatic conditions, and to predict the effect of future climate changes on peatland carbon-balance processes.

Remote sensing data, available at a range of spatial and spectral resolutions, can complement ground-based observations that are often expensive, logistically difficult, time consuming, and typically of limited temporal resolution and spatial extent. Although currently available remote sensing instrumentation has not been designed to directly monitor the peatland carbon cycle, sensors can derive information related to relevant land-surface properties [*Sitch et al.*, 2007]. One of the traditional uses of remotely sensed data in carbon cycle research has been the production of thematic maps to provide information on land cover. Land cover maps are often used to quantify carbon stocks [e.g., *Bridgham et al.*, 2007; *Gorham*, 1991], to extrapolate in situ point-based field observations [e.g., *Bubier et al.*, 2005; *Johansson et al.*, 2006; *Roulet et al.*, 1994], to calibrate and extrapolate carbon models [e.g., *Sonnentag et al.*, 2008], and to inform and direct peatland restoration and management efforts [e.g., *MacKay et al.*, 2009; *Poulin et al.*, 2002, 2006]. The potential use of remote sensing data in carbon cycle applications extends beyond land cover mapping and toward the development of empirical or physically based algorithms

which relate spectral information to inherent physiological and biophysical properties of vegetation [e.g., LAI, effective fraction of absorbed photosynthetically active radiation (fPAR), and chlorophyll content]. These data are highly relevant to carbon flux estimations and are commonly assimilated into relatively simple remote sensing models, which attempt to describe spatial and temporal carbon dynamics [e.g., gross primary productivity (GPP), net photosynthesis (NPP); e.g., *Turner et al.*, 2006] or are used to parameterize and validate complex process-based carbon models [e.g., *Chen et al.*, 2003; *Sonnentag et al.*, 2008]. However, the successful application of remotely sensed data in carbon exchange monitoring and modeling is reliant on the development of appropriate methodologies for the repeated retrieval of key variables, as well as an understanding of the uncertainties involved in their estimation [*Prieto-Blanco et al.*, 2009]. This is especially true when applying conventional remote sensing approaches in northern peatland environments where vegetation composition is spatially heterogeneous because of complex hummock-hollow microtopography, and the canopy structure and spectral properties of dominant plant species, such as *Sphagnum* mosses, differ markedly from vascular terrestrial vegetation for which many conventional remote sensing algorithms have been developed.

In this paper, we provide a review of remote sensing approaches that have been used to retrieve northern peatland land-surface data of relevance to carbon exchange research, grouping the work into two main categories: (1) mapping northern peatland land cover; and (2) retrieval of biophysical properties. We discuss the theoretical and methodological challenges of these approaches, focusing on the uncertainty involved in these measurements and the potential to deliver repeatable dynamic information. We also discuss promising new methodological developments that are emerging, which may improve the utility of remote sensing data for understanding, modeling, and monitoring northern peatland carbon exchange.

2. REMOTE SENSING APPROACH

The remote sensing approach is broadly based upon principles surrounding the transfer of energy (radiance: W sr^{-1} m^{-2}) from a surface to a sensor [see *Campbell*, 2002; *Jensen*, 2005; *Lillesand et al.*, 2008]. Over much of the last century, the principal sensor used to study peatlands has been an aircraft-mounted camera. However, since the 1970s, the number and sophistication of sensors and platforms available has increased significantly, offering enhanced capabilities for mapping of peatland surfaces over a range of spatial and temporal scales. At the same time, an increased choice necessitates a clear understanding of any sensor/platform-specific considerations when selecting the optimal data type for peatland applications.

2.1. Remote Sensing Systems

Most remote sensing systems have sensors which capture radiance data in one or more discrete wavelength ranges; often with narrow, broad, or continuous bandwidths. Data are typically collected within wavelengths which range from the visible (VIS; 0.4–0.7 μm) to the microwave (0.1–1 m). Remote sensing systems can either be composed of photographic sensors, used in aerial photography, or more commonly digital sensors. Generally, digital sensors can be categorized as either: (1) multispectral, using few channels (1–10) and broad bands (50–100 nm), or (2) hyperspectral, with the capability to measure in numerous (up to 250 bands), narrowly defined bands (1–10 nm) or continuous parts of the electromagnetic spectrum. Those sensors that operate in the microwave region (0.1–1 m) operate in a different manner to those at shorter wavelengths in that they are normally (1) active sensors, that both generate and detect electromagnetic radiation, and can be used during the day or night, (2) have a spatial resolution that is a function of antenna design and wavelength, and (3) have an all-weather capability, in that they are unaffected by most cloud and weather systems, an important consideration for some northern peatlands. Most sensors can be placed on a number of different platform types. These can range from a satellite to aircraft or even kites and balloons. However, for sensors operating outside the microwave region, important systematic relationships exist between the sensor platform combination used, and the spectral, spatial, and temporal resolution of the data produced. In general, for most photographic, multispectral, and hyperspectral systems, the distance between target and sensor (often either expressed as altitude or orbital height) impacts on the upwelling radiant flux, and therefore the pixel size. This, along with an appreciation of the sensitivity and optical setup of the sensor, can impact on factors such as the bandwidth available to the sensor, the number of bandwidths that can be used at specific wavelengths, and the size of the resulting image. For most existing and forthcoming satellite platform/sensor combinations, these factors can mean that direct and predictable relationships exist between the pixel size (0.6 m to 1 km), repeat-period (12 h to 44 days), and the bandwidth (1–100 nm). To a certain degree, the use of airborne sensors can allow greater spatial, spectral, and temporal flexibility, but at the expense of cost and complexity (Table 1). Nevertheless, the choice of sensor and platform can have significant implications for the spatial extent and timing of remote sensing monitoring programs and, therefore, can ultimately influence the ability to successfully generate useful data about peatland surfaces. The ranges of spatial and temporal scales at which peatland studies can operate are outlined in Figure 1. Details of case-studies associated with these examples are outlined in section 3.

2.2. Turning Numbers Into Data

In general, assuming some form of data has been successfully collected, further interpretation relies upon initial correction of data for: (a) atmospheric effects (e.g., air molecules, aerosols, and water vapor/droplets), (b) geometric distortion (e.g., movement of the ground surface relative to the sensor and platform), and (c) conversion of digital data to standard radiance units ($W\ sr^{-1}\ m^{-2}$). In each case, failure to correct sufficiently for each of these effects can have significant impacts on data interpretation later in the processing chain, particularly for studies that may involve parameter estimation and any form of monitoring or change detection. Most satellite sensor/platform systems operate with predictable orbital parameters, and onboard/in-flight calibration, which enable routine geometric correction [e.g., Moderate Resolution Imaging Spectroradiometer (MODIS) Level 1B; *Xiong and Barnes*, 2006; *Xiong et al.*, 2006]), and allowance for changes in illumination geometry and sensor calibration. However, these systems initially detect upwelling radiation from the top of the atmosphere, and require correction to remove the effects of scattering and absorbing components of the atmosphere [*Kaufman*, 1989]. In most instances, this is achieved through implementation of either: (1) simple or empirical approaches (e.g., histogram matching [e.g., *Richter*, 1996], empirical line correction [e.g., *Slater et al.*, 1987], and darkest pixel subtraction [e.g., *Chavez*, 1975]), or (2) atmospheric radiative transfer models (e.g., MODTRAN [*Berk et al.*, 1998], 6S [*Vermote et al.*, 1997b]). In most cases, standardized products produced from satellite systems (e.g., MODIS surface reflectance and vegetation indices (see http://modis-land.gsfc.nasa.gov/index.htm)) have undergone careful correction [e.g., *Vermote et al.*, 1997a]. Nevertheless, most satellite data (e.g., Landsat, SPOT, and IKONOS) are often delivered without initial correction for radiometric/atmospheric effects and require, at the very least, empirical correction before use. Airborne platforms and sensors, by comparison, require complex control systems to allow accurate geometric correction [e.g., *Schläpfer and Richter*, 2002], and radiometric/atmospheric correction often requires initial off-site calibration of the sensor array before data collection, alongside the application of empirical or atmospheric radiative transfer correction approaches using data collected at the time of

Table 1. Typical Remote Sensing Types and Platforms Used for Peatland Applications[a]

	Photography	Imaging Spectrometer	MSS	High Resolution MSS	Medium Resolution MSS	Low Resolution MSS/ Spectrometer	SAR
Platform	Airborne	Airborne	Airborne	Spaceborne	Spaceborne	Spaceborne	Spaceborne
Example	Wild RC10 Survey Camera	CASI, AVIRIS	Daedalus 1268 ATM	Quickbird, IKONOS	Landsat ETM, SPOT, ASTER	AVHRR, MODIS, SPOT Vegetation	ERS-1, RADARSAT
Specification							
Spatial resolution (m)	<0.5[b] (A)	0.5–20[b] (B)	0.5–20[b] (B)	0.6–4 (C)	2.5–80 (D)	250–1150 (E)	10–100 (D)
Spectral bandwidth (nm)	approx. 10	1.8	5–20[c]	80–110[c]	60–200[c]	10–500[c]	NA[d]
Temporal repetition (days)	Upon request	Upon request	Upon request	2–11 (G)	16–26 (F)	4 per day to 26 (H)	24–35 (F)
Swath width (km)	1–20[b]	0.25–12[b]	0.5–20[b]	11–16	60–185	2.2–2.7	35–100
Number of spectral bands	1–4 (if digital)	288	1–15	5	4–16	5–36	NA[d]
Logistics							
Mission targeting	upon request	upon request	upon request	none or limited	none or limited	none	none or limited
Flight time	upon request	upon request	upon request	fixed	fixed	fixed	fixed
Repetitive coverage	low cost	highest cost	high cost	low cost	low cost	lowest cost	low cost
Cost per km²	low	high	high	medium	low	lowest	low
Archive available from	1920s–present	1980s–present	1980s–present	1999–present	1970s–present	1970s–present	1990s–present
Archive coverage	high	low	low	medium	high	highest	high
Clear sky conditions	preferable	preferable	preferable	critical	critical	critical	not affected
Data volume	lowest	highest	high	high	medium	low	medium
Pre-Processing effort	high	high	high	low	low	low	medium
Vegetation parameter extraction (e.g., LAI, fAPAR)	not possible	possible	possible	possible	possible	standard product	not possible

[a]Letters in parentheses relate to Figure 2.
[b]Depends on the flying height (and focal length for air photography).
[c]Bandwidths normally increase at longer wavelengths.
[d]NA indicates not applicable. AR systems can operate with a range of wavelengths and polarization modes.

overpass [e.g., *Harris et al.*, 2006; *Richter and Schläpfer*, 2002].

Following initial data correction, subsequent analysis of data relies on there being some form of relationship between the radiation measurements collected by the sensor and some variable of interest on the ground [*Verstraete et al.*, 1996]. Such relationships can be derived using an array of approaches, which differ in their complexity (Figure 2). As the link between the measured radiation and the variable of interest becomes more complex, uncertainty in the variable of interest will generally increase. Consequently, the remote sensing data analysis approach adopted has implications for

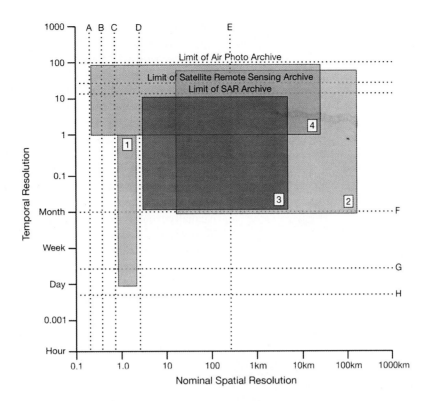

Figure 1. Examples of the spatial and temporal constrains of the different forms of data extraction for peatlands from remote sensing data. The dotted lines (A to H) indicate the temporal and spatial characteristics of remote sensing data types outlined in Table 1. Limits to remote sensing data archives are also noted. The boxes (1 to 4) provide a guide to the spatial and temporal characteristics of typical peatland studies involving: (Box 1) the extraction of species level data [e.g., *Thomas et al.*, 2002], (Box 2) the broadscale mapping of peatlands [e.g., *Frey and Smith*, 2007], (Box 3) the mapping of functional peatland vegetation using Landsat data [e.g., *Sonnentag et al.*, 2007] or studies involving the broadscale mapping of peatland types using similar data [e.g., *Bronge and Naslund-Landenmark*, 2002], and (Box 4) a typical study mapping peatland types and surfaces from aerial photographs [e.g., *Malmer et al.*, 2005]. Modified after *Jensen* [2000].

our ability to extract useful information relating to the mapping and assessment of peatland vegetation.

3. MAPPING NORTHERN PEATLAND LAND COVER WITH REMOTELY SENSED DATA

Within the last 20 years, a number of studies have utilized remotely sensed data to map northern peatlands at a range of spatial scales (i.e., from global to community level), using an array of methods, which range in complexity. Because peatland environments, plant communities, and species exhibit spatial heterogeneity as well as unique spectral reflectance properties associated with the mixture of vascular and nonvascular species [*Bubier et al.*, 1997; *Schaepman-Strub et al.*, 2008], classical approaches that are commonly used in remote sensing to map large, relatively homogenous patches of land dominated by vascular plants, such as forests and

grasslands, are not necessarily suitable for mapping northern peatlands. In the section that follows, we discuss the advantages and limitations of the most conventional approaches used for mapping northern peatlands and discuss how advances in algorithm development, and sensor and platform technology may enable improved mapping of peatland ecosystems in the future.

3.1. Conventional Approaches to Mapping Peatland Land Cover

3.1.1. Visual interpretation. Visual interpretation of aerial photographs is one of the most widely adopted approaches for mapping peatland vegetation and has been used extensively to delineate major plant communities across North American continental peatlands [e.g., *Glaser*, 1987, 1992]. A number of criteria have been developed to map vegetation

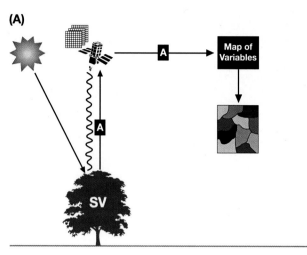

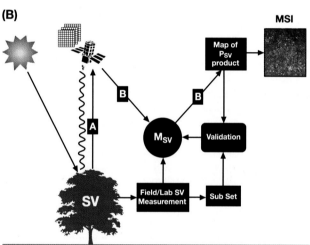

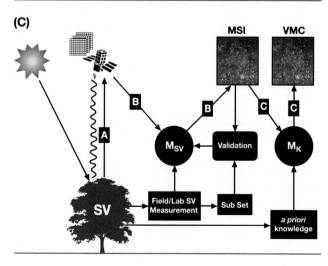

composition based on image tones, textures, shadows, and spatial associations with other objects. Color infrared (CIR) films are particularly useful, since they convey information contained in both the visible and the near-infrared wavelengths which are appropriate for vegetation mapping [*Lillesand et al.*, 2008]. For example, *Glaser* [1992] illustrated how lightly treed *Sphagnum* raised bogs could be separated from dense forest based on color representation in CIR photographs. *Malmer et al.* [2005] demonstrated the use of CIR images in order to differentiate surface moisture contents using the blue band signal, whereas *Pelletier et al.* [2007] used aerial photographs to map surface bog patterns and spatially weight methane emission estimates across several Canadian peatlands. A number of studies have also used visual interpretation techniques to analyze satellite imagery, often to refine automated image classification procedures [e.g., *Muller et al.*, 1999; *Bronge and Naslund-Landenmark*, 2002]. Although useful, visual interpretation techniques have a number of drawbacks. First, the visual classification of images can be somewhat subjective because of the difficulty in generating standardized measurement procedures. Thus, the quality and accuracy of a given classification can differ between interpreters and within or between peatlands. Difficulties also occur at the interpretation stage when features of interest are too small for the analyst to define or identify [*Murphy et al.*, 2007]. Another significant issue with this approach is that the manual extraction of fine-scale detail is prohibitively time consuming [*Kadmon and Harari-Kremer*, 1999]. As a consequence, where possible, most peatland vegetation mapping efforts utilize automated or semiautomated digital image classification methods.

3.1.2. Automated and semiautomated digital image classification. A large proportion of automated and semiautomated peatland mapping efforts make use of conventional

Figure 2. Three alternate strategies for the generation of data from remote sensing imagery, involving: (a) simple direct mapping from image data to produce maps of variables of interest (e.g., supervised classification or direct interpretation of aerial photographs) via path A (indicated by arrows), (b) the use of field knowledge and measurements of surface state variables to derive and test a simple invertible, often empirical model (M_{SV}) for the production of some variable of interest (e.g., P_{SV} = canopy moisture content using the moisture stress index; MSI) via a combination of paths A and B (indicated by arrows), and (c) the conversion, based upon further field knowledge, of this variable of interest to another variable relating to peatland surface condition (e.g., surface volumetric moisture content; VMC) via paths A, B, and C (indicated by arrows) and an additional model (e.g., M_K).

per-pixel image classification approaches (e.g., supervised, unsupervised and hybrid classification), which are well documented in image processing texts [see *Campbell*, 2002; *Lillesand et al.*, 2008; *Mather*, 2006]. In essence, these methods allocate each pixel to a single thematic class (e.g., plant community) or spectral cluster based on their spectral response pattern (e.g., Figure 2, path A). Per-pixel classification approaches have been used to map peatland vegetation across a range of different scales and with various levels of detail.

Information on peatland distribution at a global scale is commonly gained from standard categorical global land cover products derived from low-resolution imagery (~1 km), such as advanced very high resolution radiometer (AVHRR), MODIS, and SPOT vegetation (Table 1). The advantages of using these data are that they provide daily repeat global coverage of the Earth's surface and are often generated from a carefully corrected and calibrated time series of observations. However, many northern peatlands are often omitted from global land cover maps or misclassified as other land cover types because of their small size (<1 km) and spatial and spectral complexity [*Frey and Smith*, 2007; *Krankina et al.*, 2008]. Underrepresentation of northern peatlands may also be attributed to the classification schemes used by standard global land cover products (e.g., MODIS Land Cover), which do not include legend definitions that are representative of northern peatland ecosystems [e.g., *Heiskanen*, 2008]. As part of an intercomparison of land cover maps generated from medium- to low-resolution imagery for identifying peatlands in the St. Petersburg region of Russia, *Krankina et al.* [2008] noted that maps which defined "wetlands" as inundated land units, omitted a significantly larger proportion of peatlands than maps which contained more inclusive and relevant definitions in their legends.

At the landscape and community scale, peatland type and vegetation composition are commonly mapped on a per-pixel basis using data from medium to high spatial resolution sensors (Table 1). For example, *Glaser* [1989] used Landsat TM data to broadly classify a North American peatland complex into three classes (bog, fen, and standing water) using a conventional hybrid per-pixel classification approach (i.e., both supervised and unsupervised classification). In contrast, a more detailed vegetation map was produced from Landsat Enhanced Thematic Mapper (ETM+) data by *Poulin et al.* [2002] who used a supervised per-pixel classification approach to distinguish 13 different habitats within individual peatlands in Southern Quebec, Canada. In a similar vein, multitemporal SPOT-4 imagery has been utilized to map general vegetation patterns across a wetland complex in northeastern Alberta, Canada using a supervised classification [*Töyrä and Pietroniro*, 2005].

The potential of active microwave images for mapping northern peatlands using standard per-pixel approaches has also been investigated. Synthetic aperture radar (SAR) systems have a similar spatial and temporal resolution to Landsat data, but are not constrained by cloudy conditions and low sun angle. In addition, under certain conditions, these data are able to penetrate plant canopies to gain information on the below canopy ground conditions and hydrology [e.g., *Moghaddam et al.*, 2000; *Sokol et al.*, 2004]. Traditionally, the use of SAR data for per-pixel classification has been directed at the broad discrimination of peatland and nonpeatland vegetation based on water extent characterization [e.g., *Li and Chen*, 2005; *Töyrä and Pietroniro*, 2005], although promising preliminary investigations suggest that data from the fully polarized SAR sensor onboard the recently launched Radarsat-2 satellite may provide increased peatland vegetation discrimination capabilities [*Touzi et al.*, 2007]. The synergistic use of microwave (e.g., Radarsat-1) and optical data (such as Landsat ETM) has also shown potential for effective discrimination of different types of peatlands [e.g., *Grenier et al.*, 2007].

While medium-resolution sensors (e.g., Landsat, SPOT, and SAR systems) can be used to map general peatland vegetation classes, their spatial resolution often remains too coarse to effectively represent the heterogeneous nature of many peatland plant communities using traditional per-pixel approaches. Aerial photographs and high spatial resolution satellite sensors (e.g., IKONOS and Quickbird) can overcome the spatial limitations of medium-resolution sensors, but they lack sufficiently detailed spectral information needed to be able to differentiate between peatland species [e.g., *Bubier et al.*, 1997; *Malmer et al.*, 2005; *Mehner et al.*, 2004]. Recent advances in remote sensing have led the way for the development of hyperspectral sensors, also known as imaging spectrometers [see *Lillesand et al.*, 2008]. When placed on an airborne platform, these sensors often have the requisite spatial and spectral resolutions to enable land cover types to be resolved based on subtle differences in their spectral signatures. For example, *Sphagnum* mosses have a higher reflectance in the red portion of the visible range and lower reflectance in the near-infrared region compared with vascular plants [*Bubier et al.*, 1997]. While hyperspectral data have been used to map vegetation across coastal and shoreline wetlands [e.g., *Schmidt et al.*, 2004; *Belluco et al.*, 2006], there are relatively few examples of their use in complex northern peatland landscapes. Nevertheless, *Thomas et al.* [2002] demonstrated the use of imaging spectrometers [e.g., Compact Airborne Spectrographic Imager (CASI)] for mapping fen and bog vegetation across a northern peatland complex using conventional per-pixel classification methods. Despite their potential for discriminating complex plant communities, airborne imaging spectrometers currently have a limited geographic coverage (e.g., 1 to 10 km), although high spatial resolution imagery can be used as a form of supporting data for larger-scale mapping efforts [*Aplin*, 2005]. For example, *Quinton et al.* [2003] utilized classified high-resolution IKONOS data to assist in the supervised classification of Landsat data in a study of connectivity and storage functions in peatlands. A similar approach was taken by *Bubier et al.* [2005] who utilized a combination of CASI and Landsat TM data to map vegetation composition across a northern peatland. However, using multiple data sets in this way requires that the data are collected in a synchronous manner and that careful correction (geometric, radiometric, and atmospheric) and initial intercalibration of the airborne and spaceborne data needs to be undertaken before use. In addition, the direct comparison, processing, and use of data with contrasting spectral and spatial resolution requires a detailed understanding of the sensor spectral response functions in each case [e.g., *Schmid et al.*, 2005].

More computationally advanced pixel-based classification algorithms, such as the spectral angle mapper (SAM), have also been used to map peatland vegetation [e.g., *Jollineau and Howarth*, 2008]. The SAM algorithm differs from standard classification approaches in that it compares unknown pixel spectra to predefined representative spectra for each class (i.e., end-member spectra) by calculating the angle, in radians, between them in feature space. If angles are smaller than a user-defined threshold, then the pixels are thought to match and are assigned to that class [*Kruse et al.*, 1993]. An advantage of SAM is that because the algorithm utilizes data related to the spectral angle, and not reflectance values, the classification is largely unaffected by varying illumination conditions and albedo effects [*Kruse et al.*, 1993; *Leckie et al.*, 2005]. Although the SAM algorithm was initially developed for use with hyperspectral data, the approach has been used to map peatland vegetation from both hyperspectral [e.g., CASI; e.g., *Jollineau and Howarth*, 2008] and multispectral imagery, e.g., Landsat [e.g., *Sonnentag et al.*, 2007].

Despite the importance of land cover data in carbon exchange research and the proliferation of studies utilizing remotely sensed data to map northern peatlands using per-pixel approaches, few studies have specifically utilized the resultant data to improve the monitoring and estimation of peatland carbon fluxes [e.g., *Bubier et al.*, 2005; *Sonnentag et al.*, 2008]. There are several reasons for this: (1) First, thematic maps have often been created with an alternative research focus (e.g., for site-specific conservation or management purposes). While often achieving the objectives of a given study, the subsequent use of such data for carbon exchange research may be constrained by the original choice of vegetation classes and mapping scale, both of which may not be aligned with the requirements of carbon exchange research (see section 1); (2) Classification accuracy is a critical issue where land cover maps are used for the extrapolation or interpolation of point-based carbon flux observations or estimates. Low classification accuracies can therefore bias carbon exchange estimates [e.g., *Bubier et al.*, 2005; *Becker et al.*, 2008]. Mixed pixels are a major source of classification error in thematic maps created using per-pixel classification approaches. One of the primary causes is the presence of more than one land cover class within a single pixel. It is thus difficult to identify a distinct spectral signature that can be clearly related to a single vegetation class. Mixed pixels are a problem when mapping peatlands using mid- to low-resolution data (20 m to 1 km pixels) because the patterning of plant communities in relation to environmental gradients often occurs at a scale finer than the resolution of the pixel. However, classification performance does not necessarily improve with the use of higher spatial resolution imagery, such as IKONOS and CASI, because the variability within a single class tends to increase at finer spatial scales [e.g., *Thomas et al.*, 2002; *Bubier et al.*, 2005; *Jollineau and Howarth*, 2008]. Though errors of overall classification accuracy and per-class accuracy are often derived from the use of conventional approaches (i.e., the use of a confusion matrix [see *Lillesand et al.*, 2005]), to the authors' knowledge, there have been no attempts to derive information on the spatial variation in thematic error and the uncertainty associated with the allocation of a particular class to a pixel across peatland environments. Such information would enable more accurate modeling of classification errors and potentially reduce errors when monitoring peatland vegetation dynamics [*Brown et al.*, 2009]; (3) Finally, the vast majority of the classification approaches not only ignore the issue of mixed pixels, but often rely on the existence of statistical or empirical relationships between upwelling radiance and the land cover class of interest (Figure 2, path A). Generally, no direct physical understanding of how radiance interacts with the different components of the land surface is incorporated into a per-pixel classification approach. This is an important issue, as we know that the radiative transfer process from one media to another is often influenced by a number of state variables (i.e., the smallest set of variables needed to fully describe the remote sensing data). In the case of vegetation canopies, there are three broad state variable parameters (SV; Figure 2). These are (1) the radiation scattering, absorbing, and transmission properties of the plant canopy (e.g., leaves, stems, flowers, and fruit), (2) geometric properties of the canopy (e.g., aboveground biomass, leaf area index, and arrangement of foliage), and (3) illumination angle (i.e., is the sun the only significant source of illumination?;

is the direction of view toward the hot spot or nadir?). In essence, SV 1 and 2 are stochastic in nature and can change over a wide range of timescales (e.g., hour/day/season/year). Consequently, a per-pixel classification algorithm trained at one time of the year or in a particular locality may not work at a different time of the year or in a different region, which makes many classification techniques labor intensive and unreliable over large regions or for multidate sequences of images [Hall et al., 1997].

3.2. Contemporary Approaches to Mapping Peatland Land Cover

The limitations of visual interpretation and per-pixel mapping approaches have led to continued research into the development and novel use of more complex image processing techniques, which have potentially greater capabilities for mapping spatially and spectrally heterogeneous peatland vegetation. Contemporary approaches focus on either: (1) improved classification accuracy through the adaptation of existing or development of new automated or semiautomated classification algorithms, or (2) the combination of data from multiple sources and multiple angles and the use of multiresolution data, which can enable continuous mapping of peatlands at increasing spatial scales. These approaches are reviewed in turn below.

3.2.1. Complex classification approaches
3.2.1.1. Linear mixture modeling. Much of the recent work on the development of innovative approaches to map northern peatlands has focused on the retrieval of subpixel level data. One such approach, linear mixture modeling [or spectral mixture analysis (SMA)], assumes that the spectral signature of a pixel is a linear mixture of different land-surface types and further assumes that it is possible to identify each of the surface types that contribute to the mixing [see *Campbell*, 2002]. Reference spectra for every known land-surface type are therefore a required prerequisite. These are known as end-members and represent the response that would be observed from a homogenous pixel containing a single land-surface type. The pure end-members are then used to linearly unmix each mixed pixel within the image to produce a continuous map of proportional vegetation coverage [*Colwell*, 1974; *Settle and Drake*, 1993; *Smith et al.*, 1990]. However, the underlying assumptions associated with this approach (e.g., linear and additive mixing between components), coupled with a reliance on the availability of pure pixels or spectral signatures to define every land-surface type, and the mathematical constraint which limits the number of image end-members to one less than the available number of wavebands, mean that linear mixture modeling is often ill-suited for mapping large numbers of vegetation classes across wide geographic areas. For these reasons, the standard linear mixture modeling technique and variants of the standard model [e.g., multiple end-member spectral mixture analysis (MESMA), which allows for the use of different end-member combinations to model different pixels within a scene; *Roberts et al.*, 1998], are often used for localized mapping applications. While linear mixture modeling of this nature is not a new technique, its utility for mapping the subpixel fractions of northern peatland vegetation has only recently been explored. For example, *Sonnentag et al.*, [2007] applied MESMA to a three-end-member model (sunlit vascular canopy, *Sphagnum* moss and shadow) to produce tree, shrub, and *Sphagnum* moss distribution maps from a Landsat TM image of a Canadian ombrotrophic peatland. The research shows that the distinctive spectral characteristics of *Sphagnum* mosses must be accounted for in any northern peatland linear mixture model due to the strong influence of moss ground cover on the spectral response of the shrub canopy [*Sonnentag et al.*, 2007]. Recently, field spectroradiometry has also been used in an attempt to investigate the potential of MESMA for mapping the fractional cover of peatland plant functional types [e.g., shrub, graminoids, and *Sphagnum* moss; *Schaepman-Strub et al.*, 2008]. While such an approach was able to map open patches of *Sphagnum* moss, partitioning of the vascular plant fractional cover into shrub and graminoid proved difficult, highlighting the reliance of linear mixture models on the availability of appropriate and well defined end-members.

Alternative subpixel mapping approaches that search for specific materials of interest within a mixed pixel [for example mixed tuned matched filtering (MTMF)] can also be utilized to map peatland vegetation. This method has a significant advantage over traditional mixture modeling approaches in that the composition of a pixel is not constrained to be a combination of all of the defined end-members. Thus, the spectral signature of every end-member in the image need not be known, and only the spectral signature of the material/surface of interest (e.g., *Sphagnum*) is required (i.e., a single end-member). The output of an MTMF classification is therefore a fraction image detailing the spatial distribution and proportion of the single material of interest. A separate classification must be undertaken for each material of interest; thus, the approach is best suited for mapping of target plant species or communities. For example, *Harris et al.* [2005] used MTMF to map the distribution of *Sphagnum* moss across a peatland as part of a broader study, which utilized the unique spectral characteristics of *Sphagnum* to determine near-surface hydrological conditions (Figure 3).

a) ATM false colour composite   b) *S. pulchrum* fraction image

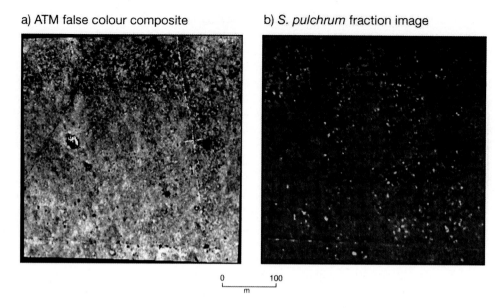

0   100
m

Figure 3. (a) An Airborne Thematic Mapper (ATM) subset image and (b) Mixed Tuned Match Filtering (MTMF) fraction image used to identify *Sphagnum pulchrum* at an ombrotrophic peatland in Wales, UK. The light regions in the fraction image represent patches of *S. pulchrum*. From *Harris et al.* [2006].

3.2.1.2. Nonlinear mixture modeling. It is evident that linear mixture models (or subpixel approaches in general) can provide improved classification accuracy compared to conventional per-pixel classification approaches. However, the underlying assumption that changes in the proportion of the surface types within a pixel is the only control on a pixel's observed response is often false. In reality, nonlinear relationships between parameters are common in remote sensing, particularly where land-surface heterogeneity is high relative to the resolution of the sensor, as is often the case when mapping northern peatland vegetation. Under these conditions, radiation mixing will often occur within and between land-surface types before the radiation reaches the sensor (e.g., between branches and leaves of the same and different species), making it impossible to fully de-convolve the land-surface components of the pixel using linear mixture models. Consequently, much of the focus in contemporary image classification is on the use of nonlinear classifiers such as artificial neural networks (ANN), which do not make assumptions about the nature of the mixing within a pixel. A particular advantage of neural networks is that they are able to directly accommodate mixed pixels, thus removing the need for end-member spectra, which are difficult to obtain in heterogeneous environments [*Foody et al.*, 1997]. However, their use for mapping peatland vegetation has thus far been limited and somewhat unsuccessful [e.g., *Brown et al.*, 2007]. Problems again arise from the extremely heterogeneous nature of peatland vegetation, where high spatial and

spectral variability within mixed pixels make it difficult for ANNs to generalize (i.e., to classify outputs which are not encountered in the training data). *Mills et al.* [2006] reported similar issues when attempting to use ANNs to map a complex mosaic of upland plant communities using high spatial resolution imagery. The ANNs outperformed conventional classifiers when mapping the same regions as those used for network training, but classification accuracies were low when applied to unseen data from a remote location [*Mills et al.*, 2006]. ANNs may be best suited to mapping peatland vegetation at coarse resolutions and with simple classification schemes (e.g., the differentiation of peatland or nonpeatland surfaces). The use of support vector machines (SVM) [*Brown et al.*, 1999]), may be a potential alternative to neural network classifiers for peatland mapping. Like ANNs, SVM algorithms also use information contained within mixed pixels. However, the main advantage of SVMs is that they require substantially less training data than other approaches, and are able to achieve similar, if not better, classification accuracies [*Foody and Mathur*, 2006; *Koetz et al.*, 2008]. The requirement for only a small number of training samples is a particular advantage for hyperspectral image processing, where meeting the requirements of recommended training sample sizes for conventional classifiers can be problematic (e.g., 10–30 times the number of wavebands [*Mather*, 2006]). SVMs are a relatively new tool for image classification, and there are currently no examples of their use for mapping peatland vegetation. Evidently, future research efforts should focus on the potential use and development of nonlinear classification approaches for mapping peatland vegetation across a range of spatial scales.

Soft classification approaches, such as linear mixture modeling and some forms of ANNs and SVMs, may be preferable to conventional per-pixel approaches when more detailed information on the composition of mixed pixels is desired. However, the major drawback of all subpixel classification approaches lies in the difficulty of assessing the classification accuracy and the need to collect suitable (often quite detailed) reference data [*Lu and Weng*, 2007].

3.2.1.3. Synergizing data sources. While the spectral characteristics of plant communities are most commonly used to map peatland vegetation distribution using remote sensing, spectral confusion between different vegetation classes can lower the accuracy of the final output map. Improved classification results can sometimes be achieved with the use of additional ancillary data sets. For example, decision tree and rule-based classification approaches often utilize data from several sources and multiple dates to improve mapping accuracy. These approaches are particularly useful when integrating optical and radar imagery to maximize the benefits of each sensor (see Table 1). *Li and Chen* [2005] developed a rule-based model in which a digital elevation model DEM, multitemporal SAR, and Landsat imagery were used to map several peatland classes (e.g., swamp, marsh, treed and open fen, treed and open bog), achieving greater classification accuracy compared to a classical approach based on Landsat data alone.

The synergistic use of elevation data, derived from lidar, and optical data can also be exploited for mapping peatland land cover at higher spatial resolutions [e.g., *Milton et al.*, 2004]. However, the vertical information provided by lidar is unlikely to provide much additional information on more discrete vegetation types (e.g., shrubs) due to issues related to the vertical separability of laser returns and vegetation density [*Hopkinson et al.*, 2005; *Koetz et al.*, 2008]. Instead, recent innovative approaches attempt to utilize lidar data to establish links between peatland spatial structure and plant communities using geostatistical approaches [*Anderson and Bennie*, 2008].

Object-oriented classification approaches are also emerging in the remote sensing literature. These methods also enable the use of several different sources of data and involve two main steps. The first is the application of a segmentation algorithm used to merge neighboring pixels to form spatial objects (or segments) that are homogenous with regard to spatial or spectral characteristics [*Ryherd and Woodcock*, 1996; *Jensen*, 2005]. The second stage involves the clas-

sification of those objects either using traditional "hard" classification methods of spectral proximity such as nearest neighbor or "soft" methods based on fuzzy logic with membership functions (MFs, or attribute values). For example, *Grenier et al.* [2007] used an object-based approach to map peatland classes in two regions of Quebec using a combination of Landsat ETM and Radarsat-1 imagery. One of the advantages of object-based approaches over per pixel techniques is that homogenous groups of pixels (or objects) are often more closely related to ecological classes than individual pixels [*Jensen,* 2005]. However, selection of MFs and attributes which relate objects to thematic classes can be subjective, and care must be taken to choose MFs which are repeatable and easily transferred to other ecologically comparable sites [*Grenier et al.*, 2007].

3.2.1.4. Utilizing the angular domain. Although not widely utilized within land cover mapping studies at present, the incorporation of multiangular data into classification approaches could also lead to improvements in peatland vegetation mapping. Multiangular observations, collected by viewing the target from a number of different angles, can be used to determine the bidirectional reflectance distribution function (BRDF), which describes the dependence of observed reflectance on illumination and sensor viewing angles. Differences in the structure and optical properties of vegetation canopies mean that the BRDF of vegetation is highly anisotropic [*Asner et al.*, 1998]. The information on angular dependence of different vegetation classes can be used in conjunction with multispectral and multitemporal data to improve classification accuracy, particularly when spectral signatures are insensitive to vegetation composition [*Barnsley et al.*, 1997]. For example, *Heiskanen and Kivinen* [2008] were able to discriminate between open peatlands and forest classes more easily using multitemporal, multispectral, and multiangular MODIS data than by using either the spectral or temporal domains in isolation.

3.2.2. Scaling peatland land cover maps using multiresolution data. Another area of research of increasing importance in the field of peatland remote sensing is the development of scaling algorithms, which enable upscaling of high-resolution land cover data. Scaling is necessary due to the trade-offs that exist between the spectral, spatial, and temporal resolution, and geographic coverage of remotely sensed data (see section 2.1 for further details). This is particularly important when mapping northern peatlands because of the stated high levels of vegetation heterogeneity, which often prevent the direct application of subpixel processing approaches to moderate or coarse resolution imagery (≤250 m). Consequently, high-resolution data are often used as a basis for mapping subpixel proportions of peatland land cover at lower spatial resolutions. Several approaches have been used to upscale peatland land cover data. The first is based on the development of a statistical model to relate spectral measurements in the coarse resolution data to the proportion of the desired vegetation type mapped using high-resolution imagery. *Pflugmacher et al.* [2007] utilized this regression approach to map the fractional cover of mined and unmined peatlands in the St. Petersburg area of Russia from MODIS data using a combination of high-resolution forest inventory maps and Landsat imagery. Although the approach indicated a potential improvement in peatland mapping performance based on MODIS data, statistical models of this nature are empirical, and thus, more research is required to ascertain whether the approach is applicable to other locations or periods in time. Alternative, physically based approaches utilize linear mixture models. *Takeuchi et al.* [2003] used linear mixture modeling to map methane emissions from a Siberian wetland by scaling directly between AVHRR and SPOT HRV data (see Table 1). A land cover map was generated from the higher resolution SPOT data and used to estimate the proportional coverage of land cover in overlapping 1-km AVHRR pixels. The liner mixture model was then used to determine the spectral properties of each end-member within the image given that the proportional coverage of every land-surface type within a given pixel is known. The scaling model was then applied to the entire AVHRR scene to map subpixel land cover. Another viable physically based approach is the use of a pixel scaling-based method. Although not yet used for peatland vegetation mapping, this approach identifies the location of a number of extreme pixels in the coarse resolution data. Information on the fractional coverage of the land cover of each pixel, derived from the higher resolution data, is then used to select the purest pixels which represent the end-members for each land-surface type [e.g., *Zeng et al.*, 2008].

While scaling approaches show much promise for subpixel mapping of vegetation at moderate to low spatial resolutions, more research is required regarding their application to northern peatlands. Additionally, scaling algorithms must be developed in conjunction with improved classification algorithms (outlined in section 3.2.1), since the accuracy of scaled data is often inherently reliant on effective characterization of the land cover at higher spatial resolutions.

4. RETRIEVAL OF VEGETATION PHYSIOLOGICAL AND BIOPHYSICAL PROPERTIES FROM REMOTELY SENSED DATA

Although land cover mapping is the most common use of remotely sensed data in the study of northern peatlands, remote sensing data can be used to retrieve more detailed information related to peatland carbon exchange processes (e.g., GPP, LAI, *f*PAR, light use efficiency (LUE), chlorophyll content).

Once image preprocessing has been completed (section 2.2), successful retrieval of anything other than class-level information from remote sensing data hinges on the availability of SV models (M_{SV}; typical state variables for vegetation are outlined in section 3.1.2) that can be inverted against the measurements of radiation at the sensor to retrieve the information or parameter of interest [(a subset of SV) parameters (P_{SV}); *Pinty et al.*, 2004]. In essence, in order to derive detailed measurements of a variable of interest (e.g., LAI, *f*PAR), it is often essential to initially measure (e.g., through field and laboratory measurement), model (i.e., develop M_{SV}) and/or minimize the effect of a range of relevant SV parameters. In Figure 2, this approach is represented by path B, where a model is used to convert radiance measurements to a variable of interest which is based upon SV or P_{SV}. A priori knowledge can subsequently be used to link these variables to those peatland carbon exchange processes which cannot be directly measured from remotely sensed data (e.g., GPP, NPP, LUE), either through models (e.g., Boreal Ecosystem Productivity Simulator (BEPS) [*Liu et al.*, 1997] and MOD17 GPP product) or the derivation of statistical relationships. In Figure 2, this approach is represented by path C.

Retrieval of vegetation physiological and biophysical properties from remotely sensed imagery is commonly achieved either through: (1) development of empirical or statistical models which relate spectral information to vegetation parameters of utility in carbon flux estimations, or (2) by physically modeling the radiative transfer (RT) process. These approaches are reviewed below.

4.1. Empirical or Statistical Modeling Approaches

The importance of ecosystem level carbon dynamics in global carbon cycle research has led to the development of a number of medium- to coarse-resolution operational products specifically designed for use in carbon exchange research (e.g., AVHRR LAI, and *f*PAR; MODIS LAI, *f*PAR, and GPP products; and vegetation indices). A number of these data sets, notably those derived from long-term AVHRR observations, have either been derived empirically from spectral reflectance, using vegetation indices such as the normalized difference vegetation index (NDVI) [*Los et al.*, 2005], or used to formulate new empirical relationships between reflectance and carbon exchange parameters. For example, spectral indices that are related to vegetation greenness such as the NDVI or the Enhanced Vegetation Index (EVI) have been correlated with flux tower measures of GPP with varying degrees of success [*La Puma et al.*, 2007; *Rahman et al.*, 2005; *Sims et al.*, 2006a, 2006b; *Wylie et al.*, 2003]. Regression relationships have also been formulated between the MODIS-derived *f*PAR product and ground-based flux-tower measurements to derive light use efficiency estimates across Canadian peatlands [*Connolly et al.*, 2008].

Spectral indices have also been used to derive empirical relationships between spectral properties and biophysical attributes of peatland vegetation at higher spatial resolutions. Most have concentrated on understanding dynamic changes in *Sphagnum* mosses because of their dominance across northern peatland landscapes and their importance in the carbon cycle. Vegetation indices have primarily been used to monitor relative changes in moss water content [*Vogelmann and Moss*, 1993; *Bryant and Baird*, 2003; *Harris et al.*, 2005, 2006; *van Gaalen et al.*, 2007] and photosynthetic capacity [*van Gaalen et al.*, 2007; *Harris*, 2008]. If sufficient field data (or similar a priori information) exist, these data can, in turn, be processed further to produce additional information relating to the condition of peatland surfaces. For example, the conversion of moss canopy moisture content (derived from a P_{SV} model) to surface volumetric water content (VMC) may be possible (e.g., via a knowledge-based model M_K, represented as path C in Figure 2).

High-resolution laboratory and field spectrometry studies have utilized statistical models (e.g., P_{SV} or M_K), to successfully retrieve plant physiological and biophysical information from small peatland plots (e.g., 10^0–10^1 m). However, as the scale of observation increases, i.e., from in situ to airborne to satellite-based remote sensing, additional variations in reflection can occur as a result of atmospheric conditions, sun-angle effects (e.g., shadowing) and the influence of reflectance from nontarget surfaces [*Harris and Bryant*, 2009]. Thus, empirical relationships of this nature will only hold true at increasing spatial scales if the variable of interest (e.g., water content) has a greater impact on reflectance than that of the extraneous background effects and/or variability. Currently, only a small number of studies have attempted to assess the potential of regression models, specifically adapted for use across northern peatlands, at increasing spatial and temporal scales [e.g., *Harris et al.*, 2006; *Harris and Bryant*, 2009]. A further issue related to all statistical modeling approaches is that of repeatability. Empirical approaches derived at a given location or point in time may not be applicable to other locations, in different seasons or years, or applicable for use with different sensors. This is because the variable of interest is not the only state variable which affects the remote sensing data. Consequently, the successful retrieval of biophysical information from empiri-

cal models relies, to some extent, on a good understanding of the other RT state variables present within the scene and some knowledge of how they affect the variable of interest. If the extraneous factors can be minimized (e.g., through careful data collection, preprocessing, and sample design) then empirical models can be used effectively to quantify dynamic change in peatland plant communities in a manner that can inform carbon-balance studies.

4.2. Physically Based Approaches

In contrast to the empirical approach, physically based models such as RT and geometric optical (GO) models try to predict remote sensing data based on the complete set of RT state variables for the surface in question, e.g., the optical and geometric properties of the canopy, the leaves, and the soil background [*Goel*, 1988; *Meroni et al.*, 2004]. These models are able to explain the scattering and absorption within the vegetation canopy and thus can offer an explicit connection between canopy reflectance and plant physiological and biophysical properties. Consequently, the physical approach offers several advantages over the use of empirical models, in that: (1) physical models have more general validity since they are not governed by specific empirical relationships, (2) they are able to exploit multiangular data sets to characterize the full dimensionality of the vegetation canopy, and (3) they can retrieve information on several biophysical properties simultaneously by making use of the full spectral range of the sensor.

A number of the available medium- to coarse-resolution operational products [e.g., MODIS LAI and *f*PAR and MODIS GPP] are developed using physically based RT models. These products are freely available, provide good temporal and spatial coverage, and efforts have been made to test their validity across a range of biomes [e.g., *Morisette et al.*, 2006; *Pisek and Chen*, 2007; *Tan et al.*, 2005]. Products such as MODIS GPP have proved useful for tracking the pattern of dynamic change in photosynthesis across northern peatlands [*Moore et al.*, 2006], although the standard MODIS GPP product commonly overestimates growing season GPP across these environments particularly when rainfall levels are below the long term average (A. Harris and J. Dash, unpublished data, 2009).

Physically based models have also been developed to retrieve information on peatland vegetation structures (e.g., LAI and degree of vegetation clumping), particularly over portions of peatlands covered by monospecific tree stands where the distribution of foliage within the canopy is spatially confined and can be effectively modeled. *Sonnentag et al.* [2007] used a geometric optical (GO) RT model [four-scale; *Chen and Leblanc*, 1997] combined with linear spectral mixture modeling to map tree LAI across a Canadian peatland. The GO model was used to derive an empirical relationship between simulated shadow fraction and tree LAI. The equation was subsequently used to describe the empirical relationship between shadow fractions of the forested portions of the image, obtained from linear spectral mixture modeling, and tree LAI. A number of recent studies have also utilized physically based modeling approaches to understand the utility of combined multiangular and multispectral sensing capabilities for the retrieval of tree canopy biophysical parameters [e.g., *Chen et al.*, 2005; *Leblanc et al.*, 2005; *Prieto-Blanco et al.*, 2009; *Simic and Chen*, 2008]. The compact high-resolution imaging spectrometer satellite (CHRIS) on board the project for onboard autonomy (PROBA) has shown particular promise for deriving high-resolution vegetation biophysical parameters from northern environments [*Simic and Chen*, 2008].

Physically based models are able to fully express the underlying relationships between various biophysical properties and sensor measurements of spectral reflectance in a mathematical way and, thus, offer the greatest potential for accurate parameter retrieval and monitoring of vegetation dynamics [*Verstraete et al.*, 1996]. However, this can only be achieved if the models are able to represent reality sufficiently [*Liang*, 2007]. A commonly cited reason for mismatches between the MODIS GPP product and ground-based tower GPP data is the accepted difficulty in parameterizing the MODIS GPP model so that it can accurately represent conditions observed on the ground [*Heinsch et al.*, 2006; *Turner et al.*, 2003; *Zhao et al.*, 2006]. A large proportion of current RT models are based on monospecific stands where the radiative transfer process may be better understood and more effectively modeled. Heterogeneous multispecies canopies, such as those found in northern peatlands, are extremely difficult to model using practical, economically invertible physically based models because of the need to account for and simulate interactions between a large number of state variables [*Darvishzadeh et al.*, 2008]. Inversion of the model becomes difficult as: (1) there are often more unknowns in the RT process than there are independent measurements, (2) the more free parameters that there are in the model, the more difficult it is to guarantee that the inversion solution is unique [*Weiss and Baret*, 1999], and (3) the natural variability in vegetation structure and composition within a pixel may be such that the RT model cannot realistically simulate the canopy reflectance [*Hall et al.*, 1997]. Future research should therefore focus on the challenges associated with the development of physically based models that can explicitly take into account or minimize the effects of heterogeneity in peatland vegetation canopies.

5. SUMMARY

This paper set out to evaluate and review many of the existing and emerging remote sensing approaches that can be used to retrieve northern peatland land-surface data of relevance to peatland carbon exchange research. Within this context, a number of important research themes, methodological challenges, and research priorities associated with our continued use of remote sensing data in this manner have been discussed, and are summarized below.

We have identified a significant number of studies that involve the generation of land cover maps from remotely sensed data. In some instances, the land cover data have been used to quantify carbon stocks, extrapolate in situ field observations, calibrate and extrapolate carbon models, and inform peatland management efforts. In general, the level of detail obtained from imagery is largely a function of the spatial resolution of the sensor and the size and homogeneity of the vegetation patches being studied. As a result, traditional visual interpretation approaches based on the use of air photographs remain extremely useful for the derivation of detailed land cover information for localized peatland studies for a particular time. In the same vein, automated and semiautomated classification approaches are more suited for the routine collection of data over greater spatial extents. A large number of studies have applied conventional per-pixel remote sensing approaches for mapping peatlands. These approaches are often rapid, require limited technical expertise, and can provide useful information regarding vegetation composition and spatial distribution. However, there are a number of drawbacks with these approaches due in part to: (1) pixel heterogeneity associated with peatland surfaces and the associated problem of mixed pixels, and (2) the lack of a direct understanding of how radiance interacts with the land surface. Both factors reduce classification accuracy, making it difficult to identify temporal changes in land cover, and enhancing bias in carbon exchange estimates that rely directly on land cover data. Nevertheless, if accompanied by a full assessment of classification error and uncertainty, conventional land cover classification approaches can provide important information on the distribution and composition of peatland plant communities at a range of spatial scales.

In an attempt to directly accommodate or address the problems associated with the per-pixel approaches, a range of contemporary classification approaches for mapping peatland vegetation have been developed, including linear and nonlinear mixture modeling, the use of synergistic data sources, and utilization of the angular domain. Of these, nonlinear mixture modeling and the combined use of multiangular and multispectral data perhaps offer the greatest promise for mapping northern peatland vegetation. Another important line of research is the development of measures that can be used to upscale land cover data. This is particularly relevant for northern peatlands where the complexity in vegetation composition is such that it often becomes very difficult (if not impossible) to derive detailed information relating to vegetation composition at increasing spatial scales.

The potential of remote sensing in carbon cycle applications also clearly extends beyond the generation of simple land cover products toward the development of empirical, semiempirical, or physically based algorithms, which relate spectral information to inherent biophysical properties of vegetation relevant to the peatland carbon cycle. Although some a priori knowledge is often required to link remote sensing-derived variables to those carbon-balance processes, which cannot be directly measured by remote sensing, these approaches offer significant advantages. In particular, empirical approaches have been successfully used to monitor temporal changes in peatland photosynthetic efficiency, gross primary productivity, and near-surface hydrology. However, the inability of empirical relationships to physically model radiation interactions with the vegetation canopy mean that relationships derived at a given location, point in time, or spatial area, are often not applicable to other locations, to different seasons or years, or to alternative sensors. Nevertheless, empirical approaches have been used effectively where data are collected using an appropriate sampling design and preprocessed in such a way as to minimize the impact of extraneous factors (e.g., scattering, absorption, transmission, and geometric properties of the canopy, and illumination conditions).

Physically based RT models can be used to overcome some of the limitations of empirical approaches and also have the added advantage of being able to retrieve information on several biophysical properties simultaneously, as well as exploit multiangular data sets. However, these models require effective parameterization and cannot easily cope with the heterogeneity of peatland surfaces. Nevertheless, these models offer hope, in that they have been used to successfully derive information on the vegetation structure (e.g., LAI) of monospecific stands (e.g., treed portions of peatlands), where the distribution of the foliage is spatially confined and can be effectively modeled.

Over the last 20 years, there has been considerable progress in the development and implementation of remote sensing approaches that provide data relating to peatland carbon processes. At the same time, this review highlights a number of methodological challenges, which may limit the effectiveness of remote sensing data in some instances. Consequently, there are several ways in which future research can be directed to inform peatland carbon-balance research.

In particular, we suggest that the following lines of research are pursued:

1. **Improved** communication among multidisciplinary research groups (e.g., ecologists, hydrologists, remote sensors) is needed to ensure that: (1) products derived from remotely sensed data are of use to carbon cycle scientists, and (2) that process-based models are developed to maximize the utility and applicability of such data sets.

2. Continued development, testing, and validation of nonlinear mixture modeling approaches (such as SVM) for peatland mapping are required across a range of spatial scales.

3. Utilization of global and local-scale statistical approaches to provide more information as to how the quality of remote sensing output data (e.g., thematic maps) varies spatially. Such information would be used to improve error modeling and prevent the propagation of uncertainty when using such data for carbon modeling.

4. Effective use and integration of combined multispectral and multiangular data sets is required to improve both peatland mapping efforts and the retrieval of biophysical parameters from remotely sensed data.

5. Development of appropriate scaling algorithms is required to: (1) allow peatlands to be mapped in more detail at lower spatial resolutions, and (2) improve our understanding of the impacts of scale on the retrieval of peatland biophysical properties using empirical or statistical approaches.

6. The continued development of radiative transfer models has to take place in order to generate models that can be applied to heterogeneous peatland vegetation assemblages (e.g., specific PFTs).

Acknowledgments. The reviewers are thanked for their help in improving an earlier version of this manuscript.

REFERENCES

Anderson, K., and J. J. Bennie (2008), LiDAR-derived spatial indicators of peatland eco-hydrological condition, Proceedings of the Remote Sensing and Photogrammetry Society Conference: Measuring change in the Earth system, University of Exeter, 15–17 September 2008.

Aplin, P. (2005), Remote sensing: Ecology, *Prog. Phys. Geogr.*, *29*, 104–113, doi:10.1191/030913305pp437pr.

Asner, G. P., B. H. Braswell, D. S. Schimel, and C. A. Wessman (1998), Ecological research needs from multiangle remote sensing data, *Remote Sens. Environ.*, *63*, 155–165.

Aurela, M., T. Laurila, and J.-P. Tuovinen (2001), Seasonal CO_2 balances of a subarctic mire. *J. Geophys. Res.*, *106*, 1623–1637.

Barnsley, M. J., D. Allison, and P. Lewis (1997), On the information content of multiple view angle (MVA) images, *Int. J. Remote Sens.*, *18*, 1937–1960.

Becker, T., L., Kutzbach, I. Forbrich, J. Schneider, D. Jager, B. Thees, and M. Wilmking (2008), Do we miss the hot spots?—The use of very high resolution aerial photographs to quantify carbon fluxes in peatlands, *Biogeosciences, 5*, 1387–1393.

Belluco, E., M. Camuffo, S. Ferrari, L. Modenese, S. Silvestri, A. Marani, and M. Marani (2006), Mapping salt-marsh vegetation by multispectral and hyperspectral remote sensing, *Remote Sens. Environ., 105*, 54–67, doi:10.1016/j.rse.2006.06.006.

Belyea, L. R. and N. Malmer (2004), Carbon sequestration in peatland: Patterns and mechanisms of response to climate change, *Global Change Biol., 10*, 1043–1052.

Berk, A., L. S. Bernstein, G. P. Anderson, P. K. Acharya, D. C. Robertson, J. H. Chetwynd, and S. M. Adler-Golden (1998), MODTRAN cloud and multiple scattering upgrades with application to AVIRIS, *Remote Sens. Environ., 65*, 367–375.

Bridgham, S. D., J. P. Megonigal, J. K. Keller, N. B. Bliss, and C. Trettin (2007), Chapter 13, Wetlands (The First State of the Carbon Cycle Report), US Climate Change Science Program, Synthesis and Assessment Product 2.2.

Bronge, L. B., and B. Naslund-Landenmark (2002), Wetland classification for Swedish CORINE Land Cover adopting a semiautomatic interactive approach, *Can. J. Remote Sens., 28*, 139–155.

Brown, E., M. Aitkenhead, R. Wright, and I. H. Aalders (2007), Mapping and classification of peatland on the Isle of Lewis using Landsat ETM, *Scott. Geogr. J., 123*, 173–192, doi:10.1080/14702540701786912.

Brown, K. M., G. M. Foody, and P. M. A. Atkinson (2009), Estimating per-pixel thematic uncertainty in remote sensing classifications, *Int. J. Remote Sens., 30*, 209–229.

Brown, M., S. R. Gunn, and H. G. Lewis (1999), Support vector machines for optimal classification and spectral unmixing, *Ecol. Model., 120*, 167–179.

Bryant, R. G., and A. J. Baird (2003), The spectral behaviour of *Sphagnum* canopies under varying hydrological conditions, *Geophys. Res. Lett., 30*(3), 1134, doi:10.1029/2002GL016053.

Bubier, J., T. Moore, K. Savage, and P. Crill (2005), A comparison of methane flux in a boreal landscape between a dry and a wet year, *Global Biogeochem. Cycles, 19*, GB1023, doi:10.1029/2004GB002351.

Bubier, J. L., B. N. Rock, and P. M. Crill (1997), Spectral reflectance measurements of boreal wetland and forest mosses, *J. Geophys. Res., 102*, 29,483–29,494.

Bubier, J. L., T. R. Moore, and L. A. Bledzki (2007), Effects of nutrient addition on vegetation and carbon cycling in an ombrotrophic bog, *Global Change Biol., 13*, 1168–1186.

Campbell, J. B. (2002), *Introduction to Remote Sensing*, Taylor and Francis, New York.

Chavez, P. S. (1975), Atmospheric, solar, and MTF corrections for ERTS digital imagery, Proceedings of the American Society of Photogrammetry Fall Technical Meeting, Phoenix, Ariz.

Chen, J. M., and S. G. Leblanc (1997), A 4-Scale bidirectional reflection model based on canopy architecture, *IEEE Trans. Geosci. Remote Sens., 39*, 1061–1071.

Chen, J. M., J. Liu, S. G. Leblanc, R. Lacaze, and J. L. Roujean (2003), Multi-angular optical remote sensing for assessing vegetation structure and carbon absorption, *Remote Sens. Environ., 84*, 516–525.

Chen, J. M., C. H. Menges, and S. G. Leblanc (2005), Global mapping of foliage clumping index using multi-angular satellite data, *Remote Sens. Environ., 97*, 447–457.

Colwell, J. E. (1974), Vegetation canopy reflectance, *Remote Sens. Environ., 3*, 175–83.

Connolly, J., N. T. Roulet, J. W. Seaquist, N. M. Holden, P. M. Lafleur, E. R. Humphreys, B. W. Heumann, and S. M. Ward (2008), Using MODIS derived *f*PAR with ground based flux tower measurements to derive the light use efficiency for two Canadian peatlands, *Biogeosci. Discuss., 5*, 1765–1794.

Darvishzadeh, R., A. Skidmore, M. Schlerf, and C. Atzberger (2008), Inversion of a radiative transfer model for estimating vegetation LAI and chlorophyll in heterogeneous grassland, *Remote Sens. Environ., 112*, 2592–2604, doi:10.1016/j.rse.2007.12.003.

Dorrepaal, E., J. H. C. Cornelissen, R. Aerts, B. Wallen, and R. S. P. Van Logtestijn (2005), Are growth forms consistent predictors of leaf litter quality and decomposability across peatlands along a latitudinal gradient?, *J. Ecol., 93*, 817–828, doi:10.1111/j.1365-2745.2005.01024.x.

Dorrepaal, E., J. H. C. Cornelissen, and R. Aerts (2007), Changing leaf litter feedbacks on plant production across contrasting sub-arctic peatland species and growth forms, *Oecologia, 151*, 251–261, doi:10.1007/s00442-006-0580-3.

Foody, G. M., and A. Mathur (2006), The use of small training sets containing mixed pixels for accurate hard image classification: Training on mixed spectral responses for classification by a SVM, *Remote Sens. Environ., 103*, 179–189.

Foody, G. M., R. M. Lucas, P. J. Curran, and M. Honzak (1997), Non-linear mixture modelling without end-members using an artificial neural network, *Int. J. Remote Sens., 18*, 937–953.

Foster, D. R., H. E. Wright Jr., M. Thelaus, and G. A. King (1993), Bog development and landform dynamics in central Sweden and south-eastern Labrador, Canada, *J. Ecol., 76*, 1164–1185.

Frey, K. E., and L. C. Smith (2007), How well do we know northern land cover? Comparison of four global vegetation and wetland products with a new ground-truth database for West Siberia, *Global Biogeochem. Cycles, 21*, GB1016, doi:10.1029/2006GB002706.

Glaser, P. H. (1987), The development of streamlined bog islands in the continental interior of North-America, *Arctic Alpine Res., 19*, 402–413.

Glaser, P. H. (1989), Detecting biotic and hydrogeochemical processes in large peat basins with Landsat imagery, *Remote Sens. Environ., 28*, 109–119.

Glaser, P. H. (1992), *Peat Landforms*, Univ. of Minnesota Press.

Goel, N. S. (1988), Models of vegetation canopy reflectance and their use in estimation of biophysical parameters from reflectance data, *Remote Sens. Rev., 4*, 1–121.

Gorham, E. (1991), Northern peatlands: Role in the carbon cycle and probable responses to climatic warming, *Ecol. Appl., 1*, 182–195.

Grenier, M., A. M. Demers, S. Labrecque, M. Benoit, R. A. Fournier, and B. Drolet (2007), An object-based method to map wetland using RADARSAT-1 and Landsat ETM images: Test case on two sites in Quebec, Canada, *Can. J. Remote Sens.*, *33*, S28–S45.

Hall, F. G., D. E. Knapp, and K. F. Huemmrich (1997), Physically based classification and satellite mapping of biophysical characteristics in the southern boreal forest, *J. Geophys. Res.*, *102*, 29,567–29,580.

Harris, A. (2008), Spectral reflectance and photosynthetic properties of *Sphagnum* mosses exposed to progressive drought, *Ecohydrology*, *1*, 35–42.

Harris, A., and R. G. Bryant (2009) A multi-scale remote sensing approach for monitoring northern peatland hydrology: Present possibilities and future challenges, *J. Environ. Manage.*, doi:10.1016/j.jenvman.2007.06.025, in press.

Harris, A., R. G. Bryant, and A. J. Baird (2005), Detecting water stress in *Sphagnum* spp, *Remote Sens. Environ.*, *97*, 371–381.

Harris, A., R. G. Bryant, and A. J. Baird (2006), Mapping the effects of water stress on *Sphagnum*: Preliminary observations using airborne remote sensing, *Remote Sens. Environ.*, *100*, 363–378.

Heinsch, F. A., M. S. Zhao, S. W. Running, J. S. Kimball, R. R. Nemani, K. J. Davis, P. V. Bolstad, B. D. Cook, A. R. Desai, D. M. Ricciuto, et al. (2006), Evaluation of remote sensing based terrestrial productivity from MODIS using regional tower eddy flux network observations, *IEEE Trans. Geosci. Remote Sens.*, *44*, 1908–1925.

Heiskanen, J., (2008), Evaluation of global land cover data sets over the tundra-taiga transition zone in northernmost Finland, *Int. J. Remote Sens.*, *29*, 3727–3751, doi:10.1080/01431160701871104.

Heiskanen, J., and S. Kivinen (2008), Assessment of multispectral, -temporal and -angular MODIS data for tree cover mapping in the tundra-taiga transition zone, *Remote Sens. Environ.*, *112*, 2367–2380, doi:10.1016/j.rse.2007.11.002.

Hopkinson, C., L. E. Chasmer, G. Sass, I. F. Creed, M. Sitar, W. Kalbfleisch, and P. Treitz (2005), Vegetation class dependent errors in lidar ground elevation and canopy height estimates in a boreal wetland environment, *Can. J. Remote Sens.*, *31*, 191–206.

IPCC (2007), Climate Change 2007: The Physical Science Basis, Contribution of Working Group I to the fourth assessment report of the intergovernmental panel on climate change, edited by S. Solomon et al., Cambridge, UK, http://www.ipcc.ch/ipccreports/ar4-wg1.htm.

Jensen, J. R. (2000), *Remote Sensing of the Environment: An Earth Resource Perspective*, John Wiley, New York.

Jensen, J. R. (2005), *Introductory Digital Image Processing: A Remote Sensing Perspective*, John Wiley, New York.

Johansson, T., N. Malmer, P. M. Crill, T. Friborg, J. H. Akerman, M. Mastepanov, and T. R. Christensen (2006), Decadal vegetation changes in a northern peatland, greenhouse gas fluxes and net radiative forcing, *Global Change Biol.*, *12*, 2352–2369.

Jollineau, M. Y., and P. J. Howarth (2008), Mapping an inland wetland complex using hyperspectral imagery, *Int. J. Remote Sens.*, *29*, 3609–3631.

Kadmon, R., and R. Harari-Kremer (1999), Studying long-term vegetation dynamics using digital processing of historical aerial photographs, *Remote Sens. Environ.*, *68*, 164–176.

Kaufman, Y. J. (1989), The atmospheric effect on remote sensing and its correction, *Theory and Applications of Optical Remote Sensing*, edited by G. Asrar, pp. 336–428, John Wiley, New York.

Koetz, B., F. Morsdorf, S. van der Linden, T. Curt, and B. Allgöwer (2008), Multi-source land cover classification for forest fire management based on imaging spectrometry and LiDAR data, *For. Ecol. Manage.*, *256*, 263–271.

Krankina, O. N., D. Pflugmacher, M. Friedl, W. B. Cohen, P. Nelson, and A. Baccini (2008), Meeting the challenge of mapping peatlands with remotely sensed data, *Biogeosciences*, *5*, 1809–1820.

Kruse, F. A., A. B. Lefkoff, J. W. Boardman, K. B. Heidebrecht, A. T. Shapiro, P. J. Barloon, and A. F. H. Goetz (1993), The spectral image-processing system (SIPS)—Interactive visualization and analysis of imaging spectrometer data, *Remote Sens. Environ.*, *44*, 145–163.

La Puma, I. P., T. E. Philippi, and S. F. Oberbauer (2007), Relating NDVI to ecosystem CO_2 exchange patterns in response to season length and soil warming manipulations in arctic Alaska, *Remote Sens. Environ.*, *109*, 225–236.

Leblanc, S. G., J. M. Chen, H. P. White, R. Latifovic, R. Lacaze, and J. L. Roujean (2005), Canada-wide foliage clumping index mapping from multiangular POLDER measurements, *Can. J. Remote Sens.*, *31*, 364–376.

Leckie, D. G., E. Cloney, C. Jay, and D. Paradine (2005), Automated mapping of stream features with high-resolution multispectral imagery: An example of the capabilities, *Photogramm. Eng. Remote Sens.*, *71*, 145–155.

Li, J., and W. Chen (2005), A rule-based method for mapping Canada's wetlands using optical, radar and DEM data, *Int. J. Remote Sens.*, *26*, 5051.

Liang, S. L. (2007), Recent developments in estimating land surface biogeophysical variables from optical remote sensing, *Prog. Phys. Geogr.*, *31*, 501–516, doi:10.1177/0309133307084626.

Lillesand, T. M., R. W. Kiefer, and J. W. Chipman (2008), *Remote Sensing and Image Interpretation*, John Wiley, New York.

Liu, J., J. M. Chen, J. Cihlar, and W. M. Park (1997), A process-based boreal ecosystem productivity simulator using remote sensing inputs, *Remote Sens. Environ.*, *62*, 158–175.

Los, S. O., P. R. J. North, W. M. F. Grey, and M. J. Barnsley (2005), A method to convert AVHRR Normalized Difference Vegetation Index time series to a standard viewing and illumination geometry, *Remote Sens. Environ.*, *99*, 400–411, doi:10.1016/j.rse.2005.08.017.

Lu, D., and Q. Weng (2007), A survey of image classification methods and techniques for improving classification performance, *Int. J. Remote Sens.*, *28*, 823–870, doi:10.1080/01431160600746456.

MacKay, H., C. M. Finlayson, D. Fernández-Prieto, N. Davidson, D. Pritchard, and L. M. Rebelo, (2009) The role of Earth Observation (EO) technologies in supporting implementation of the Ramsar Convention on Wetlands. *J. Environ. Manage.*, in press.

Malmer, N. T. Johansson, M. Olsrud, and T. R. Christensen (2005), Vegetation, climatic changes and net carbon sequestration in a North-Scandinavian subarctic mire over 30 years, *Global Change Biol.*, *11*, 1895–1909, doi:10.1111/j.1365-2486.2005.01042.x.

Mather, P. M. (2006). *Computer Processing of Remotely-Sensed Images: An Introduction*, John Wiley, Chichester.

Mehner, H., M. Cutler, D. Fairbairn, and G. Thompson (2004), Remote sensing of upland vegetation: The potential of high spatial resolution satellite sensors, *Global Ecol. Biogeogr.*, *13*, 359–369.

Meroni, M., R. Colombo, and C. Panigada (2004), Inversion of a radiative transfer model with hyperspectral observations for LAI mapping in poplar plantations, *Remote Sens. Environ.*, *92*, 195–206, doi:10.1016/j.rse.2004.06.005.

Mills, H., M. E. J. Cutler, and D. Fairbairn (2006), Artificial neural networks for mapping regional-scale upland vegetation from high spatial resolution imagery, *Int. J. Remote Sens.*, *27*, 2177–2195, doi:10.1080/01431160500396501.

Milton, E. J., P. D. Hughes, K. Anderson, J. Schultz, C. T. Hill, and R. Lindsay (2004), Remote sensing condition categories on lowland raised bogs in the United Kingdom. Part 1: Development and testing methods. Proceedings of the Peterborough Remote Sensing Workshop: Working Today for Nature Tomorrow, Peterborough, UK, English Nature.

Minkkinen, K., R. Korhonen, I. Savolainen, and J. Laine (2002), Carbon balance and radiative forcing of Finnish peatlands 1900–2100—The impact of forestry drainage, *Global Change Biol.*, *8*, 785–799.

Moghaddam, M., S. Saatchi, and R. Cuenca (2000), Estimating subcanopy soil moisture with radar, *J. Geophys. Res.*, *105*, 14,899–14,911.

Moore, T. R., and R. Knowles (1989), The influence of water table levels on methane and carbon dioxide emissions from peatland soils, *Can. J. Soil Sci.*, *69*, 33–38.

Moore, T. R., P. M. Lafleur, D. M. I. Poon, B. W. Heumann, J. W. Seaquist, and N. T. Roulet (2006), Spring photosynthesis in a cool temperate bog, *Global Change Biol.*, *12*, 2323–2335.

Morisette, J. T., F. Baret, J. L. Privette, R. B. Myneni, J. E. Nickeson, S. Garrigues, N. V. Shabanov, M. Weiss, R. A. Fernandes, S. G. Leblanc, et al. (2006), Validation of global moderate-resolution LAI products: A framework proposed within the CEOS Land Product Validation subgroup, *IEEE Trans. Geosci. Remote Sens.*, *44*, 1804–1817, doi:10.1109/tgrs.2006.872529.

Muller, S. V., A. E. Racoviteanu, and D. A. Walker (1999), Landsat MSS-derived land-cover map of northern Alaska: Extrapolation methods and a comparison with photo-interpreted and AVHRR-derived maps, *Int. J. Remote Sens.*, *20*, 2921–2946.

Murphy, P. N. C., J. Ogilvie, K. Connor, and P. A. Arpl (2007), Mapping wetlands: A comparison of two different approaches for New Brunswick, Canada, *Wetlands*, *27*, 846–854.

Oechel, W. C., S. J. Hastings, G. Vourlitis, M. Jenkins, G. Riechers, and N. Grulke (1993), Recent change of arctic tundra ecosystems from a net carbon dioxide sink to a source, *Nature*, *361*, 520–523.

Pelletier, L., T. R. Moore, N. T. Roulet, M. Garneau, and V. Beaulieu-Audy (2007), Methane fluxes from three peatlands in the La Grande Riviere watershed, James Bay lowland, Canada, *J. Geophys. Res.*, *112*, G01018, doi:10.1029/2006JG000216.

Pflugmacher, D., O. N. Krankina, and W. B. Cohen (2007), Satellite-based peatland mapping: Potential of the MODIS sensor, *Global Planet. Change*, *56*, 248–257, doi:10.1016/j.glopacha.2006.07.019.

Pinty, B., et al. (2004), Radiation Transfer Model Intercomparison (RAMI) exercise: Results from the second phase, *J. Geophys. Res.*, *109*, D06210, doi:10.1029/2003JD004252.

Pisek, J., and J. M. Chen (2007), Comparison and validation of MODIS and VEGETATION global LAI products over four BigFoot sites in North America, *Remote Sens. Environ.*, *109*, 81–94, doi:10.1016/j.rse.2006.12.004.

Poulin, M., D. Careau, L. Rochefort, and A. Desrochers (2002), From satellite imagery to peatland vegetation diversity: How reliable are habitat maps? *Cons. Ecol.*, *6*, 3016.

Poulin, M., M. Beslisle, and M. Cabeza (2006), Within-site habitat configuration in reserve design: A case study with a peatland bird, *Biol. Conserv.*, *128*, 55–66, doi:10.1016/j.biocon.2005.09.016.

Prieto-Blanco, A., P. R. J. North, M. J. Barnsley, and N. Fox (2009), Satellite-driven modelling of Net Primary Productivity (NPP): Theoretical analysis. *Remote Sens. Environ.*, *113*, 137–147.

Quinton, W. L., M. Hayashi, and A. Pietroniro (2003), Connectivity and storage functions of channel fens and flat bogs in northern basins, *Hydrol. Process.*, *17*, 3665–3684, doi:10.1002/hyp.1369.

Rahman, A. F., D. A. Sims, V. D. Cordova, and B. Z. El-Masri (2005), Potential of MODIS EVI and surface temperature for directly estimating per-pixel ecosystem C fluxes, *Geophys. Res. Lett.*, *32*, L19404, doi:10.1029/2005GL024127.

Richter, R. (1996), Atmospheric correction of satellite data with haze removal including a haze/clear transition region, *Comp. Geosci.*, *22*, 675–681.

Richter, R., and D. Schläpfer (2002), Geo-atmospheric processing of airborne imaging spectrometry data. Part 2: atmospheric/topographic correction, *Int. J. Remote Sens.*, *23*, 2631–2649, doi:10.1080/01431160110115834.

Roberts, D. A., M. Gardner, R. Church, S. Ustin, G. Scheer, and R. O. Green (1998), Mapping chaparral in the Santa Monica Mountains using multiple endmember spectral mixture models, *Remote Sens. Environ.*, *65*, 267–279.

Roulet, N. T., A. Jano, C. A. Kelly, L. F. Klinger, T. R. Moore, R. Protz, J. A. Ritter, and W. R. Rouse (1994), Role of the Hudson Bay lowland as a source of atmospheric methane, *J. Geophys. Res.*, *99*, 1439–1454.

Rydin, H., and J. K. Jeglum (2006), *The Biology of Peatlands*, Oxford Univ. Press, Oxford, UK.

Ryherd, S., and C. Woodcock (1996), Combining spectral and texture data in the segmentation of remotely sensed images, *Photogramm. Eng. Remote Sens.*, *62*, 181–194.

Schaepman-Strub, G., J. Limpens, M. Menken, H. M. Bartholomeus, and M. E. Schaepman (2008), Towards spatial assessment of carbon sequestration in peatlands: spectroscopy based estimation of fractional cover of three plant functional types, *Biogeosci. Discuss.*, *5*, 1293–1317.

Schläpfer, D., and R. Richter (2002), Geo-atmospheric processing of airborne imaging spectrometry data. Part 1: parametric orthorectification, *Int. J. Remote Sens.*, *23*, 2609–2630, doi:10.1080/01431160110115825.

Schmid, T., M. Koch, and J. Gumuzzio (2005), Multisensor approach to determine changes of wetland characteristics in semi-arid environments (Central Spain), *IEEE Trans. Geosci. Remote Sens.*, *43*, 2516–2525, doi:10.1109/tgrs.2005.852082.

Schmidt, K. S., A. K. Skidmore, E. H. Kloosterman, H. Van Oosten, L. Kumar, and J. A. M. Janssen (2004), Mapping coastal vegetation using an expert system and hyperspectral imagery, *Photogramm. Eng. Remote Sens.*, 70, 703–715.

Settle, J. J., and N. A. Drake (1993), Linear mixing and the estimation of ground cover proportions, *Int. J. Remote Sens.*, 14, 1159–1177.

Simic, A., and J. M. Chen (2008), Refining a hyperspectral and multiangle measurement concept for vegetation structure assessment, *Can. J. Remote Sens.*, 34, 174–191.

Sims, D. A., H. Y. Luo, S. Hastings, W. C. Oechel, A. F. Rahman, and J. A. Gamon (2006a), Parallel adjustments in vegetation greenness and ecosystem CO_2 exchange in response to drought in a Southern California chaparral ecosystem, *Remote Sens. Environ.*, 103, 289–303.

Sims, D. A., et al. (2006b), On the use of MODIS EVI to assess gross primary productivity of North American ecosystems, *J. Geophys. Res.*, 111, G04015, doi:10.1029/2006JG000162.

Sitch, S., A. D. McGuire, J. Kimball, N. Gedney, J. Gamon, R. Engstrom, A. Wolf, Q. Zhuang, J. Clein, and K. C. McDonald (2007), Assessing the carbon balance of circumpolar Arctic tundra using remote sensing and process modeling, *Ecol. Appl.*, 17, 213–234.

Slater, P. N., S. F. Biggar, R. G. Holm, R. D. Jackson, Y. Mao, M. S. Moran, J. M. Palmer, and B. Yuan (1987), Reflectance-based and radiance-based methods for the in-flight absolute calibration of multispectral sensors, *Remote Sens. Environ.*, 22, 11–37.

Smith, L. C., G. M. MacDonald, A. A. Velichko, D. W. Beilman, O. K. Borisova, K. E. Frey, K. V. Kremenetski, and Y. Sheng (2004), Siberian peatlands a net carbon sink and global methane source since the early Holocene, *Science*, 303, 353–356, doi:10.1126/science.1090553.

Smith, M. O., S. L. Ustin, J. B. Adams, and A. R. Gillespie (1990), Vegetation in deserts.1. A regional measure of abundance from multispectral images, *Remote Sens. Environ.*, 31, 1–26.

Sokol, J., H. Ncnairn, and T. J. Pultz (2004), Case studies demonstrating hydrological applications of C-band multi-polarized and polarimetric SAR, *Can. J. Remote Sens.*, 30, 470–483.

Sonnentag, O., J. M. Chen, D. A. Roberts, J. Talbot, K. Q. Halligan, and A. Govind (2007), Mapping tree and shrub leaf area indices in an ombrotrophic peatland through multiple endmember spectral unmixing, *Remote Sens. Environ.*, 109, 342–360, doi:10.1016/j.rse.2007.01.010.

Sonnentag, O., J. M. Chen, N. T. Roulet, W. Ju, and A. Govind (2008), Spatially explicit simulation of peatland hydrology and carbon dioxide exchange: Influence of mesoscale topography, *J. Geophys. Res.*, 113, G02005, doi:10.1029/2007JG000605.

Strack, M., and J. M. Waddington (2007), Response of peatland carbon dioxide and methane fluxes to a water table drawdown experiment, *Global Biogeochem. Cycles*, 21, GB1007, doi:10.1029/2006GB002715.

Takeuchi, W., M. Tamura, and Y. Yasuoka (2003), Estimation of methane emission from West Siberian wetland by scaling technique between NOAA AVHRR and SPOT HRV, *Remote Sens. Environ.*, 85, 21–29.

Tan, B., J. Hu, P. Zhang, D. Huang, N. Shabanov, M. Weiss, Y. Knyazikhin, and R. B. Myneni (2005), Validation of Moderate Resolution Imaging Spectroradiometer leaf area index product in croplands of Alpilles, France, *J. Geophys. Res.*, 110, D01107, doi:10.1029/2004JD004860.

Thomas, V., P. Treitz, D. Jelinski, J. Miller, P. Lafleur, and J. H. McCaughey (2002), Image classification of a northern peatland complex using spectral and plant community data, *Remote Sens. Environ.*, 84, 83–99.

Touzi, R., A. Deschamps, and G. Rother (2007), Wetland characterization using polarimetric RADARSAT-2 capability, *Can. J. Remote Sens.*, 33, S56–S67.

Töyrä, J., and A. Pietroniro (2005), Towards operational monitoring of a northern wetland using geomatics-based techniques, *Remote Sens. Environ.*, 97, 174–191.

Turner, D., S. Urbanski, D. Bremwe, S. C. Wofsy, T. Meyers, S. T. Gower, and M. Gregory (2003), A cross-biome comparison of daily light use efficiency for gross primary production, *Global Change Biol.*, 9, 383–395.

Turner, D. P., W. D. Ritts, W. B. Cohen, S. T. Gower, S. W. Running, M. S. Zhao, M. H. Costa, A. A. Kirschbaum, J. M. Ham, S. R. Saleska, et al. (2006), Evaluation of MODIS NPP and GPP products across multiple biomes, *Remote Sens. Environ.*, 102, 282–292, doi:10.1016/j.rse.2006.02.017.

van Gaalen, K. E. v., L. B. Flanagan, and D. R. Peddle (2007), Photosynthesis, chlorophyll fluorescence and spectral reflectance in *Sphagnum* moss at varying water contents, *Oecologia*, 153, 19–28, doi:10.1007/s00442-007-0718-y.

Vermote, E. F., N. El Saleous, C. O. Justice, Y. J. Kaufman, J. L. Privette, L. Remer, J. C. Roger, and D. Tanré (1997a), Atmospheric correction of visible to middle-infrared EOS-MODIS data over land surfaces: Background, operational algorithm and validation, *J. Geophys. Res.*, 102, 17,131–17,141.

Vermote, E. F., D. Tanre, J. L. Deuze, M. Herman, and J. J. Morcrette (1997b), Second Simulation of the Satellite Signal in the Solar Spectrum, 6S: An overview, *IEEE Trans. Geosci. Remote Sens.*, 35, 675–686.

Verstraete, M. M., B. Pinty, and R. B. Myneni (1996), Potential and limitations of information extraction on the terrestrial biosphere from satellite remote sensing, *Remote Sens. Environ.*, 58, 201–214.

Vogelmann, J. E., and D. M. Moss (1993), Spectral reflectance measurements in the genus *Sphagnum, Remote Sens. Environ.*, 45, 273–279.

Waddington, J. M., T. J. Griffis, and W. R. Rouse (1998), Northern Canadian wetlands: Net ecosystem CO_2 exchange and climate change, *Clim. Change*, 40, 267–275.

Weiss, M., and F. Baret (1999), Evaluation of canopy biophysical variable retrieval performances from the accumulation of large swath satellite data, *Remote Sens. Environ.*, 70, 293–306.

Wiedermann, M. M., A. Nordin, U. Gunnarsson, M. B. Nilsson, and L. Ericson (2007), Global change shifts vegetation and plant-parasite interactions in a boreal mire, *Ecology*, 88, 454–464.

Wylie, B. K., D. A. Johnson, E. Laca, N. Z. Saliendra, T. G. Gilmanov, B. C. Reed, L. L. Tieszen, and B. B. Worstell (2003), Calibration of remotely sensed, coarse resolution NDVI to CO_2 fluxes in a sagebrush-steppe ecosystem, *Remote Sens. Environ.*, 85, 243–255.

Xiong, X., and W. Barnes (2006), MODIS calibration and characterization, in *Earth Science Satellite Remote Sensing. Volume 2 Data, Computational Processing and Tools*, edited by J. Qu et al., pp. 77–97, Springer, New York.

Xiong, X., A. Isaacman, and W. Barnes (2006). MODIS Level-1B Products, in *Earth Science Satellite Remote Sensing, Vol.1: Science and Instruments*, edited by J. Qu et al., pp. 33–49, Springer, New York.

Zeng, Y., M. E. Schaepman, B. Wu, J. G. P. W. Clevers, and A. K. Bregt (2008), Scaling-based forest structural change detection using an inverted geometric-optical model in the Three Gorges region of China, *Remote Sens. Environ.*, *112*, 4261–4271.

Zhao, M., S. W. Running, and R. R. Nemani (2006), Sensitivity of Moderate Resolution Imaging Spectroradiometer (MODIS) terrestrial primary production to the accuracy of meteorological reanalyses, *J. Geophys. Res.*, *111*, G01002, doi:10.1029/2004JG000004.

R. G. Bryant, Department of Geography, University of Sheffield, Winter Street, Sheffield S10 2TN, UK.

A. Harris, School of Geography, University of Southampton, Highfield, Southampton SO17 1BJ, UK. (a.harris@soton.ac.uk)

Plant Litter Decomposition and Nutrient Release in Peatlands

Luca Bragazza,[1] Alexandre Buttler,[2,3,4] Andy Siegenthaler,[2] and Edward A. D. Mitchell[2,5,6]

Decomposition of plant litter is a crucial process in controlling the carbon balance of peatlands. Indeed, as long as the rate of litter decomposition remains lower than the rate of above- and belowground litter production, a net accumulation of peat and, thus, carbon will take place. In addition, decomposition controls the release of important nutrients such as nitrogen, phosphorus, and potassium, the availability of which affects the structure and the functioning of plant communities. This chapter describes the role of the main drivers in affecting mass loss and nutrient release from recently deposited plant litter. In particular, the rate of mass loss of *Sphagnum* litter and vascular plant litter is reviewed in relation to regional climatic conditions, aerobic/anaerobic conditions, and litter chemistry. The rate of nutrient release is discussed in relation to the rate of mass loss and associated litter chemistry by means of a specific case study.

1. INTRODUCTION

Peatlands are characterized by a significantly large storage of soil organic matter per unit of surface compared to the other major ecosystems of the world [*Batjes*, 1996]. Although net primary production (NPP; i.e., the net amount of biomass accumulated by phototrophic biosynthesis per unit area and time) is relatively small in peatlands [*Ito and Oikawa*, 2004], the imbalance between NPP and decomposi-

tion is strong enough to cause significantly high rates of soil organic matter accumulation [*Schlesinger*, 1997; *Clymo et al.*, 1998].

At global scale, the largest fraction of NPP is delivered to the soil as dead organic matter through above- and belowground litter. If we estimate a global annual aboveground litter fall of 54.8×10^{15} g [*Meentemeyer et al.*, 1982] over a worldwide land area of about 121×10^6 km^2, and if we estimate a mean bulk density of litter fall of about 0.2 g cm^{-3}, then each year, a litter layer of about 2 mm would be deposited on the world's land so that this layer would theoretically reach a thickness of 2 m in 1,000 years. However, decomposition prevents such a rate of organic matter accumulation by breaking down the litter into carbon dioxide (CO_2), dissolved organic carbon (DOC), inorganic and organic nutrients, as well as stable humus.

The accumulation of soil organic matter in terrestrial ecosystems is universal, in the sense that higher rates of NPP compared to decomposition take place also in those ecosystems apparently characterized by very low soil organic matter content [*Batjes*, 1996]. Indeed, it is the "magnitude" of the long-term imbalance between NPP and decomposition that allowed peatlands to build up impressive stores of soil organic matter: the peat [*Wieder*, 2006].

In the light of the role played by decomposition in controlling the ability of peatlands to act as C sinks, in this chapter,

[1]Department of Biology and Evolution, University of Ferrara, Ferrara, Italy.

[2]Laboratory of Ecological Systems, École Polytechnique Fédérale de Lausanne, Lausanne, Switzerland.

[3]Also at Restoration Ecology Research Group, Swiss Federal Research Institute, WSL, Lausanne, Switzerland.

[4]Also at Laboratoire de Chrono-Environnement, UMR CNRS 6249, Université de Franche-Comté, Besançon, France.

[5]Also at Wetlands Research Group, Swiss Federal Research Institute, WSL, Lausanne, Switzerland.

[6]Also at Institute of Biology, University of Neuchâtel, Neuchâtel, Switzerland.

Carbon Cycling in Northern Peatlands
Geophysical Monograph Series 184
Copyright 2009 by the American Geophysical Union.
10.1029/2008GM000815

we will review the main factors controlling decomposition of recently deposited plant litter, paying particular attention to mass loss and nutrient release from aboveground litter. For a discussion of peat decomposition and associated release of dissolved organic carbon (DOC), carbon dioxide (CO_2), and methane (CH_4), the reader is referred to the chapters by *Fenner et al.* [this volume], *Moore* [this volume], and *Nilsson and Öquist* [this volume] in the present monograph.

Although litter decomposition is a complex process simultaneously affected by multiple chemical, physical, and biological drivers, we will try to highlight some general trends. To this aim, we first discuss the roles of the main drivers affecting litter mass loss and, second, we describe the trends in C, nitrogen (N), phosphorus (P), and potassium (K) releases based on a 3-year-long field study of litter decomposition in a peatland of the Italian Alps.

2. MASS LOSS

Decomposition of plant litter "involves a complex set of processes including chemical, physical, and biological agents acting upon a variety of organic substrates that are themselves constantly changing" [*Berg and McClaugherty*, 2003]. It is clear from the above definition that decomposition can be highly variable in relation to the spatial and temporal diversity of interacting abiotic and biotic factors.

A simple way to estimate decomposition rate is by measuring litter mass loss, an estimate typically obtained by means of litter bags. The litter bag technique consists in confining fresh litter in mesh bags that are placed on the ground and periodically collected so as to measure the remaining litter mass and associated litter chemistry [*Singh and Gupta*, 1977]. This simple and cheap technique has been, however, criticized because (1) confined litter bags may create their own microenvironment different from surrounding bulk soil; (2) litter bags may exclude specific faunal groups in relation to the mesh size [*Nieminen and Setala*, 1997; *Bradford et al.*, 2002]; and (3) litter bags are usually filled with litter from a single species [*Gartner and Cardon*, 2004]. Nevertheless, the litter bag technique is widely applied to monitor temporal mass loss in both terrestrial and aquatic environments.

Different mathematical models have been proposed to describe litter mass loss based on information obtained from litter bags [*Wieder and Lang*, 1982; *Berg and McClaugherty*, 2003]. The most commonly applied model, particularly for early stages of decomposition, is the single exponential model [*Olson*, 1963], which is supposed to work well for a variety of litter types until about 80% of initial litter is lost.

In the following, we will try to review how and to what extent mass loss of recently deposited litter is affected by regional and local drivers, an essential background to understand the potential response of peatlands to global change.

2.1. Climate

The role of climatic factors in affecting litter mass loss has received much attention in ecological studies so that, at a global scale, mean annual temperature and actual evapotranspiration were shown to have the strongest influence [*Meentemeyer*, 1978; *Aerts*, 1997; *Gholz et al.*, 2000; *Liski et al.*, 2003].

For assessing the role of climate on litter mass loss in peatlands, we related the annual k decomposition constant [*Olson*, 1963] from seven major terrestrial ecosystem types to corresponding mean annual temperature and mean total annual precipitation. To this aim, we selected data where (1) the litter bag technique was applied to estimate litter mass loss; (2) the k values were calculated on the basis of the single exponential model; (3) the decomposition was monitored at least for 2 years or longer, so as to include changes associated both with physical leaching and with microbial activity [*Siegenthaler et al.*, 2001].

We were able to select a total of 179 k values from 40 published and unpublished papers fulfilling the above criteria (data are available on request). Particularly for peatlands, most of the selected papers monitored litter mass loss during a burial period varying from 2 to 5 years, except for the paper by *Latter et al.* [1998] where mass loss was monitored for 22 years.

Across the seven ecosystem types, k values were positively correlated with the mean annual temperature (Pearson's $r = 0.37$; $p < 0.001$; $n = 179$) but not with mean total annual precipitation (Figure 1). In the specific case of peatlands, the mean k value of vascular plant litter (0.225 yr^{-1}) did not differ significantly from the mean k values of marshlands, Mediterranean shrublands, and deserts (Figure 2), whereas the mean k value of *Sphagnum* litter (0.081 yr^{-1}) was significantly lower compared to the k values of all the other ecosystem types including the mean k value of vascular plant litter (Figure 2). In addition, k values of vascular plant litter in peatlands were positively related with both the mean annual temperature (Pearson's $r = 0.37$; $p < 0.001$; $n = 69$) and the mean total annual precipitation (Pearson's $r = 0.39$; $p < 0.001$; $n = 69$), whereas the k values of *Sphagnum* litter were not correlated to either of these climatic factors. Increased decomposition rates of vascular plant litter in peatlands with increasing temperature were also reported by other authors [*Hobbie*, 1996; *Thormann et al.*, 2004; *Moore et al.*, 2005; *Breeuwer et al.*, 2008]. Furthermore, many laboratory and field experiments have shown that higher temperature resulted in increasing decomposition of vascular plant and

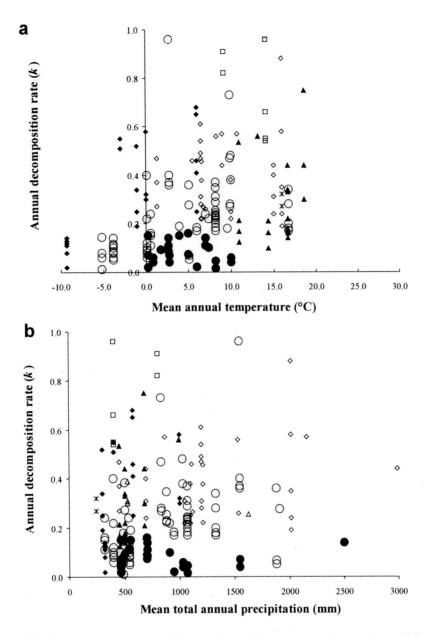

Figure 1. Relationship between annual decomposition k value and (a) mean annual temperature and (b) mean total annual precipitation for different plant species in seven major ecosystem types. Decomposition k values were obtained from studies lasting more than 1 year. For peatlands, k values of *Sphagnum* species (solid circles) and vascular plant species (open circles) were reported separately. Other symbols are as follows: open diamonds, forests; solid triangles, Mediterranean shrublands; open squares, grasslands; asterisks, deserts; open triangles, marshlands; solid diamonds, tundra. Sources are *Latter and Cragg* [1967], *Rochefort et al.* [1990], *Johnson and Damman* [1991], *Gallardo and Merino* [1993], *Aerts and De Caluwe* [1997], *Foote and Reynolds* [1997], *Wrubleski et al.* [1997], *Latter et al.* [1998], *Gholz et al.* [2000], *Moro and Domingo* [2000], *Scheffer et al.* [2001], *Seastedt and Adams* [2001], *Thormann et al.* [2001], *Aerts et al.* [2003], *Bridgham and Richardson* [2003], *Kemp et al.* [2003], *Koukoura et al.* [2003], *Quested et al.* [2003], *Albers et al.* [2004], *Heim and Frey* [2004], *Hobbie and Gough* [2004], *Palviainen et al.* [2004], *Asada and Warner* [2005], *Fioretto et al.* [2005], *Moore et al.* [2005], *Quideau et al.* [2005], *Sariyildiz et al.* [2005], *Aerts et al.* [2006], *Weerakkody and Parkinson* [2006], *Bokhorst et al.* [2007], *Bubier et al.* [2007], *Cortez et al.* [2007], *Lindo and Winchester* [2007], *Moore et al.* [2007], *Breeuwer et al.* [2008], *Buttler* [1987], *Miyamoto and Hiura* [2008], *Sariyildiz* [2008], *Turetsky et al.* [2008], Bragazza, unpublished data [2008].

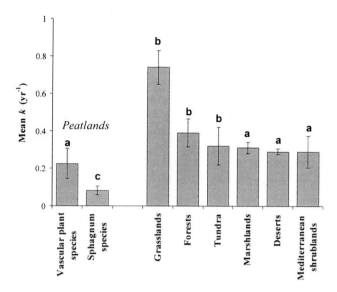

Figure 2. Mean annual *k* values (plus/minus SD) of plant litter in seven major ecosystem types. The analysis of variance was applied to compare mean *k* values of *Sphagnum* litter and vascular plant litter in peatlands with other ecosystems. The different superscripts indicate significant differences based on LSD Fisher post hoc test ($p < 0.05$). For sources, see Figure 1.

Sphagnum litter under nonlimiting moisture conditions [*Dioumaeva et al.*, 2003; *Domisch et al.*, 2006; *Glatzel et al.*, 2006] and that under excess drought the decomposition of organic matter is hampered [*Laiho et al.*, 2004; *Aerts*, 2006; *Gerdol et al.*, 2008]. On the other hand, considering the sole *Sphagnum* litter, the absence of relationship between its decay and climatic factors seems to highlight the prevailing role of litter quality in controlling *Sphagnum* decomposition in peatlands (see below).

2.2. Litter Decomposition in Aerobic Versus Anaerobic Conditions

The intensity of water flow and the degree of hydric saturation of the peat substrate are supposed to affect the rate of plant litter decomposition in peatlands. Indeed, laboratory and field experiments have demonstrated that mass loss is higher under aerobic conditions than under anaerobic conditions [*Belyea*, 1996; *Scanlon and Moore*, 2000; *Yavitt et al.*, 2000; *Blodau et al.*, 2004; *Laiho*, 2006; *Moore and Basiliko*, 2006; *Jaatinen et al.*, 2008; *Wickland and Neff*, 2008]. In Figure 3, mass losses from vascular plant litter and *Sphagnum* litter decomposing in parallel under aerobic and anaerobic conditions are compared. The percentage of mass loss was on average about 2.3 times higher in the aerobic layer

both for vascular plant and *Sphagnum* litter. It follows that the faster the decomposing plant litter enters the anaerobic zone to become saturated by water, the higher the rate of peat accumulation will be as a consequence of reduced mass loss. It is then clear that any disturbance lowering the water table can enhance, at least temporarily, the decomposition of previously water-saturated peat which becomes exposed to aerobic conditions [*Laine et al.*, this volume].

If aerobic conditions enhance litter decomposition, this raises the question of why decomposition rates of *Sphagnum* litter is generally lower in hummocks, i.e., in habitats well aerated for most of the year, compared to the litter produced by *Sphagnum* species inhabiting moister habitats such as hollows [*Johnson and Damman*, 1991; *Belyea*, 1996; *Asada and Warner*, 2005; *Dorrepaal et al.*, 2005; *Bragazza et al.*, 2007; *Moore et al.*, 2007; *Breeuwer et al.*, 2008]. This can be explained taking into account the role of litter chemistry, in particular the relatively higher content of structural carbohydrates in *Sphagnum* species forming high hummocks [*Turetsky et al.*, 2008]. Thus, the complex interactions between physical conditions at habitat scale and *Sphagnum* litter chemistry play an important role in controlling the topographical variability of peatland surface and associated rates on peat accumulation [*Ohlson and Dahlberg*, 1991; *Belyea and Malmer*, 2004].

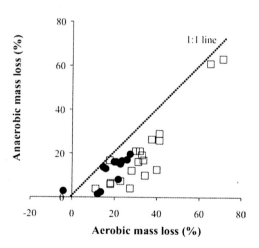

Figure 3. Comparison of mass loss (percent of initial dry weight) of vascular plant litter (open squares) and *Sphagnum* litter (solid circles) buried simultaneously both in aerobic and anaerobic conditions. Data were obtained using litter bags with different burial time intervals and different mesh size. The dotted line shows the 1:1 ratio. Sources are *Clymo* [1965], *Johnson and Damman* [1991], *Asada and Warner* [2005], *Moore et al.* [2007], *Trinder et al.* [2008], Bragazza, unpublished data [2008].

2.3. Litter Chemistry

The role of initial litter chemistry in affecting decomposition in peatlands will be separately reviewed for vascular plant litter and *Sphagnum* litter in the light of their significantly different *k* values (Figure 2).

Considering the most common stoichiometric parameters [*Hessen et al.*, 2004], it appears that mass loss of vascular plant litter is negatively correlated with the initial value of the lignin/N quotient (Pearson's $r = -0.46$; $p < 0.05$; $n = 27$), but positively correlated with the initial value of the holocellulose/lignin quotient (Pearson's $r = 0.59$; $p < 0.05$; $n = 13$; Figure 4). It is here necessary to underline that the sources selected in Figure 4 did not apply the same protocol for lignin quantification in initial litter. Indeed, more often the term "lignin" is operationally defined as the acid unhydrolyzable residue (AUR) of a sequential fractionation analysis so that AUR incorporates both true lignin as well as compounds such as cutins and tannins. Nevertheless, on the basis of our results, it seems that higher initial values of the AUR/N quotient hamper litter decomposition as a consequence of the lower chemical quality of plant litter [*Melillo*

et al., 1982; *Aerts and De Caluwe*, 1997; *Thorman et al.*, 2001; *Siegenthaler et al.*, 2001; *Cornelissen et al.*, 2004]. On the other hand, a higher initial holocellulose/AUR quotient in vascular plant litter seems to enhance decomposition because microbes have access to more easily decomposable C compounds [*Melillo et al.*, 1989; *Bridgham and Richardson*, 2003; *Comont et al.*, 2006; see also *Artz*, this volume].

Decomposition of *Sphagnum* litter has been demonstrated to be enhanced, at least after 1 year of field burial, by low initial values of phenolics/nutrient and C/nutrient quotients [*Szumigalski and Bayley*, 1996; *Limpens and Berendse*, 2003; *Dorrepaal et al.*, 2005; *Bragazza et al.*, 2007]. On the basis of selected data, we observed that *k* values of *Sphagnum* litter were negatively correlated with initial AUR/N quotients (Pearson's $r = -0.66$; $p < 0.05$; $n = 11$; Figure 4). Anyway, we must underline that (1) Sphagna do not contain true lignin as do vascular plants, so that the AUR in *Sphagnum* litter mainly refers to lignin-like polymeric phenolics [*Bland et al.*, 1968; *van der Heijden*, 1994], and (2) as mentioned above, the methods used to quantify lignin-like compounds, and therefore the results, vary among authors. The role of the phenolics/N quotient in affecting *Sphagnum* litter decomposition is confirmed by higher decomposability of litter deposited by minerotrophic *Sphagnum* species compared to ombrotrophic *Sphagnum* litter [*Bragazza et al.*, 2007], the latter being characterized by a higher phenolic content [*Rudolph and Samland*, 1985]. It has also been suggested that a better predictor of moss decomposition rate in peatlands is represented by the quotient between fructose/pentose carbohydrates in initial litter [*Turetsky et al.*, 2008], so that moss species investing relatively more in structural carbohydrates (i.e., pentosans) such as hummock-forming Sphagna are more resistant than moss species investing relatively more in metabolic carbohydrates (i.e., fructosans), such as hollow inhabiting Sphagna. In this review, the limited data did not reveal a significant role for C/nutrient quotients on *Sphagnum* litter decomposition, although the C/P and C/N quotient have been reported to significantly explain mass loss of *Sphagnum* litter after 1 year of field burial [*Hogg et al.*, 1994; *Aerts et al.*, 2001; *Limpens and Berendse*, 2003; *Bragazza et al.*, 2007].

From a global change perspective, increasing atmospheric N deposition and global climate change can have contrasting effects on litter decomposition in peatlands. Indeed, increasing N availability can be expected to enhance short-term *Sphagnum* litter decomposition through an increase of N content [*Williams et al.*, 1999; *Bragazza et al.*, 2006; *Gerdol et al.*, 2007] and a decrease of (soluble) phenolics [*Bragazza and Freeman*, 2007], thereafter reducing the phenolics/N and the C/nutrient quotients. On the other hand, *Sphagnum* litter decomposition is strongly affected by its water content,

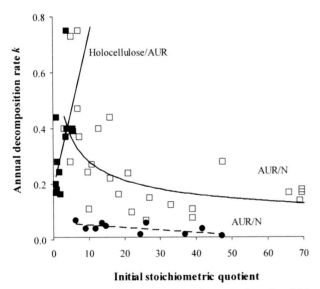

Figure 4. Relationship between initial values of selected stoichiometric quotients and annual *k* values of vascular plant litter (open and solid squares) and *Sphagnum* litter (solid circles) based on studies in which mass loss was monitored for more than 1 year. AUR is acid unhydrolyzable residue. Sources are *Aerts and De Caluwe* [1997], *Scheffer et al.* [2001], *Thormann et al.* [2001], *Bridgham and Richardson* [2003], *Aerts et al.* [2006], *Bubier et al.* [2007], *Breeuwer et al.* [2008], *Turetsky et al.* [2008], Bragazza, unpublished data [2008].

so that below a certain level of litter moisture, decomposition initially slows down and then stops completely [*Crow and Wieder*, 2005; *Aerts*, 2006; *Gerdol et al.*, 2007]. For vascular plants, increasing N availability had variable species-specific effects on mass loss depending on the chemistry of initial litter [*Aerts et al.*, 2006; *Breeuwer et al.*, 2008]. At ecosystem level, increased N deposition can provoke a shift in species composition, with vascular plants or brown mosses displacing *Sphagnum* species as a result of efficient use of nitrogen [*Berendse et al.*, 2001; *Mitchell et al.*, 2002; *Bubier et al.*, 2007], and this ultimately will change the litter quality of both individual species and the plant community.

As previously underlined, this chapter centers on aboveground litter decomposition because most of the studies on litter decomposition in peatlands deal with leaf decay. Nevertheless, belowground litter derived from roots and rhizomes can play an important role in C sequestration and nutrient cycling, particularly in fens, i.e., peatlands receiving mineral inputs from surrounding mineral soil, where belowground production represents the greatest part of total NPP [*Buttler*, 1992; *Wieder*, 2006]. At the global scale, it has been suggested that leaf litter decomposition is more controlled by (micro)climate factors, whereas belowground litter decomposition is more affected by root and rhizome chemical quality [*Whendee and Miya*, 2001]. Particularly in peatlands, the very few studies on belowground litter decomposition [e.g., *Hartmann*, 1999; *Scheffer and Aerts*, 2000; *Thormann et al.*, 2001; *Moore et al.*, 2007] seem to indicate that mass loss of roots and/or rhizomes is lower than the corresponding leaf mass loss, particularly for vascular plant species adapted to waterlogged soils. In addition, *Scheffer and Aerts* [2000] reported a greater net loss of N and P from decomposition of roots and rhizomes compared to leaves. Although more studies are necessary on belowground litter decomposition, the few available results highlight the important role of belowground litter for C accumulation and nutrient cycling, particularly in fens.

3. NUTRIENT RELEASE

The supply of nutrients in peatlands is provided by external and internal sources. External sources are represented by atmospheric wet and dry deposition as well as by groundwater influxes from the surrounding mineral soil. In bogs (i.e., ombrotrophic peatlands), atmospheric deposition is the primary external source of nutrients, whereas atmospheric deposition and groundwater inputs represent the external sources supplying fens (i.e., minerotrophic peatlands) with nutrients. Internal sources are, instead, associated with nutrient release during organic matter decomposition. The rate of nutrient release is not necessarily equal to the rate

of mass loss because some nutrients can be immobilized by microbes and incorporated in humic compounds rather than being mineralized during mass loss [*Jonasson and Shaver*, 1999]. The degree of nutrient release during litter decomposition can then affect nutrient availability for plant growth and, ultimately, the structure and the functioning of plant communities [*Aerts et al.*, 1999; *Parton et al.*, 2007].

In this section, we will assess the rates of release of C, N, P, and K during (aboveground) litter decomposition in peatlands. To this aim, we will use data from a 3-year long litter decomposition experiment carried out in a peatland of the Italian Alps.

3.1. Study Area, Material, and Methods

The Marcesina peatland (45°57′N; 11°37′E) is located at 1300 m above sea level under climatic conditions characterized by a mean annual temperature of 3°C and total annual precipitation of 1550 mm. On the basis of the floristic composition, the study site can be classified as an ombrotrophic peatland or bog.

For the aims of this study, two *Sphagnum* species (i.e., *S. magellanicum* and *S. fuscum*) and four vascular plant species (i.e., *Potentilla erecta*, *Eriophorum vaginatum*, *Calluna vulgaris*, and *Carex rostrata*) were selected. Litter samples from each plant species were collected in mid-September 2004 so as to prepare litter bags (mesh size = 0.5 mm), which were then buried at the beginning of October 2004 just below the bog surface in the typical habitat of each plant species. The total number of litter bags was 48 for *Sphagnum* litter and 96 for vascular plant litter.

At the beginning of October 2005, 2006, and 2007, eight litter bags for each species were retrieved, and mass losses as well as N, P, K, and C concentrations were determined. For technical details concerning litter bag preparation, cleaning, and chemical analyses, see *Bragazza et al.* [2007].

For each litter bag, nutrient release was calculated as: $[(M_0C_0 - M_tC_t) / M_0C_0] \times 100$ where M_0 and M_t are, respectively, the oven-dry mass of litter in the bag at the beginning of the experiment and at time t (i.e., years 1, 2, and 3), whereas C_0 and C_t are the corresponding nutrient concentrations in the litter. Positive values of release indicate net loss of the nutrient, whereas negative values indicate immobilization of the nutrient.

3.2. Results and Discussion

During 3 years of field decomposition, the cumulative release, i.e., at time t, of C, N, P, and K was always positive, therefore indicating a net loss of these nutrients from decomposing litter (Figure 5). However, at the end of the third year

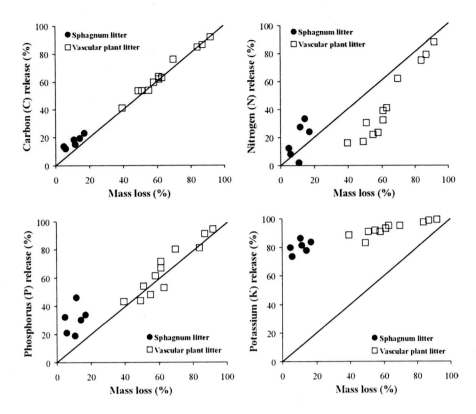

Figure 5. Relationship between mass loss after 1, 2, and 3 years of decomposition and corresponding release of C, N, P, and K from two *Sphagnum* species (i.e., *S. fuscum* and *S. magellanicum* litter) and four vascular plant species (i.e., *Potentilla erecta*, *Carex rostrata*, *Eriophorum vaginatum*, and *Calluna vulgaris* litter). Data represent, for each plant species, mean values calculated on eight annually retrieved litter bags. The line corresponds to the 1:1 ratio.

of burial, the release of C, N, P, and K was significantly lower in *Sphagnum* litter than in vascular plant litter, suggesting a relatively lower rate of nutrient loss from *Sphagnum* litter (Figure 5). Similarly, the mean annual mass loss of *Sphagnum* litter (11.2%, standard deviation (SD) = 4.7; $n = 48$) was significantly lower ($p < 0.001$) than the mean annual mass loss of vascular plant litter (64.5%, SD = 16.3; $n = 96$).

Independently from litter type, the mean annual release of K (89% ± 7%) was significantly greater ($p < 0.01$) compared to all other nutrients (C = 49.5% ± 27%; N = 35% ± 24%; P = 54% ± 22%). The high leaching of K from decomposing litter was already reported [*Brock and Bregman*, 1989; *Sundstrom et al.*, 2000; *Bragazza et al.*, 2007], and it seems to explain the peaks of K concentration in peatland waters typically found outside the growing season, when plant growth has not yet started, but K is released by decomposing litter [*Buttler*, 1992; *Proctor*, 1992; *Vitt et al.*, 1995; *Bragazza et al.*, 1998].

In the case of vascular plant litter, the cumulative release of C and P paralleled the mass loss (i.e., the nutrient release/ mass loss quotient was always around 1 during the decom-

position period), whereas the cumulative release of N and K was, respectively, lower and higher than the corresponding mass loss (Fig. 5). Instead, in the case of *Sphagnum* litter, the cumulative release of N, P, and K was greater than the corresponding mass loss (i.e., the nutrient release/mass loss quotient was always > 1), whereas the cumulative C release tended to parallel the mass loss (Figure 5). Different trends in nutrient release between *Sphagnum* litter and vascular plant litter can initially be explained by the absence of protective tissues and waxes on *Sphagnum* leaves allowing a rapid physical leaching of N and P [*Scheffer et al.*, 2001].

Vascular plant litter had initial C/N quotients ranging from 38 to 48, and the absence of N immobilization is in accordance with findings of *Parton et al.* [2007], who reported net N loss with leaf litter C/N quotient < 40. For *Sphagnum* litter, during the first year of decomposition, immobilization is absent as already reported by *Brock and Bregman* [1989] and *Vehoeven et al.* [1990], but N release tended to decrease over time (Figure 6). More precisely, with increasing mass loss, the AUR/N quotient increased significantly for both *Sphagnum* litter and vascular plant litter, but the corresponding

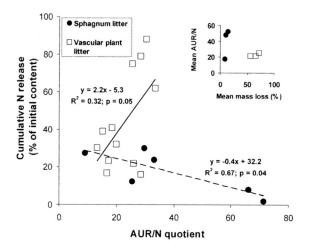

Figure 6. Relationship between AUR/N quotient after 1, 2, and 3 years of decomposition and corresponding N release for *Sphagnum* litter and vascular plant litter. Each value for each species is the mean based on eight litter bags annually retrieved. The inner graph represents the relationship between cumulative mean mass loss and corresponding AUR/N quotient for *Sphagnum* litter and vascular plant litter (in this case the individual species have been merged for *Sphagnum* litter and vascular plant litter).

cumulative N release tended to decrease for *Sphagnum* litter and to increase for vascular plant litter, therefore indicating that, over time, *Sphagnum* litter tends to retain more N (Figure 6).

Carbon released from both *Sphagnum* and vascular plant litter was much more linearly related to mass loss than for P, K, and N (Figure 5). Accordingly, the rate of C release effectively indicates the rate of mass loss [*Bragazza et al.*, 2007].

Phosphorus release was enhanced by low initial C/P quotient (Figure 7), suggesting that litter with greater initial C/P quotient tends to lose P at lower rates [*Aerts and De Caluwe*, 1997; *Bragazza et al.*, 2007]. It is generally assumed that P release in bogs is higher than in fens because of a geochemical P immobilization in fen habitats through the formation of complexes with Fe, Ca, or Al, which allows a faster turnover of P in ombrotrophic habitats compared to minerotrophic habitats [*Verhoeven et al.*, 1990; *Bridgham et al.*, 1998; *Scheffer et al.*, 2001; *Bragazza et al.*, 2007].

4. CONCLUSIONS

The low rate of plant litter decomposition in actively growing peatlands is not only a peculiarity of this type of ecosystem, but it is also the key process to ensure peat accumulation and, consequently, to guarantee that active peat-

lands act as C sinks. Litter quality and anoxic soil conditions appear as the major factors affecting decomposition rates in peatlands. It has been shown that the peculiar chemistry of *Sphagnum* litter is fundamental for enhancing long-term peat accumulation. Accordingly, any environmental change favoring vascular plants at the expense of *Sphagnum* plants can potentially reduce peat accumulation by increasing the rates of litter decomposition. This is particularly true in bogs where *Sphagnum* litter form the bulk of dead biomass. At the same time, changes in litter chemistry can also affect nutrient release and therefore nutrient availability in peatlands, thus potentially altering the competitive balance between different plant species.

Our review has also identified three important current limitations in studies of litter decomposition and associated nutrient release in peatlands: (1) There are relatively few studies monitoring decomposition in peatlands for periods longer than 1 year, so that the patterns and mechanisms of longer-term litter decomposition are poorly known. (2) There are few data on belowground litter decomposition in peatlands, so that we currently miss the functioning of an important component for peatland biogeochemistry, especially for fens where belowground litter production is highest. (3) For some chemical analyses, such as for lignin (AUR) and polyphenols, the methodologies are not standardized. Hence, we suggest that future studies should extend the burial time of litter bags and pay more attention to belowground

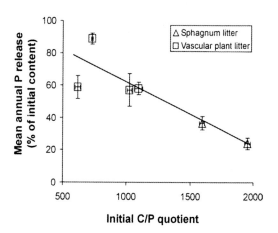

Figure 7. Relationship between mean (plus/minus SD) annual release of phosphorus (P) during 3 years of decomposition and corresponding mean (plus/minus SD) initial C/P quotient in four vascular plant litter types and two *Sphagnum* litter types ($Y = -0.04X + 97.8$; $r^2 = 0.76$; $p = 0.02$). The values of annual P release were calculated on 24 litter bags for each plant species, whereas the mean initial C/P quotient was calculated on three initial litter samples for each plant species.

litter decomposition. In addition, we call for the definition of a standard protocol for the chemical characterization and quantification of plant litter components, particularly for lignin (AUR) and polyphenols, so as to make comparisons among studies more reliable.

REFERENCES

Aerts, R. (1997), Climate, leaf litter chemistry and leaf litter decomposition in terrestrial ecosystems: A triangular relationship, *Oikos*, *79*, 439–449.

Aerts, R. (2006), The freezer defrosting: Global warming and litter decomposition in cold biomes, *J. Ecol.*, *94*, 713–724.

Aerts, R., and H. De Caluwe (1997), Nutritional and plant-mediated controls on leaf litter decomposition of *Carex* species, *Ecology*, *78*, 244–260.

Aerts, R., J. T. A. Verhoeven, and D. F. Whigham (1999), Plant-mediated controls on nutrient cycling in temperate fens and bogs, *Ecology*, *80*, 2170–2181.

Aerts, R., B. Wallen, N. Malmer, and H. De Caluwe (2001), Nutritional constraints on *Sphagnum*-growth and potential decay in northern peatlands, *J. Ecol.*, *89*, 292–299.

Aerts, R., H. De Caluwe, and B. Beltman (2003), Plant community mediated vs. nutritional controls on litter decomposition rates in grasslands, *Ecology*, *84*, 3198–3208.

Aerts, R., R. S. P. van Logtestijn, and P. S. Karlsson (2006), Nitrogen supply differentially affects litter decomposition rates and nitrogen dynamics of sub-arctic bog species, *Oecologia*, *146*, 652–658.

Albers, D., S. Migge, M. Schaefer, and S. Scheu (2004), Decomposition of beech leaves (*Fagus sylvatica*) and spruce needles (*Picea abies*) in pure and mixed stands of beech and spruce, *Soil Biol. Biochem.*, *36*, 155–164.

Artz, R. R. E. (2009), Microbial community structure and carbon substrate use in northern peatlands, *Geophys. Monogr. Ser.*, doi:10.1029/2008GM000806, this volume.

Asada, T., and B. G. Warner (2005), Surface peat mass and carbon balance in a hypermaritime peatland, *Soil Sci. Soc. Am. J.*, *69*, 549–562.

Batjes, N. H. (1996), Total carbon and nitrogen in the soils of the world, *Eur. J. Soil Sci.*, *47*, 151–163.

Belyea, L. R. (1996), Separating the effects of litter quality and microenvironment on decomposition rates in patterned peatland, *Oikos*, *77*, 529–539.

Belyea, L. R., and N. Malmer (2004), Carbon sequestration in peatland: Patterns and mechanisms of response to climate change, *Global Change Biol.*, *10*, 1043–1052.

Berendse, F., et al. (2001), Raised atmospheric CO_2 levels and increased N deposition cause shifts in plant species composition and production in *Sphagnum* bogs, *Global Change Biol.*, *7*, 591–598.

Berg, B., and C. McClaugherty (2003), *Plant Litter: Decomposition, Humus Formation, Carbon Sequestration*, 286 pp., Springer, Heidelberg, Germany.

Bland, D. E., A. Logan, M. Menshum, and S. Sternhell (1968), The lignin of *Sphagnum*, *Phytochemistry*, *7*, 1373–1377.

Blodau, C., N. Basiliko, and T. R. Moore (2004), Carbon turnover in peatland mesocosms exposed to different water table levels, *Biogeochemistry*, *67*, 331–351.

Bokhorst, S., A. Huiskes, P. Convey, and R. Aerts (2007), Climate change effects on organic matter decomposition rates in ecosystems from Maritime Arctic and Falkland Islands, *Global Change Biol.*, *13*, 2642–2653.

Bradford, M. A., G. M. Tordoff, T. Eggers, T. H. Jones, and E. Newington (2002), Microbiota, fauna, and mesh size interactions in litter decomposition, *Oikos*, *99*, 317–323.

Bragazza, L., and C. Freeman (2007), High nitrogen availability reduces polyphenol content in *Sphagnum* peat, *Sci. Total Environ.*, *377*, 439–443.

Bragazza, L., R. Alber, and R. Gerdol (1998), Seasonal chemistry of pore water in hummocks and hollows in a poor mire in the southern Alps (Italy), *Wetlands*, *18*, 320–328.

Bragazza, L., et al. (2006), Atmospheric nitrogen deposition promotes carbon loss from peat bogs, *Proc. Natl. Acad. Sci. U. S. A.*, *103*, 19,386–19,389.

Bragazza, L., C. Siffi, P. Iacumin, and R. Gerdol (2007), Mass loss and nutrient release during litter decay in peatlands: The role of microbial adaptability to litter chemistry, *Soil Biol. Biochem.*, *39*, 257–267.

Breeuwer, A., M. Heijmans, B. J. M. Robroek, J. Limpens, and F. Berendse (2008), The effect of increased temperature and nitrogen deposition on decomposition in bogs, *Oikos*, *117*, 1258–1268.

Bridgham, S. D., and C. J. Richardson (2003), Endogenous versus exogenous nutrient control over decomposition and mineralization in North Carolina peatlands, *Biogeochemistry*, *65*, 151–178.

Bridgham, S. D., K. Updegraff, and J. Pastor (1998), Carbon, nitrogen, and phosphorus mineralization in northern wetlands, *Ecology*, *79*, 1545–1561.

Brock, T. M. C., and R. Bregman (1989), Periodicity in growth, productivity, nutrient content and decomposition of *Sphagnum recurvum* var *mucronatum* in a fen woodland, *Oecologia*, *80*, 44–52.

Bubier, J. L., T. R. Moore, and L. A. Bledzki (2007), Effects of nutrient addition on vegetation and carbon cycling in an ombrotrophic bog, *Global Change Biol.*, *13*, 1168–1186.

Buttler, A. (1987), Etude écosystémique des marais non boisés de la rive sud du lac de Neuchâtel (Suisse): phytosociologie, pédologie, hydrodynamique et hydrochimie, production végétale, cycles biogéochimiques et influence du fauchage sur la végétation, Ph.D. thesis, 284 pp., Univ. of Neuchâtel, Neuchâtel, Switzerland.

Buttler, A. (1992), Hydrochimie des nappes des prairies humides de la rive sud du lac de Neuchâtel, *Bull. Ecol.*, *23*, 35–47.

Clymo, R. S. (1965), Experiments on breakdown of *Sphagnum* in two bogs, *J. Ecol.*, *53*, 747–758.

Clymo, R. S., J. Turunen, and K. Tolonen (1998), Carbon accumulation in peatland, *Oikos*, *81*, 368–388.

Comont, L., F. Laggoun-Défarge, and J.-R. Disnar (2006), Evolution of organic matter indicators in response to major environmental changes: The case of a formerly cut-over peat bog (Le Russey, Jura Mountains, France), *Org. Geochem.*, *37*, 1736–1751.

Cornelissen, J. H. C., H. M. Quested, D. Gwyynn-Jones, R. S. P. van Logtestijn, M. A. H. De Beus, A. Kondratchuk, T. V. Callaghan, and R. Aerts (2004), Leaf digestibility and litter decomposability are related in a wide range of subarctic plant species and types, *Funct. Ecol.*, *18*, 779–786.

Cortez, J., E. Garnier, N. Perez-Harguindeguy, M. Debussche, and D. Gillon (2007), Plant traits, litter quality and decomposition in a Mediterranean old-field succession, *Plant Soil*, *296*, 19–34.

Crow, S. E., and K. Wieder (2005), Sources of CO_2 emission from a northern peatland: Root respiration, exudation, and decomposition, *Ecology*, *86*, 1825–1834.

Dioumaeva, I., S. Trumbore, E. A. G. Schuur, M. L. Goulden, M. Litvak, and A. I. Hirsch (2003), Decomposition of peat from upland boreal forest: Temperature dependence and sources of respired carbon, *J. Geophys. Res.*, *108*(D3), 8222, doi:10.1029/2001JD000848.

Domisch, T., L. Finer, J. Laine, and R. Laiho (2006), Decomposition and nitrogen dynamics of litter in peat soils from two climatic regions under different temperature regimes, *Eur. J. Soil Biol.*, *42*, 74–81.

Dorrepaal, E., J. H. C. Cornelissen, R. Aerts, B. Wallen, and R. S. P. van Logtestijn (2005), Are growth forms consistent predictors of leaf litter quality and decomposability across peatlands along a latitudinal gradient?, *J. Ecol.*, *93*, 817–828.

Fenner, N., C. Freeman, and F. Worrall (2009), Hydrological controls on dissolved organic carbon production and release from UK peatlands, *Geophys. Monogr. Ser.*, doi:10.1029/2008GM000823, this volume.

Fioretto A., C. Di Nardo, S. Papa, and A. Fuggi (2005), Lignin and cellulose degradation and nitrogen dynamics during decomposition of three leaf litter species in a Mediterranean ecosystem, *Soil Biol. Biochem.*, *37*, 1083–1091.

Foote A. L., and K. A. Reynolds (1997), Decomposition of saltmeadow cordgrass (*Spartina patens*) in Louisiana Coastal Marshes, *Estuaries*, *20*, 579–588.

Gallardo, A., and J. Merino (1993), Leaf decomposition in two Mediterranean ecosystems of southwest Spain: influence of substrate quality, *Ecology*, *74*, 152–161.

Gartner, T. B., and Z. G. Cardon (2004), Decomposition dynamics in mixed-species leaf litter, *Oikos*, *104*, 230–246.

Gerdol, R., A. Petraglia, L. Bragazza, P. Iacumin, and L. Brancaleoni (2007), Nitrogen deposition interacts with climate in affecting production and decomposition rates in *Sphagnum* mosses, *Global Change Biol.*, *13*, 1810–1821.

Gerdol, R., L. Bragazza, and L. Brancaleoni (2008), Heatwave 2003: High summer temperature, rather than experimental fertilization affects vegetation and CO_2 exchange in an alpine bog, *New Phytol.*, *179*, 142–154.

Gholz, H. L., D. A. Wedin, S. M. Smitherman, M. E. Harmon, and W. J. Parton (2000), Long-term dynamics of pine and hardwood litter in contrasting environments: Toward a global model of decomposition, *Global Change Biol.*, *6*, 751–765.

Glatzel, S., S. Lemke, and G. Gerold (2006), Short-term effects of an exceptionally hot and dry summer on decomposition of surface peat in a restored temperate bog, *Eur. J. Soil Biol.*, *42*, 219–229.

Hartmann, M. (1999), Species dependent root decomposition in re-wetted fen soils, *Plant Soil*, *213*, 93–98.

Heim, A., and B. Frey (2004), Early stage litter decomposition rates for Swiss forests, *Biogeochemistry*, *70*, 299–313.

Hessen, D. O., G. I. Agren, T. R. Anderson, J. J. Elser, and P. C. De Ruiter (2004), Carbon sequestration in ecosystems: The role of stoichiometry, *Ecology*, *85*, 1179–1192.

Hobbie, S. (1996), Temperature and plant species control over litter decomposition in Alaskan tundra, *Ecol. Monogr.*, *66*, 503–522.

Hobbie, S. E., and L. Gough (2004), Litter decomposition in moist acidic and non-acidic tundra with different glacial histories, *Oecologia*, *140*, 113–124.

Hogg, E. H., N. Malmer, and B. Wallen (1994), Regional and microsite variation in the potential decay rate of *Sphagnum magellanicum* in south Swedish raised bog, *Ecography*, *17*, 50–59.

Ito, A., and T. Oikawa (2004), Global mapping of terrestrial primary productivity and light-use efficiency with a processes-based model, in *Global Environmental Change in the Ocean and on Land*, edited by M. Shiyomi et al., pp. 343–358, Terra Sci., Tokyo.

Jaatinen, K., R. Laiho, A. Vuorenmaa, U. del Castillo, K. Minkinnen, T. Pennanen, T. Penttila, and H. Fritze (2008), Responses of aerobic microbial communities and soil respiration to water-level drawdown in a northern boreal fen, *Environ. Microbiol.*, *10*, 339–353.

Johnson, L. C., and A. W. H. Damman (1991), Species-controlled *Sphagnum* decay on a South Swedish raised bog, *Oikos*, *61*, 234–242.

Jonasson, S., and G. R. Shaver (1999), Within-stand nutrient cycling in arctic and boreal wetlands, *Ecology*, *80*, 2139–2150.

Kemp, P. R., J. F. Reynolds, R. A. Virginia, and W. G. Whitford (2003), Decomposition of leaf and root litter of Chihuahuan desert shrubs: Effects of three years of summer drought, *J. Arid Environ.*, *53*, 21–39.

Koukoura, Z., A. P. Mamolos, and K. L. Kalburtji (2003), Decomposition of dominant plant species litter in a semi-arid grassland, *Appl. Soil Ecol.*, *23*, 13–23.

Laiho, R. (2006), Decomposition in peatlands: Reconciling seemingly contrasting results on the impacts of lowered water levels, *Soil Biol. Biochem.*, *38*, 2011–2024.

Laiho, R., J. Laine, C. C. Trettin, and L. Finer (2004), Scots pine litter decomposition along drainage succession and soil nutrient gradients in peatland forests, and the effects of inter-annual weather variation, *Soil Biol. Biochem.*, *36*, 1095–1109.

Laine, J., K. Minkkinen, and C. Trettin (2009), Direct human impacts on the peatland carbon sink, *Geophys. Monogr. Ser.*, doi:10.1029/2008GM000808, this volume.

Latter, P. M., and J. B. Cragg (1967), The decomposition of *Juncus squarrosus* leaves and microbiological changes in the profile of *Juncus* moor, *J. Ecol.*, *55*, 465–482.

Latter, P. M., G. Howson, D. M. Howard, and W. A. Scott (1998), Long-term study of litter decomposition on a Pennine peat bog: Which regression?, *Oecologia*, *113*, 94–103.

Limpens, J., and F. Berendse (2003), How litter quality affects mass loss and N loss from decomposing *Sphagnum*, *Oikos*, *103*, 537–547.

Lindo, Z., and N. N. Winchester (2007), Oribatid mite communities and foliar litter decomposition in canopy suspended soils and forest floor habitats of western redcedar forests, Vancouver Island, Canada, *Soil Biol. Biochem.*, *39*, 2957–2966.

Liski, J., A. Nissenen, M. Erhard, and O. Taskinen (2003), Climatic effects on litter decomposition from arctic tundra to tropical rainforest, *Global Change Biol.*, *9*, 575–584.

Meentemeyer, V. (1978), Macroclimate and lignin control of litter decomposition rates, *Ecology*, *59*, 465–472.

Meentemeyer, V., E. O. Box, and R. Thompson (1982), World patterns and amounts of terrestrial plant litter production, *BioScience*, *32*, 125–128.

Melillo, J. M., J. D. Aber, and J. F. Muratore (1982), Nitrogen and lignin control on hardwood leaf litter decomposition dynamics, *Ecology*, *63*, 621–626.

Melillo, J. M., J. D. Aber, A. E. Linkins, A. Ricca, B. Fry, and K. J. Nadelhoffer (1989), Carbon and nitrogen dynamics along the decay continuum: Plant litter to soil organic matter, *Plant Soil*, *115*, 189–198.

Mitchell, E. A. D., P. Grosvernier, A. Buttler, A. Rydin, A. Siegenthaler, and J.-M. Gobat (2002), Contrasted effects of increased N and CO_2 supply on two keystone species in peatland restoration and implications for global change, *J. Ecol.*, *90*, 529–533.

Miyamoto, T., and T. Hiura (2008), Decomposition and nitrogen release from the foliage litter of fir (*Abies sachalinensis*) and oak (*Quercus crispula*) under different forest canopies in Hokkaido, Japan, *Ecol. Res.*, *23*, 673–680.

Moore, T. R. (2009), Dissolved organic carbon production and transport in Canadian peatlands, *Geophys. Monogr. Ser.*, doi:10.1029/2008GM000816, this volume.

Moore, T. R., and N. Basiliko (2006), Decomposition in Boreal Peatlands, in *Boreal Peatland Ecosystems*, edited by R. K. Wieder and D. H. Vitt, pp. 125–143, Springer, Heidelberg, Germany.

Moore, T. R., J. A. Trofymow, M. Siltanen, C. Prescott, and CIDET Working Group (2005), Patterns of decomposition and carbon, nitrogen, and phosphorus dynamics of litter in upland forest and peatland sites in central Canada, *Can. J. For. Res.*, *35*, 133–142.

Moore, T. R., J. L. Bubier, and L. Bledzki (2007), Litter decomposition in temperate peatland ecosystems: The effect of substrate and site, *Ecosystems*, *10*, 949–963.

Moro, M. J., and F. Domingo (2000), Litter decomposition in four woody species in a Mediterranean climate: Weight loss, N and P dynamics, *Ann. Bot.*, *86*, 1065–1071.

Nieminen, J. K., and H. Setala (1997), Enclosing decomposer food web: Implications for community structure and function, *Biol. Fertil. Soils*, *26*, 50–57.

Nilsson, M., and M. Öquist (2009), Partitioning litter mass loss into carbon dioxide and methane in peatland ecosystems, *Geophys. Monogr. Ser.*, doi:10.1029/2008GM000819, this volume.

Ohlson, M., and B. Dahlberg (1991), Rate of peat increment in hummock and lawn communities on Swedish mires during the last 150 years, *Oikos*, *61*, 369–378.

Olson, J. S. (1963), Energy storage and the balance of producers and decomposers in ecological systems, *Ecology*, *44*, 322–331.

Palviainen, M., L. Finer, A. M. Kurka, Il. Mannerkoski, S. Pirainen, and M. Starr (2004), Decomposition and nutrient release from logging residues after clear-cutting of mixed boreal forest, *Plant Soil*, *263*, 53–67.

Parton, W., et al. (2007), Global-scale similarities in nitrogen release patterns during long-term decomposition, *Science*, *135*, 361–364.

Proctor, M. C. F. (1992), Regional and local variation in the chemical composition of ombrogenous mire waters in Britain and Ireland, *J. Ecol.*, *80*, 719–736.

Quested, H. M., J. H. C. Cornelissen, M. C. Press, T. V. Callaghan, R. Aerts, F. Trosien, P. Riemann, D. Gwynn-Jones, A. Kondratchuk, and S. E. Jonasson (2003), Decomposition of sub-arctic plants with differing nitrogen economies: A functional role of hemiparasites, *Ecology*, *84*, 3209–3221.

Quideau, S. A., R. C. Graham, S.-W. Oh, P. F. Hendrix, and R. E. Wasylishen (2005), Leaf litter decomposition in a chaparral ecosystem, Southern California, *Soil Biol. Biochem.*, *37*, 1988–1998.

Rochefort, L., D. H. Vitt, and S. E. Bayley (1990), Growth, production, and decomposition dynamics of *Sphagnum* under natural and experimentally acidified conditions, *Ecology*, *71*, 1986–2000.

Rudolph, H., and J. Samland (1985), Occurrence and metabolism of sphagnum acid in the cell walls of bryophytes, *Phytochemistry*, *24*, 745–749.

Sariyildiz, T. (2008), Effects of gap-size classes on long-term litter decomposition rates of beech, oak and chestnut species at high elevations in northeast Turkey, *Ecosystems*, *11*, 841–853.

Sariyildiz, T., J. M. Anderson, and M. Kucuk (2005), Effects of tree species and topography on soil chemistry, litter quality and decomposition in Northeast Turkey, *Soil Biol. Biochem.*, *37*, 1695–1706.

Scanlon, D., and T. Moore (2000), Carbon dioxide production from peatland soil profiles: The influence of temperature, oxic/anoxic conditions and substrate, *Soil Sci.*, *165*, 153–160.

Scheffer, R. A., and R. Aerts (2000), Root decomposition and soil nutrient and carbon cycling in two temperate fen ecosystems, *Oikos*, *91*, 541–549.

Scheffer, R. A., R. S. P. van Logtestijn, and J. T. A. Verhoeven (2001), Decomposition of *Carex* and *Sphagnum* litter in to mesotrophic fens differing in dominant plant species, *Oikos*, *92*, 44–54.

Schlesinger, W. H. (1997), *Biogeochemistry: An Analysis of Global Change*, pp. 588, Academic, San Diego, Calif.

Seastedt, T. R., and G. Adams (2001), Effects of mobile tree islands on alpine tundra soils, *Ecology*, *82*, 8–17.

Siegenthaler, A., E. van der Heijden, E. A. D. Mitchell, A. Buttler, P. Grosvernier, and J.-M. Gobat (2001), Effects of elevated atmospheric CO_2 and mineral nitrogen deposition on litter quality, bioleaching and decomposition in a *Sphagnum* peat bog, in *Global Change and Protected Areas*, edited by G. Visconti et al., pp. 311–321, Springer, Dordrecht, Netherlands.

Singh, J. S., and S. R. Gupta (1977), Plant decomposition and soil respiration in terrestrial ecosystems, *Bot. Rev.*, *43*, 449–528.

Sundstrom, E., T. Magnusson, and B. Hånell (2000), Nutrient conditions in drained peatlands along a north-south climatic gradient in Sweden, *For. Ecol. Manage.*, *126*, 149–161.

Szumigalski, A. R., and S. E. Bayley (1996), Decomposition along a bog to rich fen gradient in central Alberta, Canada, *Can. J. Bot.*, *74*, 573–581.

Thormann, M. N., S. E. Bayley, and R. S. Currah (2001), Comparison of decomposition of belowground and aboveground plant litters in peatlands of boreal Alberta, *Can. J. Bot.*, *79*, 9–22.

Thormann, M. N., S. E. Bayley, and R. S. Currah (2004), Microcosm tests of the effects of temperature and microbial species number on the decomposition of *Carex aquatilis* and *Sphagnum fuscum* litter from southern boreal peatlands, *Can. J. Microbiol.*, *50*, 793–802.

Trinder, C. J., D. Johnson, and R. E. Artz (2008), Interactions among fungal community structure, litter decomposition and depth of water table in a cutover peatland, *FEMS Microbiol. Ecol.*, *64*, 433–448.

Turetsky, M. R., S. E. Crow, R. J. Evans, D. H. Vitt, and R. K. Wieder (2008), Trade-off in resource allocation among moss species control decomposition in boreal peatlands, *J. Ecol.*, *96*, 1297–1305.

Van der Heijden, E. (1994), A combined anatomical and pyrolysis mass spectrometric study of peatified plant tissues, Ph.D. thesis, 157 pp., Univ. of Amsterdam, Amsterdam.

Verhoeven, J., E. Maltby, and M. B. Schmidt (1990), Nitrogen and phosphorus mineralization in fens and bogs, *J. Ecol.*, *78*, 713–726.

Vitt, D. H., S. E. Bayley, and T. L. Jin (1995), Seasonal variation in water chemistry over a bog-rich fen gradient in continental western Canada, *Can. J. Fish. Aquat. Sci.*, *52*, 587–606.

Weerakkody, J., and D. Parkinson (2006), Leaf litter decomposition in an upper montane rainforest in Sri Lanka, *Pedobiologia*, *50*, 387–395.

Whendee, L. S., and R. K. Miya (2001), Global patterns in root decomposition: Comparisons of climate and litter quality effects, *Oecologia*, *129*, 407–419.

Wickland, K. P., and J. C. Neff (2008), Decomposition of soil organic matter from boreal black spruce forest: Environmental and chemical controls, *Biogeochemistry*, *87*, 29–47.

Wieder, R. K. (2006), Primary Production in Boreal Peatlands, in *Boreal Peatland Ecosystems*, edited by R. K. Wieder and D. H. Vitt, pp. 145–164, Springer, Heidelberg, Germany.

Wieder R. K., and G. E. Lang (1982), A critique of the analytical methods used in examining decomposition data obtained from litter bags, *Ecology*, *63*, 1636–1642.

Williams, B. L., A. Buttler, P. Grosvernier, A.-J. Francez, D. Gilbert, M. Ilomets, J. Jauhiainen, Y. Matthey, D. J. Silcock, and H. Vasander (1999), The fate of NH_4NO_3 added to *Sphagnum magellanicum* carpets at five European mire sites, *Biogeochemistry*, *45*, 73–93.

Wrubleski, D. A., H. R. Murkin, A. G. van der Valk, and C.B. Davis (1997), Decomposition of litter of three mudflat annual species in a northern prairie marsh during drawdown, *Plant Ecol.*, *129*, 141–148.

Yavitt, J. B., C. J. Williams, and R. K. Wieder (2000), Controls on microbial production of methane and carbon dioxide in three *Sphagnum*-dominated peatland ecosystems as revealed by a reciprocal field peat transplant experiment, *Geomicrobiol. J.*, *17*, 61–88.

L. Bragazza, Department of Biology and Evolution, University of Ferrara, Corso Ercole I d'Este 32, I-44100 Ferrara, Italy. (luca.bragazza@unife.it)

A. Buttler, Restoration Ecology Research Group, Swiss Federal Research Institute, WSL, Station 2, CH-1015 Lausanne, Switzerland.

E. A. D. Mitchell, Institute of Biology, University of Neuchâtel, Rue Emile-Argand 11, Case postale 158, CH-2009 Neuchâtel, Switzerland.

A. Siegenthaler, Laboratory of Ecological Systems, École Polytechnique Fédérale de Lausanne, CH-1015 Lausanne, Switzerland.

Microbial Community Structure and Carbon Substrate Use in Northern Peatlands

Rebekka R. E. Artz

The Macaulay Land Use Research Institute, Aberdeen, UK

The net C sink function of many northern peatlands is a product of an imperfect balance between net primary production and net respiration. This review summarizes our current knowledge of the microbial pathways of carbon flow in such ecosystems, the key players in the aerobic and anaerobic decomposition routes, and the drivers of microbial community structure and substrate use. The review identifies several areas in need of future research, notably on the key players and their activities in the periodically oxic and the permanently anoxic zones of peatlands.

1. INTRODUCTION

Peatlands form where the sequestration of photosynthetically fixed carbon (C) exceeds losses through decomposition. Conditions suitable to peatland formation have occurred over the last 11,000 years in the relatively cool but wet climatic zones in the northern regions of the world and have led to sometimes spectacular accumulations of peat deposits. In the boreal and subarctic regions, an average depth of 1.7 m of peat is often cited [*Turunen*, 2003], but figures of up to 6 or 7 m [e.g., *Beilman et al.*, 2008] have been reported. Together, these deposits of organic carbon in Northern peatlands are currently thought to harbor up to 25% of the global terrestrial C stocks [*Maltby and Immirzi*, 1993; *Gruber et al.*, 2004], although the figures carry a high level of uncertainty due to extrapolation from a limited database of peat depths and bulk density values [*Vasander and Kettunen*, 2006]. The current trends toward continuing increases in nitrogen deposition in many northern regions, as well as a growing awareness of the potential changes in climatic patterns caused by the cumulative emissions from fossil fuel use have sparked an interest into the likely fate of this vast C store.

Carbon Cycling in Northern Peatlands
Geophysical Monograph Series 184
Copyright 2009 by the American Geophysical Union.
10.1029/2008GM000806

Decomposition in peatlands is thought to be constrained by a combination of factors that can vary with the geographical location and source of aqueous inputs. First, available O_2 declines rapidly with depth due to permanent waterlogging (Figure 1), thus placing energy constraints on the metabolism of chemically complex molecules. Decomposition does partly proceed via pathways utilizing alternative electron acceptors (e.g., $Fe(III)$, SO_4^{3-}, NO_3^-) until low biological availability limits further decomposition (Figure 1). The relative importance of decomposition pathways utilizing alternative electron acceptors in the cycling of C depends on the trophic status of the peatland: as nutrient inputs in ombrotrophic peatlands (bogs) are entirely dependent on precipitation, these pathways are constrained by the low concentrations of such alternative electron acceptors (as long as deposition has not been altered, more about this below) and depend on recharge via oxidizing pathways within the soil. In minerotrophic (also sometimes termed rheotrophic) peatlands, i.e., fens, recharge of such pathways occurs also through inputs of external groundwater or overland flow. In addition, the low pH values of ombrotrophic peatlands, but also the generally low temperature, energetic constraints on bioavailable N and/or P from organic sources and the potential energy yield from available carbon substrates in peatlands add to constraints on microbial activity (Figure 1). Therefore, these factors may also be important in determining the size and composition of the microbial community involved in carbon

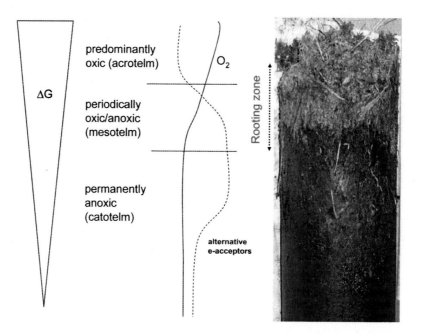

Figure 1. Schematic representation of the energy constraints placed by the intrinsic decomposability of C inputs, the location and temporal variation of the water table, and plant root extension. These define the division into the predominantly oxic surface peat above the water table (often referred to as "acrotelm"), the periodically oxic/anoxic interface zone (referred to as "mesotelm" by *Clymo and Bryant* [2008]) and the permanently anoxic zone ("catotelm").

and nutrient cycling. This review will address the current state of our knowledge of the community structure of peatland microbiota, their involvement in C cycling through their substrate use abilities, and the likely effects of anthropogenic changes as well as gaps in our knowledge of microbial community structure and activity, and ultimately, the fate of C, in Northern peatlands.

2. DIRECT DRIVERS OF MICROBIAL COMMUNITY STRUCTURE AND SUBSTRATE USE

2.1. Vertical Niche Differentiation Caused by Energetic Constraints on Decomposition

That the structure of the microbial community of peatlands is altered through the depth profile has been observed in almost all studies over the last century [e.g., *Latter and Cragg*, 1967; *Williams and Crawford*, 1983; *Morales et al.*, 2006]. In particular, fungi numerically dominate the predominantly oxic, upper, layers (generally referred to as the "acrotelm") of peatland ecosystems, although yeast cells are often found throughout peat profiles [*Polyakova et al.*, 2001; *Golovchenko et al.*, 2002; *Artz et al.*, 2007]. With increasing oxygen limitation with depth, the community changes toward more and more specialized microbial taxa that utilize

metabolic intermediates of the higher trophic level microbiota. For practical reasons, these niches will be discussed separately at first, as transfer of C between these compartments will be addressed at a later stage.

2.1.1. C flow pathways in the predominantly oxic surface layer. The oxic layer ("acrotelm") receives a large proportion of plant litter inputs, consisting of large molecular weight polymers that cannot be taken up directly by microbial cells, but also a large proportion of easily assimilated C. The first stage of litter or root decay is generally associated with loss of such labile components, followed by turnover of the more structural components such as lignocellulose. Other sources of easily assimilated C substrates to the microbial communities in this zone consist of the exudates of vascular plant roots (rhizoexudates), leachates from bryophyte stems, and faunal fecal inputs. As a consequence, the turnover times of easily assimilated C substrates in the soil tend to be in the order of hours to days. Hence, decomposition of the bulk of more structural plant litter inputs in this zone depends to some degree on the production of extracellular hydrolytic enzymes (Figure 2).

The ability of many fungi to produce such enzymes has historically focused attention on identifying fungi from peatlands and assessing their enzymatic capacities. Our current

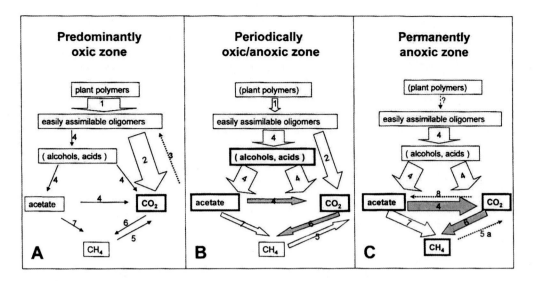

Figure 2. Schematic representation of the major pathways of C flow in the (a) predominantly oxic zone, (b) zone characterized by fluctuations between oxic and anoxic conditions, and (c) permanently anoxic zone. Microbial reactions are shown as arrows with numbers representative of the decomposition pathway: (1) hydrolysis via release of extracellular enzymes, (2) aerobic respiration, (3) autotrophic CO_2 fixation, (4) fermentation or anaerobic respiration with alternative electron acceptors (except H_2), (5) methane oxidation, (6) hydrogenotrophic (autotrophic) methanogenesis, (7) acetoclastic methanogenesis, (8) acetogenesis. Arrow size indicates relative sink strength (not to scale). Dotted arrows are pathways requiring further investigation (e.g., 5a which indicates anaerobic methane oxidation). Shaded arrows indicate syntrophic pathways (see text). End products/intermediates of the microbially mediated decomposition pathways are indicated in boxes, with products that are likely to accumulate shown in bold. Undegraded or partially degraded plant C that accumulates as peat and recycling of microbial necromass have been omitted for clarity.

knowledge of the array of enzyme activities possessed by the microbiota associated with peatland ecosystems stems almost entirely from cultivation-based studies, and may thus be severely biased by the limited range of microbiota studied, to date, and the artificial conditions (e.g., lack of competition for resources with other microbiota) in laboratory incubations. While the many studies of fungal, and also bacterial, polymer-decomposing isolates from peatland ecosystems have noted a high level of saprophytic ability and unusual physiological traits [*Schulz and Thormann*, 2005; *Thormann et al.*, 2002, 2004, 2006; *Pankratov et al.*, 2006, 2007; *Artz et al.*, 2007], it is unclear how this translates to in situ decomposition potential. Recent advances in proteomics have enabled identification of extracellularly produced enzymes [*Schulze et al.*, 2005; *Maron et al.*, 2007], although these approaches have not yet been applied to peatland ecosystems.

Similarly, there are no reports that specifically investigated how or if genes for extracellular enzymes are differentially expressed under varying environmental conditions in peatland environments. This is primarily due to a lack of knowledge of the gene sequences encoding for such enzymes, although progress has recently been made, for example, in

the case of fungal laccases [e.g., *Luis et al.*, 2005; *Kellner et al.*, 2007]. This has led to a few recent studies that have investigated laccase genetic diversity in environmental samples [e.g., *Hofmockel et al.*, 2007], although such approaches have not yet been applied to peatlands. Some recent intriguing findings in this respect, however, include the discovery that certain ascomycete fungi within the Myxotrichaceae (often mycorrhizal) are able to decompose *Sphagnum* tissues to a degree that resembles white rot in wood [*Rice et al.*, 2006]. It would be interesting to further investigate the metabolic potential and its genetic origins of such fungi. Similarly, according to *Caldwell et al.* [2000], dark septate endophytic fungi (DSE; commonly found in the roots of ericaceous species) are capable of producing the extracellular enzymes necessary to process major detrital polymers. Such activities by DSE would also allow the host plant of such endophytic fungi access to nutrients sequestered in organic matter.

Even the respiration of hydrolytic products appears to be dominated by fungi and/or actinobacteria in this layer, as *Bergman et al.* [2000] found that the main biomass sink of a ^{13}C-glucose label was found in mannitol and triglycerides, storage compounds of these microbial groups. It is possible that the production of extracellular enzymes is tightly controlled to uptake of the hydrolysis products, to limit cheating [*Allison and Vitousek*, 2005], yet this would require further study. Little is known about the diversity and degradative capacities of actinobacteria specific to peatlands, although they are known to possess the capacity to produce various extracellular hydrolytic enzymes. Several clones forming two groups and a novel lineage containing peat-derived sequences have recently been described by *Dedysh et al.* [2006] and *Rheims and Stackebrandt* [1999]. In summary, the dominant decomposition pathway in the permanently oxic zone is via aerobic, exo-enzyme-driven polymer decomposition (Figure 2), except for minor contributions from other pathways in anoxic microsites. However, more detailed knowledge about the types of enzymes that are produced and an understanding of the regulation of these would be a major step forward in understanding microbially mediated C and nutrient cycling in this zone.

In recent years, many studies have used stable isotope or radioisotope tracers to locate the pathways of C in soil environments. Very few studies, to date, have attempted to quantify the strength of the flux of photosynthetically fixed C into microbial biomass and respiration in peatlands. In tundra ecosystems, *Loya et al.* [2002] calculated that 6% of the fixed C was allocated to the oxic zone, a figure similar to other ecosystems. Approximately 30% of this C was subsequently cycled through microbial biomass [*Loya et al.*, 2002], although there were differences between microsite types both in the total assimilation and turnover within the microbial community. The sum of respiration of such photosynthetically fixed C within the plant (autotrophic plant respiration) and via microbial turnover (heterotrophic respiration) has been assessed in various recent pulse chase experiments [e.g., *Loya et al.*, 2002; *Olsrud and Christensen*, 2004; *Trinder et al.*, 2008b] in tundra or peatland ecosystems, yet it has been impossible to distinguish the relative contribution of the two processes. In addition, our understanding of the fate of more complex carbon (e.g., cellulose, lignin) in peatland ecosystems, so far, comes entirely from litter decomposition experiments without the use of C tracers. All studies conducted on peatlands using carbon tracers, to date, have been relatively short in duration. Hence, predominantly the pathways of rhizoexudate C and root mucilage turnover are likely to have been followed. The results of *Loya et al.* [2002], who studied C tracer fixation into various pools of soil organic matter, did suggest slow but detectable increase in incorporation into acid unhydrolyzable residue pools. Similarly, there was incorporation into nonpolar compounds such as fat, lipid, or wax pools, although the authors were unable to distinguish between pools that may turn over

relatively rapidly (e.g., storage lipids) or those that may be recalcitrant (e.g., wax). However, together with the data of *Latter et al.* [1998], who showed that the isotopic composition of structural biomarkers (*n*-alkanes) of plant cuticular waxes after extensive litter decomposition did not change, this suggests that there is a small, but significant pool of fixed C that does not undergo any decomposition even in the permanently oxic layer.

2.1.2. C flow pathways in the periodically oxic zone. The zone where the water table fluctuates is generally anoxic, but is periodically either completely (in microsites with root penetration) or partially oxygenated. This zone has been referred to as the mesotelm by *Clymo and Bryant* [2008]. In this zone, there is a proliferation of both obligate and facultative anaerobic microorganisms, and as will become clear, it is here where there is little information available on in situ microbial community structure and their contribution to the various possible C transformation pathways. In addition to the aerobic respiration pathways (Figure 2), which are periodically impeded in this zone, further transformation of C can proceed via a multitude of microbial reactions. A likely pathway of major importance in this zone is the fermentation of relatively recent plant assimilates such as rhizoexudates. Indeed, *Coles and Yavitt* [2004] found a close association of higher rates of anaerobic respiratory activity with the zones of belowground allocation of relatively recent plant C in a forested peatland. Under states of anoxia, fermentative processes result in intermediate metabolites such as volatile fatty acids and alcohols as well as CO_2. *Bergman et al.* [2000] showed substantial accumulation of a ^{13}C label originating from glucose into intermediate metabolite pools, yet little accumulation in newly formed biomass. It is therefore possible that many of these reactions are carried out primarily by facultative anaerobic microbes. For example, an increase in the relative proportion of yeast cells with depth has been noted by many studies [e.g., *Polyakova et al.*, 2001]. Complex polymers, such as cellulose and polyphenolic polymers such as lignin, are not degraded by the most common yeast species [e.g., *Latter et al.*, 1967; *Barnett et al.*, 1983], and this is thought to preclude yeasts, in general, from the latter stages of polymer decomposition except as secondary saprobes of relatively easily assimilated C [*Thormann et al.*, 2007]. There are several interesting examples of basidiomycetous yeasts isolated from cold and/or waterlogged environments that are capable of utilizing protein, aromatic intermediates of lignin degradation, hemicellulose, and pectin under aerobic conditions [e.g., *Schönwalder*, 1958; *Vishniac*, 1996 and references therein]. However, the energetic constraints of decomposition of aromatic compounds like lignin under anaerobic conditions are severe, thus generally leading to accumulation of such compounds in this zone. Indeed, the activity of oxygen-requiring enzymes that catalyze the decomposition of phenolics, phenol oxidases, have been hypothesized to be the "enzyme latch" that controls further hydrolytic reactions [*Freeman et al.*, 2001; more on this in later paragraphs]. Thus, under anoxia, it is likely that the primary fermentative pathways proceed through facultative anaerobic bacterial taxa. Very few investigations have so far been carried out to study the microbial diversity within this layer or, indeed, of their net contribution to C cycling. In the study of *Dedysh et al.* [2006] in the periodically oxic zone of a Siberian peatland, only 16 of the 84 bacterial 16S sequences were closely related to previously described organisms. One third of all 16S rRNA gene sequences were found to be affiliated with the Acidobacteria [*Dedysh et al.*, 2006]. The catabolic capacities of this globally distributed group of bacteria are still relatively unstudied. In a follow-up study that surveyed the proportional abundance of Acidobacteria in a wider set of peatlands in Russia, *Pankratov et al.* [2008] found that these comprised up to 4% of the bacterial cells in the periodically oxic zone. Amendments of peat microcosms with various substrates revealed that glucose, xylan, ethanol, or methanol (depending on incubation temperature) were preferred substrates for acidobacterial growth. Interestingly, the morphology of the detected acidobacterial types in the peat microcosms by fluorescence in situ hybridization (FISH) was related to the type of added substrate, although this was not observed in pure cultures obtained from these microcosms. All isolates were able to utilize galacturonic acid, a cell wall component of *Sphagnum*. Interestingly, *Pankratov et al.* [2008] observed no correlation of acidobacterial relative abundance with porewater pH in their FISH-based studies of total populations, in contrast with the results of *Hartmann et al.* [2008] who used a nonexhaustive clone library approach on samples collected within the surface 10 cm. It is possible that this is due to the potentially different pH optima for acidobacterial subgroups; *Pankratov et al.* [2008] showed that subdivision I isolates have a lower pH range than subdivision 3 isolates and speculated that their lack of an association of relative abundance with pore water pH may have been due to the inclusive nature of their FISH probe which does not report on subdivision affiliation. Although there are currently only four isolated and described species [*Pankratov et al.*, 2008 and references therein], acidobacteria seem to have the capacity to utilize various organic acids, including, interestingly, potentially humic acids.

Other charismatic bacterial groups that have been recently found in the periodically oxic zone include the Verrucomicrobia, which have been detected as clones and by FISH in a Siberian peatland [*Dedysh et al.*, 2006]. Some of the clones

formed a cluster with the anaerobic polysaccharide-utilizing *Opitutus terrae*. FISH results by *Dedysh et al.* [2006] and *Kulichevskaya et al.* [2006] suggest that *Planctomyces* spp are one of the numerically dominant groups in the periodically oxic zone. Novel acidophilic isolates described by *Kulichevskaya et al.* [2007, 2008] have been shown to have the capacity to degrade various biopolymers under microaerophilic conditions. Even more interesting is the recent discovery that some related Verrucomicrobiales are able to carry out methane oxidation (methanotrophy, Figure 2) under aerobic conditions [e.g., *Dunfield et al.*, 2007; *Pol et al.*, 2007]. Also, in some newly identified species within this taxonomic group, anaerobic oxidation of ammonium (the "anammox" process) has been recently discovered [e.g., *Strous et al.*, 1999; *Schmid et al.*, 2003]. Whether this latter process takes place in peatlands, however, is as yet unknown.

Aerobic methanotrophs fall within two groups of Proteobacteria: the alpha and gamma Proteobacteria. These microorganisms, also known as type I (gamma) and type II (alpha) methanotrophs, have been detected in many acidic peatlands [e.g., *Dedysh et al.*, 2006], including in the hyaline cells of *Sphagnum* [*Raghoebarsing et al.*, 2005], and their involvement in methane oxidation in peatlands (Figure 2) was recently confirmed by stable isotope probing (SIP) using a ^{13}C tracer [*Morris et al.*, 2002; *Chen et al.*, 2008a]. Recent work appears to indicate that the relative activity of type I and type II aerobic methanotrophs may be dependent on soil methane production and mixing ratios [*Knief et al.*, 2006].

As briefly mentioned earlier, many anaerobic respiratory processes utilize alternative electron acceptors such as sulfate, nitrate, or Fe(III). Many of the microbial groups that can carry out such processes are members of the Deltaproteobacteria, which have often been described as utilizers of common fermentation products such as primary organic acids and alcohols, although many also have known capacities to decompose aromatic compounds, halogenated aromatics, and long chain-length fatty acids [e.g., *Mohn and Kennedy*, 1992; *Chakraborty and Coates*, 2004]. Clone sequences for deltaproteobacterial species with affiliation to Fe(III)-reducing species have been found in the periodically oxic zone by *Dedysh et al.* [2006]. Similarly, putative sulfate-reducing species related to, e.g., *Desulfomonile* spp and the family Syntrophobacteriaceae, and also a wide variety of species with no cultured relatives within the Deltaproteobacteria, have been described from many peatlands [e.g., *Loy et al.*, 2004; *Juottonen et al.*, 2005; *Dedysh et al.*, 2006; *Schmalenberger et al.*, 2007]. Interestingly, *Juottonen et al.* [2008] observed a higher diversity of bacterial lineages in fens than bogs along a gradient from mesotrophy to oligotrophy, as might be expected given the higher potential inputs of alternative electron acceptors as well as the potential for recharge of these via oxygenation provided by the more extensive root systems of fen plant communities. Syntrophy, whereby the H_2 generated by the acid- or alcohol-degrading microorganism is passed to methanogenic microbiota (interspecies hydrogen transfer), is another likely form in which such taxa exist in peatlands [*Horn et al.*, 2003; *Loy et al.*, 2004]. Evidence of this was recently provided by inhibitor experiments by *Metje and Frenzel* [2005]. This leads us neatly to the final pathway of anaerobic decomposition, which results in formation of methane (methanogenesis). This process, the key microbial groups involved and some of the drivers are described in more detail in the next section of this chapter, as they are the predominant processes in C flow in permanently anoxic environments.

Studies of quantitative assessment of C flow in the periodically oxic zone are relatively scarce. For example, *Metje and Frenzel* [2005] published one of the few studies, to date, that quantified the flow of C from ethanol. Even less information is available on whether any of the observed potential pathways for anaerobic degradation and their microbial operators are of significance in this zone. *Hamberger et al.* [2008] conducted the first such study using a 16S rRNA stable isotope probing approach in anoxic incubations of samples taken below the winter water table of an acidic fen. They identified a diverse suite of fermentative microbiota, including members of the families Acidaminococcaceae, Aeromonadaceae, Clostridiaceae, Enterobacteriaceae, Pseudomonadaceae, and the order Actinomycetales, as well as methanogenic and nonmethanogenic archaea as the primary recipients of common root exudate constituents (xylose and glucose) as substrates. As it is this zone that is the most likely to be impacted by changes in site hydrology, for example, through the potential changes in climatic conditions forecast, it is imperative that further studies are carried out to ascertain the flow of C in this zone.

2.1.3. C flow pathways in the permanently anoxic zone.

Because the availability of O_2 is constrained by diffusion below the water table, the lower layers of peat are permanently anoxic. Hence, recharge of terminal electron acceptors such as sulfate and Fe(III) proceeds through electron shuttling via chemical reduction of organic matter [*Lovley et al.*, 1998; *Heitmann and Blodau*, 2006]. Therefore, as the energetic constraints for decomposition are increasing, the predominant pathways of C flow in this zone are methanogenesis and fermentative processes as discussed above.

Methanogenesis is one of the terminal steps of anaerobic decomposition (Figure 2). Peatlands are a large source of atmospheric methane: consequently, the production and consumption of methane in peatlands and the microbial

drivers of such processes have generated much research interest. Although methane dynamics are covered in detail in other chapters of this monograph, a short overview of the processes and microbial players are presented here for completeness. The net efflux of methane to the atmosphere is controlled by a variety of factors including the availability of bypass pathways (ebullition, transfer through aerenchymatous tissues of certain vascular species) but also by the rates of microbial production and consumption. Further factors include the relative location of these processes in relation to the water table and the surface of the peat as this influences net methane production and diffusion processes. As one of the terminal pathways of C metabolism in peat, it has been noted to be greatly influenced by, for example, water table fluctuations and temperature, as these have both direct effects on methanogenesis but can also alter the fate of C in processes at other trophic levels.

Methane is produced either through the reduction of CO_2 with H_2 (termed autotrophic or hydrogenotrophic methanogenesis), the reduction of acetate (acetoclastic methanogenesis), or the reduction of methanol or other methyl amines or sulfides (methylotrophic methanogenesis). Only the first two of these processes have been noted to be of importance in peatlands. The stable isotopic carbon signature of net methane emissions from peatlands has been used frequently to ascribe whether its origins are likely from predominantly acetoclastic or autotrophic pathways [e.g., *Whiticar*, 1999; *Hornibrook et al.*, 1997]. Although it was thought until fairly recently that around two thirds of global methane production originated from acetoclastic methanogenesis [*Conrad*, 1999], in many Northern peatlands, acetoclastic methanogenesis does not appear to be a major contributor to net methane emissions and acetate accumulates [e.g., *Hines et al.*, 2001; *Duddleston et al.*, 2002]. In support of this, in a recent study of North American peatlands, the methanogenic communities were found to contain few, if any, members that may be capable of producing methane via the acetoclastic pathway [*Rooney-Varga et al.*, 2007]. The currently cultured and phenotypically described members of the methanogenic archaea suggest that there may be an association of taxonomic affiliation with substrate utilization. Cultured members of the Methanosarcinaceae have been described as capable of all three types of methanogenic pathways. In contrast, currently described members of the Methanosaetaceae have so far only been described as acetoclastic, while members of the, e.g., Methanobacteriaceae appear to only use autotrophic pathways [*Jetten*, 1992]. Indeed, a number of recent reports suggest that the trophic status of a peatland indicates which process is likely to dominate [*Hornibrook and Bowes*, 2007], with autotrophic methanogenesis dominating in ombrotrophic peatlands [*Hornibrook et al.*, 1997] and possibly in the permanently anoxic peat layers, generally [e.g., *Hornibrook et al.*, 2000; *Clymo and Bryant*, 2008]. However, a few exceptions [e.g., *Kotsyurbenko et al.*, 2004] do exist.

A few potential pathways of C flow that may or may not occur in the deep anoxic layers remain to be discussed. Dimethyl sulfide (DMS) and its oxidation products, as well as other sulfur-containing gases have been reported as important players in the regulation of cloud formation above the world's oceans [*Ayers and Gillett*, 2000] and have thus earned the reputation of acting as anti-greenhouse gases. *Kiene and Hines* [1995] first observed the potential for significant production of DMS in anoxic peat slurries. *De Mello and Hines* [1994] reported peatlands as a source of DMS while net consuming carbonyl sulfides. The mechanism of DMS production in peatlands is as yet unclear; it is possible that homoacetogenic bacteria and/or methanogens are involved [*Kiene and Hines*, 1995; *Stets et al.*, 2004], but direct evidence is lacking. It is proposed that DMS is produced via methylation of sulfide via an intermediate (methanethiol). As the intermediate has been found to be relatively biologically unavailable, the final step in DMS production may be biologically limited [*Kiene and Hines*, 1995; *Stets et al.*, 2004].

Anaerobic methane oxidation has been reported in many, mostly marine, ecosystems [*Thauer and Shima*, 2008], and recently observed in minerotrophic peatlands [*Smemo and Yavitt*, 2007]. The recently discovered anaerobic methanotrophic archaea appear to be phylogenetically close relatives of the methanogenic Methanosarcinales [e.g., *Hinrichs et al.*, 1999]. Both sulfate- and nitrate-dependent anaerobic methane oxidation via "reverse methanogenesis" have been described. In at least one group of anaerobic methane oxidizing archaea, this process involves genes typically associated with methane production, for example, methyl coenzyme-M reductase A (mcrA), which catalyzes the terminal step in biogenic methane production [*Hallam et al.*, 2004]. Hence, group-specific primers for the detection of the mcrA genes of anaerobic methane-oxidizing archaea have been applied successfully to distinguish methanogenic and anaerobic methane-oxidizing groups [*Hallam et al.*, 2003]. In the only related species thus far detected in peat, *Juottonen et al.* [2008] identified a single mcrA clone sequence as distantly related, but phylogenetically separate, to the anaerobic methane-oxidizing archaeal cluster. Similarly, potential anaerobic members of the Planctomycetes (see previous section) were recently described by *Ivanova and Dedysh* [2006] in the permanently anoxic layers of peat. Finally, acetogenic microorganisms (producing acetate from CO_2 reduction) of the genus *Clostridium* have been recently isolated from peat [*Gossner et al.*, 2008]. However, no direct evidence exists of the occurrence of acetogenesis in peat, and it has been cal-

culated to be thermodynamically unfavorable in peat [*Metje and Frenzel*, 2007].

2.2. Vegetation Composition as a Direct Driver of Microbial Community Composition and Substrate Use

The vast majority of carbon and nutrient inputs to microbiota originate from primary production in peatlands. Hence, some degree of correlation between the microbial community composition and either the aboveground vegetation structure or the composition of litter inputs, would be reasonably expected.

2.2.1. Fungal community composition in relation to vegetation. The community structure of fungal assemblages in the upper parts of peatlands has been shown to relate to some degree to vegetation composition or the chemical composition of litter [*Thormann et al.*, 2004, 2006; *Artz et al.*, 2007; *Trinder et al.*, 2008a]. Direct plant:fungal associations may be explained by some level of resource partitioning or symbiotic relationships with plant hosts. Similarly, a certain level of specificity is generally attached to the distribution of mycorrhizal fungi, although specificity to a particular plant genus is generally low with the exception of certain ectomycorrhizal species. Mycorrhizal fungi have traditionally been classified on the basis of the structure of the interface (i.e., mantels) they produce between the host plant root and the remainder of their vegetative structures (for a thorough introduction to mycorrhizal classification and their functions, see *Smith and Read* [2008]). It is, however, generally accepted that some type of mycorrhiza are formed by specific taxonomic groups of fungi with physiologically different plant types. For example, ectomycorrhizal Basidiomycota generally associate with the root tips of coniferous trees. Traditionally, very few ectomycorrhizal (ECM) fungi have been isolated in culture-dependent studies from forested peatlands [*Thormann*, 2006; *Thormann and Rice*, 2007; *Ludley and Robinson*, 2008], yet their predominance within the fungal community of tree-dominated peatlands has been shown using cloning and sequence analysis [*Jaatinen et al.*, 2008]. Similarly, ericoid mycorrhizal (ERM) fungi, generally members of the Helotiales or *Myxotrichaceae*, are thought to predominantly associate with ericaceous plant species [*Read et al.*, 2004]. Clone sequences with high homology to both ecto- and ericoid mycorrhiza were more commonly found in clone libraries from more advanced stages of regeneration that included substantial cover of *Calluna vulgaris* from a chronosequence on a raised peatland recovering after peat harvesting operations [*Artz et al.*, 2007]. Arbuscular mycorrhiza form with most plant species globally, yet the fungi forming these associations (Glomeromycota) appear not to have been reported much in peatland ecosystems. This is perhaps partly due to the specialized vegetation of peatlands; for example, sedges, which can be the dominant monocots in wetlands, are nonmycorrhizal [*Davies et al.*, 1972] and instead rely on sophisticated physiological adaptations of their root systems to access nutrients. Due to the intimate nature of the mycorrhizal symbiosis, which relies on microbial mobilization of nutrients in exchange for host plant C, it may be expected that mycorrhizal fungi play a significant part in C flow within the permanently oxic layer. In a recent ^{13}C pulse-chase study on a previously cutover peatland, *Trinder et al.* [2008a] showed the highest enrichment in microbial biomass under *C. vulgaris*, presumably due to the direct allocation pathway of photosynthate to their mycorrhizal fungi. Unfortunately, the levels of ^{13}C incorporation did not allow for stable isotope probing. Aside from the function of mycorrhiza in peatlands as mobilizers of nutrients from organic matter, mycorrhizal colonization may also, curiously, increase root turnover rates as photosynthate C is sequestered into recalcitrant forms of C such as fungal cell wall chitin or melamin [*Langley et al.*, 2006]. *C. vulgaris*, for example, produces vast quantities of fine "hair" roots, which are heavily colonized by ericoid mycorrhizal fungi [*Smith and Read*, 2008]. These fungi contain a high concentration of chitin, and this is likely to be a significant source of carbon and nitrogen in the rhizosphere for saprobic microorganisms on root death [*Kerley and Read*, 1997]. *Yan et al.* [2008] did not, however, observe a direct effect of presence of *C. vulgaris* on the utilization of *N*-acetylglucosamine (chitin monomer), although the utilization of this substrate discriminated between peat horizons.

Other species that specialize as endophytes also have preferential access to plant C. Some further evidence of potential plant specificity of a nonmycorrhizal fungal species has recently been provided by the repeated occurrence of a clone related to *Sarcoleotia* (=*Ascocoryne*) *turficola* (a member of the Helotiales) from vegetated peatland sites but not from areas devoid of vegetation [*Artz et al.*, 2007]. This species has usually been found in the vicinity of *Sphagnum* spp. (The fungal records database of Britain and Ireland, British Mycological Society, 2006, available at http://www.fieldmycology. net/FRDBI/FRDBI.asp). However, more detailed study is required to ascertain if this particular species is endophytic in *Sphagnum*. Another group of fungal endophytes, the dark septate endophytes (DSE), are the most widely distributed and abundant fungal endophytes found in ecto-, ectendo-, endo-, and nonmycorrhizal plant roots in many boreal peatlands [*Grunig and Sieber*, 2005; *Mandyam and Jumpponen*, 2005]; however, their ecological functions are little understood [*Jumpponen and Trappe*, 1998]. An interesting ques-

tion is whether competitive or mutualistic interactions result from differences in nutrient acquisition, including extracellular enzyme production, by DSE and mycorrhizal fungi in the same host plant.

2.2.2. Bacterial and archaeal community composition in relation to vegetation. The composition of bacterial communities in peatlands shows a similar stratification with depth as fungi [*Morales et al.*, 2006], but there is less evidence of a relationship with vegetation structure in the bacterial community. In *Morales et al.*'s [2006] study of 24 peatlands, for example, a ribotype that was classified as a *Planctomyces* spp was present in 94% of the samples. Only a small number of ribotypes were site-specific, often these were associated with Alphaproteobacterial ribotypes. No significant correlations with site chemistry, vegetation composition, or geographical characteristics were found [*Morales et al.*, 2006]. Some level of plant species specificity within *Sphagnum* bacterial endophytic communities was recently reported by *Opelt et al.* [2007a] where *Sphagnum* species association was much stronger than geographical origin. This may, however, be related to nutrient status or pH range of the *Sphagnum* species under investigation [*Opelt et al.*, 2007a]. Soil pH was also shown to be a major driver of bacterial community composition in a study of natural, restored, and agricultural wetlands [*Hartmann et al.*, 2008] with a strong increase in the abundance of Acidobacteria at lower pH ranges and likely pH optima for Actinobacteria and Alphaproteobacteria. In tandem, betaproteobacterial groups were found in greater abundance in agriculturally used wetlands, in agreement with the finding in other ecosystems that eutrophication increases their relative abundance.

Similarly, the structure of microbial communities that process secondary microbial metabolites may be affected by changes in vegetation structure. In tandem with temperature and water table controls, both the methanogenic and crenarchaeal community composition has also been found to be related to pH, hydrology, and vegetation composition [*Galand et al.*, 2003, 2005; *Jaatinen et al.*, 2005; *Rooney-Varga et al.*, 2007]. *Galand et al.* [2005] showed similar vegetation-related differences in the composition of methanogenic archaea. Sedge-dominated mesotrophic fen sites were dominated by acetoclastic archaeal species, while nutrient-poor fen (containing few vascular species) and bog sites were characterized by archaeal species clustering with species previously described as exclusively autotrophic. These results were confirmed in a more exhaustive study of an ecohydrological gradient at the same site ranging from mesotrophic fen to ombrotrophic bog, which also demonstrated a general decrease in archaeal diversity with increasing ombrotrophy [*Juottonen et al.*, 2005]. Both *Nicol et al.* [2007] and

Bomberg and Timonen [2007] reported shifts in the community structure of archaea upon forestation of moorland or in mycorrhizal Scots pine roots compared to forest humus, respectively. Similar associations between vegetation structure in peatlands and the relative importance of methanogenesis finally led *Hines et al.* [2008] to conclude that vegetation change toward a more vascular plant-dominated community due to climate change-induced warming may cause a positive feedback via increased methane contributions toward net respiration pools. Broad vegetation type classification, in terms of, e.g., *Sphagnum*-dominated versus sedge-dominated vegetation types, therefore appears to hold some promise in terms of identifying the dominant pathways of methane production although the picture is not complete. Very little is known about any direct effects of vegetation changes on methanotroph diversity and functioning. Removal of *C. vulgaris* from peat cores was shown to lower methane oxidation potential and numbers of copies of a gene encoding the α subunit of the particulate methane monooxygenase (pmoA), yet no overall effect on pmoA diversity [*Chen et al.*, 2008b].

Variation in rooting extent and depth due to different vegetation types may also drive microbial community structure and substrate use. Due to the limited penetration of live roots in peat, the inputs of rhizoexudates decline rapidly with depth. This manifests itself in measures of both biomass C and substrate-induced respiration assays as a decline with depth [*Artz et al.*, 2006; *Yan et al.*, 2008]. Differences in plant traits, such as rooting depths, spatial distribution of fine roots, composition of root exudates, and finally, the roots and their symbionts themselves in the case of root turnover, are all potentially important factors in determining both the composition and distribution of the labile C pool in peatlands [*de Deijn et al.*, 2008]. This is supported by evidence from a study in which dissolved organic carbon (DOC) concentrations were quantified on a previously cutover peatland and found to be higher in vegetated areas, with significantly higher values reported under *Eriophorum* species than *C. vulgaris* [*Trinder et al.*, 2008b]. Rhizoexudate composition has also been shown to differ between peatland vegetation types [e.g., *Ström et al.*, 2005; *Crow and Wieder*, 2005]. The rhizosphere microbial communities that access labile C sources can be targeted using multiple substrate-induced respiration assays. Different vegetation types have recently been noted to support functionally different rhizosphere microbial communities in a recent study using such techniques [*Yan et al.*, 2008]. Seasonal variations in the rhizoexudate amino acid composition between *Eriophorum vaginatum* and *E. angustifolium* were observed in a subarctic wetland, which in turn affected ecosystem level methane emissions [*Ström and Christensen*, 2007]. Similarly,

Wright and Reddy [2007] observed differences in substrate-induced respiration rates as well as differentiation in terms of the substrate types used between wetland sites of differing levels of P enrichment. In a comparison of five European peatlands at different stages of regeneration following peat harvest, *Artz et al.* [2008] showed strong relationships between the vegetation succession and the functional diversity of the microbiota that utilize simple carbon sources. *Artz et al.* [2006] showed that a high proportion of the variation in substrate-induced respiration in a peat profile could be explained by the changes in the chemical composition of key components of soil organic matter, notably the ratio of polysaccharidic to polymeric carboxylic compounds. When soils are sampled and stored, for example, photosynthetic input of labile substrates stops. Due to inherent differences in the turnover times of different carbon pools, labile substrates are progressively depleted during the course of storage. This should be noticeable as a change in carbon substrate utilization patterns over time. Indeed, this is commonly seen with soil incubation experiments. In a storage experiment with peat samples from different depths, representing a gradient of humification (Figure 3), while storage effects notably altered substrate utilization profiles in upper horizons, the fingerprints were not; however, significantly different in highly humified peat despite prolonged storage in aerated conditions. Presumably this is due to the low residual availability of labile C at this depth and may indicate predominance of a microbial community within this horizon that is adapted to C substrates of low energy yields. Similarly, *Andersen et al.* [2006] showed that, while microbial biomass recovery in a peatland restoration sequence followed the chronosequence, this was not the case for glucose-induced respiration which remained lower than expected in the restored sites. The authors attributed this to the poor substrate quality of the restored sites. These findings highlight the need to understand more clearly the quantity and quality of rhizosphere carbon inputs from vegetation colonizing cutover peatland and the changes this may induce in the overall community structure of peat microbial communities.

2.3. Direct Effects of Fluctuations in the Physicochemical Environment

2.3.1. Water table fluctuations. The effect of water table on decomposition rates and the community structure of the decomposing microbiota has been a subject of long debate. *Laiho* [2006] reviewed the often contradictory findings of the effects of water table fluctuations, and therefore, this topic will only briefly be considered here. While in many cases, a distinct increase in decomposition rates and/or peat respiratory fluxes have been reported [e.g., *Riutta et al.*, 2007]

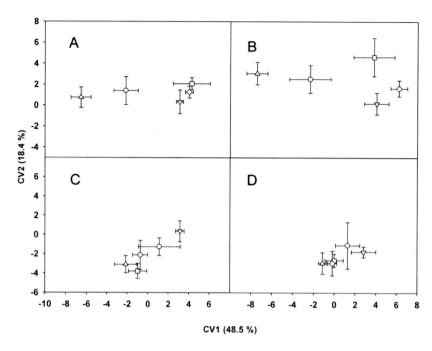

Figure 3. Results of canonical variate analysis after dimension-reducing principal components analysis on the substrate utilization patterns of microbial communities utilizing labile C sources in peat of varying levels of decomposition for (a) fresh moss litter, (b) decaying litter, (c) peat of low von Post humification index, (d) peat of high von Post humification index) after storage at 4°C for extended periods (circles = 7 days, downward triangles = 28 days, squares = 49 days, diamonds = 70 days, and upward triangles = 103 days). There was no significant difference between the substrate utilization patterns at the five incubation lengths in the most decomposed peat (999 MonteCarlo permutation replicates). For further information about the substrate-induced respiration assay methodology and statistical tests, refer to *Artz et al.* [2006]. Data from R. Artz (unpublished, 2007).

there are also reports of lowered decomposition leading to higher C accumulation rates [e.g., *Minkkinen et al.*, 2002]. Root-associated ("autotrophic" respiration contains fluxes from mycorrhizal fungi and other root-associated microorganisms) respiration was shown to be insensitive to drought in a study in temperate fen microcosms [*Knorr et al.*, 2008]. In terms of enzyme production associated with decomposition processes, increased oxygenation due to water table drawdown in peatlands has been consistently shown to increase the activity of extracellular phenol oxidases [e.g., *Freeman et al.*, 1996, 1997], which play a pivotal role in the breakdown of complex organic matter and the cycling of soil phenolic compounds that may inhibit other extracellular hydrolytic enzymes. Through this "enzymic latch mechanism" [*Freeman et al.*, 2001], environmentally mediated changes in the activity of extracellular phenol oxidases may directly affect the net decomposition rates of organic matter in soil [*Freeman et al.*, 2001, 2004]. As the first reported example of this mechanism on microbial community structure and functions, bacterial populations with potential phenolic-decomposing abilities were significantly increased in drought-

manipulated peatlands, which correlated with increased soil phenol oxidase activity and net CO_2 fluxes [*Fenner et al.*, 2005]. *Laiho* [2006], in common with many others, concluded that the effect of water table fluctuations on decomposition rates depends on whether the change occurs in a zone where conditions (nutrient concentrations, remaining labile C, and decomposer community) allow for further decomposition. Indeed, *Basiliko et al.* [2007] observed that aerobic and anaerobic decomposition of peat was constrained by the organic matter quality (including P availability), which was closely linked to microbial biomass. Furthermore, *Beer et al.* [2008] showed that through the slow rate of vertical diffusion through peat, there was little release of carbohydrate-rich dissolved organic matter into deeper peat layers, while aromatic, phenolic, and inorganic carbon accumulates. Hence, decomposition in deeper peat layers is effectively slowed due to shifts in the energetically favorable biological pathways, and water table drawdown in these layers would therefore likely have only limited effects on decomposition.

It might be expected that, at least in more minerotrophic peatlands, an alteration of the physicochemical environment that affects substrate availability (such as shifts in redox potential resulting in direct and indirect effects on substrate and alternative electron acceptor availability) may shift the balance of pathways of methane production. For example, some methanogenic archaea may be capable of switching to Fe(III) reduction [*van Bodegom et al.*, 2004; *Reiche et al.*, 2008] when stocks of Fe(III) are replenished in minerotrophic peatlands following water table reduction or oxygenation via plant roots. In addition, competition with other users of alternative electron acceptors (e.g., ethanol released by roots under anoxic conditions or as metabolite of fermentative reactions) can, in theory, influence the balance of net methane production, although *Reiche et al.* [2008] did not find evidence of such competition for common substrates.

2.3.2. Temperature fluctuations. Some isotopic studies on methane in peatlands have found some evidence of altered pathways in relation to seasonality [*Avery et al.*, 1999; *Kotsyurbenko et al.*, 2004; *Juottonen et al.*, 2008], while others have not observed such changes [*Lansdown et al.*, 1992]. *Metje and Frenzel* [2007] observed shifts in methanogenic pathways according to temperature in microcosm experiment from Siberian peat, with the autotrophic pathway dominating at very low or high temperatures. Archaeal populations have been noted to be altered by temperature in arctic peat, with (some) methanogenic archaeal populations increasing in line with methane production, while nonmethanogenic, crenarchaeotal groups declined [*Hoj et al.*, 2008]. Interestingly, *Hoj et al.* [2008] also observed a shift in the community structure of the methanogens, with some observed increases in the Methanosaetaceae (acetoclastic) population with increasing temperature. A recent study of methanogenesis in fen peat subjected to experimental drought, however, concluded that drought and re-wetting did not alter the generally autotrophic pathway of methane production in their system; rather, it more likely induced a shift in the trade-off between methanogenesis and methanotrophy [*Knorr et al.*, 2008].

2.3.3. Effects of increased atmospheric deposition. Increased nitrogen deposition is affecting many areas of Northern peatlands, with reported effects on vegetation structure through initial increases in foliar N concentration. *Sphagnum* and other bryophytes initially accumulate additional tissue N until a threshold is reached, and N reaches soil porewater directly. In such N saturated peatlands, competition between vascular plants and denitrifiers for N ensues [*Glatzel et al.*, 2008]. Long-term effects on litter quality from both vascular plants and bryophytes have therefore generally been reported, with a notable decrease in the C:N ratio. As a consequence, higher decomposition rates of vascular plant and

Sphagnum litters with increased N deposition have frequently been reported [e.g., *Gerdol et al.*, 2007]. *Bragazza et al.* [2007] demonstrated the correlation between litter quality of peatland species, specifically their polyphenol/nutrient and C/nutrient ratios, and their decomposition rates. In addition, *Bragazza and Freeman* [2007] concluded that the observed reduced polyphenol content of *Sphagnum* mosses with increased N deposition may also aid enhanced decomposition rates due to the lowered inhibitory effect of such phenolics. The impact of increased N deposition on the microbial communities is thus far largely unknown. The endophytic bacterial community of *Sphagnum* mosses was recently investigated by *Opelt et al.* [2007b] and found to contain substantial diversity of nitrogen-fixing bacteria. Similarly, the nitrogen-fixing cyanobacterial endophytes of the pleurocarpous mosses that characterize the understorey of many boreal forests are crucial to the ecosystem N budget [*DeLuca et al.*, 2008]. It is likely that such communities would be affected by higher N uptake by the plant under elevated N deposition scenarios. In a study of N addition effects in an Alaskan boreal forest, *Allison et al.* [2007] showed evidence of altered fungal community composition. This was also found by *Allison et al.* [2008] in tandem with changes in the activities of an array of extracellular enzyme, yet no overall effects on microbial biomass or soil respiration rate and signature were found. Mycorrhizal fungi are involved in >70% of the N and P uptake by plants in boreal forests and other peatland ecosystems [*van der Heijden et al.*, 2008], and thus, increased mineral N inputs may lead to direct changes in N cycling and C allocation to mycorrhizal fungi if it results in changes in the suite of extracellularly produced hydrolytic enzymes. *Lucas and Casper* [2008] noted decreases in extracellular laccase activity under increased N deposition scenarios with concomitant increases in proteolytic activity, combined with a shift in ECM community composition. This would fit a scenario whereby increased competitive advantage by the plant to take up mineral N directly results in lower requirements for N mobilization from recalcitrant organic matter, hence, resulting in lower laccase production. N excreted in the rhizoexudate pool as amino acids, which has been shown to be correlated with protease activity [*Kielland et al.*, 2007], may explain increases in proteolytic activity to aid nutrient recycling in the mycorrhizosphere. Deposition of S as acid rain has declined drastically over the last few decades and is predominantly a feature of the more industrialized temperate areas, but is included for completeness. This may induce sulfate reduction which can compete strongly with methanogenesis in the periodically oxic layers. Indeed, sulfate additions from acid rain but also volcanic eruptions have been shown to lead to a marked reduction in net methane efflux from peatlands [e.g., *Gauci et al.*, 2008].

2.3.4. Effects of fire. Depending on the severity and frequency of fires, symbiotic microbiota may be at a disadvantage due to complete removal of their host or a reduction in plant-derived C due to the reduced photosynthetic ability of the plant. In forest ecosystems, there are known fungal species that fruit prolifically after fire and frequent fires have been shown to significantly affect microbial and, specifically, ectomycorrhizal fungal community composition [*Cairney and Bastias*, 2006]. In boreal forests, for example, the loss of ECM fungi due to fire can shift the dominance of mycorrhizal groups to AM fungi. This can have effects on net C and N cycling, as C storage into glomalin is carried out at the expense of organic matter turnover [*Bergner et al.*, 2004]. *Treseder et al.* [2007] only observed a decline in total mycorrhizal C pools in one of a series of fertilized sites from a chronosequence of recovery after fire. In other trophic levels, burning appears to reduce type I methanotrophic pmoA

even though there is no significant difference in methane oxidation potential [*Chen et al.*, 2008b].

3. FEEDBACKS AND THEIR EFFECTS ON THE FATE OF PEATLAND C STOCKS

Many long-term alterations of physicochemical conditions that peatland microbiota are adapted to, such as increases in long-term temperatures, altered rainfall patterns, increased fire frequency, and long-term elevated atmospheric deposition may also have indirect effects on the peatland microbiota (Figure 4). In some cases, these changes manifest themselves in altered vegetation structure, which may cause additional alterations in the microbial community structure (Figure 4).

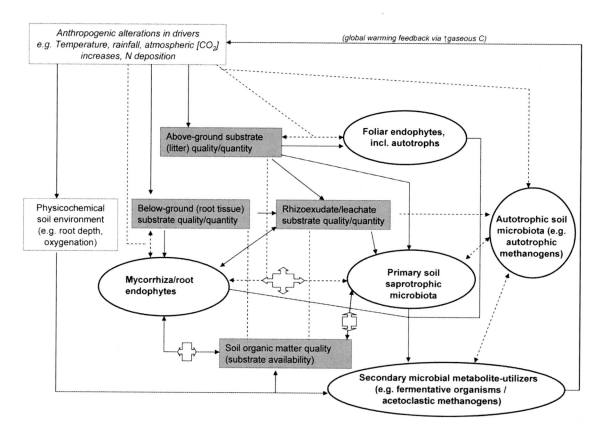

Figure 4. Direct and indirect effects of various drivers on microbial community structure and their catabolic activities from a microbiocentric perspective. Generalized groups of functionally different microbiota are shown in bold, while "energy sources" (C and nutrients, alternative electron acceptors) are shown in gray boxes. Direct effects of various drivers (in dotted boxes), that may be altered by human intervention, on energy flow through the microbial community are shown with solid arrows in the direction of the effect. Indirect effects that could cause competition for energy sources are indicated with dotted arrows. Finally, involvement of extracellularly produced enzymes is indicated with block arrows. Complex synergistic effects or within-group competitive interactions have not been considered.

For example, in the long term, increased N deposition results in increases in cover of graminoid species and dwarf shrubs at the expense of *Sphagnum* [e.g., *Bubier et al.*, 2007], while repeated burning also alters vegetation structure. In a recent study by *Gunnarsson et al.* [2008], it was concluded that the vegetation change associated with enhanced N deposition in a Southern Swedish bog was a primary factor in the extremely low observed carbon accumulation rates. *Nowinski et al.* [2008] showed this indirectly through radiocarbon dating of the soil carbon pools in a long-term fertilization experiment in the arctic tundra, where shifts to a shrub-dominated vegetation led to increased losses of root and soil organic matter C. Long-term water table drawdown, which resulted in a change in vegetation from a wet fen to a peatland forest, was shown to affect the fungal community composition, but not actinobacterial members [*Jaatinen et al.*, 2008]. Temperature increases have generally been reported to increase decomposition rates [*Aerts*, 2006] in the cooler climatic regions. In line with theory that the higher activation energy associated with the breakdown of recalcitrant substrates should result in a greater temperature sensitivity of decomposition, several studies in other soil types have now found that the turnover of the more recalcitrant soil C pools is likely affected more than turnover of labile C by rises in temperature [*Boddy et al.*, 2008; *Mikan et al.*, 2002; *Conant et al.*, 2008]. Increased temperature in peatland ecosystems has been noted to increase the fungal to bacterial ratio [e.g., *Thormann et al.*, 2004] or result in increases in mycorrhizal colonization in both mesocosms [e.g., *Domisch et al.*, 2002] and arctic tundra [*Clemmensen et al.*, 2006]. As yet, however, there is a paucity of studies that have addressed whether such changes in microbial community structure are reflected at a functional level in terms of their substrate use. *Schimel and Mikan* [2005] did observe alterations in microbial substrate use in response to freeze-thaw cycles in Arctic tundra soils that could have large effects on seasonal N availability.

Permafrost thaw, especially, in combination with temperature increases, may well stimulate methanogen abundance and methane production rates beyond the sole effects of warming [*Turetsky et al.*, 2008]. The projected altered patterns in snow fall and permafrost disappearance may thus lead to further feedback in C and N cycling through higher methane emissions (Figure 4). In the longer-term, temperature increases through climatic shifts are thought to affect vegetation structure by temperature-driven migration and increased competition, both of plant species and of fauna. This has the potential to affect microbial community structure through alteration in substrate quality via both the rhizoexudate and litter pools. Both *Thormann et al.* [2004] and *Trinder et al.* [2008a], for example, found litter type to

be strong determinants of fungal community composition throughout the course of litter decomposition. In turn, as proposed by *van der Heijden et al.* [2008], the ability of an altered microbial community to mobilize nutrients from the organic matter pool may influence the N available for plant growth. Hence, potential feedback loops may arise with climate change if this causes shifts in energy flow, thus altering the zones of biologically favorable conditions (Figure 4).

One of the major questions is the fate of the vast store of C in peatlands with the projected changes in climatic conditions and other anthropogenic influences. Figure 4 summarizes the various drivers discussed in this review and their likely routes of influence on the size, composition, and activity of peatland microbiota. As discussed, many of these effects have only recently begun to be investigated. For example, "priming" is the increased release of CO_2 from turnover of previously recalcitrant organic matter due to an increase in microbial biomass and/or changes in community structure following inputs of energy-rich substrates [*Kuzyakov et al.*, 2002]. Does priming occur in peatlands? *Fontaine et al.* [2007] showed that supplying fresh photosynthate to (nonpeat) subsoils stimulated mineralization of ancient C. By analogy, the quantity or composition of energy-rich material that is introduced to peat soils may determine whether priming occurs. However, other limiting factors in peatlands, such as product accumulation or nutrient/oxygen availability, may control the likelihood and size of a priming effect [e.g., *Basiliko et al.*, 2007; *Goldhammer and Blodau*, 2008]. Such studies have yet to be carried out for Northern peatland ecosystems.

4. SYNTHESIS AND RECOMMENDATIONS FOR FURTHER RESEARCH

The net accumulation of peat is a result of only a fractional difference between C inputs via photosynthetic fixation and subsequent allocation to the soil environment above (as shoot litter) and belowground (root litter and rhizoexudation) and the net decomposition of this pool of C. The net balances available for Northern peatlands [e.g., *Gorham*, 1995; *Roulet et al.*, 2007; *Bortoluzzi et al.*, 2006] suggested this difference to be <10% of net fixation. These studies also highlight large uncertainties, particularly in the process rates of fixed C during decomposition. Very little work has been carried out to understand the trajectory of C through microbial communities, especially in the lower horizons. A myriad of microbial taxa have been isolated, characterized, and/or observed in molecular studies on Northern peatlands, yet their relative contributions to net C cycling are still virtually unknown. A multidisciplinary approach is required to study the involvement of microbiota in peatland C cycling in order to identify their roles, relative contributions to the major decomposition pathways, and trophic interactions, both in direct response to alterations in their physicochemical environment and in ecosystems undergoing succession of their plant communities. In terms of understanding the drivers of microbial community structure, their activities within the carbon and nutrient cycles, and ultimately, their involvement in the long-term fate of sequestered C in Northern peatlands, exciting times lie ahead.

Acknowledgments. I wish to express my sincere thanks to many colleagues and especially Colin Campbell, Pete Millard, Stephen Chapman, Brajesh Singh, Andy Taylor, Markus Thormann, and Roxane Andersen, for valuable and lively discussions. I also wish to thank Lisa Belyea for the invitation to write this chapter.

REFERENCES

Aerts, R. (2006), The freezer defrosting, global warming and litter decomposition rates in cold biomes, *J. Ecol.*, *94*, 713–724.

Allison, S. D., and P. M. Vitousek (2005), Responses of extracellular enzymes to simple and complex nutrient inputs, *Soil Biol. Biochem.*, *37*, 937–944.

Allison, S. D., C. A. Hanson, and K. K. Treseder (2007), Nitrogen fertilization reduces diversity and alters community structure of active fungi in boreal ecosystems, *Soil Biol. Biochem.*, *39*, 1878–1887.

Allison, S. D., C. I. Czimczik, and K. K. Treseder (2008), Microbial activity and soil respiration under nitrogen addition in Alaskan boreal forest, *Global Change Biol.*, *14*, 1156–1168.

Andersen, R., A.-J. Francez, and L. Rochefort (2006), The physicochemical and microbiological status of a restored bog in Québec, identification of relevant criteria to monitor success, *Soil Biol. Biochem.*, *38*, 1375–1387.

Artz, R. R. E., S. J. Chapman, and C. D. Campbell (2006), Substrate utilisation profiles of microbial communities in peat are depth-dependent and correlate with whole soil FTIR profiles, *Soil Biol. Biochem.*, *38*, 2958–2962.

Artz, R. R. E., I. C. Anderson, S. J. Chapman, A. Hagn, M. Schloter, J. M. Potts, and C. D. Campbell (2007), Fungal diversity and community composition change in response to vegetational succession during natural regeneration of cut-over peatlands, *Microb. Ecol.*, *54*, 508–522.

Artz, R. R. E., S. J. Chapman, A. Siegenthaler, E. A. D. Mitchell, A. Buttler, E. Bortoluzzi, D. Gilbert, M. Yli-Petays, H. Vasander, and A.-J. Francez (2008), Functional microbial diversity in cutover peatlands responds to vegetation succession and is partly directed by labile carbon, *J. Appl. Ecol.*, *45*, 1799–1809.

Avery, G. B., R. D. Shannon, J. R. White, C. S. Martens, and M. J. Alperin (1999), Effect of seasonal changes in the pathways of methanogenesis on the δ^{13}C values of pore water methane in a Michigan peatland, *Global Change Biol.*, *13*, 475–484.

Ayers, G. P., and R. W. Gillett (2000), DMS and its oxidation products in the remote marine atmosphere: Implications for climate and atmospheric chemistry, *J. Sea Res.*, *43*, 275–286.

Barnett, J. A., R. W. Payne, and D. Yarrow (1983), *Yeasts: Characteristics and Identification*, Cambridge Univ. Press, Cambridge, U.K.

Basiliko, N., C. Blodau, C. Roehm, P. Bengtsson, and T. R. Moore (2007), Regulation of decomposition and methane dynamics across natural, commercially mined, and restored northern peatlands, *Ecosystems*, *10*, 1148–1165.

Beer, J., K. Lee, M. Whiticar, and C. Blodau (2008), Geochemical controls on anaerobic organic matter decomposition in a northern peatland, *Limnol. Oceanogr.*, *53*, 1393–1407.

Beilman, D. W., D. H. Vitt, J. S. Bhatti, and S. Forest (2008), Peat carbon stocks in the southern Mackenzie River Basin, uncertainties revealed in a high-resolution case study, *Global Change Biol.*, *14*, 1221–1232.

Bending, G. D., and D. J. Read (1997), Lignin and soluble phenolic degradation by ectomycorrhizal and ericoid mycorrhizal fungi, *Mycol. Res.*, *101*, 1348–1354.

Bergman, I., P. Lundberg, C. M. Preston, and M. Nilsson (2000), Degradation of C-13-U-glucose in *Sphagnum majus* litter: Responses to redox, pH, and temperature, *Soil Sci. Soc. Am. J.*, *64*, 1368–1381.

Bergner, B., J. Johnstone, and K. K. Treseder (2004), Experimental warming and burn severity alter soil CO_2 flux and soil functional groups in a recently burned boreal forest, *Global Change Biol.*, *10*, 1996–2004.

Blodau, C., B. Mayer, S. Peiffer, and T. R. Moore (2007), Support for an anaerobic sulfur cycle in two Canadian peatland soils, *J. Geophys. Res.*, *112*, G02004, doi:10.1029/2006JG000364.

Boddy, E., P. Roberts, P. W. Hill, J. Farrar, and D. L. Jones (2008), Turnover of low molecular weight dissolved organic C (DOC) and microbial C exhibit different temperature sensitivities in Arctic tundra soils, *Soil Biol. Biochem.*, *40*, 1557–1566.

Bomberg, M., and S. Timonen (2007), Distribution of cren- and euryarchaeota in scots pine mycorrhizospheres and boreal forest humus, *Microb. Ecol.*, *54*, 406–416.

Bortoluzzi, E., D. Epron, A. Siegenthaler, D. Gilbert, and A. Buttler (2006), Carbon balance of a European mountain bog at contrasting stages of regeneration, *New Phytol.*, *172*, 708–718.

Bragazza, L., and C. Freeman (2007), High nitrogen availability reduces polyphenol content in *Sphagnum* peat, *Sci. Total Environ.*, *377*, 439–443.

Bragazza, L., C. Siffi, P. Iacumin, and R. Gerdol (2007), Mass loss and nutrient release during litter decay in peatland, The role of microbial adaptability to litter chemistry, *Soil Biol. Biochem.*, *39*, 257–267.

Bubier, J. L., T. R. Moore, and L. A. Bledzki (2007), Effects of nutrient addition on vegetation and carbon cycling in an ombrotrophic bog, *Global Change Biol.*, *13*, 1168–1186.

Cairney, J. W. G., and B. A. Bastias (2007), Influences of fire on forest soil fungal communities, *Can. J. For. Res.*, *37*, 207–215.

Caldwell, B. A., A. Jumpponen, and J. M. Trappe (2000), Utilization of major detrital substrates by dark-septate, root endophytes, *Mycologia*, *92*, 230–232.

Chakraborty, R., and J. D. Coates (2004), Anaerobic degradation of monoaromatic hydrocarbons, *Appl. Microbiol. Biotechnol.*, *64*, 437–446.

Chen, Y., M. G. Dumont, J. D. Neufeld, L. Bodrossy, N. Stralis-Pavese, N. P. McNamara, N. Ostle, M. J. I. Briones, and J. C. Murrell (2008a), Revealing the uncultivated majority: Combining DNA stable-isotope probing, multiple displacement amplification and metagenomic analyses of uncultivated *Methylocystis* in acidic peatlands, *Environ. Microbiol.*, *10*, 2609–2622.

Chen, Y., N. P. McNamara, M. G. Dumont, M. G., L. Bodrossy, N. Stralis-Pavese, and J. C. Murrell (2008b), The impact of burning and *Calluna* removal on below-ground methanotroph diversity and activity in a peatland soil, *Appl. Soil Ecol.*, *40*, 291–298.

Clemmensen, K. E., A. Michelsen, S. Jonasson, and G. R. Shaver (2006), Increased ectomycorrhizal fungal abundance after long-term fertilization and warming of two arctic tundra ecosystems, *New Phytol.*, *171*, 391–404.

Clymo, R. S., and C. L. Bryant (2008), Diffusion and mass flow of dissolved carbon dioxide, methane, and dissolved organic carbon in a 7-m deep raised peat bog, *Geochim. Cosmochim. Acta*, *72*, 2048–2066.

Coles, J. R. P., and J. B. Yavitt (2004), Linking belowground carbon allocation to anaerobic CH_4 and CO_2 production in a forested peatland, New York State, *Geomicrobiol. J.*, *21*, 445–455.

Conant, R. T., R. A. Drijber, M. L. Haddix, W. J. Parton, E. A. Paul, and A. F. Plante (2008), Sensitivity of organic matter decomposition to warming varies with its quality, *Global Change Biol.*, *14*, 868–877.

Conrad, R. (1999), Contribution of hydrogen to methane production and control of hydrogen concentrations in methanogenic soils and sediments, *FEMS Microbiol. Ecol.*, *28*, 193–202.

Crow, S. E., and R. K. Wieder (2005), Sources of CO_2 emission from a northern peatland, Root respiration, exudation, and decomposition, *Ecology*, *86*, 1825–1834.

Davies, J., L. G. Bryarty, and J. O. Rieley (1972), Observations on the swollen lateral roots of the Cyperaceae, *New Phytol.*, *72*, 167–174.

De Deyn, G. B., J. H. C. Cornelissen, and R. D. Bardgett (2008), Plant functional traits and soil carbon sequestration in contrasting biomes, *Ecol. Lett.*, *11*, 516–531.

Dedysh, S. N., T. A. Pankratov, S. E. Belova, I. Kulichevskaya, and W. Liesack (2006), Phylogenetic analysis and in situ identification of *Bacteria* community composition in an acidic *Sphagnum* peat bog, *Appl. Environ. Microbiol.*, *72*, 2110–2117.

DeLuca, T. H., O. Zackrisson, M. J. Gundale, and M.-C. Nilsson (2008), Ecosystem feedbacks and nitrogen fixation in boreal forests, *Science*, *320*, 1181.

de Mello, W. Z., and M. E. Hines (1994), Application of static and dynamic enxlosures for determining dimethyl sulfide and carbonyl sulfide exchange in *Sphagnum* peatlands: Implications for the magnitude and direction of flux, *J. Geophys. Res.*, *99*(D7), 14,601–14,607.

Domisch, T., et al. (2002), Effect of soil temperature on nutrient allocation and mycorrhizas in Scots pine seedlings, *Plant Soil*, *239*, 173–185.

Duddleston, K. N., M. A. Kinney, R. P. Kiene, and M. E. Hines (2002), Anaerobic microbial biogeochemistry in a northern bog: Acetate as a dominant metabolic end product, *Global Biogeochem. Cycles*, *16*(4), 1063, doi:10.1029/2001GB001402.

Dunfield, P. F., et al. (2007), Methane oxidation by an extremely acidophilic bacterium of the phylum Verrucomicrobia, *Nature*, *450*, 879–882.

Fenner, N., C. Freeman, and B. Reynolds (2005), Hydrological effects on the diversity of phenolic degrading bacteria in a peatland: Implications for carbon cycling, *Soil Biol. Biochem.*, *37*, 1277–1287.

Fontaine, S., S. Barot, P. Barre, N. Bdioui, B. Mary, and C. Rumpel (2007), Stability of organic carbon in deep soil layers controlled by fresh carbon supply, *Nature*, *450*, 277.

Freeman, C., G. Liska, N. J. Ostle, M. A. Lock, B. Reynolds, and J. Hudson (1996), Microbial activity and enzymic decomposition processes following peatland water table drawdown, *Plant Soil*, *180*, 121–127.

Freeman, C., G. Liska, N. J. Ostle, M. A. Lock, S. Hughes, B. Reynolds, and J. Hudson (1997), Enzymes and biogeochemical cycling in wetlands during a simulated drought, *Biogeochemistry*, *39*, 177–187.

Freeman, C., N. Ostle, and H. Kang (2001), An enzymic 'latch' on a global carbon store - A shortage of oxygen locks up carbon in peatlands by restraining a single enzyme, *Nature*, *409*, 149.

Freeman, C., N. J. Ostle, N. Fenner, and H. Kang (2004), A regulatory role for phenol oxidase during decomposition in peatlands, *Soil Biol. Biochem.*, *36*, 1663–1667.

Galand, P. E., H. Fritze, and K. Yrjälä (2003), Microsite-dependent changes in methanogenic populations in a boreal oligotrophic fen, *Environ. Microbiol.*, *5*, 1133–1143.

Galand, P. E., H. Juottonen, H. Fritze, and K. Yrjälä (2005), Pathways of methanogenesis and diversity of methanogenic Archaea in three boreal peatland ecosystems, *Appl. Environ. Microbiol.*, *71*, 2195–2198.

Gauci, V., S. Blake, D. S. Stevenson, and E. J. Highwood (2008), Halving of the northern wetland CH_4 source by a large Icelandic volcanic eruption, *J. Geophys. Res.*, *113*, G00A11, doi:10.1029/2007JG000499.

Gerdol, R., A. Petraglia, L. Bragazza, P. Iacumin, and L. Brancaleoni (2007), Nitrogen deposition interacts with climate in affecting production and decomposition rates in *Sphagnum* mosses, *Global Change Biol.*, *13*, 1810–1821.

Glatzel, S., I. Forbrich, C. Kruger, S. Lemke, and G. Gerold (2008), Small scale controls of greenhouse gas release under elevated N deposition rates in a restoring peat bog in NW Germany, *Biogeosci.*, *5*, 925–935.

Goldhammer, T., and C. Blodau (2008), Desiccation and product accumulation constrain heterotrophic anaerobic respiration in peats of an ombrotrophic temperate bog, *Soil Biol. Biochem.*, *40*, 2007–2015.

Golovchenko, A. V., N. G. Dobrovol'skaya, and L. I. Inisheva (2002), Structure and stocks of microbial biomass in oligotrophic peat bogs of the southern Taiga in western Siberia, *Eurasian Soil Sci.*, *35*, 1296–1301.

Gorham, E. (1995), The biogeochemistry of northern peatlands and its possible response to global warming, in *Biotic Feedbacks in the Global Climate System: Will the Warming Feed the Warming?*, edited by G. M. Woodwell and F. T. MacKenzie, pp. 169–187, Oxford Univ. Press, New York.

Gossner, A. S., F. Picardal, R. S. Tanner, and H. R. Drake (2008), Carbon metabolism of the moderately acid-tolerant acetogen *Clostridium drakei* isolated from peat, *FEMS Microbiol. Ecol.*, *287*, 236–242.

Gruber, N., et al. (2004), The vulnerability of the carbon cycle in the 21st century: An assessment of carbon-climate-human interactions, in *Toward CO_2 Stabilization, Issues, Strategies and Consequences*, edited by C. B. Field and M. R. Raupach, pp. 45–76, Island Press, Washington, D. C.

Grunig, C. R., and T. N. Sieber (2005), Molecular and phenotypic description of the widespread root symbiont *Acephala applanata* gen. et sp nov., formerly known as dark-septate endophyte Type 1, *Mycologia*, *97*, 628–640.

Gunnarsson, U., L. B. Bronge, H. Rydin, and M. Ohlson (2008), Near-zero recent carbon accumulation in a bog with high nitrogen deposition in SW Sweden, *Global Change Biol.*, *14*, 2152–2165.

Hallam, S. J., P. R. Girguis, C. M. Preston, P. M. Richardson, and E. F. DeLong (2003), Identification of methyl coenzyme M reductase A (mcrA) genes associated with methane-oxidizing archaea, *Appl. Environ. Microbiol.*, *69*, 5483–5491.

Hallam, S. J., N. Putnam, C. M. Preston, J. C. Detter, D. Rokhsar, P. M. Richardson, and E. F. DeLong (2004), Reverse methanogenesis: Testing the hypothesis with environmental genomics, *Science*, *305*, 1457–1462.

Hamberger, A., M. A. Horn, M. G. Dumont, J. C. Murrell, and H. L. Drake (2008), Anaerobic consumers of monosaccharides in a moderately acidic fen, *Appl. Environ. Microbiol.*, *74*, 3112–3120.

Hartman, W. H., C. J. Richardson, R. Vilgalys, and G. L. Bruland (2008), Environmental and anthropogenic controls over bacterial communities in wetland soils, *Proc. Natl. Acad. Sci. U. S. A.*, *105*, 17,842–17,847.

Heitmann, T., and C. Blodau (2006), Oxidation and incorporation of hydrogen sulfide by dissolved organic matter, *Chem. Geol.*, *235*, 12–20.

Hines, M. E., K. N. Duddleston, and R. P. Kiene (2001), Carbon flow to acetate and C_1 compounds in northern wetlands, *Geophys. Res. Lett.*, *28*(22), 4251–4254.

Hines, M. E., K. N. Duddleston, J. N. Rooney-Varga, D. Fields, and J. P. Chanton (2008), Uncoupling of acetate degradation from methane formation in Alaskan wetlands: Connections to vegetation distribution, *Global Biogeochem. Cycles*, *22*, GB2017, doi:10.1029/2006GB002903.

Hinrichs, K. U., J. M. Hayes, S. P. Sylva, P. G. Brewer, and E. F. DeLong (1999), Methane-consuming archaebacteria in marine sediments, *Nature*, *398*, 802.

Hofmockel, K. S., D. R. Zak, and C. B. Blackwood (2007), Does atmospheric NO_3-deposition alter the abundance and activity of ligninolytic fungi in forest soils?, *Ecosystems*, *10*, 1278–1286.

Hogberg, M. N., P. Hogberg, and D. D. Myrold (2007), Is microbial community composition in boreal forest soils determined by pH, C-to-N ratio, the trees, or all three?, *Oecologia*, *150*, 590–601.

Hoj, L., R. A. Olsen, and V. L. Torsvik (2008), Effects of temperature on the diversity and community structure of known methanogenic groups and other Archaea in high Arctic peat, *ISME J.*, *2*, 37–48.

Horn, M. A., C. Matthies, K. Kusel, A. Schramm, and H. L. Drake (2003), Hydrogenotrophic methanogenesis by moderately acid-tolerant methanogens of a methane-emitting acidic peat, *Appl. Environ. Microbiol.*, *69*, 74–83.

Hornibrook, E. R. C., and H. L. Bowes (2007), Trophic status impacts both the magnitude and stable carbon isotope composition of methane flux from peatlands, *Geophys. Res. Lett.*, *34*, L21401, doi:10.1029/2007GL031231.

Hornibrook, E. R. C., F. J. Longstaffe, and W. S. Fyfe (1997), Spatial distribution of microbial methane production pathways in temperate zone wetland soils: Stable carbon and hydrogen isotope evidence, *Geochim. Cosmochim. Acta*, *61*, 745–753.

Hornibrook, E. R. C., et al. (2000), Carbon-isotope ratios and carbon, nitrogen and sulfur abundances in flora and soil organic matter from a temperate-zone bog and marsh, *Geochem. J.*, *34*, 237–245.

Ivanova, A. O., and S. N. Dedysh (2006), High abundance of planctomycetes in anoxic layers of a *Sphagnum* peat bog, *Microbiology*, *75*, 716–719.

Jaatinen, K., E. S. Tuittila, J. Laine, K. Yrjala, and H. Fritze (2005), Methane-oxidizing bacteria in a finnish raised mire complex: Effects of site fertility and drainage, *FEMS Microbiol. Ecol.*, *50*, 429–439.

Jaatinen, K., R. Laiho, A. Vuorenmaa, U. del Castillo, K. Minkkinen, T. Pennanen, T. Penttila, and H. Fritze (2008), Responses of aerobic microbial communities and soil respiration to water-level drawdown in a northern boreal fen, *Environ. Microbiol.*, *10*, 339–353.

Jetten, M. S. M. (1992), Methanogenesis from acetate—A comparison of the acetate metabolism in *Methanothrix soehngenii* and *Methanosarcina* spp., *FEMS Microbiol. Rev.*, *88*, 181–197.

Jumpponen, A., and J. M. Trappe (1998), Dark septate endophytes, a review of facultative biotrophic root-colonizing fungi, *New Phytol.*, *140*, 295–310.

Juottonen, H., P. E. Galand, E. S. Tuittila, J. Laine, H. Fritze, and K. Yrjala (2005), Methanogen communities and *Bacteria* along an ecohydrological gradient in a northern raised bog complex, *Environ. Microbiol.*, *7*, 1547–1557.

Juottonen, H., E. S. Tuittila, S. Juutinen, H. Fritze, and K. Yrjala (2008), Seasonality of rDNA and rRNA-derived archaeal communities and methanogenic potential in a boreal mire, *ISME J.*, *2*, 1157–1168.

Kellner, H., P. Luis, and F. Buscot (2007), Diversity of laccase-like multicopper oxidase genes in Morchellaceae, identification of genes potentially involved in extracellular activities related to plant litter decay, *FEMS Microbiol. Ecol.*, *61*, 153–163.

Kerley, S. J., and D. J. Read (1997), The biology of mycorrhiza in the Ericaceae, 19, fungal mycelium as a nitrogen source for the ericoid mycorrhizal fungus *Hymenoscyphus ericae* and its host plants, *New Phytol.*, *136*, 691–701.

Kielland, K., J. W. McFarland, R. W. Ruess, and K. Olson (2007), Rapid cycling of organic nitrogen in taiga forest ecosystems, *Ecosystems*, *10*, 360–368.

Kiene, R. P., and M. E. Hines (1995), Microbial formation of dimethyl sulphide in anoxic *Sphagnum* peat, *Appl. Environ. Microbiol.*, *61*, 2720–2726.

Knief, C., S. Kolb, P. L. E. Bodelier, A. Lipski, and P. F. Dunfield (2006), The active methanotrophic community in hydromorphic soils changes in response to changing methane concentration, *Environ. Microbiol.*, *8*, 321–333.

Knorr, K. H., M. R. Oosterwoud, and C. Blodau (2008), Experimental drought alters rates of soil respiration and methanogenesis but not carbon exchange in soil of a temperate fen, *Soil Biol. Biochem.*, *40*, 1781–1791.

Kotsyurbenko, O. R., M. W. Friedrich, M. V. Simankova, A. N. Nozhevnikova, P. N. Golyshin, K. N. Timmis, and R. Conrad (2004a), Shift from acetoclastic to H_2-dependent methanogenesis in a West Siberian peat at low pH values and isolation of a acidophilic *Methanobacterium* strain, *Appl. Environ. Microbiol.*, *73*, 2344–2348.

Kotsyurbenko, O. R., K. J. Chin, M. V. Glagolev, S. Stubner, M. V. Simankova, A. N. Nozhevnikova, and R. Conrad (2004b), Acetoclastic and hydrogenotrophic methane production and methanogenic populations in an acidic West-Siberian peat bog, *Environ. Microbiol.*, *6*, 1159–1173.

Kulichevskaya, I. S., T. A. Pankratov, and S. N. Dedysh (2006), Detection of representatives of the Planctomycetes in *Sphagnum* peat bogs by molecular and cultivation approaches, *Microbiology*, *75*, 329–335.

Kulichevskaya, I. S., A. O. Ivanova S. E. Belova, O. I. Baulina, P. L. E. Bodelier, W. I. C. Rijpstra, J. S. S. Damste, G. A. Zavarzin, and S. N. Dedysh (2007), Schlesneria paludicola gen. nov., sp nov., the first acidophilic member of the order Planctomycetales, from *Sphagnum*-dominated boreal wetlands, *Int. J. Syst. Evol. Microbiol.*, *57*, 2680–2687.

Kulichevskaya, I. S., A. O. Ivanova, O. I. Baulina, P. L. E. Bodelier, J. S. S. Damste, and S. N. Dedysh (2008), *Singulisphaera acidiphila* gen. nov., sp nov., a non-filamentous, *Isosphaera*-like planctomycete from acidic northern wetlands, *Int. J. Syst. Evol. Microbiol.*, *58*, 1186–1193.

Kuzyakov, Y. (2002), Review, factors affecting rhizosphere priming effects, *J. Plant Nutr. Soil Sci.*, *165*, 382–396.

Laiho, R. (2006), Decomposition in peatlands, Reconciling seemingly contrasting results on the impacts of lowered water levels, *Soil Biol. Biochem.*, *38*, 2011–2024.

Langley, J. A., S. K. Chapman, and B. A. Hungate (2006), Ectomycorrhizal colonization slows root decomposition, the post-mortem fungal legacy, *Ecol. Lett.*, *9*, 955–959.

Lansdown, J. M., P. D. Quay, and S. L. King (1992), CH_4 production via CO_2 reduction in a temperate bog: A source of ^{13}C-depleted CH_4, *Geochim. Cosmochim. Acta*, *56*, 3493–3503.

Latter, P. M., and J. B. Cragg (1967), The decomposition of Juncus squarrosus leaves and microbiological changes in the profile of Juncus moor, *J. Ecol.*, *55*, 465–482.

Latter, P. M., J. B. Cragg, and O. W. Heal (1967), Comparative studies on the microbiology of four moorland soils in the Northern Pennines, *J. Ecol.*, *55*, 445–464.

Latter, P. M., G. Howson, D. M. Howard, and W. A. Scott (1998), Long-term study of litter decomposition on a Pennine neat bog: Which regression?, *Oecologia*, *113*, 94–103.

Lovley, D. R., J. L. Fraga, E. L. Blunt-Harris, L. A. Hayes, E. J. P. Phillips, and J. D. Coates (1998), Humic substances as a mediator for microbially catalyzed metal reduction, *Acta Hydrochim. Hydrobiol.*, *26*, 152–157.

Loy, A., K. Kusel, A. Lehner, H. L. Drake, and M. Wagner (2004), Microarray and functional gene analyses of sulfate-reducing prokaryotes in low-sulfate, acidic fens reveal cooccurrence of recognized genera and novel lineages, *Appl. Environ. Microbiol.*, *70*, 6998–7009.

Loya, W. M., L. C. Johnson, G. W. Kling, J. Y. King, W. S. Reeburgh, and K. J. Nadelhoffer (2002), Pulse-labeling studies of carbon cycling in arctic tundra ecosystems: Contribution of photosynthates to soil organic matter, *Global Biogeochem. Cycles*, *16*(4), 1101, doi:10.1029/2001GB001464.

Lucas, R. W., and B. B. Casper (2008), Ectomycorrhizal community and extracellular enzyme activity following simulated atmospheric N deposition, *Soil Biol. Biochem.*, *40*, 1662–1669.

Ludley, K. E., and C. H. Robinson (2008), 'Decomposer' Basidiomycota in Arctic and Antarctic ecosystems, *Soil Biol. Biochem.*, *40*, 11–29.

Luis, P., H. Kellner, F. Martin, and F. Buscot (2005), A molecular method to evaluate basidiomycete laccase gene expression in forest soils, *Geoderma*, *128*, 18–27.

Maltby, E., and C. P. Immirzi (1993), Carbon dynamics in peatlands and other wetland soils, regional and global perspectives, *Chemosphere*, *27*, 999–1023.

Mandyam, K., and A. Jumpponen (2005), Seeking the elusive function of the root-colonising dark septate endophytic fungi, *Stud. Mycol.*, 173–189.

Maron, P. A., L. Ranjard, C. Mougel, and P. Lemanceau (2007), Metaproteomics, a new approach for studying functional microbial ecology, *Microb. Ecol.*, *53*, 486–493.

Metje, M., and P. Frenzel (2005), The effect of temperature on anaerobic ethanol oxidation and methanogenesis in an acidic peat from a northern wetland, *Appl. Environ. Microbiol.*, *71*, 8191–8200.

Metje, M., and P. Frenzel (2007), Methanogenesis and methanogenic pathways in a peat from subarctic permafrost, *Environ. Microbiol.*, *9*, 954–964.

Mikan, C. J., J. P. Schimel, and A. P. Doyle (2002), Temperature controls of microbial respiration in arctic tundra soils above and below freezing, *Soil Biol. Biochem.*, *34*, 1785–1795.

Minkkinen, K., R. Korhonen, I. Savolainen, and J. Laine (2002), Carbon balance and radiative forcing of Finnish peatlands 1900-2100 - the impact of forestry drainage, *Global Change Biol.*, *8*, 785–799.

Mohn, W. W., and K. J. Kennedy (1992), Reductive dehalogenation of chlorophenols by *Desulfomonile tiedjei* DCB-1, *Appl. Environ. Microbiol.*, *58*, 1367–1370.

Morales, S. E., P. J. Mouser, N. Ward, S. P. Hudman, N. J. Gotelli, D. S. Ross, and T. A. Lewis (2006), Comparison of bacterial communities in New England *Sphagnum* bogs using terminal restriction fragment length polymorphism (T-RFLP), *Microb. Ecol.*, *52*, 34–44.

Morris, S. A., S. Radajewski, T. W. Willison, and J. C. Murrell (2002), Identification of the functionally active methanotroph population in a peat soil by stable-isotope probing, *Appl. Environ. Microbiol.*, *68*, 1446–1453.

Nicol, G. W., C. D. Campbell, S. J. Chapman, and J. I. Prosser (2007), Afforestation of moorland leads to changes in crenarchaeal community structure, *FEMS Microbiol. Ecol.*, *60*, 51–59.

Nowinski, N. S., S. E. Trumbore, E. A. G. Schuur, M. C. Mack, and G. R. Shaver (2008), Nutrient addition prompts rapid destabilization of organic matter in an arctic tundra ecosystem, *Ecosystems*, *11*, 16–25.

Olsrud, M., and T. R. Christensen (2004), Carbon cycling in subarctic tundra: Seasonal variation in ecosystem partitioning based on in situ ^{14}C pulse-labelling, *Soil Biol. Biochem.*, *36*, 245–253.

Opelt, K., C. Berg, S. Schonmann, L. Eberl, and G. Berg (2007a), High specificity but contrasting biodiversity of *Sphagnum*-associated bacterial and plant communities in bog ecosystems independent of the geographical region, *ISME J.*, *1*, 502–516.

Opelt, K., V. Chobot, F. Hadacek, S. Schonmann, L. Eberl, and G. Berg (2007b), Investigations of the structure and function of bacterial communities associated with *Sphagnum* mosses, *Environ. Microbiol.*, *9*, 2795–2809.

Pankratov, T. A., I. S. Kulichevskaya, W. Liesack, and S. N. Dedysh (2006), Isolation of aerobic, gliding, xylanolytic and laminarinolytic bacteria from acidic *Sphagnum* peatlands and emended description of *Chitinophaga arvensicola* Kampfer et al. 2006, *Int. J. Syst. Evol. Microbiol.*, *56*, 2761–2764.

Pankratov, T. A., B. J. Tindall, W. Liesack, and S. N. Dedysh (2007), *Mucilaginibacter paludis* gen. nov., sp. nov. and *Mucilaginibacter gracilis* sp. nov., pectin-, xylan- and laminarin-degrading members of the family *Sphingobacteriaceae* from acidic *Sphagnum* peat bog, *Int. J. Syst. Evol. Microbiol.*, *57*, 2349. (Correction, *Int. J. Syst. Evol. Microbiol.*, *57*, 2979, 2007.)

Pankratov, T. A., Y. M. Serkebaeva, I. S. Kulichevskaya, W. Liesack, and S. N. Dedysh (2008), Substrate-induced growth and isolation of *Acidobacteria* from acidic *Sphagnum* peat, *ISME J.*, *2*, 551–560.

Pol, A., K. Heijmans, H. R. Harhangi, D. Tedesco, M. S. Jetten, and H. J. Op den Camp (2007), Methanotrophy below pH 1 by a new Verrucomicrobia species, *Nature*, *450*, 874–878.

Polyakova, A. V., I. Y. Chernov, and N. S. Panikov (2001), Yeast diversity in hydromorphic soils with reference to a grass-*Sphagnum* wetland in western Siberia and a hummocky tundra region at Cape Barrow (Alaska), *Microbiology*, *70*, 617–622.

Raghoebarsing, A. A., et al. (2005), Methanotrophic symbionts provide carbon for photosynthesis in peat bogs, *Nature*, *436*, 1153–1156.

Read, D. J., J. R. Leake, and J. Perez-Moreno (2004), Mycorrhizal fungi as drivers of ecosystem processes in heathland and boreal forest biomes, *Can. J. Bot.*, *82*, 1243–1263.

Reiche, M., G. Torburg, and K. Kuesel (2008), Competition of Fe(III) reduction and methanogenesis in an acidic fen, *FEMS Microbiol. Ecol.*, *65*, 88–101.

Rheims, H., and E. Stackebrandt (1999), Application of nested polymerase chain reaction for the detection of as yet uncultured organisms of the class *Actinobacteria* in environmental samples, *Environ. Microbiol.*, *1*, 137–143.

Rice, A. V., A. Tsuneda, and R. S. Currah (2006), In vitro decomposition of *Sphagnum* by some microfungi resembles white rot of wood, *FEMS Microbiol. Ecol.*, *56*, 372–382.

Riutta, T., J. Laine, and E .S. Tuittila (2007), Sensitivity of CO_2 exchange of fen ecosystem components to water level variation, *Ecosystems*, *10*, 718–733.

Rooney-Varga, J. N., M. W. Giewat, K. N. Duddleston, J. P. Chanton, and M. E. Hines (2007), Links between archaeal community structure, vegetation type and methanogenic pathway in Alaskan peatlands, *FEMS Microbiol. Ecol.*, *60*, 240–251.

Roulet, N. T., P. M. Lafleur, P. J. H. Richard, T. R. Moore, E. R. Humphreys, and J. Bubier (2007), Contemporary carbon balance and late Holocene carbon accumulation in a northern peatland, *Global Change Biol.*, *13*, 397–411.

Schimel, J. P., and C. Mikan (2005), Changing microbial substrate use in Arctic tundra soils through a freeze-thaw cycle, *Soil Biol. Biochem.*, *37*, 1411–1418.

Schmalenberger, A., H. L. Drake, and K. Kuesel (2007), High unique diversity of sulphate-reducing prokaryotes characterized in a depth gradient in an acidic fen, *Environ. Microbiol.*, *9*, 1317–1328.

Schmid, M., et al. (2003), Candidatus "*Scalindua brodae*", sp. nov., Candidatus "*Scalindua wagneri*", sp. nov., two new species of anaerobic ammonium oxidizing bacteria, *Syst. Appl. Microbiol.*, *26*, 529–538.

Schoenwalder, H. (1958), Über die Verwertung yon Huminsäuren als Nährstoffquelle durch Mikroorganismen, *Arch. Microbiol.*, *30*, 162–180.

Schulz, M. J., and M. N. Thormann (2005), Functional and taxonomic diversity of saprobic filamentous fungi from Typha latifolia from central Alberta, Canada, *Wetlands*, *25*, 675–684.

Schulze, W. X., G. Gleixner, K. Kaiser, G. Guggenberger, M. Mann, and E.-D. Schulze (2005), A proteomic fingerprint of dissolved organic carbon and of soil particles, *Oecologia*, *142*, 335–343.

Smemo, K. A., and J. B. Yavitt (2007), Evidence for anaerobic CH_4 oxidation in freshwater peatlands, *Geomicrobiol. J.*, *24*, 583–597.

Smith, S. E., and D. J. Read (2008), *Mycorrhizal Symbiosis*, 3rd ed., Academic, San Diego, Calif.

Stets, E. G., M. E. Hines, and R. P. Kiene (2004), Thiol methylation potential in anoxic, low-pH wetland sediments and its relationship with dimethylsulfide production and organic carbon cycling, *FEMS Microbiol. Ecol.*, *47*, 1–11.

Ström, L., and T. R. Christensen (2007), Below-ground carbon turnover and greenhouse gas exchanges in a subarctic wetland, *Soil Biol. Biochem.*, *39*, 1689–1698.

Ström, L., M. Mastepanov, and T. R. Christensen (2005), Species-specific effects of vascular plants on carbon turnover and methane emissions from wetlands, *Biogeochemistry*, *75*, 65–82.

Strous, M., J. A. Fuerst, E. H. M. Kramer, S. Logemann, G. Muyzer, T. K. Van de Pas-Schoonen, R. Webb, J. G. Kuenen, and M. S. M. Jetten (1999), Missing lithotroph identified as new planctomycete, *Nature*, *400*, 446–449.

Thauer, R. K., and S. Shima (2008), Methane as fuel for anaerobic microorganisms, *Ann. N. Y. Acad. Sci.*, *1125*, 158–170.

Thormann, M. N. (2006), Diversity and function of fungi in peatlands, a carbon cycling perspective, *Can. J. Soil Sci.*, *86*, 281–293.

Thormann, M. N., and A. V. Rice (2007), Fungi from peatlands, *Fungal Diversity*, *24*, 241–299.

Thormann, M. N., R. S. Currah, and S. E. Bayley (2002), The relative ability of fungi from *Sphagnum* fuscum to decompose selected carbon substrates, *Can. J. Microbiol.*, *48*, 204–211.

Thormann, M. N., S. E. Bayley, and R. S. Currah (2004), Microcosm tests of the effects of temperature and microbial species number on the decomposition of Carex aquatilis and *Sphagnum* fuscum litter from southern boreal peatlands, *Can. J. Microbiol.*, *50*, 793–802.

Thormann, M. N., A. V. Rice, and D. W. Beilman (2007), Yeasts in peatlands, A review of richness and roles in peat decomposition, *Wetlands*, *27*, 761–772.

Treseder, K. K., K. M. Turner, and M. C. Mack (2007), Mycorrhizal responses to nitrogen fertilization in boreal ecosystems, potential consequences for soil carbon storage, *Global Change Biol.*, *13*, 78–88.

Trinder, C. J., D. Johnson, and R. R. E. Artz (2008a), Interactions among fungal community structure, litter decomposition and depth of water table in a cutover peatland, *FEMS Microbiol. Ecol.*, *64*, 433–448.

Trinder, C. J., R. R. E. Artz, and D. Johnson (2008b), Contribution of plant photosynthate to soil respiration and dissolved organic carbon in a naturally recolonising cutover peatland, *Soil Biol. Biochem.*, *40*, 1622–1628.

Turetsky, M. R., C. C. Treat, M. P. Waldrop, J. M. Waddington, J. W. Harden, and A. D. McGuire (2008), Short-term response of methane fluxes and methanogen activity to water table and soil warming manipulations in an Alaskan peatland, *J. Geophys. Res.*, *113*, G00A10, doi:10.1029/2007JG000496.

Turunen, J. (2003), Past and present carbon accumulation in undisturbed boreal and subarctic mires, a review, *Suo*, *54*, 15–28.

van Bodegom, P. M., J. C. M. Scholten, and A. J. M. Stams (2004), Direct inhibition of methanogenesis by ferric ion, *FEMS Microbiol. Ecol.*, *49*, 261–268.

van der Heijden, M. G. A., R. D. Bardgett, and N. M. van Straalen (2008), The unseen majority, soil microbes as drivers of plant diversity and productivity in terrestrial ecosystems, *Ecol. Lett.*, *11*, 296–310.

Vasander, H., and A. Kettunen (2006), Carbon in boreal peatlands, in *Boreal Peatland Ecosystem*, edited by R. K. Wieder and D. H. Vitt, pp. 165–194, Springer, Heidelberg.

Vishniac, H. S. (1996), Biodiversity of yeasts and filamentous microfungi in terrestrial Antarctic ecosystems, *Biodivers. Conserv.*, *5*, 1365–1378

Whiticar, M. J. (1999), Carbon and hydrogen isotope systematics of bacterial formation and oxidation of methane, *Chem. Geol.*, *161*, 291–314.

Williams, R. T., and R. L. Crawford (1993), Microbial diversity in Minnesota peatlands, *Microbiol. Ecol.*, *9*, 201–214.

Wright, A. L., and K. R. Reddy (2007), Substrate-induced respiration for phosphorus-enriched and oligotrophic peat soils in an Everglades wetland, *Soil Sci. Soc. Am. J.*, *71*, 1579–1583.

Yan, W., R. R. E. Artz, and D. Johnson (2008), Species-specific effects of plant colonising cut-over peatlands on patterns of carbon source utilisation by soil microorganisms, *Soil Biol. Biochem.*, *40*, 544–549.

R. R. E. Artz, The Macaulay Land Use Research Institute, Craigiebuckler, Aberdeen AB15 8QH, UK. (r.artz@macaulay.ac.uk)

Partitioning Litter Mass Loss Into Carbon Dioxide and Methane in Peatland Ecosystems

M. Nilsson and M. Öquist

Department of Forest Ecology and Management, Swedish University of Agricultural Sciences, Umeå, Sweden

Carbon is lost from peatland ecosystems as CO_2-C, CH_4-C, or organic C. Despite the obvious contributions of all three of these forms of carbon to total mass losses arising from the degradation of organic material in peatland ecosystems, our understanding of both the relative amounts of these C forms that are emitted, and the factors controlling their partitioning, is limited. In this chapter, we summarize and evaluate current knowledge regarding partitioning of terminally mineralized carbon into CO_2 and CH_4 to identify the main factors controlling the partitioning of mass loss of plant-derived biomass into these species; gaps in knowledge regarding the processes involved, their rates and their interactions; and important issues to address in future research initiatives. We first outline the main factors that influence the partitioning into CO_2 and CH_4 of organic matter degraded in northern peatlands, then evaluate their relative importance, and related issues, using available data in the scientific literature. We conclude that current knowledge of the partitioning between CO_2 and CH_4 production during the degradation of organic material in peatlands is very limited, and values of CO_2/CH_4 reported in the literature differ by a factor of about 80,000. The main environmental factors affecting the partitioning are the composition of the plant litter entering the soil, the soil redox conditions, the age of the peat, and soil temperature. However, attempts to draw sound general conclusions supported by reliable quantitative data are hampered by the high variability reported in the literature.

1. INTRODUCTION

Carbon is lost from peatland ecosystems as CO_2-C, CH_4-C, or organic C. Numerous estimates of the annual losses of each individual component, methane, carbon dioxide, and runoff exports of organic carbon [dissolved organic carbon (DOC) plus particulate organic carbon (POC)] have been published. However, despite the obvious importance of

contributions of all three of these forms of carbon to total mass losses arising from the degradation of organic material in peatland ecosystems [*Roulet et al.*, 2007; *Nilsson et al.*, 2008], our understanding of both the relative amounts of these C forms that are emitted, and the factors influencing their partitioning, is limited.

Fundamental differences between two of these forms of C released from peatlands and the other is that CO_2 and CH_4 result from terminal carbon mineralization, while DOC largely consists of intermediates produced during various stages of organic carbon degradation. The main sources of DOC are polymers from which oligomers and monomers are released by the action of microbial exoenzymes, and fermentation processes, i.e., processes in which organic materials

Carbon Cycling in Northern Peatlands
Geophysical Monograph Series 184
Copyright 2009 by the American Geophysical Union.
10.1029/2008GM000819

serve as both donors and acceptors for electrons. Hence, as well as being an important form of carbon lost from the peatland, DOC also constitutes an important source of both CO_2 and CH_4. Thus, the factors controlling the production of carbon dioxide and methane are very different from those controlling DOC production. Considerations of the complex processes involved in DOC production are beyond the scope of this analysis. Instead, we focus solely on the partitioning of the carbon released during terminal mineralization into carbon dioxide and methane.

The production of CO_2 may be mediated by either autotrophic organisms (more specifically, plant roots) or heterotrophic organisms, while CH_4 is produced solely by heterotrophic activity. In addition, the heterotrophic production of CO_2-C and CH_4-C generally involves the degradation of two functionally and structurally distinct sources of organic carbon: exudates from moss tissues and vascular plant roots; and more complex organic material, i.e., biopolymers [*Popp et al.*, 1999; *Strom et al.*, 2003]. Here we consider only the production of CO_2 and CH_4 during the degradation of plant-derived biopolymers. In most peatlands, this degradation is totally dominated by fungi and bacteria even if other heterotrophic organisms may contribute. We do not include the decomposition of plant exudates nor autotrophic respiration by plant roots, nor do we consider processes coupled to the oxidation of CH_4 by methanotrophic organisms, which can occur during the diffusion of CH_4 from the production zone to the atmosphere.

The main objectives of this contribution are (1) to identify the controls of the partitioning of plant-derived biomass losses into CO_2 and CH_4; (2) to assess current quantitative knowledge regarding factors controlling the partitioning into CO_2 and CH_4; and (3) to identify current gaps in our knowledge and key issues that should be addressed in future research initiatives. We consider controls at two levels: the mechanisms whereby the microtopography of the peatland surface and plant community composition controls peatland redox conditions and substrate quality, and physical and microbial controls at the molecular level. These controls are then combined in a theoretical framework, the validity of which is tested against available data of processes and environmental variables related to the partitioning of mineralized peatland organic matter into the two fractions.

2. DEGRADATION OF ORGANIC MATERIAL

Consider a cohort of fresh plant litter with a specific organic chemical composition being degraded over time following a negative exponential function [e.g., *Agren and Bosatta*, 1996]. The most important factors controlling the partitioning into carbon dioxide and methane are the relative contributions of the aerobic and anaerobic degradation processes driving the mass losses (Figure 1), which depend in turn on the vertical zonation of oxic and anoxic conditions (Figure 2), and the vertical distribution of microorganisms, e.g., methanogenic organisms must be present for methane production to occur. The vertical distribution of methanogenic organisms is controlled by the average water table level during the growing season [*Sundh et al.*, 1994; *Nedwell and Watson*, 1995]. Under oxic conditions, only CO_2 will be produced as the end product of mineralization, while under anoxic conditions both CO_2 and CH_4 will be produced. The master control of the partitioning of a cohort of organic material into CO_2 and CH_4 will be the proportional mass loss occurring under oxic conditions. To simplify the conceptual framework, we assume that the water table represents a distinct border between oxic and anoxic conditions. The major factors controlling the proportional aerobic mass loss in a peatland ecosystem are then the depth of the water table in conjunction with the depth at which plant litter is added to the system (Figures 1 and 2).

In a peatland, the spatial variation in average water table depth is reflected in variations in the microtopographical characteristics of its surface. Most peatlands generally have very heterogeneous surfaces, forming a mosaic of structures varying from "dry" hummocks through lawns to "wet" hollows. Typical average water table depths for hummocks, lawns, and hollows are 20–60 (or more), 5–20, and <5 cm below the peatland surface, respectively (Figure 2) [*Belyea and Clymo*, 2001; *Rietkerk et al.*, 2004]. Thus, the relative proportions of oxic and anoxic zones in the top 50–60 cm of a peatland will reflect the microtopographical heterogeneity.

In addition to the microtopographical structure, the vertical position at which new plant litter is added also constitutes an important control of its fate during degradation (Figure 1). In terms of the depth at which plant litter is added, three main categories of plant material can be identified: aboveground vascular plant material, belowground vascular plant material (mainly roots), and moss litter. Aboveground vascular plant tissues are added at the top of the peatland surface (Figures 1a and 1b), while vascular plant root litter is added at various depths below the peatland surface (Figure 1c). The vertical distribution of root biomass may differ between microtopographical units, but the vast majority is produced within the topmost 20 cm [*Wallén*, 1986; *Saarinen*, 1997]. *Sphagnum* plant litter is added just below the vegetation surface, but the exact depth at which it is added depends on the surface microtopography. In hummocks, moss litter is produced about a centimeter below the surface, while the corresponding depth in hollows is 5–10 cm, depending on the morphology and growth characteristics of the moss.

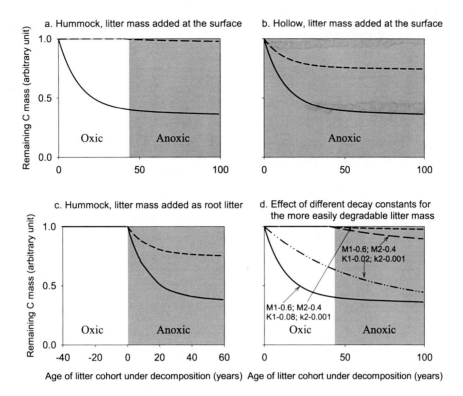

Figure 1. Illustration of how the partitioning of plant litter carbon into CO_2 and CH_4 is influenced by the degradability of the plant material and when the material is transferred into the anoxic zone. To illustrate this, we use a simple decomposition model with two carbon fractions with different decomposition rate constants (k). (fraction 1, 0.6 mass units, $k = -0.08$ year^{-1}; fraction 2, 0.4 mass units, $k = -0.001$ year^{-1}). The $CH_4:CO_2$ quotient during anaerobic decomposition is set to 1.5. Solid line, total C mass loss; dashed line, amount of mass lost as methane C: (a) mass loss in a hummock (water table depth (WT) = 40 cm), litter mass added at the surface; (b) mass loss in a hollow (WT = 0 cm), litter mass added at the surface; (c) mass loss in hummock (WT = 40 cm), litter mass added as below ground litter, i.e., dead plant roots, at the 40-year depth. (d) This graph is comparable with graph 1a, i.e., mass loss in a hummock (WT = 40 cm), mass added at the surface, but two alternative decomposition rate constants for the most easily degradable carbon fraction are used, $k = -0.08$ and -0.02 year^{-1} (arrows from the rate constants indicate the accompanying line in the graph).

Most of the decomposition of plant litter added at the surface of a hummock will proceed under oxic conditions, generating solely CO_2 (Figure 1a), before it reaches the anoxic zone, where both CO_2 and CH_4 may be produced (Figures 1a and 2). By the time it reaches the anoxic zone, however, only a small proportion of the initial amount of plant litter will remain for anaerobic decomposition, and its organic chemical composition will have changed dramatically, greatly increasing its recalcitrance [*Bohlin et al.*, 1989]. This is illustrated by the functions presented in Figure 1d describing the decomposition of two relatively easily degradable fractions of plant material, with exponential decay constants, k, of -0.08 or -0.02 year^{-1}, and a recalcitrant fraction with a decay constant of -0.001. Applying a decay constant of -0.08 for the more easily degradable material leaves only recalcitrant material for anaerobic decomposition when the mate-

rial passes into the anoxic zone after 40 years of degradation. If the material is slightly less readily degradable ($k = -0.02$ year^{-1}), more relatively easily degradable material remains for anaerobic decomposition, resulting in substantially more methane (Figure 1d). In contrast, aboveground plant litter with the same organic chemical composition in a hollow will be available for anaerobic degradation immediately after senescence, and CH_4 will be produced from a much larger fraction of the organic material (see Figure 1b).

3. TERMINAL CARBON MINERALIZATION AND ITS PARTITIONING INTO CO_2 AND CH_4

In this section, we present a brief review of the metabolic reactions involved in the mineralization of organic matter in peatlands and their specific requirements. In aerobic

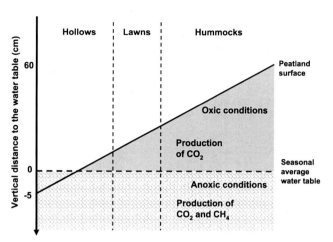

Figure 2. Relationship between mire microtopography and extension of the oxic zone.

metabolism, oxygen serves two main functions: (1) as a terminal e⁻ acceptor during respiration, and (2) as a direct oxidant of organic molecules. Oxygen-based respiration is a key component of the primary metabolism of aerobic organisms because it is essential for the re-oxidation of their key electron carriers [usually NADH (nicotinamide adenine dinucleotide)]. In the absence of oxygen, this function can be fulfilled by other oxidized compounds such as N-oxides, oxidized metal ions, organic compounds, S-oxides, carbon dioxide, or protons. These alternative inorganic and organic e⁻ acceptors serve the same basic need in microbial metabolism, i.e., regeneration of the electron carriers, but yield less (or even no) energy. As a direct oxidant of organic molecules, oxygen cannot be replaced by other compounds because it involves the enzyme-driven insertion of the oxygen molecule into the substrate molecule by oxygenases. However, similar reactions to those catalyzed by oxygenases can also proceed in the absence of oxygen and can occur in strictly anaerobic organisms [cf. *Zehnder and Svensson*, 1986].

In general, the microorganisms responsible for the mineralization of organic matter are unable to take up particulate organic material, so the first step of mineralization involves breaking down solid, insoluble biopolymers by exoenzymes into soluble compounds. This hydrolyzation is the first essential step in the microbial utilization of organic matter (OM) and is ultimately the main source of DOC. However, even if the DOC formed via the hydrolyzation of particulate OM is soluble, it may require further breakdown by (exo)enzymes before it can be used in metabolic processes.

Microbial metabolism can be divided into catabolic and anabolic processes. The former involve (1) oxidation reactions leading from the substrate to intermediate products [such as pyruvate, acetyl-Co-A, reduced co-factors, and ad-

enosine tri phosphate (ATP)], and (2) reduction reactions in which the reduced co-factors (e.g., NADH) are reoxidized (and a terminal e⁻ acceptor is reduced) and regenerated (i.e., respiration or fermentation). Anabolic processes are biosynthetic reactions whereby the organisms produce cell-building materials and are generally ATP consuming.

If the composition of the substrate is known, and it is completely mineralized under methanogenic conditions, the partitioning between CO_2 and CH_4 arising from its degradation can be described by the following equation [*Symons and Buswell*, 1933]:

$$C_n H_a O_b + (n - a/4 - b/2) H_2O \rightarrow$$
$$(n/2 + a/8 - b/4) CH_4 + (n/2 - a/8 + b/4) CO_2. \quad (1)$$

Thus, the oxidation state of the carbon should determine the apportionment between CH_4 and CO_2 in the produced gas. The oxidation state of carbon ranges from -4 (CH_4) to $+4$ (e.g., CO_2), but for most compounds present in most substrates available to microorganisms (e.g., carbohydrates, proteins and lipids), the carbon oxidation state typically ranges from -2 to 0 (Figure 3). Thus, when OM is subjected to terminal mineralization under optimal methanogenic conditions (Figure 3), the theoretical expected CO_2/CH_4 quotient will range from 0.3 (fat) to 1 (carbohydrates). Another important conclusion that can be drawn from Figure 3 that has profound implications for the partitioning of organic C into CO_2 and CH_4 during mineralization under methanogenic conditions is that the quotient decreases with reductions in the oxidation state of the C compound used as an e⁻ donor. Degradation of organic material normally proceeds sequentially, starting with the more readily degradable carbohydrate polymers (although *Sphagnum* carbohydrate polymers are resistant to degradation) followed by degradation of polymers of fatty acids and aromatic compounds. Thus, according to equation (1), greater proportions of CH_4, relative to CO_2, can be produced during the mineralization of old, more decomposed peat than during the mineralization of younger, less recalcitrant material.

During degradation of organic matter under anoxic conditions, and in the absence of alternative e⁻ acceptors for respiration, and under methanogenic conditions, methane production is integral for terminal mineralization of organic matter because it serves a H_2 sink. Many fermentative bacteria form H_2 as a terminal electron sink, as this gives a higher potential energy yield than passing electrons to carbon compounds and forming reduced organic end products. However, the anoxic oxidation of, e.g., volatile fatty acids and other fermentation products by acetogenic bacteria, which is an important step toward complete degradation of organic matter, is only favorable at low partial pressures of H_2 ($<10^{-3}$ atm) [*Wolin and Miller*, 1982]. This is because acetogenic bacteria rely on

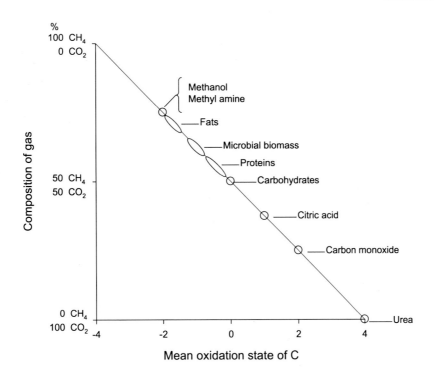

Figure 3. Theoretical CO_2 and CH_4 yields from the complete mineralization of various organic compounds under optimal methanogenic conditions. Modified after *Gujer and Zhender* [1983], reprinted with permission from the copyright holders, IWA.

NADH to maintain their electron flow, and the regeneration of this electron carrier is only exergonic at low hydrogen partial pressures. Thus, H_2 consumption during methanogenesis is important for driving the anoxic degradation of organic matter, as they are capable of sustaining the low partial pressures that the fermenting microbial communities require. In peatlands, and other ecosystems where fermentation of organic matter proceeds, microbial communities other than methanogens (e.g., sulfate and iron-reducing bacteria) can also suppress the partial pressure of H_2. Such competition for electrons by other functional groups will divert the electron flow away from methanogenesis increasing the relative yield of CO_2 at the expense of CH_4. Such processes are not accounted for in the theoretical concept of equation (1).

From a mass balance perspective, the main control of the partitioning between CO_2 and CH_4 is clearly the proportional mass loss during oxic conditions. Further, since degradation proceeds according to a negative exponential (or close to exponential) function, even after a quite limited time under oxic conditions, most of the relatively easily degradable material will have been decomposed by aerobic organisms, resulting solely in the production of CO_2. Even if the organic material that is transferred into the anoxic zone is in a more reduced oxidation state that is more favorable for the production of CH_4 than CO_2, it often constitutes a minor proportion of the original amount of dead plant litter. The remaining fraction is also normally most recalcitrant to degradation and thus yields small amounts of CH_4 and CO_2 per unit mass. The relatively large mass loss under oxic conditions of plant litter added to peatland surfaces implies that root litter, produced in the anoxic zone or the adjacent periodically oxic/anoxic zone, makes a major contribution to the production of methane from plant biomass. The importance of exudates of readily degradable compounds from plant roots is widely recognized, but the addition of undecomposed or relatively undecomposed plant root material provides an important source of organic carbon that is available for anaerobic degradation and from a C mass perspective probably is more important than the exudates. However, the terminal carbon mineralization of any cohort of peat litter under methanogenic conditions, i.e., no competing terminal e^- acceptors available, should yield CO_2:CH_4 ratios lower than 1.

4. EVALUATION OF LITERATURE DATA ON THE PARTITIONING OF MINERALIZED CARBON FROM PEAT TO CO_2 AND CH_4

In our analysis, we compiled results from investigations in which simultaneous measurements of CO_2 and CH_4 production rates were obtained during anoxic incubations of peat soil samples in order to elucidate the potential range of expected CO_2/CH_4 quotients and to identify environmental variables that control the partitioning. In the first step, we excluded any values originating from peat sampled in the uppermost oxic zone (the "acrotelm"), since anoxic incubation of samples from oxic zones will yield biased results with respect to the partitioning between CO_2 and CH_4. Samples from the oxic, or the transitional, zone incubated under anoxic conditions will not be at a steady state with respect to the anaerobic microbial community, and the partitioning of emissions into CO_2 and CH_4 strongly depends on the abundance and metabolic status of this community. In this context, it should be noted that CO_2/CH_4 quotients obtained from anoxic incubations of peat from above the groundwater table are manyfold larger than those obtained from incubations of peat from below the groundwater table, primarily because the former lack established or viable methanogenic communities. This first selection resulted in data on CO_2/CH_4 quotients ($n = 213$, Figure 4a) from 16 peer-reviewed journal articles [*Basiliko and Yavitt*, 2001; *Bergman et al.*, 1999; *Blodau et al.*, 2004; *Clymo and Bryant*, 2008; *Clymo and Pearce*, 1995; *Coles and Yavitt*, 2002; *Dettling et al.*, 2006; *Duddleston et al.*, 2002; *Glatzel et al.*, 2004; *Magnusson*, 1993; *Moore and Dalva*, 1997; *Rinnan et al.*, 2003; *Updegraff et al.*, 1995; *Yavitt and Seidmann-Zager*, 2006; *Yavitt et al.*, 1987, 2000, 2005].

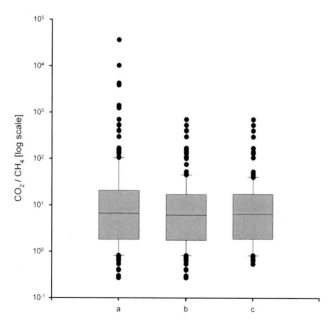

Figure 4. The distribution of CO_2/CH_4 quotients available in the literature (a) before and (b) after exclusion based on methodological aspects and (c) after sample manipulation or amendments. (See text for explanation.)

4.1. Statistical Distribution of CO_2/CH_4 Quotients

Literature values for the quotient of CO_2/CH_4 production rates derived from laboratory incubations of peat under anoxic conditions reported in the literature show tremendous variation, ranging from 0.5 to 36,000 (Figure 4a). The distributions of the quotients are also highly skewed, with a median of 7 and a mean of 305. However, the methodological procedures used to derive CO_2 and CH_4 production rates in some of the investigations raised concerns, including homogenization of peat samples using blender devices prior to incubation [*Duddleston et al.*, 2002] and rotating CH_4 production assay mixtures during incubation [*Moore and Dalva*, 1997]. Both of these procedures are known to affect the production of methane either through disruption of the microbial consortia or through changes in the substrate signature of the sample. Disruption of methanogenic consortia is well known to reduce the rate of methane production (M. Öquist, personal communication, 2008). Values from these investigations were excluded from further analysis. This constraint reduced the sample size to $n = 208$ and mainly removed high values representing the most extreme outliers (Figure 4b, median = 6, average = 28). The high excluded CO_2/CH_4 values emanated mainly from low rates of methane production, probably caused by the physical disturbance.

Finally, to limit the analysis to data obtained from incubations of samples that had not been subjected to any experimental treatment, we removed values obtained from incubations of material that had been subjected to any form of amendment or other experimental disturbance, such as addition of C substrates or changes in pH [*Bergman et al.*, 1999; *Dettling et al.*, 2006; *Yavitt and Seidmann-Zager*, 2006; *Yavitt et al.*, 1987]. However, values obtained from any control samples used in these investigations were kept in the analysis. This constraint removed many of the extreme values observed at the low end of the range (Figure 4c). We believe that the final dataset ($n = 120$, median = 6, average = 29) includes all of the data providing close indications of potential CO_2/CH_4 quotients generated during boreal peatland peat mineralization. However, the values discarded in this final selection step may provide information on the environmental variables affecting CO_2/CH_4 quotients during mineralization and are evaluated separately.

We can conclude, from the retained data, that the range of CO_2/CH_4 quotients arising from the degradation of peat from northern peatlands is extremely variable. Further, this variability is presumably linked to differences in the inherent properties of the peat matrix, including the structure of its microbial populations and their ability to utilize available substrates. The availability of alternative e^- acceptors for respiration may be another factor that influences the CO_2/CH_4

quotients, since their presence diverts some of the electron flow away from methanogenesis under anoxic conditions. Since peatland ecosystems are highly redox-sensitive, highly site-specific variations in the availability of oxidized forms of potential e^- acceptors (such as nitrogen, iron, and sulfur) may explain some of the variation in observed quotients, and the presence of alternative e^- acceptors always causes increases in the CO_2/CH_4 quotients. It is likely that the presence of such alternative e^- acceptors is most important in the unsaturated or periodically oxic/anoxic zone of the soil profile, where fluctuating water tables induce fluctuations between oxic and anoxic conditions. This was a reason why we removed data obtained from incubations of samples from this zone from the analysis.

4.2. Influence of Depth From the Peatland Surface on the CO_2/CH_4 Quotients

Results of studies in which a depth gradient in the catotelm peat was included in the experimental design indicate that the CO_2/CH_4 quotients increases with depth below the soil surface (Figure 5) [*Yavitt et al.*, 1987; *Magnusson*, 1993; *Moore and Dalva*, 1997; *Blodau et al.*, 2004]. This implies that the potential for peat as a methanogenic substrate, relative to a substrate for CO_2 production, decreases with increases in its age and degree of recalcitrance. This implication is corroborated by results from investigations (albeit very few) in which ratios have been obtained for deep peat deposits (>1m), which can yield very high CO_2/CH_4 quotients, up to 700 [*Updegraff et al.*, 1995]. However, it is conceivable that this pattern may be disrupted at sites where fresh organic material is introduced at depth through the activity of plant roots (Figures 1a, 1c, and 2) or if alternative inorganic e^- acceptors are available. In addition, it must be noted that although this pattern appears to be relatively consistent for individual sites, it is not possible to generalize among sites with respect to the overall size in CO_2/CH_4 quotient.

The general increase in the CO_2/CH_4 quotients with depth conflicts with expectations based on theoretical considerations of methane formation from substrates in relation to the oxidation state of C [equation (1)]. Older, more recalcitrant C is generally in a more reduced form than fresh plant litter

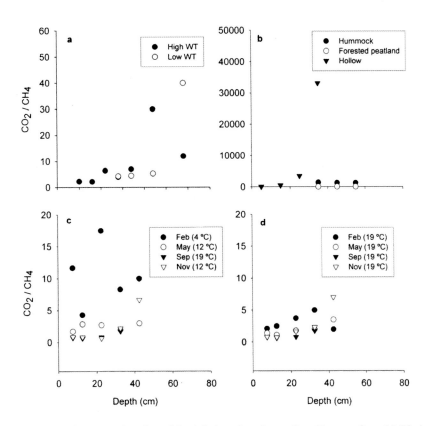

Figure 5. The CO_2/CH_4 quotients as a function of depth below the mire surface. Data are from (a) *Blodau et al.* [2004], (b) *Moore and Dalva* [1997], and (c and d), *Yavitt et al.* [1987].

and thus should theoretically yield lower CO$_2$/CH$_4$ quotients (Figure 3). However, such relationships assume complete mineralization, while microbial degradation of peat under anoxic conditions involves, to a large extent, fermentation processes in which there is no final reduction of a terminal e$^-$ acceptor, resulting in deviations from theoretically expected CO$_2$/CH$_4$ quotients. In addition, the growth of plant roots deeper in the profile may supply fresher organic material. Further, although most relevant investigations indicate that the CO$_2$/CH$_4$ quotients increases with depth, there have been few such studies to date, and the samples used have originated from widely varying types of peatlands and sub-habitats. This, in combination with the factors that obscure relationships among the pertinent variables outlined above, makes attempts to draw definitive general conclusions premature.

4.3. Relationship Between Net C Mineralization and the CO$_2$/CH$_4$ Quotients

The influence of CO$_2$ and CH$_4$ production rates on the resulting partitioning into CO$_2$ and CH$_4$ can be evaluated from studies in which data on their production have been ex-

pressed per unit mass of dry peat or present data that can be used to calculate production rates in this form. The production of both CO$_2$ and CH$_4$ are clearly positively correlated, although there is substantial variation among available data with respect to overall rates (Figure 6). It is also possible to deduce a negative relationship between CO$_2$/CH$_4$ rate quotients and total C mineralization rates (Figure 7). However, this negative relationship is highly dependent on the investigation by *Yavitt et al.* [1987], and the relationship is not statistically significant if data from this investigation are removed from the analysis.

4.4. Effect of Organic C Source on the CO$_2$/CH$_4$ Quotient

The effects of adding various carbon substrates to peat samples on production rates of CO$_2$ and CH$_4$ or both can provide insights regarding the biogeochemistry involved in mineralization processes. The examined literature contains results on responses of both CO$_2$ and CH$_4$ production to glucose amendments [*Yavitt et al.*, 1987; *Bergman et al.*, 1999; *Dettling et al.*, 2006; *Yavitt and Seidmann-Zager*, 2006], starch [*Bergman et al.*, 1999], H$_2$/CO$_2$ [*Yavitt et al.*, 1987], and ethanol [*Yavitt and Seidmann-Zager*, 2006] under an-

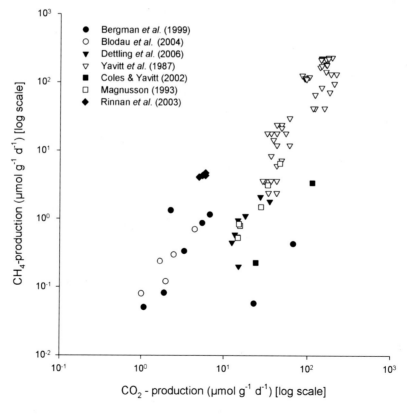

Figure 6. Relationship between CO$_2$ and CH$_4$ production rates in peat samples incubated under anoxic conditions.

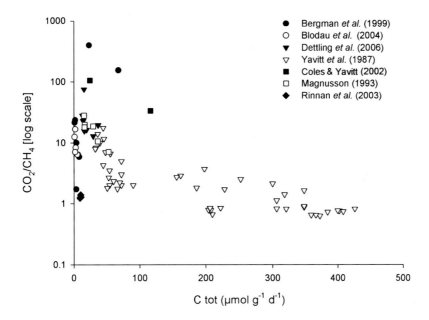

Figure 7. The quotients of rates of production of CO_2 and CH_4 as a function of the net C mineralization rate ($CO_2 + CH_4$) observed in peat samples.

oxic conditions. The results, however, are ambiguous. *Yavitt et al.* [1987] found that glucose additions resulted in, at most, very modest reductions in CO_2/CH_4 quotients in anoxic incubations. On a similar note, *Bergman et al.* [1999] observed decreases in ratios in most cases except at temperatures >15 C and pH > 6.8, when ratios increased, and similar patterns were observed when samples were amended with starch rather than glucose. *Dettling et al.* [2006] found different responses depending on peatland types. In peat samples from a dry bog and a sedge fen, the ratios decreased after glucose amendments, while they increased in samples from a wet bog and a forested fen. *Yavitt and Seidmann-Zager* [2006] found that glucose amendments could either increase or decrease the quotients of CO_2/CH_4. In all these investigations, the addition of glucose enhanced both CO_2 production and CH_4 production, compared to unamended controls, although the effect of the additions on the CO_2/CH_4 quotients differed among studies. These differences in responses of terminal mineralization to the addition of readily available C sources highlight the variability of the processes involved. Probable contributors to this variability include variations in the availability of alternative terminal e^- acceptors and in the microbial populations.

As far as we know, the effects of amending peat samples with H_2/CO_2 on both CO_2 and CH_4 production have been monitored only in one investigation [*Yavitt et al.*, 1987], in which the amendments slightly decreased CO_2 production and increased CH_4 production, leading to a concomitant decrease in the CO_2/CH_4 quotients. This is consistent with expectations since H_2 and CO_2 can serve as an e donor and e acceptor, respectively, in methane formation. Furthermore, additions of H_2 can inhibit complete fermentation of organic compounds because the re-oxidation of metabolic co-factors is inhibited by raised partial pressures of H_2, thereby decreasing CO_2 production [e.g., *Gujer and Zhender*, 1983]. In combination with the potential adverse effects on CO_2 production from fermentation resulting from high partial pressures of H_2, it is conceivable that CO_2/CH_4 quotients generally will decrease as a result.

Ethanol was used to amend peat samples in an investigation by *Yavitt and Seidmann-Zager* [2006], and its effects on CO_2 and CH_4 production resulted in a decrease in CO_2/CH_4 quotients both in the short term (2–5 days) and long term (>50 days). Ethanol is a known precursor of methane formation [e.g., *McCarty*, 1981] and is also a compound of ecological relevance for peatland ecosystems because it can be an important end product of anaerobic carbohydrate catabolism in wetland plant roots [*Gibbs and Greenway*, 2003].

Due to the limited number of relevant investigations, it is not yet possible to draw definitive conclusions regarding the effect of adding C substrates on the partitioning of mineralized carbon under anoxic conditions into CO_2 and CH_4. However, the effects of such amendments on CO_2/CH_4 quotients are clearly highly site specific and are presumably influenced by inherent peat characteristics. It is also evident that other factors, such as temperature and pH, exert strong effects on the mineralization of the added compounds [*Bergman et al.*, 1999]. Thus, these variables should be included in the experimental matrices of studies intended to resolve related issues.

4.5. Effect of Temperature on the CO$_2$/CH$_4$ Quotient

Available literature data suggest that the CO$_2$/CH$_4$ quotient decreases with increasing temperature. *Yavitt et al.* [1987] collected peat samples from different depths ($n = 5$) at regular intervals ($n = 4$) over a year and incubated them at prevailing in situ temperatures (4°–19°C). The reported CO$_2$/CH$_4$ quotients correlated negatively (and significantly) with temperature ($r = 0.61$, $p = 0.05$; values for CH$_4$ and CO$_2$ production deduced from graphs). Moreover, from the work of *Updegraf et al.* [1995] and *Bergman et al.* [1999], we can draw similar conclusions. Negative correlations between CO$_2$/CH$_4$ quotients and temperature can be explained by the high values for temperature responses (e.g., Q_{10} values) reported for CH$_4$ production relative to CO$_2$ production [e.g., *Bergman et al.*, 2000; *Svensson*, 1984]. Thus, the factor by which CH$_4$ production increases over a given temperature interval will be larger than the corresponding increase in CO$_2$ production, leading to a reduction in the CO$_2$/CH$_4$ quotient. However, it is important to note that the variability in possible peat soil temperatures is several orders of magnitude lower than the possible variability in reported CO$_2$/CH$_4$ quotients (Figure 4). Thus, for any given site, the above generalization may be valid, but it is not possible to draw conclusions with similar validity for between-site comparisons.

4.6. Effect of pH on the CO$_2$/CH$_4$ Quotient

Results of experiments by *Yavitt et al.* [1987] and *Bergman et al.* [1999] in which the pH of incubated peat samples was varied (from 3.1 to 6.4 and 4.3 to 6.8, respectively) indicate that raising the pH of the soil matrix in laboratory incubations enhances CH$_4$ production relative to CO$_2$ production and thus reduces the CO$_2$/CH$_4$ quotient. These findings are consistent with the fact that most isolated methanogens are known to have pH optima around 7, even though the environments where they are usually encountered are more acidic. Thus, increasing the pH may have a more positive effect on methane production than CO$_2$ production. In addition, at higher pH, the organisms can spend less energy on maintaining the proton gradient between them and their surroundings, with consequent reductions in rates of CO$_2$ production. In contrast, *Dettling et al.* [2006] found that the CO$_2$/CH$_4$ quotient increased with increases in native pH in incubations of peat samples from various types of peatland with a natural pH range of 4.3–6.7. However, natural variations in pH

are likely to be accompanied by variations in other relevant environmental variables (e.g., substrate availability, redox conditions, and microbial population structure), confounding the influence of pH per se on the CO$_2$/CH$_4$ quotient.

The full factorial experimental design used by *Bergman et al.* [1999] facilitates a closer analysis of the combination of factors (temperature, pH, and C additions) that influence CO$_2$/CH$_4$ quotients (Table 1). The table presents the outcome from a multiple linear regression analysis with log-transformed values of production rates or CO$_2$/CH$_4$ quotients as response variables. It is evident that C additions increase both CH$_4$ production and CO$_2$ production. However, temperature mainly controls the CH$_4$ production, while pH is the main control for CO$_2$ production. Thus, variations in both pH and temperature strongly contribute to variations in CO$_2$/CH$_4$ quotients among incubated samples.

4.7. Comparison of CO$_2$/CH$_4$ Quotients Obtained From Laboratory Incubations and Field Measurements

Obtaining reliable in situ measurements of CO$_2$ and CH$_4$ production rates in peatland soils is difficult. One approach that has been used to obtain such measurements is to calculate production rates based on detailed steady state concentration profiles obtained from field measurements. To our knowledge, this in situ approach has not been applied in parallel with laboratory incubations in any published investigation. However, in a study by *Blodau et al.* [2004], rates were derived from both concentration profiles in mesocosms and samples of the same material from corresponding depths incubated in laboratory bottles. The results from the cited study raise concerns with respect to the validity of laboratory bot-

Table 1. Output from Three Different Stepwise Multiple Linear Regression Models with Rates of CH$_4$ Production, CO$_2$ Production, and CO$_2$/CH$_4$, Respectively, as Response Variables, Based on Data Presented by *Bergman et al.* [1999]

Y^a	X_1^b	X_2^b	R^{2c}	p
CH$_4$ production	Temp. (0.81)	C add (0.54)	0.54; 0.83	<0.01
CO$_2$ production	pH (0.68)	C add (0.38)	0.42; 0.55	0.02
CO$_2$:CH$_4$	pH (0.67)	Temp (−0.43)	0.37; 0.54	0.01

^{a}Response variable in respective (SMLR) model, log-transformed rates and quotients.

^{b}X_1 and X_2 represents the independent variables and appear in order of importance in the SMLR model. Values in parentheses represent standardized regression coefficients; that is, they are directly comparable within each model.

^{c}Accumulating R^2, the first value is the variance proportion explained by X_1, the second value is the cumulative variance proportion explained by both X variables.

tle incubations in terms of their representativeness of field conditions, since the CO_2/CH_4 quotients based on calculations from the concentration profiles in the mesocosms were three- to fourfold lower than those obtained from the laboratory bottle incubations. It is conceivable that the disturbance involved in both sampling peat and preparing the samples for bottle incubations introduces artifacts in the measurements. Anoxic mineralization is carried out by consortia of various microbial populations and relies on well established syntrophy both between and among different groups of microbes [cf. *Conrad*, 1999]. Even slight disturbance of these consortia may have significant effects on their degradation capacity, and the impact may conceivably be even greater if production rates obtained from bottle incubations are compared with rates derived from in situ profiles. Although only one available reference supports this conclusion, as yet, we believe that this concern should not be neglected.

It may also be noted that quotients of CO_2/CH_4 rates inferred from field measurements of concentration depth profiles at an English peat bog yields values of 14 [*Clymo and Pearce*, 1995] and 10 [*Clymo and Bryant*, 2008], respectively. Even if these field-derived values cannot be directly compared with quotients on rates derived from laboratory incubations, they are in the low range of the data compiled in this study.

4.8. Comparison with Flux Data

Even if it is impossible to deduce the partitioning of plant litter mass losses into CO_2 and CH_4 simply from gas exchange measurements of these species between peat surfaces and the atmosphere, it is of interest to compare the magnitude of reported CO_2/CH_4 quotients from laboratory incubations with the magnitude of peatland-atmosphere gas exchange rates. To illustrate this, measured and estimated flux data obtained from Degerö Stormyr on CH_4 emissions and ecosystem respiration rates, respectively, can be used [cf. *Nilsson et al.*, 2008]. The average seasonal ecosystem respiration rate at Degerö Stormyr amounts to ~2 g CO_2-C m^{-2} d^{-1}, while corresponding CH_4 emissions are 0.075 g CH_4-C m^{-2} d^{-1}, yielding a CO_2/CH_4 quotient for C emissions of ~27. However, some of the ecosystem respiration is due to autotrophic respiration, and the net emission of CH_4 is the difference between production and oxidation. Although we do not know the exact contributions of these processes, we assume that root respiration contributes 25–50% of total CO_2 emissions, while around 5–25% of the produced CH_4 is oxidized prior to emission, based on estimates obtained from a similar peatland complex in the same region provided by *Sundh et al.* [1994]. The field gas exchange data, given the above assumptions, indicate that CO_2/CH_4

quotients of the emissions from Degerö Stormyr range over about 11 to 19. This range is similar to those obtained from bottle incubations in the literature, albeit slightly higher than the median (Figure 4). In addition, it agrees well with the ratio of 14 [*Clymo and Pearce*, 1995] and 10 [*Clymo and Bryant*, 2008] estimated for a whole peatland system based on concentration profiles, assuming steady state conditions (down to 5- and 7-m depths, respectively).

4.9. What Can We Learn From Available Data?

Based on available data, it seems difficult to make many generalizations about environmental controls on the CO_2: CH_4 partitioning under anoxic conditions in peatlands. The median value after filtering for methodological concerns and experimental treatments was about 6–7 clearly deviating from any expected value, assuming anaerobic terminal mineralization of plant material under methanogenic conditions. Published articles do not give much guidance to what extent this deviation emanates from either the occurrence of alternate terminal e acceptors, favoring CO_2-production, or fermentation processes. However, electron transfer of dissolved organic matter is suggested as a source of alternative pathways that contribute significantly to the oxidation of reduced organic substrates by anoxic respiration or by maintaining the respiratory activity of sulfate reducers via provision of thiosulfate [*Heitman et al.*, 2007]. Thus, it is likely that environmental processes directing the electron flow during organic matter mineralization are highly important for determining the partitioning of its end products. The occurrence of such compounds would be highly dependent on prevailing redox conditions and also on factors linked to the genesis and development of individual peatland sites as well as the chemical composition of groundwater entering the peatland. Such differences may explain the large variations in the observed CO_2/CH_4 quotient both within and among peatlands reported in the literature.

A few generalizations, however, can be justified based on published data, one being that the CO_2/CH_4 quotient increases with depth. This is contrary to what would be expected based on the fact that decomposition of organic material leads to enrichments of more reduced organic chemical compounds; that is, the relative proportion of carbon increases while the proportion oxygen decreases. Decomposition of more reduced organic compounds, under methanogenic conditions, would result in a lowered CO_2/CH_4 quotient (see Figure 3).

Experimentally increased pH as well as increased temperature seem to decrease the CO_2/CH_4 quotient. Increased pH increases both the production of CO_2 and CH_4, but the effect is relatively larger for methane production than for production CO_2. This can be explained by the fact that optimal pH for methanogens is around 7. The decrease in CO$_2$/CH$_4$ quotient with increasing temperature might reflect the higher Q_{10} values for the production of methane relative to the production of CO$_2$. The basis for these generalizations is, though, still very limited and should therefore be used with caution.

5. RESEARCH NEEDS

A major conclusion that can be drawn from our assessment is that there are large gaps in our knowledge and understanding of the partitioning of terminal peat mineralization into CO$_2$ and CH$_4$. Most notably, there is a clear lack of obvious explanations for the large range of CO$_2$/CH$_4$ quotients reported in the literature. This presents a challenge for future research aimed at resolving the controls of the partitioning of terminally mineralized C in plant biopolymers in peatlands into CO$_2$ and CH$_4$. Clearly, production rates of both CH$_4$ and CO$_2$ in incubated samples have been measured in a few studies, and thus, additional studies would be highly valuable. However, we would like to give a selection of recommendations for such initiatives, regarding both methodology and experimental design, which we think would facilitate attempts by the scientific community to resolve issues related to CO$_2$:CH$_4$ partitioning.

5.1. Methodological Aspects

Some of the variations in CO$_2$/CH$_4$ quotients reported in the examined studies probably originate from differences in the experimental protocols used to prepare and incubate samples. We recommend sampling and sample handling procedures that minimize the disturbance of the peat because of the adverse effects this can have on both fermentation and methanogenesis. In addition, systematic uses of abiotic variables on production rates are crucial, but it is unclear whether or not such controls were included in many of the cited investigations.

Also a more detailed description of the sampled peatland and the sampled peat would be appreciated. That would allow for a more detailed analysis from published articles on how the peat substrate influences the CO$_2$/CH$_4$ quotient. A more detailed description of the plant community composition at the peatland surface as well as of the plant remains in the sampled peat represents two important parameters in this respect. In future studies, it is also recommended that the influence of alternative e$^-$ acceptors (e.g., oxidized Fe and S compounds) is assessed by measuring these components in incubation samples.

We would also like to encourage further attempts to determine production rates in situ and calculations based on depth concentration profiles, such as those presented by *Clymo and Pearce* [1995], *Blodau et al.* [2004], and *Clymo and Bryant* [2008]. Comparison of such calculations with data obtained from laboratory bottle incubations can highlight possible discrepancies between results obtained using these two methodological approaches and, hence, indications of the validity of the data. However, there is a need for such procedures to be able to distinguish between CO$_2$ produced in situ by heterotrophic and root respiration. This requirement, however, becomes obsolete with increasing depth due to the lack of viable roots deeper down in the profile.

Another key concern is the periodically oxic/anoxic zone situated between completely oxic and anoxic zones, which may be very important for the overall partitioning of mineralized C into CO$_2$ or CH$_4$. In the theoretical framework presented earlier in this text (Figure 1), we have not considered the periodically oxic/anoxic zone in order to facilitate a schematic presentation of the theoretical concept. In addition, we have systematically removed available measurements obtained from samples from the periodically oxic/anoxic zone because we judge that such results cannot be representative of the measured variables in either completely oxic or anoxic conditions. Estimates of production rates from concentration profiles in this zone will be very uncertain because of the absence of steady state conditions in it, and in laboratory incubations of periodically oxic/anoxic zone samples, the experimental design must correspond to the potential (and probably nonlinear) range of variables such as the redox state. Again, the need for separating CO$_2$ production from heterotrophic activity from that of root respiration in sampling protocols applied in situ is large because the periodically oxic/anoxic zone is where root biomass is most abundant.

5.2. Experimental Design

Laboratory investigations have an advantage over in situ measurements in that they facilitate systematic evaluation of the main environmental controls on CO$_2$/CH$_4$ quotients. Variables such as temperature, pH, and substrate quality clearly affect the decomposition of plant litter and are major controls of its partitioning into CO$_2$ and CH$_4$. Moreover, we can conclude that there are strong interactions between them [*Bergman et al.*, 1999], and thus, rationally designed full factorial (or reduced factorial) experiments in which they are systematically varied are encouraged. Attempts should also be made to validate any laboratory-based experiments by assessing their representativeness of in situ conditions, which can be done by field manipulation experiments.

An integral aspect of elucidating the partitioning of peat decay into CO$_2$ and CH$_4$ is its implications for overall C balances of northern peatlands and their greenhouse gas dynamics. Thus, it is advisable to carry out investigations on peat material or at peatland sites that are representative with respect to their spatial distribution in the northern hemisphere. Choosing sites where investigations of other aspects of C balances are being carried out can be advantageous. This approach facilitates synthesis and validation of data across spatial and temporal scales and can also be beneficial with respect to comprehensive modeling initiatives of northern peatland C balances.

REFERENCES

Agren, G., and E. Bosatta (1996), *Theoretical Ecosystem Ecology, Understanding Element Cycles*, 234 pp., Cambridge Univ. Press, New York.

Basiliko, N., and J. B. Yavitt (2001), Influence of Ni, Co, Fe, and Na additions on methane production in *Sphagnum*-dominated Northern American peatlands, *Biogeochemistry*, *52*, 133–153.

Belyea, L. R., and R. S. Clymo (2001), Feedback control of the rate of peat formation, *Proc. R. Soc. London, Ser. B*, *268*, 1315–1321.

Bergman, I., P. Lundberg, and M. Nilsson (1999), Microbial carbon mineralisation in an acid surface peat: Effects of environmental factors in laboratory incubations, *Soil Biol. Biochem.*, *31*, 1867–1877.

Bergman, I., M. Klarqvist, and M. Nilsson (2000), Seasonal variation in rates of methane production from peat of various botanical origins: Effects of temperature and substrate quality, *FEMS Microbiol. Ecol.*, *33*, 181–189.

Blodau, C., N. Basiliko, and T. R. Moore (2004), Carbon turnover in peatland mesocosms exposed to different water table levels, *Biogeochemistry*, *67*, 331–351.

Bohlin, E., M. Hemalainen, and T. Sunden (1989), Botanical and chemical characterization of peat using multivariate methods, *Soil Sci.*, *147*, 252–263.

Clymo, R. S., and C. L. Bryant (2008), Diffusion and mass flow of dissolved carbon dioxide, methane, and dissolved organic carbon in a 7-m deep raised peat bog, *Geochim. Cosmochim. Acta*, *72*, 2048–2066.

Clymo, R. S., and D. M. E. Pearce (1995), Methane and carbon dioxide production in, transport through, and efflux from a peatland, *Philos. Trans. R. Soc. London, Ser. A*, *350*, 249–259.

Coles, J. R. P., and J. B. Yavitt (2002), Control of methane metabolism in a forested northern wetland, New York state, by aeration, substrates and peat size fractions, *Geomicrobiol. J.*, *19*, 293–315.

Conrad, R. (1999), Contribution of hydrogen to methane production and control of hydrogen concentrations in methanogenic soils and sediments, *FEMS Microbiol. Ecol.*, *3*, 193–202.

Dettling, M. D., J. B. Yavitt, and S. H. Zinder (2006), Control of organic carbon mineralization by alternative electron acceptors in four peatlands, central New York state, USA, *Wetlands*, *26*, 916–927.

Duddleston, K. N., M. A. Kinney, R. P. Kiene, and M. E. Hines (2002), Anaerobic microbial biogeochemistry in a northern bog: Acetate as a dominant metabolic end product, *Global Biogeochem. Cycles*, *16*(4), 1063, doi:10.1029/2001GB001402.

Gibbs, J., and H. Greenway (2003), Mechanisms of anoxia tolerance in plants. I. Growth, survival and anaerobic catabolism, *Funct. Plant Biol.*, *30*, 1–47.

Glatzel, S., N. Basiliko, and T. Moore (2004), Carbon dioxide and methane production potentials of peats from natural, harvested and restored sites, eastern Quebec, Canada, *Wetlands*, *24*, 261–267.

Gujer, W., and A. J. B. Zehnder (1983), Conversion processes in anaerobic digestion, *Water Sci. Technol.*, *15*(8–9), 127–167.

Heitmann, T., T. Goldhammer, J. Beer, and C. Blodau (2007), Electron transfer of dissolved organic matter and its potential significance for anaerobic respiration in a northern bog, *Global Change Biol.*, *13*, 1771–1785.

Magnusson, T. (1993), Carbon-dioxide and methane formation in forest mineral and peat soils during aerobic and anaerobic incubations, *Soil Biol. Biochem.*, *25*, 877–883.

McCarty, P. L. (1981), One hundred years of anaerobic treatment, in *Anaerobic Digestion*, edited by D. E. Hughes, pp. 3–22, Elsevier, New York.

Moore, T. R., and M. Dalva (1997), Methane and carbon dioxide exchange potentials of peat soils in aerobic and anaerobic laboratory incubations, *Soil Biol. Biochem.*, *29*, 1157–1164.

Nedwell, D. B., and A. Watson (1995), CH_4 production, oxidation and emission in a UK ombrotrophic peat bog—Influence of SO_4^{2-} from acid-rain, *Soil Biol. Biochem.*, *27*, 893–903.

Nilsson, M., J. Sagerfors, I. Buffam, H. Laudon, T. Eriksson, A. Grelle, L. Klemedtsson, P. Weslien, and A. Lindroth (2008), Contemporary carbon accumulation in a boreal oligotrophic minerogenic mire—A significant sink after accounting for all C-fluxes, *Global Change Biol.*, *14*, 2317–2332, doi:10.1111/j.1365-2486.2008.01654.x

Popp, T. J., J. P. Chanton, G. J. Whiting, and N. Grant (1999), Methane stable isotope distribution at a *Carex* dominated fen in north central Alberta, *Global Biogeochem. Cycles*, *13*, 1063–1077.

Rietkerk, M., S. C. Dekker, M. J. Wassen, A. W. M. Verkroost, and M. F. P. Bierkens (2004), A putative mechanism for bog patterning, *Am. Nat.*, *163*, 699–708.

Rinnan, R., M. Impiö, J. Silvola, T. Holopainen, and P. J. Martikainen (2003), Carbon dioxide and methane fluxes in boreal peatland microcosms with different vegetation cover—Effects of ozone or ultraviolet-B exposure, *Global Change Biol.*, *137*, 475–483.

Roulet, N. T., P. M. Lafleur, P. J. H. Richard, T. Moore, E. R. Humphreys, and J. Bubier (2007), Contemporary carbon balance and late Holocene carbon accumulation in a northern peatland, *Global Change Biol.*, *13*, 397–411.

Saarinen, T. (1997), Biomass and production of two vascular plants in a boreal mesotrophic fen, *Can. J. Bot.*, *74*, 934–938.

Strom, L., A. Ekberg, M. Mastepanov, and T. R. Christensen (2003), The effect of vascular plants on carbon turnover and methane emissions from a tundra wetland, *Global Change Biol.*, *9*, 1185–1192.

Sundh, I., M. Nilsson, G. Granberg, and B. H. Svensson (1994), Depth distribution of microbial production and oxidation of methane in Northern boreal peatlands, *Microb. Ecol.*, *27*, 253–265.

Svensson, B. H. (1984), Different temperature optima for methane formation when enrichments from acid peat are supplied with acetate or hydrogen, *Appl. Environ. Microbiol.*, *48*, 389–394.

Symons, G. E., and A. M. Buswell (1933), The methane formation of carbohydrates, *J. Am. Chem. Soc.*, *55*, 2028–2036.

Updegraff, K., J. Pastor, S. D. Bridgham, and C. A. Johnston (1995), Environmental and substrate controls over carbon and nitrogen mineralization in northern wetlands, *Ecol. Appl.*, *5*, 151–163.

Wallén, B. (1986), Above and below ground dry mass of three main vascular plants on hummocks on a sub-arctic peat bog, *Oikos*, *46*, 51–56.

Wolin, M. J., and T. L. Miller (1982), Interspecies hydrogen transfer: 15 years later, *ASM News*, *48*, 561–565.

Yavitt, J. B., and M. Y. Seidmann-Zager (2006), Methanogenic conditions in northern peat soils, *Geomicrobiol. J.*, *23*, 119–127.

Yavitt, J. B., G. E. Lang, and K. Wieder (1987), Control of carbon mineralization to CH₄ and CO₂ in anaerobic, *Sphagnum*-derived peat from Big Run Bog, West Virginia, *Biogeochemistry*, *4*, 141–157.

Yavitt, J. B., C. J. Williams, and R. K. Wieder (2000), Controls on microbial production of methane and carbon dioxide in three *Sphagnum*-dominated peatland ecosystems as revealed by a reciprocal field peat transplant experiment, *Geomicrobiol. J.*, *17*, 61–88.

Yavitt, J. B., C. J. Williams, and R. K. Wieder (2005), Soil chemistry versus environmental controls on production of CH₄ and CO₂ in northern peatlands, *Eur. J. Soil. Sci.*, *56*, 169–178.

Zehnder, A. J. B., and B. H. Svensson (1986), Life without oxygen: What can and what cannot, *Experimenta*, *42*, 1197–1205.

_____

M. Nilsson and M. Öquist, Department of Forest Ecology and Management, Swedish University of Agricultural Sciences, SE-901 83 Umeå, Sweden. (mats.b.nilsson@sek.slu.se)

Methane Accumulation and Release From Deep Peat: Measurements, Conceptual Models, and Biogeochemical Significance

Paul H. Glaser

Department of Geology and Geophysics, University of Minnesota, Minneapolis, Minnesota, USA

Jeffrey P. Chanton

Department of Oceanography, Florida State University, Tallahassee, Florida, USA

Northern peatlands account for more than half the world's wetlands but are currently estimated to contribute only about a third of the total methane emissions from all wetlands. Increasing data on the dynamics of methane gas bubbles in peat deposits now suggest that these estimates may need to be scaled upward. Rates of methanogenesis may remain high in deep peat strata because of the downward transport of labile root exudates permitting the widespread production of gas bubbles. Recent investigations using an array of methods have reported free-phase gas volumes of 10–20% within both deep and shallow peat strata and episodic ebullition fluxes exceeding 35 g CH_4 m^{-2} per event. Gas bubbles accumulate in overpressured pockets that episodically rupture in response to steep declines in atmospheric pressure or declining water tables. Although these ebullition fluxes are highly variable in both time and space, they appear to dominate the annual methane emissions from northern peatlands and represent a major and underappreciated element of the global methane cycle.

1. INTRODUCTION

During the Holocene an estimated 455 Pg of carbon was transferred from the atmosphere to the organic soils accumulating in northern peatlands [*Harden et al.*, 1992; *Gorham*, 1991]. Northern peatlands now cover more than 350 million ha and account for about one third of the carbon stored in global soils [*Wigley and Schimel*, 2000; *Kivinen and Pakarinen*, 1981]. Although the formation of this large carbon sink would tend to damp climatic warming, peatlands are also important sources for methane a potent greenhouse gas [*Khalil*, 2000]. As a result different models provide contrary predictions on the response of these carbon reservoirs to climatic change. Some models predict that climatic warming will accelerate greenhouse gas emissions from peatlands and that effect will amplify continued warming, whereas other models predict the reverse effect if a warmer but moister climate stimulates peat growth and CO_2 uptake [*Moore et al.*, 1998; *Mathews*, 2000]. One key element of uncertainty in modeling the carbon balance of peatlands is the production, storage, and emissions of free-phase gas, which account for a previously underappreciated source of methane and carbon dioxide emissions to the atmosphere.

Peat accumulation is driven by a high water table, which restricts the zone of rapid decomposition to a thin veneer of organic soil directly above the fluctuating water table. Rates

Carbon Cycling in Northern Peatlands
Geophysical Monograph Series 184
Copyright 2009 by the American Geophysical Union.
10.1029/2008GM000840

of decomposition decline dramatically below the water table where anoxic conditions prevail and the breakdown of organic matter is restricted to the less thermodynamically efficient processes of fermentation and methanogenesis. The biodegradation of organic matter within this deeper anoxic zone may be further reduced by the increasing fraction of organic matter resistant to decay, low temperatures, limited supply of nutrients, and the accumulation of toxic metabolites. Although this deeper anoxic zone represents the bulk of peat deposits, the transformations of organic matter in deeper peat strata are poorly characterized despite their importance for understanding the carbon dynamics of peatlands and the response of peatlands to climate change.

2. INITIAL MODELS OF METHANE PRODUCTION AND TRANSPORT IN PEATLANDS

Peatlands accumulate thick deposits of partially decomposed organic matter under anaerobic conditions and should therefore be important sources of atmospheric methane. However, early models proposed that most of the methane emitted from peatlands was produced just below the water table where (1) the solid phase peat still contains a large fraction of labile carbon compounds and (2) anaerobic metabolism is also stimulated by the release of simple carbon compounds exuded from plant roots [*Andreae and Schimel*, 1989; *Schütz et al.*, 1991]. Methane production was assumed to decline with depth because longer exposure to microbial

metabolism should produce peat with a higher fraction of recalcitrant organic compounds. This model was inspired by detailed studies of methane production in rice paddies but was also supported by the strong linear relationship between methane emissions and net primary production [*Whiting and Chanton*, 1993; *Bellisario et al.*, 1999] and by the modern radiocarbon signature of methane emissions sampled by chambers [*Chanton et al.*, 1995]. Incubation experiments combined with chemical determination of the peat also supported the notion that methane production declined dramatically with depth [*Moore and Dalva*, 1997; *Dunfield et al.*, 1993].

Methane produced in the shallow anaerobic peat can be transported to the atmosphere by three principal mechanisms (1) diffusion, (2) ventilation through the aerenchyma tissue (intracellular air chambers) of vascular plants, and (3) ebullition [*Schütz et al.*, 1991; *Shannon and White*, 1994]. The dominant pathway of methane emissions to the atmosphere was believed to be mass flow through the intracellular air spaces in plant roots and shoots, whereas diffusion through the soil or ebullition (bubbling) were considered secondary transport processes [*Shannon et al.*, 1996; *Sorrell et al.*, 1997; *King et al.*, 1998]. This model was subsequently challenged by measurements of pore water methane in peatlands from Canada and northern Minnesota. These studies showed significant concentrations of methane within the deeper pore waters of both raised bogs and also fens (Figure 1). In northern Minnesota, for example methane concentrations ranged

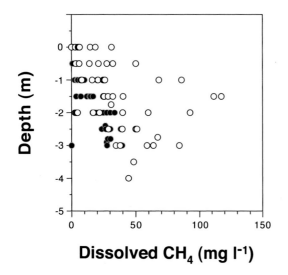

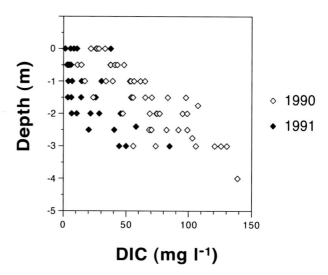

Figure 1. Pore water profiles of dissolved methane and DIC from 34 bogs and fens in the glacial Lake Agassiz peatlands, northern Minnesota. Very high concentrations of methane and DIC accumulated in the deeper pore waters of both bogs and fens at the height of a 3-year drought in 1990. However, the following year, concentrations of biogenic gases dropped by a factor of 2 or 3 after the drought ended. The open symbols indicate data from 1990, whereas the solid symbols are from 1991. Modified from *Glaser et al.* [1997a].

from 15 to 155 mg L^{-1} during the height of a 3-year drought and then declined at the same sampling sites the next year in response to the onset of normal precipitation patterns [*Romanowicz et al.*, 1995; *Glaser et al.*, 1997a]. Data also began to accumulate indicating that northern peatlands contain pockets of free-phase gas in both shallow and deeper (>1 m) peat strata that episodically degass to the atmosphere too rapidly to permit oxidation by methanotrophs [*Strack and Waddington*, 2007; *Rosenberry et al.*, 2006; *Strack et al.*, 2005; *Glaser et al.*, 2004a]. As a result new methods needed to be developed to determine the budgets for free-phase gas within peat deposits with respect to its production, storage, transport, and basin-scale distribution [e.g., *Comas et al.*, 2007, 2008; *Comas and Slater*, 2007].

3. DETECTION OF GAS BUBBLES IN PEAT

The idea that peat deposits contain gas bubbles is not new. Evidence for inflammable gas bubbling up from the bogs of northern Europe was once fairly common, inspiring reports of "will-o'-the-wisp" or ignis fatuus from the 1600s until electricity replaced flames as the primary source for nighttime illumination [*Simpson and Weiner*, 1989]. Peatland ecologists also noted gas bubbling from raised bogs sometimes producing mud volcanoes or rifts on the peat surface [*Aario*, 1932; *Gams and Ruoff*, 1929; *Thomson*, 1924; *Weber*, 1902]. More recently probes were developed to sample gas pockets in deep peat [*Mastepanov and Christensen*, 2008; *Brown*, 1998; *Brown et al.*, 1989; *Dinel et al.*, 1988]. These probes, recently reviewed by *Mastepanov and Christensen* [2008] have detected scattered pockets of gas in the peatlands of Canada and Scandinavia but not Great Britain [*Clymo and Pierce*, 1995]. One drawback of these probes is that they measure the rate that gas flows from the surrounding porous media into the artificial zone of low pressure within the probe. The amount of gas tapped will therefore vary according to (1) the effective porosity and hydraulic conductivity of the surrounding peat, (2) the mechanical strength of the peat, which resists deformation by moving gas bubbles, and (3) the distribution and abundance of gas. As a result probes probably tap different reference volumes (in the sense of *Bear* [1972]) of peat within the same peatland, and the rates of gas uptake will be difficult to translate into in situ concentration values.

An alternative approach involves indirect measurements of the free-phase gas based on geodetic, hydrogeological, or geophysical methods. In the Red Lake Peatland of northern Minnesota instrument stations were set up in 1997 to continuously monitor changes in peat volume (elevation of GPS antennas mounted on tree stumps) and peat pore pressures (instrumented piezometer nests connected to data loggers)

to calculate changes in free-phase gas volumes by three independent methods based on (1) ideal gas laws and Henry's law, (2) barometric efficiency of the wells, and (3) changes in specific storage [*Glaser et al.*, 2004a; *Rosenberry et al.*, 2003]. These independent methods produced comparable values for ebullition fluxes exceeding 35 g CH_4 m^{-2} per event. In addition the later two methods were used to show gas volumes between 10 and 20%, values that agree with those obtained by the analysis of large diameter peat cores from 2- to 3-m depth using magnetic resonance imaging (MRI) [*Glaser et al.*, 1997b, 2006].

Similar values for gas volume and ebullition fluxes have been obtained by recently developed geophysical methods such as time domain reflectometry (TDR) and ground-penetrating radar that rely on the close relationship between volumetric soil moisture and the dielectric permittivity (i.e., ability of a material to be polarized by an electromagnetic field) of soils [*Beckwith and Baird*, 2001; *Strack et al.*, 2005; *Comas et al.*, 2005a, 2005b, 2007]. TDR has the disadvantage of being an invasive technique since the probes inserted into the peat will disrupt the peat fabric and gas distribution to some extent depending upon peat composition. Other types of invasive techniques include soil capacitance probes [*Kellner et al.*, 2005; *Tokida et al.*, 2005a, 2005b] and deformation sensor rods [*Price*, 2003]. All of these methods except the latter yield values for gas content and/or ebullition fluxes similar to those reported for northern Minnesota. These studies also showed the strong relationship between episodic ebullition fluxes and steep declines in atmospheric pressure. Medical imaging methods, such as MRI have also been used to image gas bubbles from large diameter cores recovered from the deeper peat of the glacial Lake Agassiz peatlands [*Glaser et al.*, 1997b, 2006]. Gas volumes ranged from 10 to 20% supporting indirect measurements of gas volumes made from this same site with the geodetic and hydrogeologic methods described above. Although most of these methods cannot measure gas volumes directly, they seem to provide consistent results that peatlands contain large volumes of gas that episodically escape to the atmosphere.

4. MODELS OF METHANE PRODUCTION IN DEEP PEAT

Bubbles may form in a saturated porous media when the concentration of a dissolved gas exceeds its saturation value as defined by Henry's law [*Hemond and Fechner*, 1994]. The partition coefficient in Henry's law is dependent on temperature and to a lesser extent pore pressures, but bubble formation is also influenced by the production rate of that gas, its ability to diffuse to bubble nucleation points, and the material properties affecting the total porosity, hydraulic conductivity, and mechanical strength of the peat fabric. It is therefore not surprising that insoluble gases, such as methane with low mass transfer coefficients, have been reported to exceed their theoretical saturation limits by several factors in bioreactors [*Pauss et al.*, 1990] or peat pore waters [*Glaser et al.*, 1997a; *Romanowicz et al.*, 1995]. Bubbles in anaerobic sediments including peatlands consist primarily of nitrogen, methane, and to a lesser extent, carbon dioxide. The rate of methane production in anaerobic sediments is related to the methane content of the bubbles and inversely to the nitrogen content [*Chanton et al.*, 1989; *Walter et al.*, 2008]. At higher methane production rates, bubble stripping removes nitrogen from pore waters reducing its composition in bubbles equilibrated with pore water [*Martens and Chanton*, 1989; *Kipphut and Martens*, 1982].

The first hydrogeologic investigation in the glacial Lake Agassiz peatlands of Minnesota provided indirect evidence for high production rates of methane in the deeper peat. A zone of abnormal high pressure was observed at 1 m depth under the raised bog at Lost River that persisted for the duration of a 3-year study [*Siegel and Glaser*, 1987]. Although abnormal pore pressures were known in deep sedimentary basins [*Hunt*, 1996; *Neuzil*, 1995], the existence of such a zone of overpressure at only 1 m depth in a peatland was unexpected. During the course of this study benchmarks (spikes driven into trees) along a survey line rose by over 10 cm on the bog indicating significant changes in peat volume and pore pressure at depth [*Almendinger et al.*, 1986]. A mechanism for these overpressures and topographic oscillations only became clearer after a regional hydrogeologic survey was completed of 34 bogs and fens across the glacial Lake Agassiz region of northern Minnesota in 1990 and 1991 [*Glaser et al.*, 1997a; *Siegel et al.*, 1995; *Romanowicz et al.*, 1993, 1995].

During the height of a 3-year drought, zones of overpressure were found in the deeper peat of many raised bogs of this region [*Glaser et al.*, 1997a; *Romanowicz et al.*, 1993, 1995]. These zones of abnormal high pressure were usually associated with pockets of free-phase gas that bubbled vigorously from wide diameter boreholes or ejected mixtures of gas and water from piezometer standpipes. The deeper pore waters of both bogs and fens contained exceptionally high concentrations of dissolved methane and dissolved inorganic carbon (DIC) under drought conditions in 1991 with the concentrations of methane ranging as high as 150 mg L^{-1} (Figure 1). The following year, however, pore water concentrations of methane and DIC fell by a factor of 3, only 1 month after the drought ended. *Romanowicz et al.* [1995] hypothesized that the end of this drought triggered a massive release of free-phase methane trapped under hydraulic confining layers within the deeper peat.

Zones of overpressure were next discovered in the Hudson Bay Lowland where they were again associated with pockets of free-phase gas [*Glaser et al.*, 2004b]. Pore water profiles also showed significant concentrations of dissolved methane that at many sites increased with depth to values as high as 33 mg L^{-1} [*Glaser et al.*, 2004b; *Reeve*, 1996; *Reeve et al.*, 1996]. These findings indicated that rates of methanogenesis must be sufficiently high in the deeper peat to support the formation of gas bubbles despite the reputedly more recalcitrant nature of the peat fabric. Methanogenesis may also be affected by low peat temperatures, but this effect is mitigated by the stability of the thermal regime at deeper depths in any soil profile. In the Red Lake Peatland of northern Minnesota, for example, the annual range in peat temperature may be greater than 25°C at the peat surface, but this range decreases to 6°C at 1 m, 3.5°C at 2 m, 1.5°C at 3 m, and less than 1°C at 4 m depth [*McKenzie et al.*, 2007]. At depths greater than 2 m, peat temperatures lag behind those in the air by as much as 8 months because of the geothermal heat flux and the low thermal diffusivity and conductivity of peat. Nevertheless, peat temperatures never drop below 4°C within the deeper peat at Red Lake, which should favor microbial metabolism throughout the year, although the cooler temperatures may have some effect on the production and partitioning of methane into bubbles.

However, *Siegel et al.* [2001] showed that rates of methane production in the deeper peat of the Red Lake Peatland and Lost River Peatland were sufficiently high to enrich the surrounding pore waters in deuterium via the carbon dioxide reduction pathway for methanogenesis. Using an isotopic mass balance approach, they calculated that local hot spots for methanogenesis could produce over 438 g CH$_4$ m^{-3} a^{-1}, indicating considerable spatial heterogeneity with regard to microbial metabolism at depth. Methanogenesis may be favored by the higher pH of the deeper pore waters in the large peat basins of North America, which are supplied by bicarbonate ions transported upward into the peat profile from the underlying calcareous glacial deposits by advection or transverse dispersion [*Glaser et al.*, 1997a, 2004a, 2004b; *Reeve et al.*, 2001; *Romanowicz et al.*, 1995; *Siegel et al.*, 1995; *Siegel and Glaser*, 1987].

High rates of methanogenesis are also apparently driven by the availability of labile carbon compounds dissolved in the deeper pore waters since the radiocarbon age of dissolved methane, dissolved organic carbon (DOC) and DIC in peat profiles is significantly younger than that of the surrounding peat (Figure 2). Root exudates may be transported downward from the rhizosphere into the deeper peat by advection or transverse dispersion as suggested by *Charman et al.* [1999, 1994, 1993]; *Aravena et al.* [1993], *Chanton et al.* [1995], and *Chasar et al.* [2000a, 2000b] or by the deep

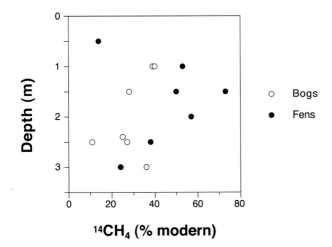

Figure 2. Fraction of modern ¹⁴CH₄ as a function of depth. An isotopic mixing model indicates that a large fraction of the methane dissolved in the pore waters of both bogs and fens from the glacial Lake Agassiz peatlands is derived from modern carbon substrates. Modified from *Chanton et al.* [1995].

penetration of sedge roots [*Chanton et al.*, 2008]. Woody plant roots cannot tolerate long periods of submergence, but the roots of various types of aquatic and emergent, herbaceous vascular plants may contain internal air cavities (aerenchyma tissue) that permit survival in anaerobic soils. Little documentary evidence exists for the depth that living roots can penetrate in peatlands despite their significance for biogeochemical cycling. The high rates of methanogenesis at depth may also be products of unknown microbial taxa that are adapted to the low temperatures, low pH, and poor substrate quality of the deeper portions of peat profiles that have yet to be characterized.

5. MODELS OF GAS TRANSPORT, STORAGE, AND EBULLITION IN PEATLANDS

The dynamics of biogenic gas bubbles in peat deposits has recently been clarified by a combination of field studies using arrays of sensors and by laboratory experiments on peat cores. Early column experiments on peat and sediment cores showed that gas bubbles produced by anaerobic microbes could occlude soil pores [e.g., *Reynolds et al.*, 1992] and thereby lower the hydraulic conductivity of the sediment. These studies using permeameters were extended by *Beckwith and Baird* [2001] by examining small diameter peat cores.

In situ investigations of gas dynamics within the large 1200 km² Red Lake Peatland of northern Minnesota provided a broader-scale conceptual model of ebullition fluxes linked to climate-driven ruptures in deep overpressured gas compartments (Figure 3). These studies simultaneously monitored changes in peat volume (by mounting GPS antennas on tree boles) and pore pressure at 1 m depth intervals (by instrumenting nests of piezometers connected to data loggers) at a bog and fen site [*Glaser et al.*, 2004a; *Rosenberry et al.*, 2003]. The piezometers detected a zone of overpressure at 2 m depth that persisted across an area greater than 0.4 ha for the duration of the study (Figure 4). The piezometer nests also recorded abrupt depressuring cycles at 2 m depth when a semiconfining layer episodically ruptured in response to seasonal droughts and sharp declines in atmospheric pressure (Figures 4, 5, and 6) [*Glaser et al.*, 2004a; *Rosenberry et al.*, 2003]. During each depressuring cycle the peat surface oscillated both vertically and horizontally by as much as 36 cm in 12–18 hours as the bubble mass rose to the surface, degassed to the atmosphere, and pore waters moved in horizontally to fill the voids (Figure 6). Three of the largest of 19 oscillations in August 1997 were associated with the release of 136 g CH₄ m⁻² [*Glaser et al.*, 2004a], which was an order of magnitude higher than the annual flux estimated by chambers at the same site [*Chasar*, 2002]. During a 31-day release period in 1998, 75 g CH₄ m⁻² were released at this same site indicating that ebullition fluxes remained high and dominated the total methane flux from this site during the study period [*Rosenberry et al.*, 2003].

These data indicate that large volumes of gas are produced within the deeper peat of the glacial Lake Agassiz peatlands despite the cool peat temperatures and the presumably more recalcitrant nature of the solid-phase peat. The gas bubbles appear to rise through the peat profile by deforming the local peat fabric [*Glaser et al.*, 1997b, 2006] or through the interconnected pore space. The bubbles are then trapped under more competent layers of peat particularly where the peat fabric is reinforced with woody root systems and branches (Figure 3). The gas accumulates under these competent peat strata that act as transient semiconfining layers until the pore pressures exceed the hydraulic fracture threshold of the seal, whereupon the seal ruptures and the excess pressure is bled off by the release of a bubble mass. The confining layer can then elastically reseal itself beginning the cycle again. The trigger for hydraulic fracturing of the confining layer can be the buildup of excess gas volume under the seal or a lowering of hydrostatic pressure acting on the peat column from above in response to seasonal drawdown of the water table or sharp drops in atmospheric pressure [*Glaser et al.*, 2004a; *Rosenberry et al.*, 2003]. The large vertical and horizontal oscillations of the peat mass indicate that the peat fabric deforms elastically in response to large bubble movements and that the transient passageways are sealed after the passage of bubbles.

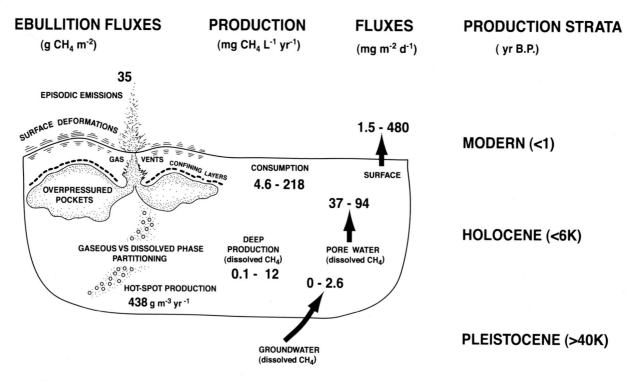

Figure 3. Conceptual model of methane fluxes in the glacial Lake Agassiz peatlands [from *Glaser et al.*, 2004a]. Methane emissions from the glacial Lake Agassiz peatlands appear to be dominated by (1) high rates of methanogenesis in deep peat strata [*Siegel et al.*, 2001], (2) trapping of gas bubbles under transient elastic, semiconfining layers in deep woody strata [*Glaser et al.*, 2004a; *Rosenberry et al.*, 2003], (3) episodic elastic rupturing of the confining layer releasing a mass of bubbles to the peat surface, (4) significant surface oscillations as the bubble mass degasses to the atmosphere [*Glaser et al.*, 2004a]. The transit time for an escaping bubble mass is too short for significant oxidation by methanotrophs. In contrast, as much as 90% of the methane produced in shallow peat strata is oxidized to carbon dioxide [*Chasar*, 2003], and the average annual fluxes for methane measured by chambers are over an order of magnitude lower than those associated with deep ebullition [*Glaser et al.*, 2004a; *Rosenberry et al.*, 2003; *Chasar*, 2003; *Chasar et al.*, 2000a, 2000b; *Chanton et al.*, 1995; *Romanowicz et al.*, 1995, 1993; *Crill et al.*, 1992, 1988].

A slightly different mechanism for ebullition fluxes in peatlands was developed in a shallow partially mined peatland in eastern Canada [*Kellner et al.*, 2005; *Strack et al.*, 2005; *Price*, 2003]. In this case pockets of overpressuring were highly localized in shallow peat strata, and degassing continued until the zone of overpressuring dissipated The release of a bubble mass apparently followed preferred pathways through the existing interconnected pore space that remain open. This plastic-type response did not appear to be the case in the Red Lake Peatland since the horizontal oscillations of the GPS antennas at Red Lake indicated an elastic expansion and contraction of the voids created by the escaping bubble mass with the passageways closing after the bubble mass passed by. These differences could be due to regional changes in the composition of peat profiles under raised bogs across eastern North America [*Glaser and Janssens*, 1986], the relatively shallow (>1.5 m) disturbed nature of the eastern Canadian study site, which had been mined, or local examples of the variable behavior of gas bubbles in peat deposits.

6. REGIONAL COMPARISONS

Carbon flux models, whether conceptual or mathematical, assume simple spatial relationships, uniform parameters, and idealized boundary conditions that are inconsistent with the variability of natural environments. In order to properly assess the limitations of such models, the regional and local variability of peatlands also should to be considered. Peatlands are widely distributed across northerly latitudes spanning major climatic and physiographic gradients. No single peatland site can therefore be considered as typical of all northern peatlands although distinct regional patterns are readily apparent. The bulk of northern peatlands, for ex-

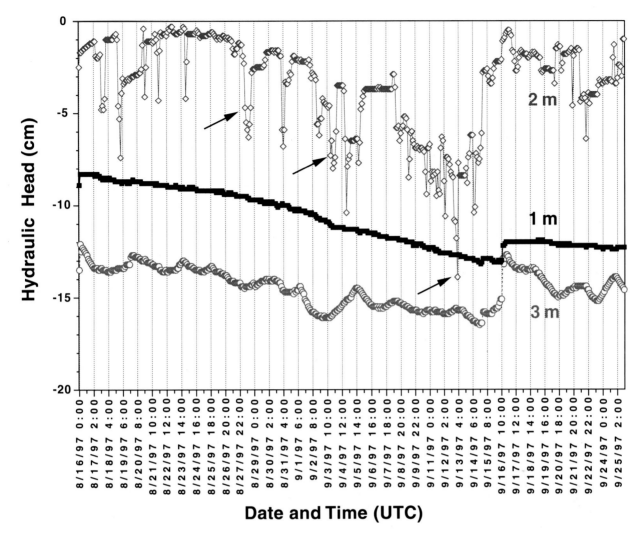

Figure 4. Changes in hydraulic head (relative to peat surface) from the Red Lake II bog during a seasonal drought in 1997. The presence of a transient, semielastic, hydraulic confining layer below 1 m depth is indicated by (1) the overpressures recorded at 2 m depth and (2) the head at 3 m depth, which varied inversely with atmospheric pressure. The short depressuring cycles at 2 m depth were produced by the episodic rupturing of the confining layer and the release of large volumes of trapped gas bubbles, which rose toward the peat surface. The arrows mark the three largest depressuring cycles, which occurred synchronously with significant oscillations of the peat surface. Each of the data points represent one instantaneous measurement taken every 2 hours. Modified from *Glaser et al.* [2004a].

ample, are concentrated within the continental interiors of North America and Eurasia where peat has spread across broad lowlands under a cool monsoonal climate [*Neustadt*, 1984; *Zoltai and Pollet*, 1983]. These large peat basins contain the largest raised bogs, bog complexes, and fen water tracks in existence, and often exhibit intricate surface patterning [*Glaser et al.*, 1997a, 2004b]. Although peatlands tend to be smaller and confined to depressions on the rugged terrain enclosing these lowlands, continental peatlands still retain many common features. Conifer forests, for example, typically grow on better drained sites such as bog crests and hummocky fen landforms where seasonal droughts create a deep aerobic layer [*Glaser and Janssens*, 1986]. Tree cover, however, declines toward the north where the colder climate favors the aggradation of permafrost producing distinctive landform patterns such as palsas and high or low centered polygons [*Vitt et al.*, 1994; *Zoltai and Pollet*, 1983]. The gradient from continental to maritime conditions is also marked

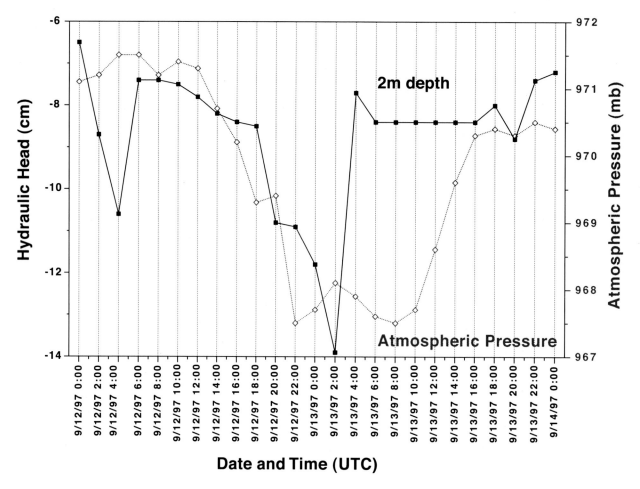

Figure 5. Synchronous changes in atmospheric pressure and depressuring cycles at 2 m depth at the Red Lake II bog site. The sampling interval covers that shown for the steepest drop in pore pressure (third arrow) in Figure 4. Deep ebullition events are apparently triggered by (1) steep declines in atmospheric pressure, (2) declines in the water table elevation, or (3) buildup of excess pressure from the accumulation of bubbles under confining layers. Modified from *Glaser et al.* [2004a].

by a decline of tree cover on peatlands because the moister and equable oceanic climate maintains perennially high water tables, whereas exposed sites are blasted by high winds. Mire pools also become more abundant toward oceanic and more northerly latitudes [*Glaser and Janssens*, 1986].

The patterns of anthropogenic disturbance also vary across North America and Eurasia. The largest peat basins of North America, for example, are located in some of the most sparsely populated regions of the world and largely escaped from anthropogenic disturbance until recently. The smaller peatlands in eastern North America and western Europe, however, are closer to population centers and have been subjected to longer and more intensive periods of anthropogenic disturbance. The peatlands of western Europe in particular were altered by burning, grazing, drainage ditches,

agricultural runoff, and peat cutting since the mid-Holocene and more recently by high loadings of atmospheric pollutants [*Roberts*, 1998; *Berglund*, 1991]. These pollutants, especially sulfate and nitrate anions, can not only alter the surface vegetation and water chemistry on peatlands but can also affect anaerobic metabolism at depth by providing alternative terminal electron acceptors [*Koster*, 2005; *Holland et al.*, 2004]. Despite the magnitude of these impacts, these smaller peatlands remain the most intensively studied sites because of their accessibility and close proximity to research institutions.

Modeling the dynamics of free-phase methane in peatlands is therefore challenging given the variability of peat deposits and the transient nature of gas bubbles in porous media. Gas bubbles are mobile in saturated sediments because of their

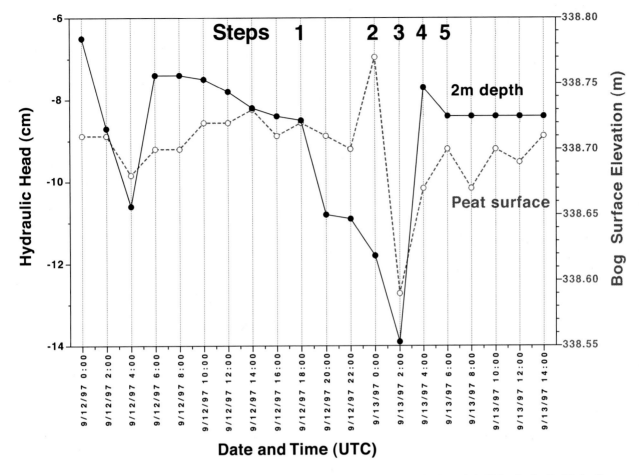

Figure 6. Synchronous changes in surface elevation and peat pore pressures indicative of ebullition in the Red Lake II bog, northern Minnesota. Note the (1) initial decline in pore pressures as a bubble mass is released from an overpressured pocket at 2 m depth followed by (2) a bulge in the peat surface as the bubble mass approaches the land surface, (3) a drop in the peat surface as the bubbles degas to the atmosphere, (4) a rise in pore pressure at 2 m depth due to the elastic closure of the confining layer after the excess pressure was bled off, and (5) elastic rebound in the peat surface as pore waters move in from the sides to fill the voids. The sampling interval is the same as that shown in Figure 5. Modified from *Glaser et al.* [2004a].

buoyancy, but they can also produce secondary porosity by deforming the fabric of weak sediments. This later effect was demonstrated by the elastic surface deformations recorded by GPS units in the Red Lake Peatland of Minnesota [*Glaser et al.*, 2004a] and also by electronic or optic leveling surveys in the nearby Lost River Peatland [*Almendinger et al.*, 1986] and elsewhere [e.g., *Roulet*, 1991; *Seppälä et al.*, 1985]. Mechanical systems equipped with pulleys on potentiometers have also been used to detect surface oscillations driven by ebullition on small peatlands [*Strack et al.*, 2006; *Fechner-Levy and Hemond*, 1996]. The three-dimensional distribution of gas bubbles should therefore vary both spatially and seasonally in response to changes in methane bubble production, transport, storage, and degassing to the

atmosphere. Although a high degree of variability would be expected in the dynamics of free methane in peat deposits, several preliminary generalizations can still be made on the basis of investigations in North American peatlands.

First, a growing body of evidence indicates that gas bubbles are commonly produced within undisturbed peat deposits although their three-dimensional distribution may be discontinuous and localized. Gas bubbles frequently appear in the surface or near surface waters of peatlands whenever an external load is applied to the peat surface. The source strata for these bubbles are uncertain since gas bubbles are buoyant and can therefore rise from any portion of a peat profile toward the land surface. Synoptic studies across several large peat basins in North America showed that gas pockets are common within the deeper portions of peat profiles that were 2 to 6 m deep [e.g., *Siegel et al.*, 2001; *Glaser et al.*, 1997a; *Romanowicz et al.*, 1995]. Gas bubbles usually erupt from wide-diameter boreholes drilled into the deeper peat, and these observations are supported by data from integrated arrays of geodetic, hydrogeologic, and geophysical instruments at specific sites [e.g., *Glaser et al.*, 2004a; *Rosenberry et al.*, 2003]. This same set of direct and indirect evidence confirmed the existence of overpressured gas pockets within the deeper strata of small raised bogs in northern Maine [e.g., *Comas et al.*, 2005a, 2005b, 2007, 2008] and the shallow strata from a cutover peatland in Quebec [e.g., *Strack et al.*, 2005, 2006; *Kellner et al.*, 2005].

Second, the distribution of gas bubbles in peat deposits may also be related to the physical properties of the peat fabric. In the glacial Lake Agassiz region, the Hudson Bay Lowland, and the Caribou bog in Maine, free-phase gas tends to accumulate in deep peat strata reinforced with woody structural elements such as buried branches, tree boles, and root mats [*Glaser et al.*, 2004a; *Rosenberry et al.*, 2003]. These structurally more competent peat strata resist deformation from rising bubbles creating traps where methane bubbles can accumulate and create local zones of overpressure. Methane bubbles can also be trapped under seasonal ice layers at sites where the water table is sufficiently close to the surface to freeze in winter [*Dise*, 1992; *Kelly et al.*, 1992]. This transient pool of gas can then degas to the atmosphere when the ice layer thaws in spring, creating a pulse in methane fluxes sometimes captured in chambers or tower-mounted sensors. Farther north permafrost sites containing ground ice may produce the same effect in peatlands where permafrost does not extend down to the basal peat or lake sediment [*Walter et al.*, 2006].

Third, rates of methanogenesis remain sufficiently high within the deeper portions of peat profiles to maintain a constant production of gas bubbles. Deep methane is consistently enriched in ^{14}C relative to the peat at any depth indicating that high rates of methanogenesis are fueled in part by the downward transport of modern root exudates [e.g., *Chanton et al.*, 1995; *Chasar et al.*, 2000a, 2000b]. These isotopic studies also indicate that methanogens metabolize recently fixed carbon compounds throughout the peat column mixed with varying fractions of old carbon derived from the adjacent peat skeleton. The formation of methane bubbles even within the deep peat therefore seems to be partly linked to net primary production. The reactivity of the dissolved organic carbon at depth depends on the hydrogeologic setting of the site that determines the direction of groundwater flow [*Siegel et al.*, 1995]. However, isotopic studies also indicate that sites with a dense cover of *Carex* sedges contain more highly reactive DOC at depth [*Chanton et al.*, 2008]. This

potentially tight coupling between the carbon dynamics of the surface and subsurface suggests an important means to upscale methane production rates using remote sensing.

Not all peatlands may contain deep overpressured gas pockets given the wide variability of peatlands across the boreal belt. Some sites will lack the buried wood layers that act as efficient traps for rising methane bubbles, whereas in permafrost regions thick layers of ground ice may inhibit methane production and transport at depth. The very low hydraulic conductivity of some peats may restrict the transport of labile DOC at depth and thus depress rates of methane production in these peatlands [e.g., *Chanton et al.*, 2008; *Clymo and Bryant*, 2008]. Anthropogenic disturbance is another factor that may inhibit the production of methane bubbles at depth by altering the hydraulics, fabric, and microbiology of deep peat profiles. The ability to detect deep gas pockets may also depend upon the design, frequency, and type of sampling employed.

Nevertheless, a growing body of evidence from both synoptic and site-intensive investigations indicates that northern peatlands trap large pools of free-phase gas that episodically escape to the atmosphere. These findings have important implications for the global methane budget by (1) extending the source strata for methane emissions to the entire peat profile, (2) providing a mechanism for rapid degassing from the deeper peat that precludes oxidation by methanotrophs, (3) identifying an alternative source for fueling methane production in the deeper peat strata, and (4) increasing the incubation time for producing methane pools from less than hundreds of years in near-surface layers to thousands of years in the deeper peat. Unlike the surface layers, which are subjected to seasonal cooling and freezing, the deeper portions of a peat profile remain relatively stable throughout the year and never freeze except in regions with permafrost. The total methane emissions from northern peatlands may therefore be 2 or 3 times higher than currently expected when the contributions from deep peat strata via ebullition are considered.

7. ROLE OF EBULLITION IN THE GLOBAL METHANE CYCLE

Northern peatlands above 45°N Latitude account for more than half the total area for global wetlands but are now considered to contribute only about a third of the 100 T g a^{-1} of methane emitted from all wetlands [*Mathews*, 2000; *Khalil*, 2000]. *Fung et al.* [1991], for example, estimated the total methane emissions from wetlands to be 115 Tg a^{-1} with only 35 Tg a^{-1} emitted from northern wetlands compared to 80 Tg a^{-1} from tropical sources. *Bartlett and Harris* [1993] also estimated total wetland emissions at 109 Tg a^{-1} with northern sources at 38 Tg a^{-1} and tropical sources at 66 Tg a^{-1}. Increasing data on the dynamics of methane gas bubbles in northern peat deposits, however, subsequently suggested that the methane emissions from northern peatlands may be much higher than previously thought. Not only must the entire depth (average of 2–3 m, range 1–8 m) as well as the aerial extent of these peatlands be considered as potential sources for methane production, but the release of gas bubbles via ebullition minimizes the exposure of methane to oxidation by methanotrophs. Moreover, high-latitude sources such as lakes and wetlands formed from degrading permafrost must also be considered as sources [e.g., *Walter et al.*, 2006; *Prater et al.*, 2007]. Unfortunately, the mobility of gas bubbles in peat deposits and the episodic nature of ebullition in both and time and space make it difficult to calculate reliable regional estimates ebullition fluxes in peatlands, lakes, and thermokarst terrain. Existing data on ebullition and free-phase gas in peatlands, however, suggest that the source strength for atmospheric methane from northern wetlands need to be scaled upward.

The major challenge for future research will be to (1) monitor ebullition fluxes over broader areas to obtain more reliable regional estimates, (2) better define the dynamics of gas formation, transport, and release in deep peat deposits, and (3) realistically characterize rates of anaerobic metabolism and organic matter transformations in peatlands. All the available techniques for monitoring free-phase gas in peatlands are to some extent invasive except for surface-based geophysical exploration methods, which still disturb subsurface pressure regimes by applying an external load to these easily deformable soils. Inserting sensors into the peat profile, however, also introduces artifacts by disrupting the peat fabric to some extent. The best approach may therefore be determinations of gas volumes and dynamics using multiple independent approaches employing different principles that sense the physical environment in slightly different but compatible ways. The rapidly evolving picture of the interplay among biogenic gas bubbles, hydraulics, and the easily deformable peat fabric provides a challenge for future research that is central for understanding the methane budget and carbon balance of global peatlands.

REFERENCES

Aario, L. (1932), Pflanzentopographische und paläogeographische Mooruntersuchungen in N-Satakunta, *Fennia*, 55, 1–179.

Almendinger, J. C., J. E. Almendinger, and P. H. Glaser (1986), Topographic fluctuations across a spring fen and raised bog in the Lost River Peatland, northern Minnesota, *J. Ecol.*, 74, 393–401.

Andreae, M. O., and D. S. Schimel (Eds.) (1989), *Exchange of Trace Gases Between Terrestrial Ecosystems and the Atmo-*
sphere: Report of the Dahlem Workshop on Exchange of Trace Gases Between Terrestrial Ecosystems and the Atmosphere, Berlin 1989, February 19–24, John Wiley, Chichester, U. K.

Aravena, R., B. G. Warner, D. J. Charman, L. R. Belyea, S. P. Mathur, and H. Dinel (1993), Carbon isotopic composition of deep carbon gases in an ombrogenous mire, northwestern Ontario, Canada, *Radiocarbon*, 35, 271–276.

Bartlett, K. B., and R. C. Harriss (1993), Review and assessment of methane emissions from wetlands, *Chemosphere*, 26(1–4), 261–320.

Bear, J. (1972), *Dynamics of Fluids in Porous Media*, Elsevier, New York.

Beckwith, C. W., and A. J. Baird (2001), Effect of biogenic gas bubbles on water flow through poorly decomposed blanket peat, *Water Resour. Res.*, 37(3), 551–558.

Bellisario, L. M., J. L. Bubier, T. R. Moore, and J. P. Chanton (1999), Controls on CH$_4$ emissions from a northern peatland, *Global Biogeochem. Cycles*, 13, 81–91.

Berglund, B. E. (1991), *The Cultural Landscape During 6,000 Years in Southern Sweden: The Ystad Project*, Munksgaard Int., Copenhagen.

Brown, A., S. P. Mathur, and D. J. Kushner (1989), An ombrotrophic bog as a methane reservoir, *Global Biogeochem. Cycles*, 3, 205–213.

Brown, D. A. (1998), Gas production from an ombrotrophic bog: Effect of climate change on microbial ecology, *Clim. Change*, 40, 277–284.

Chanton, J. P., C. S. Martens, and C. A. Kelley (1989), Gas transport from methane-saturated tidal freshwater and wetland sediments, *Limnol. Oceanogr.*, 34, 807–819.

Chanton, J. P., J. E. Bauer, P. A. Glaser, D. I. Siegel, C. A. Kelley, S. C. Tyler, E. H. Romanowicz, and A. Lazrus (1995), Radiocarbon evidence for the substrates supporting methane formation within northern Minnesota peatlands, *Geochim. Cosmochim. Acta*, 59, 3663–3668.

Chanton, J. P., P. H. Glaser, L. S. Chasar, D. J. Burdige, M. E. Hines, D. I. Siegel, L. B. Tremblay, and W. T. Cooper (2008), Radiocarbon evidence for the importance of surface vegetation on fermentation and methanogenesis in contrasting types of boreal peatlands, *Global Biogeochem. Cycles*, 22, GB4022, doi:10.1029/2008GB003274.

Charman, D. J., R. Aravena, and B. G. Warner (1993), Isotope geochemistry of gas and water samples from deep peats in boreal Canada, *Suo*, 43, 199–201.

Charman, D. J., R. Aravena, and B. G. Warner (1994), Carbon dynamics in a forested peatland in northeastern Ontario, Canada, *J. Ecol.*, 82, 55–62.

Charman, D. J., R. Aravena, C. L. Bryant, and D. D. Harkness (1999), Carbon isotopes in peat, DOC, CO$_2$ and CH$_4$ in a Holocene peatland on Dartmorre, SW England, UK, *Geology*, 27, 539–542.

Chasar, L. C. (2002), Implications of environmental change for energy flow through natural systems: Wetlands and coastal systems, Ph.D. dissertation, 320 pp., Fla. State Univ., Tallahassee.

Chasar, L. S., J. P. Chanton, P. H. Glaser, D. I. Siegel, and J. S. Rivers (2000a), Radiocarbon and stable carbon isotopic evidence for transport and transformation of dissolved organic carbon, dissolved inorganic carbon, and CH4 in a northern Minnesota peatland, *Global Biogeochem. Cycles*, *14*, 1095–1108.

Chasar, L. S., J. P. Chanton, P. H. Glaser, and D. I. Siegel (2000b), Methane concentration and stable isotope distribution as evidence of rhizospheric processes: Comparison of a fen and bog in the glacial Lake Agassiz peatland complex, *Anal. Bot.*, *86*, 655–663.

Clymo, R. S., and C. L. Bryant (2008), Diffusion and mass flow of dissolved carbon dioxide, methane, and dissolved organic carbon in a 7 m deep raised peat bog, *Geochim. Cosmochim. Acta*, *72*, 2048–2066.

Clymo, R. S., and D. M. E. Pearce (1995), Methane and carbon dioxide production in, transport through, and efflux from a peatland, *Philos. Trans. R. Soc. London, Ser. A*, *350*, 249– 259.

Comas, X., and L. Slater (2007), Evolution of biogenic gases in peat blocks inferred from noninvasive dielectric permittivity measurements, *Water Resour. Res.*, *43*, W05424, doi:10.1029/2006WR005562.

Comas, X., L. Slater, and A. Reeve (2005a), Spatial variability in biogenic gas accumulations in peat soils is revealed by ground penetrating radar (GPR), *Geophys. Res. Lett.*, *32*, L08401, doi:10.1029/2004GL022297.

Comas, X., L. Slater, and A. Reeve (2005b), Geophysical and hydrological evaluation of two bog complexes in a northern peatland: Implications for the distribution of biogenic gases at the basin scale, *Global Biogeochem. Cycles*, *19*, GB4023, doi:10.1029/2005GB002582.

Comas, X., L. Slater, and A. Reeve (2007), In situ monitoring of free-phase gas accumulation and release in peatlands using ground penetrating radar (GPR), *Geophys. Res. Lett.*, *34*, L06402, doi:10.1029/2006GL029014.

Comas, X., L. Slater, and A. Reeve (2008), Seasonal geophysical monitoring of biogenic gases in a northern peatland: Implications for temporal and spatial variability in free phase gas production rates, *J. Geophys. Res.*, *113*, G01012, doi:10.1029/2007JG000575.

Crill, P. M., K. B. Bartlett, R. C. Harriss, E. Gorham, E. S. Verry, D. I. Sebacher, L. Madzar, and W. Sanner (1988), Methane flux from Minnesota peatlands, *Global Biogeochem. Cycles*, *2*, 371–384.

Crill, P., K. B. Bartlett, and N. T. Roulet (1992), Methane flux from boreal peatlands, *Suo*, *43*, 173–182.

Dinel, H., S. P. Mathur, A. Brown, and M. Lévesque (1988), A field study of the effect of depth on methane production in peatland waters: Equipment and preliminary results, *J. Ecol.*, *76*, 1083–1091.

Dise, N. B. (1992), Winter fluxes of methane from Minnesota peatlands, *Biogeochemistry 17*, 71–83, doi:10.1007/BF00002641.

Dunfield, P., R. Knowles, R. Dumont, and T. R. Moore (1993), Methane production and consumption in temperate and subarctic peat soils: Response to temperature and pH, *Soil Biol. Biochem.*, *25*, 321–326.

Fechner-Levy, E. J., and H. F. Hemond (1996), Trapped methane volume and potential effects on methane ebullition in a northern peatland, *Limnol. Oceanogr*, *41*, 1375–1383.

Fung, I., J. John, J. Lerner, E. Mathews, M. Prather, L. P. Steele, and P. J. Fraser (1991), Three-dimensional model synthesis of the global methane cycle, *J. Geophys. Res.*, *96*, 13,033–13,065.

Gams, H., and S. Ruoff (1929), Geschichte, Aufbau und Pflanzendeckedes Zelaubruches, Monographie eines wachsenden Hochmoores in Ostpreuoen, *Schr. Physik. Ges. Konigsb.*, *LXVI*, 1–193.

Glaser, P. H., and J. A. Janssens (1986), Raised bogs in eastern North America: Transitions in landforms and gross stratigraphy, *Can. J. Bot.*, *64*, 395–415.

Glaser, P. H., D. I. Siegel, E. A. Romanowicz, and Y. P. Shen (1997a), Regional linkages between raised bogs and the climate, groundwater, and landscape features of northwestern Minnesota, *J. Ecol.*, *85*, 3–16.

Glaser, P. H., P. J. Morin, J. Kamp, N. Tsekos, and D. I. Siegel (1997b), The size and distribution of biogenic-gas bubbles in peat cores directly determined by magnetic resonance imaging (MRI), *Eos Trans. AGU*, *78*(46), Fall Meet. Suppl., F226.

Glaser, P. H., J. P. Chanton, P. Morin, D. O. Rosenberry, D. I. Siegel, O. Ruud, L. I. Chasar, and A. S. Reeve (2004a), Surface deformations as indicators of deep ebullition fluxes in a large northern peatland, *Global Biogeochem. Cycles*, *18*, GB1003, doi:10.1029/2003GB002069.

Glaser, P. H., D. I. Siegel, A. S. Reeve, J. A. Janssens, and D. R. Janecky (2004b), Tectonic drivers for vegetation patterning and landscape evolution in the Albany River region of the Hudson Bay Lowlands, *J. Ecol.*, *92*, 1054–1070.

Glaser, P. H., P. J. Morin, N. Tsekos, D. I. Siegel, D. O. Rosenberry, J. P. Chanton, and A. S. Reeve (2006), The transport of free-phase methane in hydraulically confined and unconfined peat strata analyzed by magnetic resonance imaging (MRI), *Eos Trans. AGU*, *87*(36), Fall Meet. Suppl., Abstract B31D-05.

Gorham, E. (1991), Role in the carbon cycle and probable responses to climatic warming, *Ecol. Appl.*, *1*, 182–195.

Harden, J. W., R. K. Mark, E. T. Sundquist, and R. F. Stallard (1992), Dynamics of soil carbon during deglaciation of the Laurentide Ice Sheet, *Science*, *258*, 1921–1924, doi:10.1126/science.258.5090.1921.

Hemond, H. F., and E. J. Fechner (1994), *Chemical Fate and Transport in the Environment*, Academic, San Diego, Calif.

Holland, E. A., B. H. Braswell, J. Sulzman, and J.-F. Lamarque (2004), Nitrogen deposition onto the United States and western Europe, http://www.daac.ornl.gov, Distrib. Active Arch. Cent., Oak Ridge Natl. Lab., Oak Ridge, Tenn.

Hunt, J. M. (1996), *Petroleum Geochemistry and Geology*, 2nd ed., Freeman, San Francisco, Calif.

Kellner, E., J. M. Waddington, and J. S. Price (2005), Dynamics of biogenic gas bubbles in peat: Potential effects on water storage and peat deformation, *Water Resour. Res.*, *41*, W08417, doi:10.1029/2004WR003732.

Kelly, C. A., N. B. Dise, and C. S. Martens (1992), Temporal variations in the stable carbon isotopic composition of methane emitted from Minnesota peatlands, *Global Biogeochem. Cycles*, *6*, 263–269.

Khalil, M. A. K. (Ed.) (2000), *Atmospheric Methane: Its Role in the Global Environment*, Springer, New York.

King, J. Y., W. S. Reeburgh, and S. K. Regli (1998), Methane emission and transport by arctic sedges in Alaska: Results of a vegetation removal experiment, *J. Geophys. Res.*, *103*(D22), 29,083–29,092.

Kipphut, G. W., and C. S. Martens (1982), Biogeochemical cycling in an organic-rich coastal marine basin: Dissolved-gas transport in methane-saturated sediments, *Geochim. Cosmochim. Acta, 46*, 2049–2060.

Kivinen, E., and P. Pakarinen (1981), Geographical distribution of peat resources and major peatland complex types in the world, *Ann. Acad. Sci. Fenn., Ser. A3, 132*, 5–28.

Koster, A. (2005), *The Physical Geography of Western Europe*, Oxford Univ. Press, New York.

Martens, C. S., and J. P. Chanton (1989), Radon as a tracer of biogenic gas equilibration and transport from methane-saturated sediments, *J. Geophys. Res., 94*, 3451–3459.

Mastepanov, M., and T. R. Christensen (2008), Bimembrane diffusion probe for continuous recording of dissolved and entrapped bubble gas concentrations in peat, *Soil Biol. Biochem., 40*, 2992–3003.

Mathews, E. (2000), Wetlands, in *Atmospheric Methane: Its Role in the Global Environment*, edited by M. A. K. Khalil, pp. 202–233, Springer, New York.

McKenzie, J. M., D. I. Siegel, D. O. Rosenberry, P. H. Glaser, and C. Voss (2007), Heat transport in the Red Lake Bog, glacial Lake Agassiz peatlands, *Hydrol. Processes, 21*, 369–378, doi:10.1002, doi:10.1002/hyp.6239.

Moore, T. R., and M. Dalva (1997), Methane and carbon dioxide exchange potentials of peat soils in aerobic and anaerobic laboratory incubations, *Soil Biol. Biochem., 29*, 1157–1164.

Moore, T. R., N. T. Roulet, and J. M. Waddington (1998), Uncertainty in predicting the effect of climatic change on the carbon cycling of Canadian peatlands, *Clim. Change, 40*, 229–245.

Neustadt, M. I. (1984), Holocene peatland development, in *Late Quaternary Environments of the Soviet Union*, Engl. lang. ed., pp. 201–206, Univ. of Minn. Press, Minneapolis.

Neuzil, C. E. (1995), Abnormal pressures as hydrodynamic phenomena, *Am. J. Sci., 295*, 742–786.

Pauss, A., G. Andre, M. Perrier, and S. R. Guiot (1990), Liquid-to-gas mass transfer in anaerobic processes: Inevitable transfer limitations of methane and hydrogen in the biomethanation process, *Appl. Environ. Microbiol., 56*, 1636–1644.

Prater, J. L., J. P. Chanton, and G. J. Whiting (2007), Variation in methane production pathways associated with permafrost decomposition in collapse scar bogs of Alberta, Canada, *Global Biogeochem. Cycles, 21*, GB4004, doi:10.1029/2006GB002866.

Price, J. S. (2003), Role and character of seasonal peat soil deformation on the hydrology of undisturbed and cutover peatlands, *Water Resour. Res., 39*(9), 1241, doi:10.1029/2002WR001302.

Reeve, A. S. (1996), Numerical and multivariate statistical analysis of hydrogeology and geochemistry in large peatlands, Ph.D. dissertation, Syracuse Univ., Syracuse, New York.

Reeve, A. S., D. I. Siegel, and P. H. Glaser (1996), Geochemical controls on peatland pore water from the Hudson Bay Lowland: A multivariate statistical approach, *J. Hydrol., 181*, 285–304.

Reeve, A. S., D. I. Siegel, and P. H. Glaser (2001), Simulating dispersive mixing in large peatlands, *J. Hydrol., 242*, 103–114.

Reynolds, W. D., D. A. Brown, S. P. Mathur, and R. P. Overend (1992), Effect of in-situ gas accumulation on the hydraulic conductivity of peat, *Soil Sci., 153*, 397–408.

Roberts, N. (1998), *The Holocene: An Environmental History*, 2nd ed., Blackwell, Oxford, U. K.

Romanowicz, E. A., D. I. Siegel, and P. H. Glaser (1993), Hydraulic reversals and episodic methane emissions during drought cycles in mires, *Geology, 21*, 231–234.

Romanowicz, E. A., D. I. Siegel, J. P. Chanton, and P. H. Glaser (1995), Temporal variations in dissolved methane deep in the Lake Agassiz Peatlands, Minnesota, *Global Biogeochem. Cycles, 9*, 197–212.

Rosenberry, D. O., P. H. Glaser, D. I. Siegel, and E. P. Weeks (2003), Use of hydraulic head to estimate volumetric gas content and ebullition flux in northern peatlands, *Water Resour. Res., 39*(3), 1066, doi:10.1029/2002WR001377.

Rosenberry, D. O., P. H. Glaser, and D. I. Siegel (2006), The hydrology of northern peatlands as affected by biogenic gas: Current developments and research needs, *Hydrol. Processes, 20*, 3601–3610.

Roulet, N. T. (1991), Surface level and water table fluctuations in a subarctic fen, *Arct. Alp. Res., 23*, 303–310.

Schütz, H., P. Schröder, and H. Rennenberg (1991), Role of plants in regulating methane flux to the atmosphere, in *Trace Gas Emissions From Plants*, edited by T. D. Sharkey, E. A. Holland, and H. A. Mooney, pp. 29–63, Academic, San Diego, Calif.

Seppälä, M., and L. Koutaniemi (1985), Formation of a string and pool topography as expressed by morphology, stratigraphy and current processes on a mire in Kuusamo, Finland, *Boreas, 14*, 287–309.

Shannon, R. D., and J. R. White (1994), A three-year study of controls on methane emissions from two Michigan peatlands, *Biogeochemistry, 27*, 35–60.

Shannon R. D., J. R. White, J. E. Lawson, and B. S. Gilmore (1996), Methane efflux from emergent vegetation in peatlands, *J. Ecol., 84*, 239–246.

Siegel, D. I., and P. H. Glaser (1987), Groundwater flow in a bog-fen complex, Lost River Peatland, northern Minnesota, *J. Ecol., 75*, 743–754.

Siegel, D. I., A. S. Reeve, P. H. Glaser, and E. Romanowicz (1995), Climate-driven flushing of pore water in humified peat, *Nature, 374*, 531– 533.

Siegel, D. L., J. R. Chanton, P. H. Glaser, L. S. Chasar, and D. O. Rosenberry (2001), Estimating methane production rates in bogs and landfills by deuterium enrichment of pore water, *Global Biogeochem. Cycles, 15*, 967–975.

Simpson, J. A., and E. S. C. Weiner (Eds.) (1989), *Oxford English Dictionary*, 2nd ed., Oxford Univ. Press, New York.

Sorrell, B. K., H. Brix, and P. T. Orr (1997), *Eleocharis sphacelata*: Internal gas transport pathways and modeling of aeration by pressurized flow and diffusion, *New Phytol., 136*, 433–442.

Strack, M., and J. M. Waddington (2007), Response of peatland carbon dioxide and methane fluxes to a water table drawdown experiment, *Global Biogeochem. Cycles, 21*, GB1007, doi:10.1029/2006GB002715.

Strack, M., E. Kellner, and J. M. Waddington (2005), Dynamics of biogenic gas bubbles in peat and their effects on peatland biogeochemistry, *Global Biogeochem. Cycles, 19*, GB1003 doi:10.1029/2004GB002330.

Strack, M., E. Kellner, and J. M. Waddington (2006), Effect of entrapped gas on peatland surface level fluctuations, *Hydrol. Processes*, *20*, 3611–3622.

Thomson, P. V. (1924), Influence of eruption of gas on the surface patterns of bogs, *Bot. Arch.*, *8*, 1–2.

Tokida, T., T. Miyazaki, and M. Mizoguchi (2005a), Ebullition of methane from peat with falling atmospheric pressure, *Geophys. Res Lett*, *32*, L13823, doi:10.1029/2005GL022949.

Tokida, T., T. Miyazaki, M. Mizoguchi, and K. Seki (2005b), In situ accumulation of methane bubbles in a natural wetland soil, *Eur. J. Soil Sci.*, *56*, 389–395.

Vitt, D. H., L. A. Halsey, and S. C. Zoltai (1994), The bog landforms of continental western Canada in relation to climate and permafrost patterns, *Arct. Alp. Res.*, *26*, 1–13.

Walter, K. M., S. A. Zimov, J. P. Chanton, D. Verbyla, and F. S. Chapin III (2006), Methane bubbling from Siberian thaw lakes as a positive feedback to climate warming, *Nature*, *443*, 71–75.

Walter, K. M., J. P. Chanton, F. S. Chapin III, E. A. G. Schuur, and S. A. Zimov (2008), Methane production and bubble emissions from arctic lakes: Isotopic implications for source pathways and ages, *J. Geophys. Res.*, *113*, G00A08, doi:10.1029/2007JG000569.

Weber, C. A. (1902), *Über die Vegetation und Entstehung des Hochmoors von Augstumal im Memeldelta mit vergleichenden Ausblicken auf andere Hochmoore der Erde*, Paul Parey, Berlin.

Whiting, G. J., and J. P. Chanton (1993), Primary production control of methane emissions from wetlands, *Nature*, *364*, 794–795.

Wigley, T. M. L., and D. S. Schimel (2000), *The Carbon Cycle*, Cambridge Univ. Press, Cambridge, U. K.

Zoltai, S. C., and F. C. Pollet (1983), Wetlands in Canada: Their classification, distribution, and use, in *Mires—Swamp, Bog, Fen, and Moor, Ecosyst. World*, vol. 4B, edited by A. J. P. Gore, pp. 245–268, Elsevier, Amsterdam.

J. P. Chanton, Department of Oceanography, Florida State University, Tallahassee, FL 32306, USA.

P. H. Glaser, Department of Geology and Geophysics, University of Minnesota, Pillsbury Hall, Minneapolis, MN 55455, USA. (glase001@umn.edu)

Noninvasive Field-Scale Characterization of Gaseous-Phase Methane Dynamics in Peatlands Using the Ground-Penetrating Radar Method

Xavier Comas

Department of Geosciences, Florida Atlantic University, Boca Raton, Florida, USA

Lee D. Slater

Department of Earth and Environmental Sciences, Rutgers, State University of New Jersey, Newark, New Jersey, USA

Ground-penetrating radar (GPR) is a promising technology for investigating methane cycling in peatlands over a wide range of spatial scales. Unlike most commonly applied techniques that seek to ascertain information on the vertical distribution of peat structure and/or gas content, GPR can be employed entirely noninvasively with minimal disruption to the in situ gas regime. We discuss the following applications of the GPR method in peatlands research: (1) imaging the lateral continuity of confining layers that may permit free-phase gas (FPG) accumulation and regulate methane emissions; (2) estimation of vertical profiles of FPG content using surface and cross-hole measurements; and (3) noninvasive temporal monitoring of FPG production and emissions from a peat column. We present new GPR results that demonstrate how 1-D velocity profiles from the glacial Lake Agassiz peatland (GLAP) complex support the suggestion of accumulation of FPG below confining layers at 3–4 m depth. Using cross-borehole GPR data collected from a peatland in Maine, we show how GPR tomography (previously unapplied in peatlands) can image the spatial distribution of FPG content in Caribou Bog, with minimal invasiveness. We also present results of 3-D surface reflection amplitude analysis, suggesting that reflection amplitudes, in addition to travel times, may yield insights into changes in methane production and emissions (e.g., via ebullition) from peat soils. We finish by discussing limitations of the technique (e.g., petrophysical conversion) and recommendations for further research to improve the application of this technology in studies of methane cycling in northern peatlands.

1. INTRODUCTION

It is increasingly obvious that the dynamics of biogenically produced free-phase methane (CH_4) in northern peatlands

Carbon Cycling in Northern Peatlands
Geophysical Monograph Series 184
Copyright 2009 by the American Geophysical Union.
10.1029/2008GM000810

must be better quantified to constrain the atmospheric methane burden associated with northern wetlands. Free-phase methane is released to the atmosphere from northern peatlands via both diffusive (through plant shoots and/or across the peat surface) and rapid outgassing (ebullition) pathways. Although diffusive methane fluxes from peat soils have been extensively reported (and range from 1.5 to 480 mg m^{-2} d^{-1}) [*Rosenberry et al.*, 2006], rapid releases (ebullition fluxes) are poorly quantified. Most significantly, these ebullition fluxes vary both spatially and temporally such that reliable estimates for the upscaled emission of CH_4 from northern peatlands are lacking. Temporal variations in point-scale ebullition fluxes have been captured in recent studies via high-resolution chamber techniques [*Tokida et al.*, 2005], surface deformation monitoring [*Glaser et al.*, 2004], and water level records [*Rosenberry et al.*, 2003]. These studies have suggested that short-duration ebullition events, often induced by rapid drops in atmospheric pressure, may contribute more CH_4 to the atmosphere than the diffusive flux from the same peatland for an entire year [e.g., *Glaser et al.*, 2004]. Such studies have clearly determined the importance of atmospheric pressure changes in regulating CH_4 releases, with rapid outgassing often accompanying sudden drops in atmospheric pressure [*Tokida et al.*, 2005]. Some have suggested that peat composition and stratigraphy exerts a strong control on ebullition, with woody layers acting as confining layers to trap upwardly ascending gas until some critical pressure gradient threshold is breached, after which the gas escapes [e.g., *Glaser et al.*, 2004; *Rosenberry et al.*, 2006].

It is clear that achieving accurate estimates of the atmospheric methane burden associated with northern peatlands is critically dependent on improved quantification of CH_4 emissions, specifically accounting for ebullition fluxes and taking into account the spatial and temporal variability in these fluxes. Although not within peat deposits, it is interesting to note that the importance of ebullition fluxes of CH_4 from north Siberian thermokarst lakes has recently received considerable attention [*Walter et al.*, 2006]. Studies that quantified the patchiness of ebullition from this source showed that adding emission estimates of CH_4 from north Siberian thermokarst lakes increased current estimates of total northern wetland emission by 10–63% [*Walter et al.*, 2006]. An obvious question that then arises is, To what extent will accounting for the patchiness of ebullition from northern peatlands increase current estimates of total northern wetland emission? In order to address this question, measurement methods are needed to capture the spatiotemporal variability in CH_4 releases across a peatland such that reliable upscaled estimates of total ebullition CH_4 fluxes from northern peatlands can be calculated. Whereas studies of ebullition from thermokarst lakes have been aided by the fact that ebullition hot spots are visibly identified from bubbles frozen in lake ice, ebullition hot spots in northern peatlands are invisible to the naked eye and must be detected with appropriate sensing techniques.

The commonly employed methods for studying free-phase gas dynamics in peat soils suffer from two distinct limitations. The first disadvantage pertains to those methods based on evaluating the spatial distribution of gas within the peat profile via destructive sampling or invasive insertion of probes. Such methods include moisture probes [*Kellner et al.*, 2005], time domain reflectometry (TDR) sensors [*Beckwith and Baird*, 2001], deformation rods placed at depth [*Price*, 2003], and piezometers [*Rosenberry et al.*, 2003], all of which can potentially disrupt the in situ gas regime. The second disadvantage pertains to previously employed surface-based measurements that, although noninvasive, provide little/no information on the spatial distribution of gases in the peat soil. Such methods include gas chambers [e.g., *Whalen and Reeburgh*, 1992] and surface deformation measurements using high-precision GPS measurements [*Glaser et al.*, 2004]. The latest method also suffers in that the measurements are not directly related to the free-phase gas (FPG) concentration.

There is thus a need for measurement technologies that can noninvasively quantify and monitor FPG dynamics in northern peatlands. One technology that has been recently applied in peatland studies with considerable success is the ground-penetrating radar (GPR) geophysical method [*Comas et al.*, 2005a, 2007, 2008]. The method has been used to image peatland stratigraphy and identify potential gas-trapping layers, estimate the vertical distribution of FPG within the peat deposit, and monitor seasonal changes in FPG concentration and ebullition fluxes. Previous studies have also suggested that it may be possible to map FPG hot spots rapidly within a peatland using the GPR method [*Comas et al.*, 2005b]. Although partially invasive (with disruption to the in situ gas regime) when deployed using boreholes, the GPR method is attractive as it can be applied completely noninvasively from the surface to determine the distribution of FPG within the peat profile if other sources for FPG disruption, such as instrumentation/observer weight, are avoided (e.g., by using platforms anchored to the mineral soil [*Comas et al.*, 2007, 2008]). Furthermore, the FPG content is readily upscaled to representative values for the peat column that would be more appropriate as input parameters to global climate models relative to values obtained from point source probe methods. The focus of this chapter is to discuss the application, successes, and limitations of the GPR method for investigating methane cycling in northern peatlands. We combine the most significant findings of the published studies using this method with recent new results that further highlight the potential for adopting the GPR technique in peatlands in a variety of application modes.

2. GPR METHOD

GPR is a widely used hydrogeophysical method for noninvasively measuring water content (θ) in soil layers (see, e.g., *Huisman et al.* [2003] for review) in the vadose zone, which can assist with the characterization of peatland stratigraphy and hydrology [e.g., *Slater and Reeve*, 2002]. GPR has been used widely for peat deposit profiling for almost 30 years. Classical applications include detection of peat thickness, detection of the nature of underlying peat-mineral soil interfaces, and detection of peat stratigraphical features and other internal structures [e.g., *Bjelm*, 1980; *Ulriksen*, 1981; *Warner et al.*, 1990; *Jol and Smith*, 1995]. More recent hydrological applications in peat studies include investigation of local changes in moisture content [*Theimer et al.*, 1994], water table position [*Lapen et al.*, 1996], structural features within peatlands such as natural pipes or macropores [*Holden et al.*, 2002], and stratigraphic controls on peatland hydrology [*Slater and Reeve*, 2002]. The method uses a transmitting antenna (Tx) to generate a high-frequency electromagnetic (EM) wave that, in the traditional surface-based configuration (Figure 1), penetrates the subsurface and is returned to a receiving antenna (Rx) as a sequence of reflections from stratigraphic interfaces. The velocity of this EM wave is primarily controlled by the relative

dielectric permittivity (ε_r), a geophysical property strongly dependent on water content. Changes in bulk density and organic matter content are associated with changes in moisture content within sediments, causing strong GPR reflections [e.g., *Warner et al.*, 1990]. EM wave propagation in peat soils is limited by high fluid electrical conductivity and/or high percent of clay in the underlying mineral soil, causing excessive EM wave attenuation and thus reducing the depth of penetration in the peat [*Theimer et al.*, 1994]. However, the low electrical conductivity (typically less than 200 μS cm^{-1}) of pore waters in ombotrophic northern peatlands typically facilitates investigation depths of up to 11 m, usually sufficient to reach mineral soil in most basins encountered in the northern peatlands of Maine [*Slater and Reeve*, 2002].

GPR is very well suited to studying CH$_4$ dynamics in peatlands. First, in the absence of free-phase gas, peat is saturated to within ~0.5 m of the surface and has on the order of 80–95% porosity (ϕ). Second, free-phase CH$_4$ production

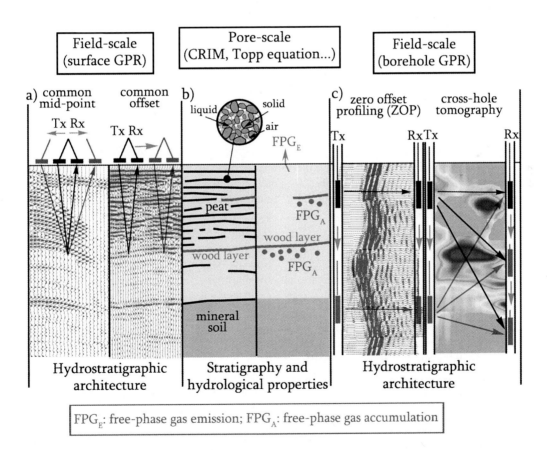

| Field-scale (surface GPR) | Pore-scale (CRIM, Topp equation...) | Field-scale (borehole GPR) |

a) common mid-point   common offset
b)   liquid   solid   air   FPG$_E$ peat   FPG$_A$ wood layer   FPG$_A$ wood layer   mineral soil
c) zero offset profiling (ZOP)   cross-hole tomography

Hydrostratigraphic architecture   Stratigraphy and hydrological properties   Hydrostratigraphic architecture

FPG$_E$: free-phase gas emission; FPG$_A$: free-phase gas accumulation

Figure 1. Overview of peatlands-related applications of GPR: characterization of (a) hydrostratigraphic architecture at the field scale using surface (common midpoint and common offset) and (c) borehole (single and cross-hole reflection) GPR surveys and characterization of (b) hydrological properties and dynamics at the pore scale combining GPR measurements and petrophysical models (e.g., complex refractive index model (CRIM) and/or Topp equation).

has been observed to account for up to ~20% changes in FPG content ($\theta_{FPG} = \phi - \theta$) in peat soils [*Rosenberry et al.*, 2006]. Considering that successful hydrogeophysical applications of GPR for monitoring vadose zone moisture dynamics have focused on detecting ~5% changes in θ [e.g., *Binley et al.*, 2002], these large, biogenically driven variations are readily detectable with the GPR technique [*Comas et al.*, 2005a, 2007, 2008]. This application of GPR is also supported by the well-constrained ε_r (of around 65–70) for saturated peat deposits [*Theimer et al.*, 1994]. Very precise estimates of peat thickness can be obtained from GPR with an assumed value of ε_r or, better still, a few common midpoint (CMP) measurements (Figure 1a) made in a peatland [*Jol and Smith*, 1995; *Slater and Reeve*, 2002].

The GPR method has been employed on studies of northern peatlands in a variety of measurement modes as summarized in Figure 1. The most common operation mode is constant-offset profiling (transmitter and receiver are separated by a fixed distance) from the surface (Figure 1a, right), that records the two-way travel time of the EM wave from the surface transmitter to reflectors at depth and returned to the surface receiver. Constant-offset surveys have been used to image the stratigraphy of the peat sequence (e.g., location of possible confining layers) and to determine depth to the mineral soil contact (depending on the nature of the mineral soil contact it may also be possible to image the mineral soil stratigraphy) [*Comas et al.*, 2005c]. Another surface-based methodology is the CMP survey, whereby the two-way travel time is again recorded but this time for transmitter and receiver pairs separated by increasingly larger distances (Figure 1a, left). This technique can be used to generate a model for the one-dimensional distribution of ε_r with depth. Using appropriate petrophysical relationships (Figure 1b and described below), it is possible to convert this ε_r depth profile into a depth profile of estimated FPG content as discussed in section 3. This approach has been successfully employed to monitor FPG production and release at sites within Maine peatlands, leading to new insights into dynamics of FPG cycling in these systems [*Comas et al.*, 2005a, 2007, 2008]. Other less commonly employed measurement modes rely on signal transmission between pairs of boreholes (Figure 1c). Here the one-way travel time between transmitter and receiver is recorded either as zero-offset profiling (ZOP) with depth (Figure 1c, left) [*Comas et al.*, 2005a] or in a tomographic mode, whereby a large number of raypaths are recorded for a variety of transmitter-receiver locations (Figure 1c, right and in section 3.3). Other variants on these basic measurement acquisition modes have included measurements on peat blocks in the laboratory during temperature-induced biogenic gas production [*Comas and Slater*, 2007].

Petrophysical relations (Figure 1b) exist to relate ε_r to moisture content and hence FPG content ($\theta_{FPG} = \phi - \theta$). Application of such relations is critical for quantitative interpretation of GPR data in terms of FPG content. By far, the most commonly applied relationship for estimating θ of soils and rocks is the empirical Topp equation [*Topp et al.*, 1980],

$$\theta = -5.3 \times 10^{-2} + 2.92 \times 10^{-2}\,\varepsilon_r \\ -5.5 \times 10^{-4}\,\varepsilon_r^2 + 4.3 \times 10^6\,\varepsilon_r^3. \qquad (1)$$

The calibration parameters in this equation have been found to satisfy measurements on mineral soils, but organic-rich soils (e.g., peat) tend to deviate from this relationship, and calibration measurements on peat soils are lacking in the published literature. Furthermore, these relationships were derived on the basis of TDR, which is typically a 500–1000 MHz measurement. Dispersion in the permittivity at low frequencies can affect ε_r-θ relationships as observed for clayey sandstone samples [*West et al.*, 2003] and clay-rich agricultural soils [*Roth et al.*, 2004]. Furthermore, dual-porosity media can exhibit abrupt changes of slope in the ε_r-θ relation when all interaggregate pores are saturated [*Blonquist et al.*, 2006]. Considering that peat is often regarded as a dual-porosity medium [*Ours et al.*, 1997; *Baird*, 1997], such effects might be important in calibrations of the ε_r-θ relation for FPG estimation. Thus it may be necessary to generate site-specific calibrations of θ for use in the interpretation of low-frequency GPR measurements.

Theoretical approaches are also frequently used to relate ε_r to θ. The simplest theoretical approaches utilize dielectric mixing models, whereby the volume fractions and ε_r of the components making up the soil are used to derive a relationship (for review, see *Lesmes and Friedman* [2005]). The complex refractive index model (CRIM) is a commonly applied mixing formula for estimating the measured ε_r from the dielectric and volumetric properties of the soil constituents (w, water; s, solid; and FPG, free-phase gas) (Figure 1b),

$$\varepsilon_r = [\theta\varepsilon_{r(w)}^\alpha + (1-\phi)\varepsilon_{r(s)}^\alpha + (\phi-\theta)\varepsilon_{r(FPG)}^\alpha]^{1/\alpha}, \qquad (2)$$

where α depends on the orientation of the electric field relative to geometry of the medium. Equation (2) applies to materials in which wavelengths are much greater than the dimensions of the dielectric or conductive inhomogenicities (typically between 0.2 and 0.4 m for 200 and 100 MHz, respectively, assuming typical velocities of 0.039 m ns^{-1} for peat soils). Note that the FPG content of interest in studies of methane cycling in peatlands is part of the third term in equation (2) ($\phi - \theta$). This approach has been applied in a number of recent studies of methane production and emission from peatlands and peat soils [*Comas et al.*, 2005a, 2007, 2008; *Comas and Slater*, 2007] with considerable success as evidenced by comparison with direct measurements of θ [*Comas and Slater*, 2007], capacitance probes [*Comas et al.*, 2005a], and surface deformation measurements [*Comas et al.*, 2007, 2008]. The limitations of this approach are discussed in section 4.

3. APPLICATIONS AND EXAMPLES

3.1. Imaging Stratigraphy and Confining Layers Trapping Methane Accumulation Using Surface GPR

A traditional application of GPR is in imaging stratigraphy. Stratigraphy can be resolved to over 10 m depth in most ombotrophic northern peatlands as a consequence of the low EM wave attenuation coefficients resulting from the low electrolytic conductivity of peat pore waters. Studies of peatland stratigraphy using GPR have yielded insights into pool formation [*Comas et al.*, 2005b] and evolution of vegetation patterning with peat growth [*Kettridge et al.*, 2008]. Monitoring of water levels and surface deformation, combined with coring, at the Red Lake peatland complex (northern Minnesota) have generated data suggesting that competent layers of peat (e.g., woody layers) act as confining layers impeding the upward transport and release of gas bubbles [*Glaser et al.*, 2004]. Although other models (e.g., *Edwards et al.* [1998] and *Daulat and Clymo* [1998] as discussed in section 3.2) have proved shallow FPG production and buildup in the absence of confining layers, *Glaser et al.* [2004] suggested that such layers regulate ebullition events at Red Lake by permitting the buildup of high FPG concentrations immediately below the layer, with episodic releases of large volumes of FPG following rapid drops in atmospheric pressure. Such competent layers should be strong GPR targets assuming that there is a large contrast in θ between the competent woody layers and surrounding less competent peat fabric.

Although the presence of woody layers trapping FPG release has been inferred from core data, the lateral continuity of such layers is not easily inferred from sparsely distributed core locations. We have conducted constant-offset measurements at sites in the Red Lake peatland complex and in raised bogs of Maine to noninvasively determine the vertical distribution of laterally continuous layers that may represent gas-trapping, confining layers. Plate 1 shows a constant-offset GPR profile at a site in Caribou Bog (Maine) where a vertically continuous core is available for comparison. The constant-offset GPR profile clearly shows two strong, laterally continuous reflectors at depths of 4.5 and 6 m, coin-

ciding with woody layers encountered in the core data. The reflector at 4.5 m is recorded as a continuous wood layer at the core location. The reflector at 6 m coincides with a zone of wood fragments and roots at the core location. The GPR data clearly confirm the lateral continuity of these woody layers over the 50 m of the survey line (Plate 1a). Results of free-phase gas sampling, performed as part of a separate study, are available for a location ~200 m away from this GPR line. Gas was sampled at multiple depths using a device, based on a water displacement concept, designed to capture pressurized gas trapped beneath competent layers. This sampler was advanced to the top of a confining layer (identified by the first sign of resistance to downward force), which, at this location, was approximately 2.7 m below the peat surface (BPS). The sampler was then entirely filled with water, and the sampler head was fitted with a gas extraction port and pushed through the confining layer to about 3.05 m BPS. Resulting rupture of the confining layer allowed FPG to enter the sampler and migrate upward to the sampler head where FPG was collected. Gas chromatography measurements identified a CH_4 concentration of 48–54%. Although not based on colocated measurements, these results support the concept that the laterally continuous layers identified with GPR (Plate 1a) may act as confining layers limiting the upward transport of FPG under buoyancy forces.

Given that GPR is a noninvasive technique that can be adopted for rapid reconnaissance surveying of peat basins (e.g., the unit can be mounted on a sled and pulled by a snowmobile during winter months), the possibility exists of applying the technique to map out laterally continuous, potential confining layers across an entire basin. Such data are needed to effectively upscale plot-scale estimates of CH_4 production and release beneath confining layers to values representative of the basin scale and thus more appropriate for input into global climate models. Furthermore, peatlands display a complex succession of vegetation patterning, and GPR has recently been used to illuminate stratigraphy across vegetation boundaries with implications for vegetation succession and persistence [*Kettridge et al.*, 2008]. The continuity of woody FPG-trapping layers across major vegetation boundaries in peatlands is largely unknown. GPR is a technology for rapidly evaluating the dependence of peat stratigraphy on vegetation and whether competent woody layers transcend vegetation boundaries or are only related to specific vegetation types.

3.2. Common Midpoint Velocity Analysis and 1-D Velocity Models

A major uncertainty in cycling of FPG in peat soils remains the vertical distribution of gas production zones and the depths contributing most CH_4 to surface emissions. Some studies, focusing on the upper 1–2 m of peat soils, have shown zones of enhanced CH_4 production immediately below the water table (see, e.g., *Kellner et al.* [2005] field studies). *Edwards et al.* [1998] report a zone of peak production immediately below the water table with a second zone of production centered around 0.15 m below the water table; *Daulat and Clymo* [1998] also observed a production peak ~0.15 m below the water table. *Slater et al.* [2007] also found that resistivity changes related to FPG production in a peat block were focused in two zones within the top 0.15 m below the water table. However, as reported elsewhere and noted above, studies of thicker peat sequences, e.g., the Red Lake Complex, have suggested that zones of extensive FPG accumulation occur at depth and are likely controlled by the location of competent woody layers that confine upward migration and release of FPG. These layers have been found at depths between 2 and 4 m in the Red Lake complex. Such layers have also been identified in northern bogs of Maine, as discussed in section 3.1 and illustrated in Plate 1a.

The CMP measurement (Figure 1a, left) can be used noninvasively to estimate the vertical distribution of FPG content within a peat soil (Plate 2). The simplest implementation of the technique involves fitting a series of hyperbolae to reflection events captured in a CMP record (Plate 2b). Assuming a 1-D velocity (v) structure (i.e., horizontal layering), the theoretical travel time between transmitter (Tx) and receiver (Rx) as a function of Tx-Rx separation is described by a hyperbola. A series of hyperbolae (Plate 2b) can then be fitted to the reflection record from a series of horizontal layers at progressively greater depths and the root-mean-square (RMS) velocities (V_{RMS}) between the surface and each reflection computed from the equation for a hyperbola. When a marked series of reflections is recorded, it is possible (depending on the velocity contrasts) to transform these RMS values into a model of interval velocities, e.g., using the Dix formulation [*Dix*, 1955], from which the interval relative dielectric permittivities can be computed (typically assuming a low loss medium, $v = \sqrt{c_0/\varepsilon_r}$, where c_0 is electromagnetic wave velocity in a vacuum, $c_0 = 3 \times 10^8$ m s^{-1}). A CMP can therefore provide a 1-D vertical FPG content model when a petrophysical relation (e.g., CRIM, equation (2)) is applied on the interval ε_r values (Plate 2c).

We have performed CMP surveys at a number of sites in the Red Lake peatland complex and in northern peatlands of Maine (primarily Caribou Bog). Plate 2b shows an example CMP reflection record that was centered at 15 m on the constant-offset profile shown for comparison in Plate 2a. This data set was collected at a clearing in a forested bog and close to the bog crest, where other measurements support the presence of FPG accumulation at depth below confining layers. Numerous near-horizontal reflectors exist within the peat soil down to time (t) = 200 ns, where a strong reflector indicates the lacustrine clay mineral soil (the GPR signal is entirely attenuated by the clay beyond this point). The CMP reflection record shown in Plate 2b shows the air wave, direct wave between Tx and Rx in the peat soil, followed by a series of reflection hyperbolae from layers within the peat down to mineral soil. Normal move out velocity estimation can be done manually (e.g., Plate 2b) or through processing algorithms, such as semblance analysis (Plate 2d), that predict the best estimate of average velocity above a given reflector [*Yilmaz*, 1987] by recalculating the arrival times for a range of velocities and adding up the normalized energy for each arrival time [e.g., *Huisman et al.*, 2003]. The resulting plot displays an arrangement of values (between 1 and 0) representing how well reflections at specific arrival times are described by that velocity (e.g., high or low values corresponding to good or bad fittings, respectively). The 1-D velocity profile estimated for this CMP is shown in Plate 2c. Higher velocities are associated with the presence of free-phase gas. We also show the estimated FPG content based on application of the CRIM assuming a 90% porosity of the peat soil at Red Lake [e.g., *McKenzie et al.*, 2006]. The relative dielectric permittivity of FPG and water are known ($\varepsilon_{r(FPG)} = 1$; $\varepsilon_{r(w)} = 80$ at 21°C), and $\varepsilon_{r(s)}$ for peat has been measured in the laboratory as 2. The exponent α can take any value between -1 and $+1$, but values of 0.35 have previously been reported for peat soils [*Kellner and Lundin*, 2001; *Kellner et al.*, 2005] and used elsewhere [*Comas et al.*, 2005a, 2007, 2008]. The velocity model in Plate 2c shows strong evidence for FPG accumulation between 2 and 3.5 m depth and lends supports to the model of FPG accumulation at depth beneath confining layers as proposed by others for this peatland complex.

3.3. Borehole Methods

The use of transmitters and receivers in boreholes (Figure 1c), instead of on the surface, has been extensively explored for monitoring of vadose zone processes in soils and rocks (for review see *Huisman et al.* [2003] and *Daniels et al.* [2005]). This technique allows the depth of exploration to be extended beyond the attenuation-limited penetration depth achievable with surface GPR. In the cross-borehole configuration, the one-way transmission time of the EM wave between a transmitter in one hole, and the receiver in another hole, is measured. The simplest implementation of the method is zero-offset profiling (Figure 1c, left), whereby the transmitter and receiver are typically kept at the same horizontal level and simultaneously lowered to generate a 1-D

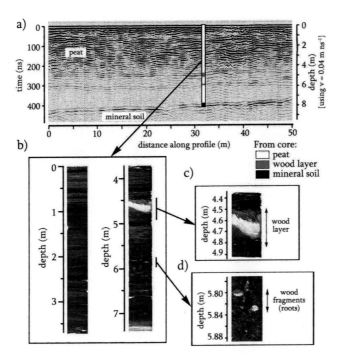

Plate 1. (a) Constant-offset GPR image (using 100 MHz antennas) of stratigraphy beneath a portion of a northern peatland (Caribou Bog, Maine) showing the correspondence between reflector attributes and presence of wood layers as inferred from coring. (b) Coring image (5 inch core diameter, courtesy of P. Glaser, University of Minnesota) showing internal peat soil structure and presence of a 30 cm wood layer between 4.5 and 4.8 m depth and wood fragments or roots at approximately 5.8 m depth. (c) Magnification of wood layer at depth shown in Plate 1b. (d) Magnification of wood fragments (or roots) at depth shown in Plate 1b.

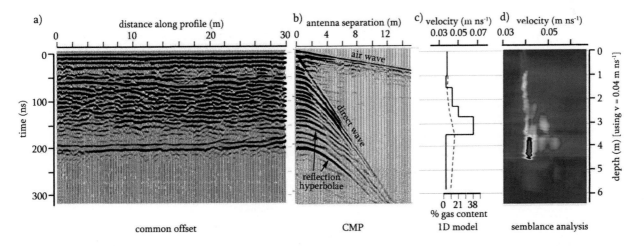

Plate 2. Multiple-offset GPR reflection survey (using 100 MHz antennas) across a northern peatland (glacial Lake Agassiz peatland, Minnesota), showing (a) constant-offset GPR survey; (b) common midpoint GPR survey (showing location of air wave, direct wave, and reflection hyperbolae); (c) 1-D vertical velocity model showing CRIM-estimated percent gas content (assuming $\varepsilon_{r(s)} = 2$, $\varepsilon_{r(w)} = 80$, $\alpha = 0.35$, and $\phi = 90\%$). Dashed and solid lines show distribution of V_{RMS} and interval velocities (e.g., Dix equation), respectively, with depth. (d) Semblance analysis. Red and blue colors indicate high and low semblance, respectively.

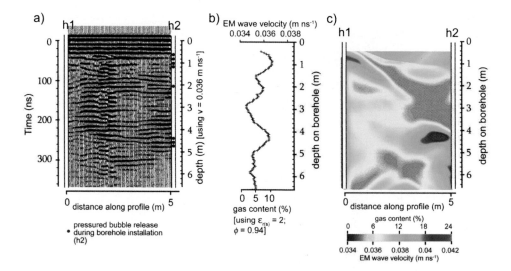

Plate 3. (a) Multiple-offset GPR reflection survey (using 250 MHz antennas) between two boreholes placed 5 m apart and to 8 m depth at Caribou Bog (Maine). (b) Borehole transmission ZOP survey results showing EM wave velocities inferred from GPR expressed as FPG content using the CRIM (assuming $\varepsilon_{r(s)} = 2$, $\varepsilon_{r(w)} = 80$, $\alpha = 0.35$, and $\phi = 94\%$). (c) Inverted tomographic image showing 2-D distribution of EM velocities and CRIM-estimated FPG contents (data in Plates 3a and 3b modified from *Comas et al.* [2005b]).

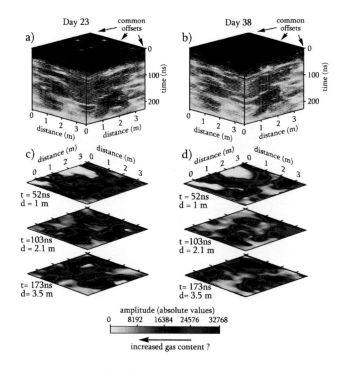

Plate 4. (a and b) Time lapse 3-D representation of the envelope over a 3.6 m × 3.6 m grid in Caribou Bog collected 15 days apart, as part of the surveys reported by *Comas et al.* [2007]. (c and d) Time slices at times $t = 52$, 103, and 173 ns (corresponding to 1, 2.1, and 3.5 m depth assuming an average velocity = 0.04 m ns^{-1}) for day 23 (Plate 4a) and day 38 (Plate 4b).

depth profile of average velocity for material between the two boreholes (Figure 1c). However, cross-borehole GPR can also be employed in a tomographic fashion, whereby the one-way travel times between transmitter and receiver for a large number of Tx-Rx configurations are recorded and used to reconstruct the best estimate of the 2-D (or 3-D if using multiple boreholes) internal velocity structure between the boreholes (Figure 1c, right).

Unlike surface GPR, borehole GPR is an invasive method because the inclinometer casing (used to survey the hole orientation required to accurately calculate the velocity from the horizontal travel time) must be installed through the peat column. Similar to FPG investigation methods that rely on insertion of probes or piezometers, the potential thus exists to disturb in situ FPG dynamics as a result of this installation. However, *Comas et al.* [2005a] performed zero-offset profiling between boreholes at a test site in Caribou Bog, Maine, and reconstructed a vertical profile of FPG content by converting the EM velocities to FPG content using the CRIM formulation (equation (2)). Their result (reconstructed in Plate 3a) predicts a zone of enhanced FPG accumulation between 3.8 and 4.8 m, with constant-offset profiles showing evidence for laterally continuous reflectors immediately above this location at about 3.6 m (Plate 3b).

We inverted tomographic data collected at this site (previously unpublished) for the 2-D distribution of velocity in between these wells and converted it to an equivalent FPG content using the CRIM (Plate 3c). Two 0.08 m diameter inclinometer casings were installed 5 m apart, from the surface to the top of the mineral soil (8.4 m), and inclinometer surveys were used to determine the deviations of the boreholes from vertical and used in the tomographic inversion (maximum separation of 5.4 m at 8 m depth). Transmitter spacing was 0.1 m starting at 1 m depth, while receiver spacing was set to 0.05 m, with both transmitter and receiver lowered to a maximum depth of 8 m. The tomographic algorithm used for the inversion (part of ReflexW by Sandmeier Scientific Software) is based on a simultaneous iterative reconstruction technique [*Gilbert*, 1972] that considers curved rays. The resulting image suggests that the FPG observed in the ZOP is locally concentrated within the image plane rather than horizontally uniform at a certain depth. Whereas the ZOP predicts average FPG of up to 10% for the 5 m section between wells, the tomography suggests that the FPG may be concentrated and localized, reaching 22% in places. This example illustrates the potential to obtain a more detailed understanding of the multidimensional distribution of FPG in peatlands using tomographic GPR methods.

Ultimately, cross-borehole GPR can yield a high-resolution vertical profile of the FPG content within a peat soil, avoiding the uncertainty associated with modeling the 1-D veloc-

ity profile using the CMP approach described in section 3.2. Furthermore, 2-D (or potentially 3-D) images of the FPG distribution between boreholes are obtainable as illustrated here. However, these methods are labor intensive relative to surface-based GPR. This extra labor effort may be significant when working at remote peatland locations. Furthermore, much of the benefit of cross-borehole GPR comes when it extends the investigation range beyond that achievable with surface GPR. This is often an important issue when attempting to explore vadose zone flow/transport when the unsaturated zone is tens of meters or more thick. In studies of methane dynamics in northern peatlands, however, our interest is in the peat sequence above the mineral soil, which is most often less than 10 m. As previously discussed, penetration depths in northern peatlands are usually sufficient to image to the mineral soil with surface reflections as a result of the low attenuation coefficients caused by the ombotrophic (low electrolytic conductivity) pore waters. We are of the view that investigations of FPG methane dynamics and cycling are best done with totally noninvasive methods such as surface GPR, such that data can be more confidently interpreted in terms of the natural system response rather than artifacts induced by disturbing the peat fabric, which may provide artificial pathways for rapid release of gas to the atmosphere. Cross-borehole measurements may be warranted when surface measurements fail either because of limited penetration or clutter in the data due to excessive reflections or diffractions from above surface objects (e.g., trees).

3.4. Temporal Monitoring of Methane Dynamics

GPR has been adopted as a tool to noninvasively monitor FPG methane dynamics in the field [*Comas et al.*, 2007, 2008] and on peat blocks in the laboratory [*Comas and Slater*, 2007]. The field-based methods have exploited the CMP survey approach and examined changes in root-mean-square EM velocity determined from the mineral soil reflector (e.g., Plate 2b). These changes have been converted to changes in FPG content integrated across the depth of the peat column using the CRIM (equation (2)). In the field-based studies, changes in GPR-estimated FPG correlate closely with measured surface deformation attributed to peat expansion during gas production (and subsequent deflation during gas release), as well as simple chamber-type measurements using gas detectors. An advantage of the technique is that it estimates the effective change in gas content for the entire peat column rather than representing values at a point in the soil as would be obtained, for example, from probes (e.g., TDR) that estimate moisture content. Depth-integrated measurements provide data at a scale more appropriate to upscaling to representative values for a peat basin, an important consideration when such data are required to constrain climate models. These studies have led to new insights into the controls of season (temperature), winter ice, and vegetation community on the production and emission of FPG from peatlands [*Comas et al.*, 2007, 2008].

Analysis of complex trace attributes such as instantaneous amplitude is often used in seismic processing to enhance the interpretation of the data set relative to an analysis of travel times alone [*White*, 1991]. One approach to amplitude analysis, known as the envelope, provides a measure of the reflectivity strength that is useful for identifying amplitude anomalies and improving signal-to-noise ratio [*Nye and Berry*, 1974]. With this approach, frequency values at every point along a trace (proportional to the square root of the complete energy of the signal at that time) are decomposed using the Hilbert transformation. Envelope attributes have been used to generate time slice profiles from three-dimensional (3-D) GPR data sets [e.g., *Nuzzo et al.*, 2002]. Plate 4 shows a 3-D representation of the envelope over a 3.6 m × 3.6 m grid in Caribou Bog (see *Comas et al.* [2007, 2008] for details) with two orthogonal data sets each containing nine equally spaced (at 0.45 m) common offset profiles with a 0.2 m trace interval. The data set was collected at two times, 15 days apart, as part of the surveys reported by *Comas et al.* [2007]. Time slice profiles are built by averaging amplitude of the radar signal within a predefined time interval. Interpretation of CMP surveys at the location generated estimates of average biogenic gas content for the 5.4 m peat column increasing from 9% on day 23 (Plate 4a) to 12% on day 38 (Plate 4b). Plates 4c and 4d show time slices at three different depths (t = 52, 103, and 173 ns, corresponding to 1, 2.1, and 3.5 m depths assuming an average velocity of 0.04 m ns^{-1}) for days 23 and 38, respectively. Anomalies within the amplitude record are characterized by marked reflectivity (purple/red shading corresponding to high-amplitude values) and appear enhanced during the period of lower gas content (day 23, Plates 4a and 4c), relative to the period of higher gas content (day 38, Plates 4b and 4d) for all depths shown. Areas of enhanced attenuation (yellow shading corresponding to low-amplitude values) show the opposite trend. Although more measurements are needed, this data set suggests that relative changes within the amplitude record may be diagnostic of FPG production and emissions as a result of backscattering induced by biogenic gas bubbles acting as small-scale heterogeneities (i.e., smaller than the propagated EM wavelengths). This data set supports (1) enhanced reflectivity during periods of low gas content (e.g., reduced backscattering) and (2) enhanced attenuation during periods of high gas content (e.g., intensified backscattering). This backscattering effect is a well known cause of amplitude attenuation in GPR surveys [*Annan*, 2006].

4. DISCUSSION

The spatially rich, noninvasive data sets obtained with GPR have provided new insights into methane cycling in northern peatlands. We have discussed here how GPR has been used to image peatland stratigraphy and identify the location of spatially continuous layers of competent peat (typically woody layers) that act to confine upward gas transport. Surface GPR profiling, 1-D velocity analysis from common midpoint surveys, and vertical EM wave velocity profiles obtained from cross-borehole measurements are all consistent with this model. In addition to proving the methodology as a unique tool for noninvasive investigation of methane dynamics, these studies have also shown how GPR can be applied to monitor the production and emission of FPG from peatlands and the dependence of production and emission rates on environmental variables (pressure, temperature, water levels, and vegetation community). For example, *Comas et al.* [2008] monitored seasonal changes in FPG content at two sites of contrasting vegetation in Caribou Bog and showed how FPG production rates and ebullition fluxes were distinctly different during the summer but followed a statistically identical trend during winter months. A significant finding of this study was the buildup of very high FPG contents (values reaching 16% integrated over the thickness of the peat column) during winter when the peat surface was frozen. In their review of the effects of biogenic gas on hydrology of northern peatlands, *Rosenberry et al.* [2006] emphasized the need for such studies of FPG accumulation under winter ice. Similar GPR monitoring studies using a higher sampling frequency have confirmed results of others (using chamber techniques, surface deformation, and hydrological sampling), suggesting that sporadic ebullition events, often triggered by sudden drops in atmospheric pressure (e.g., over few hours), can release more CH$_4$ than the diffusive contribution for an entire year [*Comas et al.*, 2007].

We summarize the main advantages of the technique as follows: (1) In the surface configuration, the technique provides an entirely noninvasive means to investigate FPG methane cycling in peatlands. (2) The scale of the measurement (from a layer interval (e.g., Plate 2b) up to the thickness of the entire peat column) is more appropriate for upscaling to basin-scale values relative to point-based measurements. (3) Spatially rich data sets are potentially obtainable; that is, it is possible to cover hundreds of meters of line in constant-offset profiling mode (e.g., for mapping continuity of confining layers) relatively rapidly.

However, the method is not without limitations. Most significantly, the GPR approach to examining methane cycling and obtaining quantitative estimates of FPG content relies on the application of an inherently uncertain petrophysical model for converting ε_r to FPG. The use of the theoretical CRIM to convert ε_r to FPG content relies on numerous assumptions. Most obviously, a number of the parameters in the model (equation (2)) may be poorly constrained. In our case, α is assumed on the basis of previous studies, and ϕ is typically somewhat constrained by limited available data, typically average porosity estimates for the peat column. Although porosity variations of only 3–4% along the peat column have been recorded in Caribou Bog [e.g., *Comas et al.*, 2005b], assuming average porosities for an entire peat column (e.g., 90–94% in Plates 2c and 3b) will generate errors in FPG content estimates. For example, our FPG content estimates in Plates 2 and 3 vary as much as 4% when considering a range of 90–94% porosity, while water content estimates vary only less than 1% using the same range. Wood layers (e.g., Plate 1) may conceivably result in much larger porosity contrasts (and therefore, larger errors in FPG content estimates) within the peat column that should be considered in future GPR work on FPG estimation in peatlands.

The CRIM presents certain advantages over other mixing models (e.g., Topp equation) by explicitly incorporating ε_r of the solid and porosity. However, another limitation of the CRIM is that it predicts ε_r on the basis of EM wave transmission through sequential layers of solid, water, and air within the soil and therefore overlooks true arrangements of soil constituents [*West et al.*, 2003] and thus may overlook the heterogeneous nature of the peat matrix. Furthermore, the scale dependency of the ε_r-θ relationship that determines how local-scale petrophysical relationships (e.g., based on TDR or neutron probe data) relate to the field-scale relationships needed to interpret GPR measurements from a peatland (Figure 1b) must be considered. The CRIM does not account for the geometric distribution of subsurface water, although this distribution is well known to impact ε_r [*Moysey and Knight*, 2004]. This dependence on geometry means that the petrophysical relationship linking ε_r to θ is scale-dependent where subsurface architecture is complex. Given that peatlands form in a very low energy environment and the dielectric properties of peat are quite uniform compared to most inorganic soils, such geometric effects may not be a major concern in peatland studies. Field-scale relationships that capture heterogeneity and account for the effect of stratigraphic complexity on the ε_r-θ relationship have been predicted using geostatistical characterizations of subsurface structure [*Moysey and Knight*, 2004]. As previously noted, other problems with the calibration of the ε_r-FPG relationship include dispersion in ε_r at low frequencies used for field GPR that may not be accounted for in empirical petrophysical relations established using higher-frequency laboratory methods (e.g., TDR). Furthermore, recent studies on inorganic soils have demonstrated how the ε_r-θ relation depends on moisture history and displays hysteresis between wetting and drying cycles [*Lai et al.*, 2006]. It is plausible that hysteresis effects may also be associated with cycles of methane production (increasing FPG) and emission (decreasing FPG). Clearly, soil (and/or site) specific calibrations of ε_r-θ are needed to improve interpretation of low-frequency GPR measurements in terms of moisture/FPG dynamics. This is a particularly pressing issue in peatland studies because solid empirical relations (e.g., equation (1)) have not been developed for peat soils. In order to advance the application of the GPR method for studies of methane cycling, we thus see a critical need for calibration measurements of the ε_r-FPG relation in peat soils, taking into account possible hysteresis effects and the dual-porosity nature of peat soils [*Ours et al.*, 1997].

Another limitation of the GPR technique is that it is not readily adaptable as an autonomous monitoring method. Autonomous data acquisition has recently been developed for other geophysical techniques, e.g., resistivity monitoring, to enhance the flexibility of the method as a hydrogeological monitoring tool. For example, *Slater et al.* [2007] used 3-D resistivity monitoring to capture FPG dynamics in a peat block. The GPR method is inherently more labor-intensive during field data acquisition, and the degree to which the method can be automated for high-resolution spatiotemporal imaging of hydrogeological processes is uncertain. Currently, an operator must be present in the field to collect CMP data sets previously used to monitor temporal evolution in FPG content. Some progress has been made regarding automation, with semiautomated acquisition systems developed that use self-tracking laser theodolites with automatic tracking capabilities [*Green et al.*, 2003]. However, such systems are not in mainstream production.

We see many opportunities for furthering studies of FPG methane dynamics in northern peatlands using the GPR method. As illustrated in Plate 2, it is possible to reconstruct the vertical distribution of FPG content from CMP data sets. Previous studies have used GPR as a tool for monitoring changes in the average FPG content integrated over the peat column, but the possibility clearly exists to monitor temporal changes in interval velocities in order to determine changes in the vertical distribution of FPG over time and its dependence on environmental parameters. For example, such measurements could help to improve understanding of what depths in the peat column contribute most to FPG production and emission via ebullition events. We also see opportunities to conduct basin-scale GPR measurements to determine the distribution of gas-trapping layers or, possibly, direct detection of gas hot spots from scattering signals [*Comas et al.*, 2005b]. The recent availability of multichannel

GPR instrumentation also provides new opportunities for studies of methane dynamics in northern peatlands. For example, methods for continuous CMP analysis using two sets of transmitters and receivers now exist. This technique could potentially be employed to map rapidly the peat column–averaged gas content across a basin. Such opportunities will likely ensure that the GPR method is increasingly adopted as a technique for exploring methane cycling in northern peatlands.

Acknowledgments. We are especially grateful to Paul Glaser (University of Minnesota) for permission to make use of the results of coring conducted at Caribou Bog during a National Science Foundation (NSF) funded workshop on Peatlands Geophysics (11–15 June 2006). The GPR data set shown in Plate 1 was collected by Harry Jol (University of Wisconsin-Eau Claire) during the same NSF workshop. Andrew Parsekian (Rutgers-Newark), Jay Nolan (Rutgers-Newark), Michael O'Brien (Rutgers-Newark), and Andrew Reeve (University of Maine) assisted with collection of field data sets shown in this chapter. We thank Kevin Hon and Mala Geoscience for the repeated loan of replacement/backup GPR equipment used in these measurements. We also thank Editor Andrew Baird and two anonymous reviewers for their suggestions to enhance the quality of an earlier version of this manuscript.

REFERENCES

Annan, A. P. (2006), Ground penetrating radar, in *Near-Surface Geophysics*, *Invest. Geophys.*, vol. 13, edited by D. K. Butler, pp. 357–438, Soc. of Explor. Geophys., Tulsa, Okla.

Baird, A. J. (1997), Field estimation of macropore functioning and surface hydraulic conductivity in a fen peat, *Hydrol. Processes*, *11*, 287–295.

Beckwith, C. W., and A. J. Baird (2001), Effect of biogenic gas bubbles on water flow through poorly decomposed blanket peat, *Water Resour. Res.*, *37*(3), 551–558.

Binley, A., P. Winship, L. J. West, M. Pokar, and R. Middleton (2002), Seasonal variation of moisture content in unsaturated sandstone inferred from borehole radar and resistivity profiles, *J. Hydrol.*, *267*(3–4), 160–172.

Bjelm, L. (1980) Geological interpretation with subsurface interface radar in peat lands, in *6th International Peat Congress, Duluth 1980*, pp. 7–8, Int. Peat Soc., Jyväskylä, Finland.

Blonquist, J. M., Jr., S. B. Jones, I. Lebron, and D. A. Robinson (2006), Microstructural and phase configurational effects determining water content: Dielectric relationships of aggregated porous media, *Water Resour. Res.*, *42*, W05424, doi:10.1029/2005WR004418.

Comas, X., and L. Slater (2007), Evolution of biogenic gases in peat blocks inferred from noninvasive dielectric permittivity measurements, *Water Resour. Res.*, *43*, W05424, doi:10.1029/2006WR005562.

Comas, X., L. Slater, and A. Reeve (2005a), Spatial variability in biogenic gas accumulations in peat soils is revealed by ground penetrating radar (GPR), *Geophys. Res. Lett.*, *32*, L08401, doi:10.1029/2004GL022297.

Comas, X., L. Slater, and A. Reeve (2005b), Stratigraphic controls on pool formation in a domed bog inferred from ground penetrating radar (GPR), *J. Hydrol.*, *315*(1–4), 40–51.

Comas, X., L. Slater, and A. Reeve (2005c), Geophysical and hydrological evaluation of two bog complexes in a northern peatland: Implications for the distribution of biogenic gases at the basin scale, *Global Biogeochem. Cycles*, *19*, GB4023, doi:10.1029/2005GB002582.

Comas, X., L. Slater, and A. Reeve (2007), In situ monitoring of free-phase gas accumulation and release in peatlands using ground penetrating radar (GPR), *Geophys. Res. Lett.*, *34*, L06402, doi:10.1029/2006GL029014.

Comas, X., L. Slater, and A. Reeve (2008), Seasonal geophysical monitoring of biogenic gases in a northern peatland: Implications for temporal and spatial variability in free phase gas production rates, *J. Geophys. Res.*, *113*, G01012, doi:10.1029/2007JG000575.

Daniels, J. J., B. Allred, A. Binley, D. LaBrecque, and A. Alumbaugh (2005), Hydrogeophysical case studies in the vadose zone, in *Hydrogeophysics*, edited by Y. Rubin and S. S. Hubbard, pp. 413–440, Springer, Dordrecht, Netherlands.

Daulat, W. E., and R. S. Clymo (1998), Effects of temperature and water table on the efflux of methane from peatland surface cores, *Atmos. Environ.*, *32*(19), 3207–3218.

Dix, C. H. (1955), Seismic velocities from surface measurements, *Geophysics*, *20*, 68–86.

Edwards, C., B. A. Hales, G. H. Hall, I. R. McDonald, J. C. Murrell, R. Pickup, D. A. Ritchie, J. R. Saunders, B. M. Simon, and M. Upton (1998), Microbiological processes in the terrestrial carbon cycle: Methane cycling in peat, *Atmos. Environ.*, *32*(19), 3247–3255.

Gilbert, H. (1972), Iterative methods for the three-dimensional reconstruction of an object from projections, *J. Theor. Biol.*, *36*, 105–117.

Glaser, P. H., J. P. Chanton, P. Morin, D. O. Rosenberry, D. I. Siegel, O. Ruud, L. I. Chasar, and A. S. Reeve (2004), Surface deformations as indicators of deep ebullition fluxes in a large northern peatland, *Global Biogeochem. Cycles*, *18*, GB1003, doi:10.1029/2003GB002069.

Green, A., R. Gross, K. Holliger, H. Horstmeyer, and J. Baldwin (2003), Results of 3-D georadar surveying and trenching the San Andreas fault near its northern landward limit, *Tectonophysics*, *368*(1–4), 7–23.

Holden, J., T. P. Burt, and M. Vilas (2002), Application of ground-penetrating radar to the identification of subsurface piping in blanket peat, *Earth Surf. Processes Landforms*, *27*, 235–249.

Huisman, J. A., S. S. Hubbard, J. D. Redman, and A. P. Annan (2003), Measuring soil water content with ground penetrating radar: A review, *Vadose Zone J.*, *2*, 476–491.

Jol, H. M., and D. G. Smith (1995), Ground penetrating radar surveys of peatlands for oilfield pipelines in Canada, *J. Appl. Geophys.*, *34*(2), 109–123.

Kellner, E., and L. C. Lundin (2001), Calibration of time domain reflectometry for water content in peat soils, *Nord. Hydrol.*, *32*, 315–332.

Kellner, E., J. M. Waddington, and J. S. Price (2005), Dynamics of biogenic gas bubbles in peat: Potential effects on water storage and peat deformation, *Water Resour. Res.*, *41*, W08417, doi:10.1029/2004WR003732.

Kettridge, N., X. Comas, A. Baird, L. Slater, M. Strack, D. Thompson, H. Jol, and A. Binley (2008), Ecohydrologically important subsurface structures in peatlands revealed by ground-penetrating radar and complex conductivity surveys, *J. Geophys. Res.*, *113*, G04030, doi:10.1029/2008JG000787.

Lai, W. L., W. F. Tsang, H. Fang, and D. Xiao (2006), Experimental determination of bulk dielectric properties and porosity of porous asphalt and soils using GPR and a cyclic moisture variation technique, *Geophysics*, *71*, K93–K102.

Lapen, D. R., B. J. Moorman, and J. S. Price (1996), Using ground-penetrating radar to delineate subsurface features along a wetland catena, *Soil Sci. Soc. Am. J.*, *60*, 923–931.

Lesmes, D. P., and S. P. Friedman (2005), Relationships between the electrical and hydrogeological properties of rocks and soils, in *Hydrogeophysics*, edited by Y. Rubin and S. S. Hubbard, pp. 87–128, Springer, Dordrecht, Netherlands.

McKenzie, J. M., D. I. Siegel, D. O. Rosenberry, P. H. Glaser, and C. I. Voss (2006), Heat transport in the Red Lake Bog, glacial Lake Agassiz peatlands, *Hydrol. Processes*, *21*, 369–378.

Moysey, S., and R. Knight (2004), Modeling the field-scale relationship between dielectric constant and water content in heterogeneous systems, *Water Resour. Res.*, *40*, W03510, doi:10.1029/2003WR002589.

Nuzzo, L., G. Leucci, S. Negri, M. T. Carozzo, and T. Quarta (2002), Application of 3D visualization techniques in the analysis of GPR data for archaeology, *Ann. Geophys.*, *45*(2), 321–337.

Nye, J. F., and M. Berry (1974), Dislocations in wave trains, *Proc. R. Soc. London, Ser. A*, *336*, 165–190.

Ours, D. P., D. I. Siegel, and P. H. Glaser (1997), Chemical dilation and the dual porosity of humified bog peat, *J. Hydrol.*, *196*(1–4), 348–360.

Price, J. S. (2003), Role and character of seasonal peat soil deformation on the hydrology of undisturbed and cutover peatlands, *Water Resour. Res.*, *39*(9), 1241, doi:10.1029/2002WR001302.

Rosenberry, D. O., P. H. Glaser, D. I. Siegel, and E. P. Weeks (2003), Use of hydraulic head to estimate volumetric gas content and ebullition flux in northern peatlands, *Water Resour. Res.*, *39*(3), 1066, doi:10.1029/2002WR001377.

Rosenberry, D. O., P. H. Glaser, and D. I. Siegel (2006), The hydrology of northern peatlands as affected by biogenic gas: Current developments and research needs, *Hydrol. Processes*, *20*, 3601–3610.

Roth, K., U. Wollschlager, Z. H. Cheng, and J. B. Zhang (2004), Exploring soil layers and water tables with ground-penetrating radar, *Pedosphere*, *14*(3), 273–282.

Slater, L., and A. Reeve (2002), Understanding peatland hydrology and stratigraphy using integrated electrical geophysics, *Geophysics*, *67*, 365–378.

Slater, L., X. Comas, D. Ntarlagiannis, and M. R. Moulik (2007), Resistivity-based monitoring of biogenic gases in peat soils, *Water Resour. Res.*, *43*, W10430, doi:10.1029/2007WR006090.

Theimer, B. D., D. C. Nobes, and B. G. Warner (1994), A study of the geoelectrical properties of peatlands and their influence on ground-penetrating radar surveying, *Geophys. Prospect.*, *42*, 179–209.

Tokida, T., T. Miyazaki, and M. Mizoguchi (2005), Ebullition of methane from peat with falling atmospheric pressure, *Geophys. Res. Lett.*, *32*, L13823, doi:10.1029/2005GL022949.

Topp, G. C., J. L. Davis, and A. P. Annan (1980), Electromagnetic determination of soil water content: Measurements in coaxial transmission lines, *Water Resour. Res.*, *16*(3), 574–582.

Ulriksen, C. P. F. (1981), Investigation of peat thickness with radar, in *6th International Peat Congress, Duluth 1980*, pp. 126–129, Int. Peat Soc., Jyväskylä, Finland.

Walter, K. M., S. A. Zimov, J. P. Chanton, D. Verbyla, and F. S. Chapin III (2006), Methane bubbling from Siberian thaw lakes as a positive feedback to climate warming, *Nature, 443*, 71–75, doi:10.1038/nature05040.

Warner, B. G., D. C. Nobes, and B. D. Theimer (1990), An application of ground penetrating radar to peat stratigraphy of Ellice Swamp, southwestern Ontario, *Can. J. Earth Sci.*, *27*, 932–938.

West, L. J., K. Handley, Y. Huang, and M. Pokar (2003), Radar frequency dielectric dispersion in sandstone: Implications for determination of moisture and clay content, *Water Resour. Res.*, *39*(2), 1026, doi:10.1029/2001WR000923.

Whalen, S. C., and W. S. Reeburgh (1992), Interannual variations in tundra methane emissions: A 4-year time series at fixed sites, *Global Biogeochem. Cycles*, *6*, 139–159.

White, R. E. (1991), Properties of instantaneous seismic attributes, *Leading Edge*, *10*(7), 26–32.

Yilmaz, Ö. (1987), *Seismic Data Processing*, pp. 240–353, 468–473, Soc. of Explor. Geophys., Tulsa, Okla.

X. Comas, Department of Geosciences, Florida Atlantic University, 777 Glades Road, Room 360, Boca Raton, FL 33431, USA. (x.comas@fau.edu)

L. D. Slater, Department of Earth and Environmental Sciences, Rutgers, State University of New Jersey, 101 Warren Street, Smith Hall, Newark, NJ 07102, USA.

Methane Dynamics in Peat: Importance of Shallow Peats and a Novel Reduced-Complexity Approach for Modeling Ebullition

T. J. Coulthard,[1] A. J. Baird,[2] J. Ramirez,[3] and J. M. Waddington[4]

Northern peatlands are one of the largest natural sources of atmospheric methane (CH$_4$), and it is important to understand the mechanisms of CH$_4$ loss from these peatlands so that future rates of CH$_4$ emission can be predicted. CH$_4$ is lost to the atmosphere from peatlands by diffusion, by plant transport, and as bubbles (ebullition). We argue that ebullition has not been accounted for properly in many previous studies, both in terms of measurement and the conceptualization of the mechanisms involved. We present a new conceptual model of bubble buildup and release that emphasizes the importance of near-surface peat as a source of atmospheric CH$_4$. We review two possible approaches to modeling bubble buildup and loss within peat soils: the recently proposed bubble threshold approach and a fully computational-fluid-dynamics approach. We suggest that neither satisfies the needs of peatland CH$_4$ models, and we propose a new reduced-complexity approach that conceptualizes bubble buildup and release as broadly similar to an upside down sandpile. Unlike the threshold approach, our model allows bubbles to accumulate at different depths within the peat profile according to peat structure, yet it retains the simplicity of many cellular (including cellular automata) models. Comparison of the results from one prototype of our model with data from a laboratory experiment suggests that the model captures some of the key dynamics of ebullition in that it reproduces well observed frequency-magnitude relationships. We outline ways in which the model may be further developed to improve its predictive capabilities.

1. METHANE LOSS FROM PEATLANDS

1.1. Mechanisms of Loss

Methane (CH$_4$) is an important greenhouse gas, and future changes in atmospheric concentrations of CH$_4$ may

have significant impacts on global climate [cf. *Gedney et al.*, 2004; *Frolking et al.*, 2006; *Intergovernmental Panel on Climate Change*, 2007; *Walter et al.*, 2001a]. Northern peatlands are one of the largest natural sources per annum of CH$_4$ emissions to the atmosphere, yet there are considerable uncertainties about how CH$_4$ is stored in, and released from, these vast ecosystems. CH$_4$ is lost to the atmosphere from peatlands via three mechanisms: (1) diffusion through pore water to the water table and thence through the zone above the water table (if one exists) to the peatland surface, (2) diffusion or active transport through vascular plants, and (3) ebullition, bubbles moving to the peatland surface. Until quite recently, ebullition was considered to be only locally important, and most attention was focused on matrix diffusion of dissolved CH$_4$ and plant-mediated transport. However, recent research, as summarized in Table 1, suggests

[1]Department of Geography, University of Hull, Hull, UK.
[2]School of Geography, University of Leeds, Leeds, UK.
[3]Miami, Florida, USA.
[4]School of Geography and Earth Sciences, McMaster University, Hamilton, Ontario, Canada.

Carbon Cycling in Northern Peatlands
Geophysical Monograph Series 184
Copyright 2009 by the American Geophysical Union.
10.1029/2008GM000811

Table 1. Examples of Recent Studies of CH_4 Ebullition From Peat Soils

Study	Rates of Ebullition (mg CH_4 m^{-2} d^{-1})	Method	Comments
Baird et al. [2004]	0–83	laboratory cores of near-surface *Sphagnum* peat ($n = 8$)	Rates are for threshold bubble content [cf. *Kellner et al.*, 2006] and are based on 2- to 4-day averages of gas collected in gas traps. No detail is available on barometric pressure in the laboratory.
Christensen et al. [2003]	36–170	laboratory cores of near-surface northern temperate/subarctic peats of various compositions ($n = 4$)	Rates appear to be for threshold bubble content and are based on continuous measurements from throughflow chambers fixed to the cores. No detail is available on barometric pressure in the laboratory.
Comas and Slater [2007]	~400–>1,200	laboratory monolith of near-surface *Sphagnum* peat	Rates are based on periodic measurements from chamber above monolith and include all transport mechanisms. However, bubbles were measured in the peat, and higher rates of CH_4 efflux seem to be associated with changes in peat bubble content.
Glaser et al. [2004]	35,000	field measurement of (1) changes in the elevation of the surface of a *Sphagnum*-dominated peatland using GPS and (2) pressure head using piezometers installed at depths of 1, 2, and 3 m	Rates are for short-lived (~4 h) ebullition events and assume (1) bubbles are lost from medium-depth peat (~2 m) and (2) a CH_4 content in the bubbles being released of 54%.
Kellner et al. [2006]	270	laboratory core of near-surface *Sphagnum* peat	Rates are for threshold bubble content and are based on 2- to 4-day averages of gas collected in gas traps.
Rosenberry et al. [2003]	4,300–10,700	field measurement of pressure head using piezometers installed at depths of 1, 2, and 3 m in *Sphagnum* peat	Rates are for short-lived (hours to days) ebullition events and assume (1) bubbles are lost from medium-depth peat (1–2 m) and (2) a CH_4 content in the bubbles being released of 50%.
Strack et al. [2005]	65	field measurement of gas trapped in collection funnels sunk into near-surface (upper meter) peat at a *Sphagnum*-dominated site	Rates are averaged over summer season. Individual events may give figures more than an order of magnitude greater.
Tokida et al. [2005]	76–1,233	laboratory core of near-surface *Sphagnum* peat	Rates are for threshold bubble content during periods of low barometric pressure and are based on high-frequency (once every 1.5–10 h) measurements using a chamber fitted to the core.
Tokida et al. [2007]	48–1,440	field measurement from two chambers installed on a *Sphagnum*-dominated site	Rates apply to periods of low barometric pressure and are based on high-frequency (once every 1.5–2 h) measurements using the chambers.
J. M. Waddington et al. (The effect of rehabilitation flooding on methane dynamics in a cutover peatland, manuscript in preparation, 2009)	49–1,090	field measurement from seven funnels installed at the surface of a flooded cutover peatland	Rates are summer averages.

that ebullition may be the dominant pathway for CH_4 losses to the atmosphere and that previous measurements and calculations of peatland CH_4 losses may be underestimates. For example, from a study of ebullition from deep (>1 m) peats, *Glaser et al.* [2004] used indirect measurements to suggest that ebullition is temporally and spatially very variable and that it can exceed diffusive fluxes by 2 orders of magnitude. Work on shallow peats (upper 1 m), both in the laboratory [*Baird et al.*, 2004] and the field [*Strack et al.*, 2005; *Tokida et al.*, 2007], also suggests that ebullition losses represent an important proportion of total CH_4 losses. Nevertheless, that ebullition is the dominant pathway for transport of CH_4 to the atmosphere in northern peatlands currently has the status of hypothesis, and more work is urgently needed on characterizing and modeling bubble buildup and losses from a range of different types of peat and plant communities.

How have (some) peatland researchers underestimated the role of ebullition in previous studies? Gas exchanges at the peatland surface are often measured using the chamber method. Permanent collars are inserted into the peat to a depth of 10–20 cm. When measurements of gas exchange are required, a chamber is fitted via a gastight seal to the collar. Typically, the following protocol is used, which we call the "normal chamber method" or NCM: (1) the chamber is fitted to its collar once/twice per week, (2) five or six sets of samples (i.e., including replicates at each time of sampling) of chamber gas are taken at regular intervals over an approximately 20- to 30-min period, (3) the samples are analyzed for CH_4 concentration, and (4) a linear regression line is fitted to the data to estimate the rate of CH_4 loss. A key assumption of the NCM is that CH_4 fluxes from peatlands are steady, at least over time frames of 1–2 weeks. Thus, if ebullition occurs continuously as a stream of small bubbles, the NCM will give accurate results. However, if ebullition is nonsteady (cyclic or episodic), the NCM could give very large errors; that is, we could be substantially underestimating the amount of CH_4 being lost from peatlands (see below). A 20-min sampling frame represents <0.2% of a week. If ebullition events are random in time and occur, on average, once a day or once every 3 days, the probability of recording a release with a chamber is 0.013 and 0.005, respectively. Recent evidence suggests that many ebullition losses are, indeed, nonsteady but that they can occur nonrandomly, with factors such as atmospheric pressure and water table decline acting as triggers for bubble release [e.g., *Comas and Slater* 2007; *Strack et al.*, 2005; *Tokida et al.*, 2005, 2007]. Many users of the NCM apparently have not used such knowledge to improve estimates of ebullition losses; that is, they have not sampled from chambers when ebullition is more likely. In this respect, the study of *Tokida et al.* [2007] is particularly notable. *Tokida et al.* [2007] measured

CH_4 efflux using two chambers placed on a temperate bog dominated by *Sphagnum* spp. but also containing vascular plants such as *Eriophorum vaginatum* L. and *Rhynchospora alba* (L.) Vahl. High-frequency measurements of CH_4 efflux were taken using the chambers every 1.5–2 h over 4 days when atmospheric pressure varied but showed a general fall from 1017 to 1000 hPa. Over this period, ebullition contributed 50–64% of the total CH_4 efflux. However, during individual events, ebullition losses exceeded the other losses combined by 1 to 2 orders of magnitude.

1.2. Conceptual Models of CH_4 Loss From Peatlands (Deep Versus Shallow Sources of Bubbles)

Interesting work has recently been undertaken suggesting that CH_4-containing bubbles may form deep (>3 m) within peat deposits and build up at middle depths (2 m) [*Glaser et al.*, 2004] below confining layers of woody peat. The mid-depth accumulations may become so large that, episodically, they break through the confining layer to the surface causing very large rates of CH_4 loss to the atmosphere (e.g., 35 g CH_4 m^{-2} per event). The evidence for the buildup and release of such large pockets or reservoirs of free-phase gas comes from changes in pore water pressures as measured using closed and open piezometers [*Rosenberry et al.*, 2003], from changes in the surface elevation of the peatland [*Glaser et al.*, 2004], and from ground-penetrating radar surveys [e.g., *Comas et al.*, 2005, 2007].

From their work on the glacial Lake Agassiz peatlands (GLAP) in Minnesota, *Glaser et al.* [2004] proposed a conceptual model of ebullition in peatlands that is shown as Figure 1. Although developed specifically for the GLAP, the model has been widely discussed and is sometimes thought of as canonical in terms of how ebullition takes place in peatlands. For example, in a study of heat transfer in peat, *McKenzie et al.* [2007, p. 369] note "Peatlands release methane, formed by anaerobic bacteria [sic] at depth, to the atmosphere. The transfer of methane from depth within the peat profile occurs either by episodic releases of large volumes of methane gas associated with the lowering of peatland water tables . . . or by continuous diffusion through the peat soil." A probably unintended consequence of the model is that it has caused the role of shallow peats as sources of bubbles escaping to the atmosphere to be somewhat overlooked. We suggest that bubbles may form more readily in shallow peats than in deep peats for at least three reasons:

1. There is a more abundant local supply of labile carbon (including exudates from the roots of vascular plants) to act as substrate for methanogens.

2. There is a greater range of temperatures near the peatland surface, with both higher and lower temperatures being

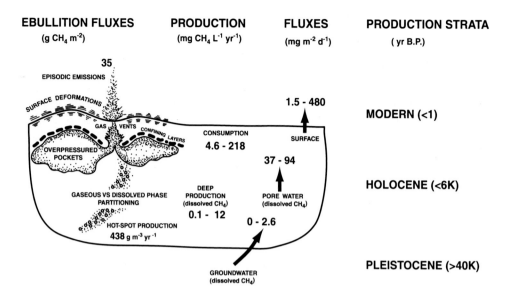

Figure 1. "Deep peat" ebullition model of *Glaser et al.* [2004].

experienced. Because CH_4 production shows a strongly nonlinear increase with temperature [e.g., *Dunfield et al.*, 1993], it will be higher near the surface than at the peatland base even if there is no difference in average temperatures (or substrate supply) between the two.

3. During water table rise after rainfall, air becomes encapsulated within peat; in other words, bubbles are immediately formed during water table rise. Once stripped of their oxygen, these bubbles may grow, as biogenic gases are produced within the peat. Thus, they may act as nuclei for free-phase CH_4 accumulation. Encapsulation during water table rise has been observed by, among others, *Beckwith and Baird* [2001] and *Baird and Waldron* [2003].

Noting reasons 1 and 2, it is not surprising that workers such as *Daulat and Clymo* [1998] and *Laing et al.* [2008] have measured intense CH_4 production within 20 cm of the water table. In addition to the points above, it has also been shown that the accumulation of bubbles within the peat profile does not require the presence of a woody confining layer (see Figure 1). Indeed, in samples of near-surface and poorly decomposed *Sphagnum* peat without confining layers, volumetric gas contents (gas volume per unit volume of peat) as high as 0.16 (16%) have been recorded [e.g., *Baird and Waldron*, 2003; *Kellner et al.*, 2006]. Gas accumulations in the absence of confining layers have also been found in the field [e.g., *Strack et al.*, 2005].

The studies cited above deal with bog or poor-fen peats. We have also found bubbles at shallow depths in rich-fen peats. For example, we installed triplicate gas trap funnels (20-cm diameter) at depths of 20, 40, and 60 cm below the

surface (all below the water table) of a 1.2-m-thick rich-fen peat in southern Ontario in Canada (Fletcher Fen) at a marl flat site devoid of vascular vegetation. Upward acting bubble fluxes of CH_4 as recorded by the traps over a 2-month period during the summer of 2006 (14 June to 21 August) were significantly greater ($p < 0.05$) in the 20-cm shallow traps (1630 ± 746 mg CH_4 m^{-2} d^{-1}) than the 40-cm (852 ± 432 mg CH_4 m^{-2} d^{-1}) and 60-cm traps (540 ± 426 mg CH_4 m^{-2} d^{-1}).

In addition to the direct evidence from Fletcher Fen, there are many studies from a range of peatlands where bubble buildup and loss have been recorded in samples of upper peat (i.e., samples excluding deep peat) [e.g., *Baird et al.*, 2004; *Beckwith and Baird*, 2001; *Christensen et al.*, 2003; *Comas and Slater*, 2007; *Kellner et al.*, 2006; *Laing et al.*, 2008; *Ström et al.*, 2005; *Tokida et al.*, 2005]. Given such evidence, we suggest that bubble formation and ebullition may be common across the upper 1 m of peat soils. Indeed, the upper 1 m may represent at least as important a source of bubble flux to the atmosphere as middle-depth and deep peats. However, direct studies of free-phase gas dynamics in deep and middle-depth peats across a range of sites are needed to confirm this assertion. We propose as Figure 2 a new conceptual model that emphasizes the importance of CH_4 production in the upper parts of a peat deposit and that formalizes the strong direct evidence that this zone is one in which bubbles may accumulate in large volumes even in the absence of woody layers of peat. We accept that more work needs to be done on bubble dynamics in shallow peats, in particular on variation between peat types. It is also important to note that the model does not exclude the possibility of

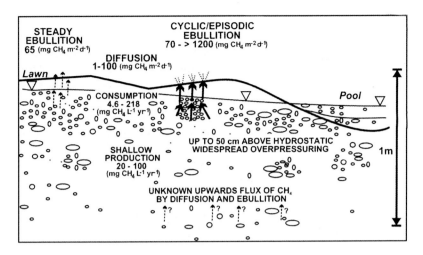

Figure 2. "Shallow peat" ebullition model.

deep and medium-depth zones of CH_4 production and bubble formation but suggests that their contribution is uncertain. A key purpose of our model is to act as a stimulus for further research, especially on the partitioning of ebullition between steady losses and episodic and cyclic events and how this may be affected by peat type.

The ranges for rates of production, consumption, and efflux that we have included in the model are necessarily wide, partly because they encompass studies done on different peat types and partly because we still lack detailed data on ebullition under a wide range of conditions such as during the passage of low-pressure weather systems, through diurnal cycles, and during prolonged drought. Noting our rationale for proposing the shallow peat conceptual model, we focus in the rest of the chapter on the upper 1 m of peat.

1.3. Vascular Plants and Ebullition

It has been suggested that the presence of vascular plants reduces the importance of ebullition. This suggestion may partly explain why the deep ebullition model has been preferred by some researchers; that is, if it is assumed that vascular plants prevent or substantially reduce bubble buildup in the rooting zone, then ebullition is only ever likely to be a deep-seated phenomenon. It should be emphasized that such an argument is not made by *Glaser et al.* [2004], but it is an argument that may explain the popularity of the model.

Chanton [2005] notes that because of the role of vascular plants as transporters of CH_4 to the atmosphere, their presence means that pore water concentrations of CH_4 may be lowered by as much as 50%, leading to the formation of fewer bubbles and lower rates of ebullition. Indeed, *Chan-*

ton [2005, p. 755] suggests that, "if vascular emergent macrophyte plants [sic] inhabit a wetland, plant transport will be the primary mechanism of CH_4 transport from the wetland." Vascular plants may also lower pore water CH_4 concentrations via rhizospheric oxidation, i.e., the diffusion of oxygen from the shoots to the roots and rhizomes, which can promote bacterial methanotrophy in pore waters around roots and rhizomes [cf. *Ström et al.*, 2005; *Waddington et al.*, 1996]. Most studies on the effects of vascular plants on ebullition have looked at inundated mineral sediments (e.g., billabong sediments and paddy fields) and it is not clear how vascular plants affect ebullition in peats. In this respect, it is worth again noting the study of *Tokida et al.* [2007] where it was found that ebullition from a peatland containing vascular plants was more important than diffusion and plant-mediated transport combined. In our study at Fletcher Fen (see section 1.2), we also examined the influence of vascular vegetation on CH_4 bubble dynamics by comparing ebullition rates and CH_4 concentrations in both pore water and gas lost via ebullition between the aforementioned marl flat site and two sedge-dominated sites (*Carex livida* (Wahlenb.) Willd. and *Scirpus cespitous* L.) in August 2007. Gas volumes collected in gas traps were not significantly different ($p < 0.05$) between the *Carex* and marl flat sites (*Scirpus* data not available). While there was no significant difference ($p < 0.05$) in bubble CH_4 concentration (12–52%) between the three sites, the 20- to 40-cm pore water CH_4 concentration at the *Scirpus* site (2.1 ± 0.4 mg L^{-1}) was significantly lower than at the *Carex* (4.9 ± 0.9 mg L^{-1}) and marl flat (5.3 ± 0.4 mg L^{-1}) sites.

Our findings suggest that the statement by *Chanton* [2005] may be too general because a key confounding factor when examining the effect of vascular plants on ebullition is that many species of vascular plants are thought to enhance CH_4 production through their root exudates, which act as substrates for methanogens [cf. *Ström et al.* 2003; *Waddington et al.*, 1996]. Thus, on the one hand, plant-mediated transport/rhizospheric oxidation might lower pore water concentrations of CH_4 and rates of bubble formation, while on the other hand, enhanced rates of CH_4 production from root exudates could increase rates of bubble production and the relative importance of ebullition [*Christensen et al.*, 2003]. Indeed, it is possible that dense vascular plant root mats may be zones of both intense CH_4 production and also bubble trapping. It would be extremely useful to separate these effects to gain a clearer picture of how vascular plants influence ebullition, and such work is underway in the second author's laboratory. Currently, the evidence from the study of *Tokida et al.* [2007] and the new data from Fletcher Fen suggest that the presence of vascular plants in peatlands does not necessarily mean that ebullition is unlikely or less likely. It is also worth noting that high rates of CH_4 flux attributed to vascular plants in some previous studies [e.g., *Waddington et al.*, 1996] may actually be partly because of an increased bubble flux (where the bubbles were lost as a continuous stream) caused by enhanced CH_4 production in the rooting zone.

1.4. Bubble Trouble: Simulating Ebullition in Peatland CH_4 Models

The way in which bubbles are trapped within peat will affect how the bubbles are released and the subsequent movement of CH_4 through the zone above the water table. Ebullition is either ignored or treated very simply in almost all existing peatland CH_4 models [cf. *Cao et al.*, 1996; *Frolking et al.*, 2002; *Gedney et al.*, 2004; *Grant and Roulet*, 2002; *Kettunen*, 2003; *Walter et al.*, 1996, 2001a, 2001b; *Zhang et al.*, 2002]. Those models that do explicitly allow for ebullition, such as that of *Walter et al.* [2001a, 2001b], assume that bubbles are not trapped in peat but are lost directly to the water table, after which the gas moves slowly via diffusion through the zone above the water table where the CH_4 within it may be consumed by methanotrophic bacteria. However, if bubbles build up and are then released in one go (i.e., episodically or cyclically), they will displace gas already present in the zone above the water table or move through that zone as mass flow so that methanotrophic processing is bypassed or overwhelmed, leading to more CH_4 being lost to the atmosphere than if the same quantity of bubbles had been lost as a steady stream over a longer time period. Therefore, the nature of bubble transport to the zone above the water table is important for how much CH_4

escapes to the atmosphere from the peatland surface [*Rosenberry et al.*, 2006]. This is a key point which apparently has not been appreciated by authors of wetland CH_4 models.

The buildup and release of bubbles from peat will depend on a range of factors, including rates of CH_4 production, the location of hot spots of production, the transport of dissolved CH_4 to and from bubbles through pore water, and the physical properties of the peat. Ebullition occurs when the buoyancy of bubbles overcomes the forces that keep them in place (in particular, surface tension) [cf. *Fechner-Levy and Hemond*, 1996; *Corapcioglu et al.*, 2004; *Strack et al.*, 2005]. It has been suggested recently that a threshold bubble volume must be reached to trigger episodic or cyclic ebullition [e.g., *Beckwith and Baird*, 2001; *Baird et al.*, 2004; *Strack et al.*, 2005]. The concept of the threshold was formalized by *Kellner et al.* [2006] in a simple model. The model is based on the idea that ebullition occurs when the bubble content rises above the threshold, the amount of gas lost in a model time step being the difference between the amount of free-phase gas in the peat and the threshold content. The model accounts for changes in bubble volume caused by changes in temperature and pressure by employing an iterative solution to Henry's law (exchange of gas between bubbles and surrounding pore water (the gaseous and aqueous/dissolved phases)) and the ideal gas equation (direct changes in bubble volume due to pressure and temperature). It also includes a production term. Comparisons of the model with laboratory data suggest it is capable of reasonable predictions of ebullition events ($r^2 = 0.66$) [*Kellner et al.*, 2006]. However, problems of the model have been noted by the model's authors. In particular, they suggest that the threshold is "fuzzy"; in other words, on occasion, the model predicted ebullition when none occurred and vice versa. Another way of thinking about this fuzziness is that ebullition sometimes occurs when the gas content is below the threshold, and sometimes it does not occur until the gas content has risen substantially above the threshold. Fuzziness can arise for a number of reasons, but perhaps the most important concerns the fact that the threshold is largely an empirical concept. In particular, the model (1) takes no account of the variations in bubble content with depth in the peat profile, (2) takes no account of bubble movement and redistribution within the peat profile, and (3) lacks a secure physical basis in terms of how different peat/pore structures cause and promote, respectively, the trapping and release of bubbles. Indeed, it may be argued that a general problem of existing ebullition studies is that the physical properties of the peat have not been described in detail; at best, these studies have looked at bulk density and degree of decomposition using an "in-the-hand" assessment. Thus, there remains a need for a better description of bubble accumulation and loss from near-surface peats.

1.5. CFD or Not CFD: That Is the Question

An alternative to the threshold approach is one where bubble dynamics (movement, break up, and coalescence) are described or modeled in great detail through a known pore structure. Somewhat surprisingly, experimental work on bubble dynamics in simple pores is ongoing, and we have found no work on bubble movement through individual pores with complicated cross-sectional shapes or through branching pores. Thus, although understanding of bubble dynamics has improved steadily over the last 20–30 years [e.g., *Borhan and Pallinti*, 1999; *Chaudhari and Hoffmann*, 1994; *Das and Pattanayak*, 1994; *Schwartz et al.*, 1986], new information on processes such as the coalescence of two moving bubbles of different sizes within uniform capillary tubes is still being discovered [e.g., *Almatroushi and Borhan*, 2006a, 2006b]. This suggests that we are some way from being able to model satisfactorily the detail of bubble dynamics in individual pores or pore networks. Nevertheless, computational fluid dynamics (CFD) models have been developed for bubble reactors and are able to predict bubble dynamics at the bubble population level. For example, the model of *Jia et al.* [2007] is able to predict local gas holdups (defined as the fractional bubble content within a section of a reactor or bubble column) and fluid velocities in bubble columns containing three phases (solid, liquid, and gaseous). However, it is clear that such models require further devel-

opment and testing; the correspondence between the model of Jia et al. and experimental data was modest at best. Additionally, to our knowledge, such approaches have not been applied to three-dimensional, two-phase fluid flow in porous media such as peats.

One hurdle to application of CFD methods to two-phase fluid transfer in peats is a lack of information on pore structures in different peat types. X-ray computed tomography (CT) promises, in part at least, to remove this hurdle. Studies by *Blais* [2005] and *Kettridge and Binley* [2008] have demonstrated that it is possible to extract information on pore size and pore continuity from X-ray images. The latter study also shows that individual bubbles can be imaged (see Figure 3) and that bubble population properties (such as the probability density function of bubble size and shape) can be derived. Despite these advances, it is unlikely that such detailed information will be used in larger models of peatland CH_4 dynamics. CFD models are computationally expensive, and their incorporation into models such as that of *Walter et al.* [2001a, 2001b] for a range of peat types would prove a huge task. In any case, there are good arguments for keeping models as simple as possible [cf. *Baird*, 1999; *Wainwright and Mulligan*, 2004a, 2004b], while avoiding the problem of simplistic (naively simple) models. Below, we present a modeling approach that we believe combines the simplicity of the threshold approach with some of the physical realism of a CFD model.

Figure 3. X-ray computed tomography images of bubbles in a sample of poorly decomposed *Sphagnum fuscum* (Schimp.) Klinggr. peat/litter. The bubbles are shown in gray against a black background, with water and peat fibers rendered transparent. Reprinted from *Kettridge and Binley* [2008], with permission of John Wiley.

2. A NEW APPROACH TO MODELING CH₄ BUBBLE BUILDUP AND EBULLITION: UPSIDE DOWN AVALANCHES

There is a significant gap between the two approaches to modeling ebullition described in section 1. The threshold approach is strongly empirical, which necessarily limits its applicability to simple peat profiles in which depth variation in, for example, bubble accumulation and the factors that affect bubble dynamics (pore structure) can be ignored. Neither does it account for depth variation in CH_4 production and solubility. Such variability is almost certainly partly responsible for the threshold exhibiting fuzziness. However, some fuzziness may be an intrinsic property of peat-bubble interactions, such that even a uniform peat might display complex ebullition behavior (i.e., a fuzzy threshold). We have also argued in section 1 that CFD models are too complex, at least for nesting within existing peatland CH_4 models.

With the problems of existing approaches in mind, we have started to explore the possibility of using "reduced complexity" models to simulate bubble dynamics in peat. What may be termed an "Occam's razor" approach involves building simple numerical approximations of bubble movement and storage within a porous medium that allow us to explore how variations in factors such as pore structure, CH_4 production, and exchanges of CH_4 between the aqueous and gaseous phases can influence ebullition.

Here we present one prototype model which is based upon a single key assumption: that a (relatively) constant production of CH_4 (or of CH_4-containing bubbles) can give rise to nonsteady releases of bubbles. We assume that (in the field) the timing of ebullition events is not regular, nor are the events necessarily of the same size, so the peatland system generates a nonlinear frequency and magnitude output from a relatively linear input. Although detailed ebullition data sets are sparse, there is evidence of such dynamics in shallow peats, as we show later (see also section 1.1).

Such behavior can be found in many other natural systems, in particular geomorphic systems where relatively steady inputs are transformed into episodic outputs. Examples include earthquakes (constant tectonic pressure leads to sudden release) and river catchment sediment systems (relatively steady rates of sediment production but episodic releases of sediment from the catchment outlet). Perhaps the best documented examples of this behavior are landslides, in particular the strikingly simple sandpile model of *Bak et al.* [1987]. *Bak et al.* [1987] developed a cellular automaton in which sand is added, grain by grain, to a surface to form a pile. When local slopes become too steep, a collapse occurs, moving sediment to neighboring cells, which too can collapse if the adjusted slopes are too steep. *Bak et al.* [1987,

1988] noticed that the addition of a single grain could cause a cascade of local collapses whose size could vary from a single cell (grain) to that of the whole length of the surface. They also found that the magnitude-frequency distribution of these cascades follows an inverse power law. After a collapse, and with the addition of more grains of sand, the system would self organize back to a critical state where the addition of a single grain of sand could again cause a cascade of collapse. The model reached its critical state through a series of positive (sand grains building up to create a pile) and negative (avalanches destroying part of the sandpile) feedbacks.

Bak et al. [1987, 1988] suggested that their model was a good analogue of natural sandpiles [cf. *Jaeger et al.*, 1989; *Held et al.*, 1990; *Rosendahl et al.*, 1993] and referred to the tendency of a system to "evolve" toward a dynamic equilibrium around a critical state as "self-organized criticality" (SOC). Here we are not concerned with SOC systems per se; however, the dynamics of the numerical sandpile appear to have some useful similarities with the dynamics of the peat-bubble system in that a small and constant input may lead to output that is highly variable in magnitude and frequency.

Our prototype model is similar to an inverted sandpile model, where, instead of grains of sand falling, we have bubbles rising. The peat profile is represented in the model as a 2-D cellular grid or lattice. Cells within this lattice may be of three types (or have one of three states): peat (solid), water (liquid), or bubble (gas). The model grid is set up as a column of water that contains obstacles or "shelves"; together, these represent the pore network of a peat soil. Therefore, by altering the size, disposition, and density of shelves, we can represent a range of different types of pore structure or network. Into the base of this column, single small "bubbles" are introduced at a constant rate (one per model iteration) but at random locations (across the base). This process represents CH_4 production; in other words, our example model assumes that pore water is already CH_4 saturated so that production translates immediately into bubbles. This is obviously oversimple, but our purpose here is to demonstrate the model's capability and simplicity rather than a realization of a particular peat type at a given time of year. Suffice it to say, the model can easily be altered so that bubble production can vary through time and model space.

During each model iteration, every bubble rises up through the column one cell until it reaches either a shelf or another bubble trapped below a shelf. When it reaches another bubble, its subsequent movement is dictated by a range of eight simple rules (as described in Figure 4) that determine whether it stops or moves around the obstacle (shelf). This allows the small bubbles to build up or coalesce into larger bubbles: in analogy, a series of upside down sandpiles. The

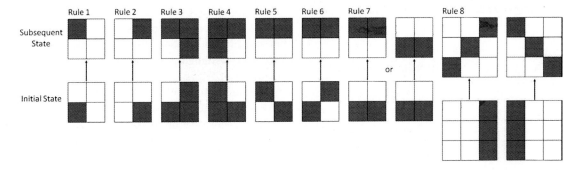

Figure 4. Rule set for the reduced-complexity ebullition model. During a model run, the cells composing the model grid are scanned from bottom to top using a 2 × 2 window, which is shifted every other iteration to operate on odd or even cells. Throughout this process the state, bubble or liquid, of the cells within the window is compared with a set of initial states in a rule set. If a match is found, new states are encoded for the cells within the window according to the rule for the initial state that has been encountered [*Chopard and Droz*, 1998]. For example, rule 1 will transport the bubble from its position in the lower left corner of the window to the upper left corner. Rule 7 is different from the preceding rules and is a remnant from the sandpile model, where there is a 50% probability that two grains of sand can become stuck together; therefore, two possible outcomes are possible. Last, rule 8 is responsible for the release of bubbles, which eventually triggers bubble "avalanches." Here the window is expanded to 3 × 3 cells, and release is dependent on three "bubbles" being stacked upon each other. Release can occur to the left or right depending on how the bubbles are arranged within the window.

bubble accumulations grow until they are large enough to shed bubbles around the side of the shelf through upward "avalanching." Similar to the sandpile model, the size of these "avalanches" bears no relation to the size of the input, so the addition of a single bubble can trigger an "avalanche" of a much larger bubble.

The model's operation is illustrated in Figure 5, where large bubbles develop beneath the shelves, leading eventually to the episodic release of different size "bubbles" repre-

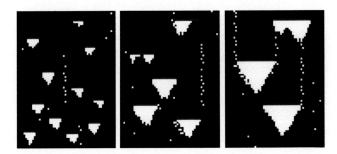

Figure 5. Example screen shots from the prototype model. Water is shown in black, bubbles are shown in white, and shelves are shown in gray. Differently sized "bubble" accumulations can be seen under the differently sized shelves representing different peat types. Also evident is the release of single and large bubbles (large bubbles being represented by a vertical stream of adjacent single bubbles).

sented by chains of individual bubbles. The model is written in Visual C# and may be obtained from the first author. The model allows the user to add or subtract shelves, which enables investigation of how different types of "pore network" affect ebullition. The size of the bubbles reaching the surface (the length of the "landslides") can be recorded by the program. To illustrate the capability of the model, we conducted three simple runs among which the width and number of shelves varied (i.e., representing different pore structures). Run 1 contained 11 shelves, each, on average, 7 cells wide. Run 2 contained five shelves, on average, 14 cells wide. Run 3 contained three shelves, on average, 22 cells wide. Bubble inputs were kept at the same rate for all three runs.

We collected output data from each of the model runs, and these are presented as unitless frequency distributions in Figure 6. They are compared to a frequency distribution of laboratory-measured ebullition events (data from *Kellner et al.* [2006]). There is a strong similarity between the frequency distribution from the laboratory experiments and the distributions from runs 2 and 3. However, run 1 produced mainly small- (size 1) and medium-sized ebullition events (size 2–3), and very few that are larger, because the shelves are not long enough to allow large accumulations of bubbles. The same volume of bubbles was introduced into the base of the "peat" in runs 1, 2 and 3. Hence, given that there were more small ebullition events in run 1, this run also produced fewer large-volume bubbles than runs 2 and 3.

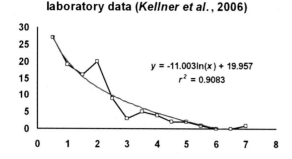

laboratory data (*Kellner et al.*, 2006)

$$y = -11.003\ln(x) + 19.957$$
$$r^2 = 0.9083$$

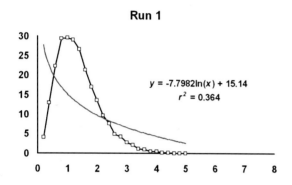

Run 1

$$y = -7.7982\ln(x) + 15.14$$
$$r^2 = 0.364$$

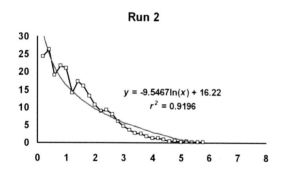

Run 2

$$y = -9.5467\ln(x) + 16.22$$
$$r^2 = 0.9196$$

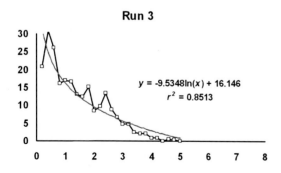

Run 3

$$y = -9.5348\ln(x) + 16.146$$
$$r^2 = 0.8513$$

Figure 6. Unitless frequency diagrams of ebullition data from *Kellner et al.* [2006] and model runs 1, 2, and 3. The data from the laboratory experiment have been scaled to allow comparison with the model runs (y axis shows frequency; x axis shows size of ebullition event). The equations in the graphs are plotted as gray lines.

Therefore, run 1 may be thought of as a poorly decomposed peat with an open pore structure in which bubbles are lost steadily (in response to CH_4 production) to the water table, as is assumed in, for example, the current version of the model of *Walter et al.* [2001a, 2001b]. We have deliberately chosen to compare frequency distributions, and the outputs from the model runs are unitless. Our aim at this stage is to show that our prototype model can produce some of the key dynamics of the real (laboratory) system, i.e., the frequency-magnitude relationship seen in the data of *Kellner et al.* [2006]. This parsimonious example model reveals how the distribution of bubble size can be radically altered by the form of the medium in which bubbles are trapped and from which they are released.

Our investigation of reduced-complexity models is in its early stages of development, and there are several obvious steps to explore. Different peat types (as represented by the size, number, and pattern of shelves) could easily be explored. Additional processes can also be incorporated to give the model greater realism, while retaining considerable simplicity compared to CFD approaches. For example, the avalanche rules could be altered to see what effect they have on both the pattern of bubble buildup and the frequency distribution of ebullition events. At the moment, the accumulations of bubbles look somewhat like inverted sandpiles, and it would be interesting to see how making the bubbles/piles more "sticky" and less "willing" to shed bubbles affects the shape of bubble accumulations and rates and patterns of release. Such an increase in stickiness could be seen as attempting to represent surface tension more accurately. It is also possible to add bubbles nonrandomly in different parts of the modeled peat profile, to reflect vertical differences in CH_4 production, for example, and to simulate exchange of CH_4 between bubbles and pore water. Finally, it is a relatively simple task to alter the model so that it operates in three dimensions. Ultimately, the power of the reduced-complexity approach is that it allows us to build parsimonious models; we can increase the level of detail only to that which is necessary to match experimental and observational data on ebullition dynamics. As part of this process, a key advantage of the reduced-complexity approach is that it is easy to conduct extensive sensitivity analyses on the range of main factors that control bubble buildup and release.

3. RESEARCH NEEDS

We have shown that ebullition needs to be accounted for properly in models of peatland CH_4 dynamics. It is not enough just to know ebullition volumes; the temporal pattern of loss also needs to be considered. Thus, we disagree with *Frolking et al.* [2002], who contend that the different modes of ebullition need not be simulated because all ebullition flux can be assumed to bypass methanotrophic processing. Such complete bypassing is unlikely to be the case in many peats. As we noted in section 1.4, CH_4 being lost via steady ebullition may, in fact, be processed in the zone above the water table, while much of that lost episodically and cyclically may not be. The fate of CH_4-containing bubbles when they reach the water table needs to be investigated empirically, but we would be surprised if there were not big differences in CH_4 efflux between a situation where a given volume of bubbles is lost via steady ebullition and one in which the same volume is lost in a single event as a large slug of gas. Although not discussed in the main body of the chapter, there is also clear evidence that bubbles may form under horizontally averaged CH_4 pore water concentrations that are substantially less than the equilibrium solubility [e.g., *Baird et al.*, 2004], and for this reason alone, more work on bubble buildup and formation is needed; that is, existing models need revision in this respect too.

We have also shown that the recently proposed bubble threshold model of ebullition [e.g., *Kellner et al.*, 2006] is probably too simple to account for the complexity of the process, while CFD approaches are still in their infancy and are unlikely ever to become part of a larger model of CH_4 dynamics in peat soils because of their extreme complexity. We believe our reduced-complexity approach offers a sensible way around the problem of how to model ebullition in a way that accounts for depth variability in key processes such as bubble accumulation. Obviously, there will still be a need for data on peat pore structures (collected using X-ray CT) and a requirement to turn such information into the right pattern of shelves found in the reduced-complexity model so that different peat types can be represented. The latter requirement is not trivial but can be addressed relatively simply using laboratory experimentation where relationships are established between different types of pore structure and bubble dynamics. Even with this requirement, the model will be parsimonious in its data requirements and setup compared with a full CFD description of the process. Finally, different versions of the prototype model will need thorough testing through both sensitivity analyses and model-data comparisons. We will report on such developments in due course.

Acknowledgments. We thank Erik Kellner and Kristen Harrison for assistance in the laboratory and field when collecting the Fletcher Fen data. Part of the research reported in this chapter was funded by a Canadian Foundation for Climate and Atmospheric Sciences (CFCAS) research grant and a NSERC Discovery Grant to J.M.W. We thank also the referees, D. Siegel and T. Tokida, whose comments on an earlier draft of the manuscript helped us improve the clarity of the explanation in this final version.

REFERENCES

Almatroushi, E., and A. Borhan (2006a), Coalescence of drops and bubbles in tube flow, *Ann. N. Y. Acad. Sci.*, *1077*, 508–526, doi:10.1196/annals.1362.025.

Almatroushi, E., and A. Borhan (2006b), Interaction and coalescence of drops and bubbles rising through a tube, *Ind. Eng. Chem. Res.*, *45*(1), 398–406, doi:10.1021/ie0505615.

Baird, A. J. (1999), Modelling, in *Eco-Hydrology: Plants and Water in Terrestrial and Aquatic Environments*, edited by A. J. Baird and R. L. Wilby, pp. 300–345, Routledge, London.

Baird, A. J., and S. Waldron (2003), Shallow horizontal groundwater flow in peatlands is reduced by bacteriogenic gas production, *Geophys. Res. Lett.*, *30*(20), 2043, doi:10.1029/2003GL018233.

Baird, A. J., C. W. Beckwith, S. Waldron, and J. M. Waddington (2004), Ebullition of methane-containing gas bubbles from near-surface *Sphagnum* peat, *Geophys. Res. Lett.*, *31*, L21505, doi:10.1029/2004GL021157.

Bak, P., C. Tang, C., and K. Wiesenfeld (1987), Self-organized criticality: An explanation of 1/f noise, *Phys. Rev. Lett.*, *59*(4), 381–384.

Bak, P., C. Tang, C., and K. Wiesenfeld (1988), Self-organized criticality, *Phys. Rev. A*, *38*(1), 364–374.

Beckwith, C. W., and A. J. Baird (2001), Effect of biogenic gas bubbles on water flow through poorly decomposed blanket peat, *Water Resour. Res.*, *37*, 551–558.

Blais, K. (2005), Measurement of physical and hydraulic properties of organic soil using computed tomographic imagery, M.Sc. thesis, 159 pp., Dep. of Geogr., Simon Fraser Univ., Burnaby, B. C., Canada.

Borhan, A., and J. Pallinti (1999), Breakup of drops and bubbles translating through cylindrical capillaries, *Physi. Fluids*, *11*, 2846–2855.

Cao, M., S. Marshall, and K. Gregson (1996), Global carbon exchange and methane emissions from natural wetlands: Application of a process-based model, *J. Geophys. Res.*, *101*(D9), 14,399–14,414.

Chanton, J. P. (2005), The effect of gas transport on the isotope signature of methane in wetlands, *Org. Geochem.*, *36*, 753–768, doi:10.1016/j.orggeochem.2004.10.007.

Chaudhari, R. V., and H. Hoffmann (1994), Coalescence of gas bubbles in liquids, *Rev. Chem. Eng.*, *2*, 131–190.

Chopard, B., and M. Droz (1998), *Cellular Automata Modeling of Physical Systems*, 353 pp., Cambridge Univ. Press, Cambridge, U. K.

Christensen, T. R., N. Panikov, M. Mastepanov, A. Joabsson, A. Stewart, M. Öquist, M. Sommerkorn, S. Reynaud, and B. Svensson (2003), Biotic controls on CO_2 and CH_4 exchange in wetlands—A closed environment study, *Biogeochemistry*, *64*, 337–354.

Comas, X., and L. Slater (2007), Evolution of biogenic gases in peat blocks inferred from noninvasive dielectric permittivity measurements, *Water Resour. Res.*, *43*, W05424, doi:10.1029/2006WR005562.

Comas, X., L. Slater, and A. Reeve (2005), Geophysical and hydrological evaluation of two bog complexes in a northern peatland:

Implications for the distribution of biogenic gases at the basin scale, *Global Biogeochem. Cycles*, *19*, GB4023, doi:10.1029/2005GB002582.

Comas, X., L. Slater, and A. Reeve (2007), In situ monitoring of free-phase gas accumulation and release in peatlands using ground penetrating radar (GPR), *Geophys. Res. Lett.*, *34*, L06402, doi:10.1029/2006GL029014.

Corapcioglu, M. Y., A. Cihan, and M. Drazenovic (2004), Rise velocity of an air bubble in porous media: Theoretical studies, *Water Resour. Res.*, *40*, W04214, doi:10.1029/2003WR002618.

Das, R. K., and S. Pattanayak (1994), Bubble to slug flow transition in vertical upward 2-phase flow-through narrow tubes, *Chem. Eng. Sci.*, *49*, 2163–2172.

Daulat, W. E., and R. S. Clymo (1998), Effects of temperature and watertable on the efflux of methane from peatland surface cores, *Atmos. Environ.*, *32*(19), 3207–3218.

Dunfield, P., K. Knowles, R. Dumont, and T. Moore (1993), Methane production and consumption in temperate and subarctic peat soils: Response to temperature and pH, *Soil Biol. Biochem.*, *25*, 321–326.

Fechner-Levy, E. J., and H. F. Hemond (1996), Trapped methane volume and potential effects on methane ebullition in a northern peatland, *Limnol. Oceanogr.*, *41*, 1375–1383.

Frolking, S., N. T. Roulet, T. R. Moore, P. M. Lafleur, J. L. Bubier, and P. M. Crill (2002), Modeling seasonal to annual carbon balance of Mer Bleue Bog, Ontario, Canada, *Global Biogeochem. Cycles*, *16*(3), 1030, doi:10.1029/2001GB001457.

Frolking, S., N. T. Roulet, and J. Fuglestvedt (2006), How northern peatlands influence the Earth's radiative budget: Sustained methane emission versus sustained carbon sequestration, *J. Geophys. Res.*, *111*, G01008, doi:10.1029/2005JG000091.

Gedney, N., P. M. Cox, and C. Huntingford (2004), Climate feedback from wetland methane emissions, *Geophys. Res. Lett.*, *31*, L20503, doi:10.1029/2004GL020919.

Glaser, P. H., J. P. Chanton, P. Morin, D. O. Rosenberry, D. I. Siegel, O. Ruud, L. I. Chasar, and A. S. Reeve (2004), Surface deformations as indicators of deep ebullition fluxes in a large northern peatland, *Global Biogeochem. Cycles*, *18*, GB1003, doi:10.1029/2003GB002069.

Grant, R. F., and N. T. Roulet (2002), Methane efflux from boreal wetlands: Theory and testing of the ecosystem model Ecosys with chamber and tower flux measurements, *Global Biogeochem. Cycles*, *16*(4), 1054, doi:10.1029/2001GB001702.

Held, G. A., D. H. Solina, H. Solina, D. T. Keane, W. J. Haaq, P. M. Horn, and G. Grinstein (1990), Experimental study of critical-mass fluctuations in an evolving sandpile, *Phys. Rev. Lett.*, *65*(9), 1120–1123.

Intergovernmental Panel on Climate Change (2007), *Climate Change 2007: Synthesis Report*, 103 pp., Geneva, Switzerland.

Jaeger, H. M., C.-H. Liu, and S. R. Nagel (1989), Relaxation at the angle of repose, *Phys. Rev. Lett.*, *62*(1), 40–43.

Jia, X., J. Wen, H. Zhou, W. Feng, and Q. Yuan (2007), Local hydrodynamics modeling of a gas-liquid-solid three-phase bubble column, *Am. Inst. Chem. Eng. J.*, *53*(9), 2221–2231, doi:10.1002/aic.11254.

Kellner, E., A. J. Baird, M. Oosterwoud, K. Harrison, and J. M. Waddington (2006), Effect of temperature and atmospheric pressure on methane (CH_4) ebullition from near-surface peats, *Geophys. Res. Lett.*, *33*, L18405, doi:10.1029/2006GL027509.

Kettridge, N., and A. Binley (2008), X-ray computed tomography of peat soils: Measuring gas content and peat structure, *Hydrol. Processes*, *22*, 4827–4837, doi:10.1002/hyp.7097.

Kettunen, A. (2003), Connecting methane fluxes to vegetation cover and water table fluctuations at microsite level: A modeling study, *Global Biogeochem. Cycles*, *17*(2), 1051, doi:10.1029/2002GB001958.

Laing, C. G., T. G. Shreeve, and D. M. E. Pearce (2008), Methane bubbles in surface peat cores: In situ measurements, *Global Change Biol.*, *14*, 916–924, doi:10.1111/j.1365-2486.2007.01534.x.

McKenzie, J. M., D. I. Siegel, D. O. Rosenberry, P. H. Glaser, and C. I. Voss (2007), Heat transport in the Red Lake Bog, glacial Lake Agassiz peatlands, *Hydrol. Processes*, *21*, 369–378, doi:10.1002/hyp.6239.

Rosenberry, D. O., P. H. Glaser, D. I. Siegel, and E. P. Weeks (2003), Use of hydraulic head to estimate volumetric gas content and ebullition flux in northern peatlands, *Water Resour. Res.*, *39*(3), 1066, doi:10.1029/2002WR001377.

Rosenberry, D. O., P. H. Glaser, and D. I. Siegel (2006), The hydrology of northern peatlands as affected by biogenic gas: Current developments and research needs, *Hydrol. Processes*, *20*, 3601–3610, doi:10.1002/hyp.6377.

Rosendahl, J., M. Vekić, and J. Kelley (1993), Persistent self-organization of sandpiles, *Phys. Rev. E*, *47*(2), 1401–1404.

Schwartz, L. W., H. M. Princen, and A. D. Kiss (1986), On the motion of bubbles in capillary tubes, *J. Fluid Mech.*, *172*, 259–275, doi:10.1017/S0022112086001738.

Strack, M., E. Kellner, and J. M. Waddington (2005), Dynamics of biogenic gas bubbles in peat and their effects on peatland biogeochemistry, *Global Biogeochem. Cycles*, *19*, GB1003, doi:10.1029/2004GB002330.

Ström, L., A. Ekberg, M. Mastepanov, and T. R. Christensen (2003), The effect of vascular plants on carbon turnover and methane emissions from a tundra wetland, *Global Change Biol.*, *9*, 1185–1192.

Ström, L., M. Mastepanov, and T. R. Christensen (2005), Species-specific effects of vascular plants on carbon turnover and methane emissions from wetlands, *Biogeochemistry*, *75*, 65–82, doi:10.1007/s10533-004-6124-1.

Tokida, T., T. Miyazaki, and M. Mizoguchi (2005), Ebullition of methane from peat with falling atmospheric pressure, *Geophys. Res. Lett.*, *32*, L13823, doi:10.1029/2005GL022949.

Tokida, T., T. Miyazaki, M. Mizoguchi, O. Nagata, F. Takakai, A. Kagemoto, and R. Hatano (2007), Falling atmospheric pressure as a trigger for methane ebullition from a peatland, *Global Biogeochem. Cycles*, *21*, GB2003, doi:10.1029/2006GB002790.

Waddington, J. M., N. T. Roulet, and R. V. Swanson (1996), Water table control of CH_4 emission enhancement by vascular plants in boreal peatlands, *J. Geophys. Res.*, *101*(D17), 22,775–22,785.

Wainwright, J., and M. Mulligan (2004a), Introduction, in *Environmental Modelling: Finding Simplicity in Complexity*, edited by J. Wainwright and M. Mulligan, pp. 1–4, John Wiley, Chichester, U. K.

Wainwright, J., and M. Mulligan (2004b), Pointers for the future, in *Environmental Modelling: Finding Simplicity in Complexity*, edited by J. Wainwright and M. Mulligan, pp. 389–396, John Wiley, Chichester, U. K.

Walter, B. P., M. Heimann, R. D. Shannon, and J. R. White (1996), A process-based model to derive methane emissions from natural wetlands, *Geophys. Res. Lett.*, *23*(25), 3731–3734.

Walter, B. P., M. Heimann, and E. Matthews (2001a), Modeling modern methane emissions from natural wetlands: 1. Model description and results, *J. Geophys. Res.*, *106*(D24), 34,189–34,206.

Walter, B. P., M. Heimann, and E. Matthews (2001b), Modeling modern methane emissions from natural wetlands: 2. Interannual variations 1982–1993, *J. Geophys. Res.*, *106*(D24), 34,207–34,219.

Zhang, Y., C. Li, C. C. Trettin, H. Li, and G. Sun (2002), An integrated model of soil, hydrology, and vegetation for carbon dynamics in wetland ecosystems, *Global Biogeochem. Cycles*, *16*(4), 1061, doi:10.1029/2001GB001838.

A. J. Baird, School of Geography, University of Leeds, Woodhouse Lane, Leeds LS2 9JT, UK. (a.j.baird@leeds.ac.uk)

T. J. Coulthard, Department of Geography, University of Hull, Cottingham Road, Hull HU6 7RX, UK.

J. Ramirez, 6370 Simmons Street, Miami, FL 33014, USA.

J. M. Waddington, School of Geography and Earth Sciences, McMaster University, 1280 Main Street West, Hamilton, ON L8S 4K1, Canada.

The Stable Carbon Isotope Composition of Methane Produced and Emitted From Northern Peatlands

Edward R. C. Hornibrook

Bristol Biogeochemistry Research Centre, Department of Earth Sciences, University of Bristol, Bristol, UK

Stable carbon isotope values, pore water concentration, and flux data for methane (CH_4) were compiled for 26 peatlands situated in the northern hemisphere to explore relationships between trophic status and CH_4 cycling. Methane produced in ombrotrophic bogs has $\delta^{13}C$ values that are significantly more negative than CH_4 formed in fens apparently because of poor dissociation of acetic acid or an absence of methanogenic archaea capable of metabolizing acetic acid under low pH conditions. The $\delta^{13}C$ values of CH_4 in pore water of ombrotrophic and minerotrophic peatlands exhibit the opposite trend: $\delta^{13}C(CH_4)$ values become more positive with depth in rain-fed bogs and more negative with depth in fens. The key zone for methanogenesis occurs at shallow depths in both types of peatland and consequently, $\delta^{13}C$ values of CH_4 emitted from ombrotrophic bogs (-74.9 ± 9.8‰; $n = 42$) are more negative than from fens (-64.8 ± 4.0‰; $n = 38$). An abundance of graminoids in fens contributes to more positive $\delta^{13}C(CH_4)$ values in pore water through (1) release of root exudates which promotes aceticlastic methanogenesis, (2) rhizosphere oxidization of CH_4 causing localized enrichment of $^{13}CH_4$, and (3) preferential export of $^{12}CH_4$ through aerenchyma, which also enriches pore water in $^{13}CH_4$. Emissions from blanket bogs and raised bogs should be attributed more negative $\delta^{13}C(CH_4)$ values relative to fens in isotope-weighted mass balance budgets. Further study is needed of bogs that have an apparently low nutrient status but exhibit a pore water distribution of $\delta^{13}C(CH_4)$ values similar to fens.

1. INTRODUCTION

Wetlands are the main perennial source of methane (CH_4) emissions to the Earth's atmosphere in the absence of human activity. Global wetland distribution is dynamic with the most significant periods of reorganization in the Quaternary Period occurring during glacial-interglacial transitions. Wetland area paradoxically is thought to be greatest during glacial maxima despite much of the high latitudes being covered in ice because the lowering of sea level exposes large areas of coastal plain which become occupied by wetlands; however, CH_4 flux per unit area of wetland probably was lower during glacial times than at present because net primary productivity (NPP) would have been attenuated by low CO_2 concentrations in the glacial period atmosphere [*Kaplan*, 2002].

The vast northern wetlands of North America, Scandinavia, and much of Asia have developed only since the retreat of continental ice sheets ~12 ka before present [*MacDonald et al.*, 2006]. Peatlands are the dominant wetland type at high latitudes during interglacial periods with differences in

Carbon Cycling in Northern Peatlands
Geophysical Monograph Series 184
Copyright 2009 by the American Geophysical Union.
10.1029/2008GM000828

topography, hydrology, and the depth of peat accumulation dictating the degree to which the ecosystems are rain-fed (ombrotrophic) or groundwater-influenced (minerotrophic). Records of succession from shallow lake → marsh → fen → bog are captured in the pollen and plant macrofossil archive of many northern peatlands and CH_4 flux on an area basis is assumed to have decreased with time because of the impact of changes in trophic status on NPP and vegetation assemblages, which in turn diminish CH_4 flux [*Whiting and Chanton*, 1993; *Smith et al.*, 2004].

Analysis of $^{13}C/^{12}C$ and $^2H/^1H$ ratios in minute quantities of CH_4 extracted from air bubbles trapped in ice cores presents an opportunity to track changes in significant global sources that have contributed to the twofold increase in atmospheric CH_4 concentration from ~350 to ~700 parts per billion (ppb) during the preindustrial Holocene [*Ferretti et al.*, 2005; *Schaefer et al.*, 2006; *Sowers*, 2006; *Whiticar and Schaefer*, 2007; *Schaefer and Whiticar*, 2008] or the dramatic rise to ~1800 ppb during the last millennium [*Ferretti et al.*, 2007; *Lassey et al.*, 2007; *Houweling et al.*, 2008]. Similarly, monitoring of the present day stable isotope composition of atmospheric CH_4 permits detection of short-term changes in CH_4 sources and sinks and, potentially, longer-term trends in ecosystems impacted by human activities or climate change [*Stevens and Engelkemeir*, 1988; *Stevens*, 1993; *Francey et al.*, 1999; *Lowe et al.*, 1999; *Quay et al.*, 1999; *Bergamaschi et al.*, 2000; *Lassey et al.*, 2000; *Allan et al.*, 2001; *Miller*, 2004]. However, effective interpretation of changes in $\delta^{13}C$ and δ^2H values of atmospheric CH_4, whether modern samples or air trapped in ice and firn, requires an accurate knowledge of the δ-signatures of CH_4 emission sources. Indeed, *Miller et al.* [2002] recently have noted that among the largest errors involved in using stable isotopes to constrain the global CH_4 budget are uncertainties associated with generalized $\delta^{13}C$ or δ^2H compositions for different CH_4 sources.

Global wetlands in early CH_4 budgets typically were assigned a single collective $\delta^{13}C(CH_4)$ value of approximately –60‰ [e.g., *Whiticar*, 1993], and a similar value has been used subsequently in many models. The $\delta^{13}C(CH_4)$ values attributed to wetlands are ^{13}C-depleted relative to atmospheric CH_4 because the source is entirely biological; however, formation of wetland CH_4 has been attributed almost exclusively to methanogenic *archaea* that utilize the acetate fermentation pathway because of a long-standing assumption that acetoclastic methanogenesis dominates freshwater systems while CO_2/H_2 methanogenesis is most prevalent in marine sediments [*Whiticar et al.*, 1986]. Despite evidence that CO_2/H_2 methanogenesis can prevail in certain types of wetlands [e.g., *Lansdown et al.*, 1992; *Hines et al.*, 2001; *Duddleston et al.*, 2002] and that differences exist in the

$\delta^{13}C$ values of CH_4 emissions from broad groups of peatlands [*Kelley et al.*, 1992; *Chanton et al.*, 1995 2000], a common $\delta^{13}C(CH_4)$ value of approximately –60‰ continues to be applied generally to wetlands as a whole in top-down and bottom-up CH_4 budgets constrained by stable isotopes. At least two models have assigned different $\delta^{13}C$ values to CH_4 emissions from low and high latitude wetlands [*Hein et al.*, 1997; *Mikaloff Fletcher et al.*, 2004], but none to date has attempted to distinguish CH_4 emissions isotopically from ombrotrophic and minerotrophic wetlands in the northern hemisphere.

Establishing $\delta^{13}C$ values for emissions from different types of wetland is not a straightforward task. Wetland CH_4 exhibits a remarkable range of stable carbon isotope values considering that the gas is derived almost exclusively from degradation of C_3 plant material having a very limited range of $\delta^{13}C$ values (e.g., approximately –30 to –24‰; *Hornibrook et al.*, 2000a). The $\delta^{13}C$ values reported to date for wetland CH_4 range from approximately –100‰ for gas transported through aerenchyma of vascular plants [*Chanton et al.*, 2002] to approximately –42‰ for residual CH_4 after preferential loss of $^{12}CH_4$ due to methanotrophy or diffusion through plant aerenchyma [*Gerard and Chanton*, 1993]. The amount of ^{13}C-enrichment in partially oxidized CH_4 will depend upon the fraction remaining and almost certainly more positive $\delta^{13}C(CH_4)$ values exist within natural wetlands than have been reported to date. These examples at the negative and positive ends of the $\delta^{13}C$ value range exhibited by wetland CH_4 are the result of secondary processes. Controls on primary $\delta^{13}C$ values of microbially formed CH_4 are similarly complex and include the biochemical pathway of methanogenesis (e.g., acetate fermentation or CO_2 reduction) [*Whiticar et al.*, 1986], differences in the $\delta^{13}C$ composition of organic matter [*Chanton et al.*, 1989], growth temperature [*Whiticar*, 1999], growth phase [*Botz et al.*, 1996], and energetics of methanogenesis [*Valentine et al.*, 2004; *Penning et al.*, 2005]. Controls on the δ^2H values of primary CH_4 are more poorly understood but include many of the same factors as for $\delta^{13}C$ values as well as the δ^2H value of ambient H_2O [*Wahlen*, 1994; *Waldron et al.*, 1999a, 1999b; *Chanton et al.*, 2006].

Despite these complexities, consistent patterns exist in the stable isotope composition of CH_4 produced and emitted from wetlands [*Chanton et al.*, 2000; *Hornibrook and Bowes*, 2007; *Prater et al.*, 2007; *Hines et al.*, 2008] that warrant further investigation with the aim of potentially refining the representation of these globally significant ecosystems in CH_4-cycle budgets and models that are constrained by isotopes. This review focuses on the $\delta^{13}C$ values of CH_4 in northern peatlands. The δ^2H values of CH_4 are not included because the availability of such data remains poor, and there have been few recent fundamental advances in our understanding of the hydrogen isotope systematics of methanogenesis in contrast to stable carbon isotopes [e.g., *Valentine et al.*, 2004; *Penning et al.*, 2005].

2. METHODS

2.1. Peatland Data

Methane emission and pore water data for northern peatlands have been compiled from the peer-reviewed literature with the exception of data from *Lansdown* [1992], which are used with permission. The basis for inclusion in this review was the availability of one or more of the following: (1) pore water $\delta^{13}C(CH_4)$ data for some range of depths (i.e., not a single depth), (2) $\delta^{13}C$ data for CH_4 emissions and pore water CH_4, or (3) $\delta^{13}C$ values for CH_4 emissions and CH_4 flux strength. Concentration and $\delta^{13}C$ data for CH_4 in soil gas bubbles were not included because of the tendency of bubbles to migrate in peat soil which precludes investigation of relationships between the depth distribution of pore water $\delta^{13}C(CH_4)$ values and peatland trophic status. Details about wetland type, location, soil pH, and dominant vegetation for each peatland are provided in Table 1. Given the differences in sampling purpose and protocols for the various studies, predictably a complete set of common parameters was not available for the peatlands. The details of sample collection and analysis methodologies can be found in the original publications and are not repeated here. The list in Table 1 is not comprehensive but does contain examples of the majority of peatland types that exist at latitudes >40°N. There is a notable dearth of published stable isotope data for CH_4 emissions and pore water from the globally significant areas of peatland in Scandinavia and Siberia.

2.2. Stable Isotope Notation

Stable isotope notation is based upon the International Union of Pure and Applied Chemistry Recommendations by T. B. Coplen (Explanatory glossary of terms used in expression of relative isotope ratios and gas ratios, submitted to *Pure and Applied Chemistry*, 2008). The delta value ($\delta^{13}C$) is defined as

$$\delta^{13}C = \frac{(^{13}C/^{12}C)_{sample}}{(^{13}C/^{12}C)_{VPDB}} - 1 \quad \text{‰}, \quad (1)$$

where $(^{13}C/^{12}C)_{VPDB}$ is 1123.75×10^{-5}, the stable carbon isotope ratio in the international standard Vienna Pee Dee Belemnite (VPDB). Note that the multiplication factor of 1000 normally present on the right-hand side of equation 1 is absent because the symbol ‰ equals 10^{-3}, and thus, express-

ing $\delta^{13}C$ values in units of ‰ makes the 1000 an extraneous numerical factor (Coplen, submitted manuscript, 2008).

Partitioning of ^{13}C and ^{12}C between CO_2 (or ΣCO_2) and CH_4 as a result of isotope effects associated with either acetoclastic or CO_2/H_2 methanogenesis is expressed using the isotope fractionation factor α (unitless) in the notation $\alpha_{CO_2-CH_4}(^{13}C)$, which is defined as

$$\alpha_{CO_2-CH_4}(^{13}C, {}^{12}C) = \frac{1000 + \delta^{13}C(CO_2)}{1000 + \delta^{13}C(CH_4)}.$$

Differences in the $\delta^{13}C$ values of two pools of CH_4 are expressed using Δ notation, which is defined as:

$$\Delta_{CH_4}(^{13}C, {}^{12}C) = \delta^{13}C(CH_4)_A - \delta^{13}C(CH_4)_B, \quad (2)$$

where A and B denote the two pools containing CH_4. Values of $\Delta_{CH_4}(^{13}C, {}^{12}C)$ also are abbreviated as $\Delta_{CH_4}(^{13}C)$ and have units of ‰.

3. RESULTS AND DISCUSSION

3.1. Pore Water CH₄

Pore water $\delta^{13}C(CH_4)$ values for 22 of the peatlands listed in Table 1 are shown in Figure 1 with data from minerotrophic and ombrotrophic peatlands grouped, respectively, in the left- and right-hand sides of the figure. The symbol size for pore water $\delta^{13}C(CH_4)$ values increases with increasing depth and thus by default, in most instances in concert also with an increasing concentration of dissolved CH_4. The depth range for samples from each site is provided in the caption to Figure 1. The minerotrophic peatlands all have a soil water pH ≥ 5 and exhibit the same trend of decreasing $\delta^{13}C(CH_4)$ values with increasing depth. Peatlands classified as "bogs" in the literature, mainly because of the acidity of surface water and types of vegetation present, are grouped as ombrotrophic in Figure 1. The sites generally have a pH < 5 although many exhibit an increase in pore water pH with depth below the *Sphagnum*-dominated acrotelm. The pH data for some sites suggest inclusion of measurements from laggs, which are fringe areas where water draining from the raised surface of an ombrotrophic peatland mixes with water from adjoining mineral soils, resulting in a higher alkalinity and greater availability of plant nutrients. Bogs that lack a raised dome or topographic position to exclude groundwater influence exhibit the same trend of pore water stable carbon isotope data as fens in which $\delta^{13}C(CH_4)$ values become more negative with increasing depth (e.g., Bakchar Bog, Bog S4, Bog 3850, Sifton Bog, and Turnagain Bog). In contrast, pore water $\delta^{13}C(CH_4)$ values in fully ombrotrophic peatlands (e.g., blanket bogs and raised bogs) show the opposite trend

Table 1. Summary Data for Peatland Sites[a]

Peatland	Type	Latitude	pH	Predominant Vegetation	Reference
Crymlyn Bog (Wales, UK)	Intermediate fen	51°38′N	5.5 ± 0.5	*Juncus squarrosum, Carex elata, Eriophorum gracile, Cladium mariscus, Molinia caerulea, Phragmites australis, Sphagnum* spp.	*Hornibrook and Bowes* [2007]; *Hornibrook et al.* [2008]
Gors Lwyd (Wales, UK)	Upland valley mire	52°15′N	4.9 ± 0.6	*Erica tetralix, Eriophorum angustifolium, Sphagnum* spp.	*Hornibrook and Bowes* [2007]; *Hornibrook et al.* [2008]
Alaskan Tundra (alkaline) (Arctic Coastal Plain, Alaska, USA)	Continuous permafrost (0.2 to 1.0 m depth)	~70°N	7–9	*Eriophorum vaginatum, Carex* spp.	*Quay et al.* [1988]; *Lansdown* [1992]
Alberta Fen (Canada)	Minerotrophic fen	54°36′N	~7.0	*Carex aquatilis, Carex rostrata*	*Popp et al.* [1999]
Upper Red Lake Fen (Minnesota, USA)	Minerotrophic fen	48°N	5.5–6.5	*Carex lasiocarpa, Scirpus cespitosus, Rhyncospora alba, Eriophorum* spp., *Scheuchzeria palustris*	*Chasar et al.* [2000]
Point Pelee Marsh (Ontario, Canada)	Freshwater marsh	41°58′N	~6.9	*Typha latifolia*	*Hornibrook et al.* [1997, 2000b, 2000c]
Sifton Bog (Ontario, Canada)	Kettle hole bog	43°00′N	3.8–5.5	*Sphagnum* spp., *Chamaedaphne calyculata, Eriophorum virginicum, Vaccinium oxycoccus, Larix laricina, Picea mariana, Kalmia polifolia, Andromeda glaucophylla*	*Hornibrook et al.* [1997, 2000b, 2000c]
Bog 3850 (Minnesota, USA)	Kettle hole bog	47°32′N	NR	NR	*Quay et al.* [1988]; *Lansdown* [1992]
Bakchar Bog (Siberia, Russia)	Ombrotrophic bog	56°51′N	4.8	*Equisetum* spp.	*Kotsyurbenko et al.* [2004]
Bog S4 (Marcel Experimental Forest, Minnesota, USA)	Nonforested kettle hole bog	47°32′N	3.9–6.0 (avg. 4.6)	*Chamaedaphne calyculata, Sphagnum capillifolium, Carex oligosperma, Eriophorum virginicum, Rhyncospora alba, Picea mariana, Larix laricinia*	*Quay et al.* [1988]; *Lansdown* [1992]; *Dise* [1993]
Turnagain Bog (Alaska, USA)	Ombrotrophic bog	60°10′N	4.7–5.1	*Sphagnum* spp., *Myrica gale*	*Chanton et al.* [2006]
Alaskan Tundra (acidic) (north slope Brooks Range, Alaska, USA)	Continuous permafrost (0.2 to 1.0 m depth)	~68°N	5–7	*Sphagnum* spp.	*Quay et al.* [1988]; *Lansdown* [1992]
Upper Red Lake Bog (Minnesota, USA)	Ombrotrophic bog	48°N	3.5–4.5	*Picea mariana, Larix laricinia, Sphagnum* spp., *Chamaedaphne calyculata, Ledum groenlandicum, Vaccinium* spp., *Kalmia polifolia*	*Chasar et al.* [2000]
Rainy River Bog (Ontario, Canada)	Ombrotrophic bog	48°47′N	NR	*Sphagnum fuscum, Sphagnum angustifolum, Picea mariana,* unspecified ericaceous shrubs	*Aravena et al.* [1993]

Table 1. (continued)

Peatland	Type	Latitude	pH	Predominant Vegetation	Reference
Blaen Fign (Wales, UK)	Blanket bog	52°16′N	4.2 ± 0.3	*Juncus squarrosus, Calluna vulgaris, Trichophorum cespitosum, Sphagnum* spp.	*Hornibrook and Bowes* [2007]; *Hornibrook et al.* [2008]
Kings Lake Bog (Washington, USA)	Lake bog	46°36′N	~3.5	*Sphagnum* spp., *Rhyncospora alba, Vaccinium oxycoccus, Ledum groenlandicum, Kalmia polifolia*	*Lansdown et al.* [1992]
Cors Caron (Wales, UK)	Raised bog	52°15′N	4.2 ± 0.1	*Sphagnum* spp., *Rhyncospora alba, Trichophorum cespitosum, Calluna vulgaris, Narcethium ossifragum*	*Hornibrook and Bowes* [2007]; *Hornibrook et al.* [2008]
Ellergower Moss (Scotland, UK)	Raised bog	55°05′N	NR	*Sphagnum* spp., *Calluna vulgaris, Erica tetralix, Trichophorum cespitosum, Eriophorum vaginatum, Rhyncospora alba*	*Waldron et al.* [1999]; *Clymo and Bryant* [2008]
Collapse Bog (CB) (Manitoba, Canada)	Open graminoid bog	55°40′N	4.0	*Sphagnum riparium, Carex paupercula, Vaccinium oxycoccus*	*Bellisario et al.* [1999]
Poor Fen (PF) (Manitoba, Canada)	Open graminoid poor fen	55°40′N	4.3–4.4	*Sphagnum riparium, Carex aquatilis, Carex limosa*	*Bellisario et al.* [1999]
Intermediate Fen (IF) (Manitoba, Canada)	Open graminoid intermediate fen	55°40′N	5.8	*Carex rostrata, Carex limosa, Sphagnum riparium*	*Bellisario et al.* [1999]
Rich Fen (RF) (Manitoba, Canada)	Open low-shrub fen	55°40′N	6.8	*Andromeda glaucophylla, Menyanthes trifoliate, Campylium stellatum, Calliergon stramineum, Carex chordorrhiza, Limprichtia revolvens, Messia triquetra, Scorpidium scorpiodes*	*Bellisario et al.* [1999]
Bena Bog	Open raised bog	47°32′N	3–5	*Sphagnum* spp.	*Crill et al.* [1988]; *Lansdown* [1992]
S2 (Marcel Experimental Forest, Minnesota, USA)	Forested ombrotrophic bog	47°32′N	4.1	*Picea mariana, Chamaedaphne calyculata, Ledum groenlandicum, Sphagnum* spp.	*Crill et al.* [1988]; *Lansdown* [1992]
S3 (Marcel Experimental Forest, Minnesota, USA)	Minerotrophic fen	47°32′N	6.9	*Salix* spp., *Alnus rugosa, Picea mariana, Fraxinus nigra, Betula papyrifera, Sphagnum* spp.	*Crill et al.* [1988]; *Lansdown* [1992]
Junction Fen (Marcel Experimental Forest, Minnesota, USA)	Poor fen transitional to bog	47°32′N	3 – 5	*Carex oligosperma, Vaccinium oxycoccus Sphagnum* spp.	*Crill et al.* [1988]; *Kelley et al.* [1992]

[a]NR is not reported.

with CH_4 in shallow peat being the most ^{13}C-depleted and $\delta^{13}C$ values becoming more positive with increasing depth (e.g., acidic Alaskan Tundra through to Ellergower Moss; Figure 1).

In ombrotrophic peatlands, the offset between $\delta^{13}C$ values for CH_4 and coexisting CO_2 (not shown) typically is constant, ranging from ~60 to 70‰. In contrast, in minerotrophic sites $\alpha_{\Sigma CO_2 - CH_4}$(^{13}C) values range from ~1.030 in anoxic shallow peat to a maximum of ~1.070 in the catotelm [*Hornibrook et al.*, 2000b]. The differences in $\delta^{13}C(CH_4)$ values with depth in the two classes of peatland conventionally are interpreted to result from a predominance of

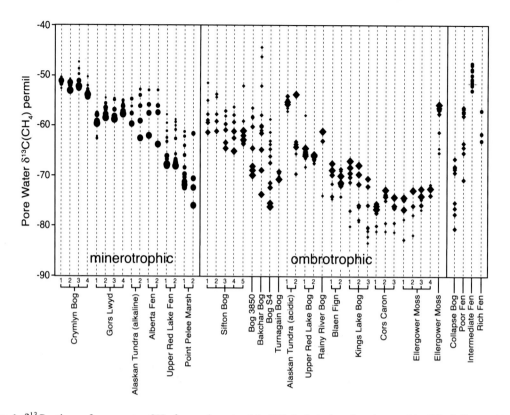

Figure 1. $\delta^{13}C$ values of pore water CH_4 from minerotrophic (filled circles) and ombrotrophic (filled diamonds) peatlands. The symbol size for each site increases with increasing depth and by default also the quantity of dissolved CH_4. Symbol size associated with absolute depths are consistent for individual peatlands for which more than one set of pore water $\delta^{13}C(CH_4)$ values was available; however, the differences in symbol size between peatlands are arbitrary because of the wide range of maximum and minimum depths and the resolution of sampling in the data set. Symbol size is constant for the four peatlands at the far right-hand side of the figure because pore water samples were collected from two depths at each site over a 4-month period. The $\delta^{13}C$ value of pore water CH_4 in minerotrophic peatlands consistently becomes more negative with increasing depth. In contrast, dissolved CH_4 in bog soils can exhibit either ^{13}C-depletion or ^{13}C-enrichment with increasing depth. The latter trend (i.e., more positive $\delta^{13}C(CH_4)$ values with depth) occurs primarily in fully ombrotrophic (i.e., blanket or raised) acidic bogs that lack significant graminoid cover. The sample depths and collection dates (and in some cases locations) for each peatland are: Crymlyn Bog 1 = 10 to 68 cm (27 May 2003), 2 = 18 to 64 cm (2 July 2003), 3 = 16 to 68 cm (28 July 2003), 4 = 14 to 68 cm (19 August 2003) [*Hornibrook and Bowes*, 2007]; (b) Gors Lwyd 1 = 14 to 66 cm (19 May 2003), 2 = 18 to 66 cm (24 June 2003), 3 = 27 to 68 cm (23 July 2003), 4 = 22 to 68 cm (11 August 2003) [*Hornibrook and Bowes*, 2007]; Alaskan Tundra (alkaline) 1 = 5 to 20 cm (km 162, August 1987), 2 = 13 to 25 cm (km 162, August 1987) [*Quay et al.*, 1988; *Lansdown*, 1992]; Alberta Fen 1 = 5 to 45 cm (1995 mean), 2 = 5 to 45 cm (1996 mean) [*Popp et al.*, 1999]; Upper Red Lake Fen 1 = 40 to 260 cm (June 1997), 2 = 20 to 280 cm (July 1997) [*Chasar et al.*, 2000]; Point Pelee Marsh 1 = 25 to 180 cm (7 September 1995), 2 = 80 to 160 cm (13 March 1996) [*Hornibrook et al.*, 1997, 2000b, 2000c]; Bakchar Bog = 10 to 120 cm [sampling date not reported (NR)] [*Kotsyurbenko et al.*, 2004]; Bog S4 = 9 to 199 cm (23 May 1986) [*Lansdown*, 1992]; Bog 3850 = 10 to 90 cm (21 June 1986) [*Lansdown*, 1992]; Sifton Bog 1 = 10 to 85 cm (July 1995), 2 = 15 to 75 cm (August 1995), 3 = 20 to 95 cm (February 1996), 4 = 40 to 150 cm (April 1996), 5 = 25 to 180 cm (August 1996) [*Hornibrook et al.*, 1997, 2000b, 2000c]; Turnagain Bog = 50 to 200 cm (August 2001) [*Chanton et al.*, 2006]; Alaskan Tundra (acidic) 1 = 4 to 20 cm (km 45, August 1987), 2 = 10 to 35 cm (km 59, August 1987) [*Quay et al.*, 1988; *Lansdown*, 1992]; Upper Red Lake Bog 1 = 20 to 280 cm (June 1997), 2 = 40 to 280 cm (July 1997) [*Chasar et al.*, 2000]; Rainy River Bog = 65 to 170 cm (date NR) [*Aravena et al.*, 1993]; Blaen Fign 1 = 10 to 45 cm (30 April 2003), 2 = 12 to 45 cm (4 June 2003) [*Bowes and Hornibrook*, 2006; *Hornibrook and Bowes*, 2007]; Kings Lake Bog 1 = 12 to 100 cm (26 May 1988), 2 = 14 to 100 cm (7 July 1990), 3 = 4 to 58 cm (16 January 1991) [*Lansdown*, 1992; *Lansdown et al.*, 1992]; Cors Caron 1 = 20 to 68 cm (7 May 2003), 2 = 18 to 68 cm (17 June 2003), 3 = 31 to 68 cm (14 July 2003) [*Hornibrook and Bowes*, 2007]; Ellergower Moss 1 and 2 = 0 to 450 cm (28 August 1992), 3 and 4 = 0 to 500 cm (26 September 1992) [*Waldron et al.*, 1999a, 1999b]; Ellergower Moss = 0 to 750 cm (nest B, October 1997) [*Clymo and Bryant*, 2008]; Collapse Bog, Poor Fen, Intermediate Fen and Rich Fen = either 15 or 50 cm (May to September 1994) [*Bellisario et al.*, 1999].

acetoclastic methanogenesis in shallow peat in minerotrophic mires with greater ^{13}C-depletion of CH_4 occurring at depth because of a transition to CO_2/H_2 methanogenesis due to lower temperature and/or a lack of labile substrate [e.g., *Hornibrook et al.*, 1997; *Popp et al.*, 1999]. The much larger and constant $\alpha_{\Sigma CO_2-CH_4}$(^{13}C) values and highly negative $\delta^{13}C(CH_4)$ values in ombrotrophic bogs are consistent with a prevalence of CO_2/H_2 methanogenesis throughout the peat profile [*Whiticar et al.*, 1986; *Lansdown et al.*, 1992].

Interpretation of $\delta^{13}C$ values of pore water CH_4 and CO_2 on this basis has been controversial [e.g., *Waldron et al.*, 1998, 1999a, 1999b; *Hornibrook et al.*, 1998, 2000b]; however, investigations employing stable isotopes in parallel with radiocarbon tracers ($H^{14}CO_3^-$ and $^{14}CH_3COO^-$) to quantify relative rates of CO_2/H_2 and acetoclastic methanogenesis have agreed well with $\delta^{13}C$ data [e.g., *Lansdown et al.*, 1992; *Avery et al.*, 1999]. *Conrad* [2005] provided an updated overview of the systematics of using natural abundance stable carbon isotope measurements to quantify methanogenic pathways, including a summary of $\delta^{13}C$ values for CH_4 generated by the two main pathways in studies that employed radiocarbon tracers. The data show clearly that CH_4 produced by CO_2/H_2 methanogenesis is significantly ^{13}C-depleted relative to CH_4 generated by acetate fermentation. While compellingly supportive of the original *Whiticar et al.* [1986] proxy model, the revised version of *Conrad* [2005] comes with several important caveats, including that carbon isotope fractionation factors associated with methanogenesis are not constant, in particular, within microbial culture experiments. *Whiticar* [1999] also made that point in an earlier re-assessment of the proxy model.

More specifically, the sizes of $\alpha_{\Sigma CO_2-CH_4}$(^{13}C) and $\alpha_{CH_3COO^--CH_4}$(^{13}C) can vary because of changes in factors (e.g., temperature or substrate availability) that impact microbial growth and metabolism. Consequently, the energetics of methanogenesis expressed via Gibbs free energy change (ΔG) correlate well with changes in isotope fractionation [*Valentine et al.*, 2004; *Penning et al.*, 2005]. The CO_2/H_2 pathway has been studied the most in this regard, and it appears that the magnitude of $\alpha_{\Sigma CO_2-CH_4}$(^{13}C) attains a maximum value as ΔG approaches the thermodynamic limit of CH_4 formation. This observation potentially has important implications for understanding the $\delta^{13}C$ composition of CH_4 in peatlands and other natural environments. The partial pressure of H_2 (pH_2) exerts a major control on ΔG of CH_4 formation and thus, variations in H_2 abundance may influence $\delta^{13}C(CH_4)$ values indirectly through the energetics of CO_2/H_2 methanogenesis (in addition to the direct influence on pathway predominance). Under highly favorable energetic conditions (i.e., very negative ΔG), the magnitude of $\alpha_{\Sigma CO_2-CH_4}$(^{13}C) is diminished to the extent that $\delta^{13}C$ values of CH_4 generated can be indistinguishable from CH_4 formed via acetate fermentation [*Penning et al.*, 2005]. Thus, an abundance of labile substrate supplied under conditions of favorable pH, moisture, and temperature could produced more positive $\delta^{13}C(CH_4)$ values giving the appearance of a shift to acetate fermentation when in reality CO_2/H_2 methanogenesis is simply more energetically feasible. This effect may occur in the rhizosphere of minerotrophic peatlands, contributing to the trend from ^{13}C-enriched CH_4 to more negative $\delta^{13}C(CH_4)$ values at depth beneath the supply of root exudates. Acetate fermentation undoubtedly is an important source of CH_4 in fens; however, it is impossible from existing data to exclude the possibility that a significant contribution of ^{13}C-enriched CH_4 is being generated by CO_2/H_2 methanogenesis because of a very negative ΔG of reaction.

Notably, a slightly different version of this conceptual model for control of $\delta^{13}C(CH_4)$ values was reported more than a decade ago by *Miyajima et al.* [1997], who suggested that substrate quality through its influence on the rate-limiting step in the anaerobic chain of decay might be responsible for differences in $\delta^{13}C(CH_4)$ values with depth and between peatlands. Increased recalcitrance of substrate results in fermentation of complex biopolymers being the rate-limiting step, which leads to a poor supply of H_2 and thus, CO_2/H_2 methanogenesis operating near its thermodynamic limit, producing CH_4 with highly negative $\delta^{13}C$ values. In contrast, an ample supply of labile compounds, such as glucose or low molecular weight volatile fatty acids from root exudates or recent plant litter, will yield an abundance of H_2 and acetic acid for methanogenesis. The rate-limiting step then becomes methanogenesis at the base of the anaerobic chain of decay, resulting in more negative values of ΔG of CH_4 formation by CO_2/H_2 methanogenesis and a greater contribution of ^{13}C-enriched CH_4 from acetate dissimilation.

In anoxic environments where significant quantities of labile organic compounds are not exuded in the subsurface by the rhizome of vascular plants, variable expression of $\alpha_{\Sigma CO_2-CH_4}$(^{13}C) values in situ is unlikely to occur because ΔG of CH_4 formation typically will be stable (i.e., methanogenesis operates as a steady state process near its thermodynamic limit). This stability normally is driven by the maintenance of very low pH_2 levels necessary for heterotrophic microorganisms to degrade higher alcohols and volatile fatty acids. It is tempting to try to interpret $\delta^{13}C(CH_4)$ values and their depth distribution in ombrotrophic bogs within this framework because of the lack of vascular vegetation; however, the cause of significant ^{13}C depletion of CH_4 in rain-fed mires cannot be attributed solely to organic matter recalcitrance and unfavorable energetics of methanogenesis. Rather, the assumption of a predominance, or even the presence, of acetoclastic methanogenesis in all freshwater systems in the

Whiticar et al. [1986] proxy model almost certainly is incorrect. In principle, CH_4 production in freshwater environments should be comprised of ~66% acetate fermentation and 33% CO_2/H_2 methanogenesis based upon the energetics of anaerobic degradation of glucose monomers [*Conrad et al.*, 1986]; however, ombrotrophic peatlands are exceptional among freshwater ecosystems because of the acidic conditions created by *Sphagnum* spp. in the absence of the buffering capacity of mineral-rich ground water or contact with alkaline bedrock or regolith. Substrate for methanogenesis in acidic mires typically is not in short supply, as commonly, there exists an abundance of acetic acid in pore water that has accumulated as an end product of anaerobic metabolism [*Hines et al.*, 2001, 2008; *Duddleston et al.*, 2002]. It appears that methanogens, for reasons that are not fully understood, are incapable of utilizing the excess acetic acid, and its eventual fate is oxidation to CO_2 once it has diffused upward through pore water to surface horizons containing SO_4^{2-} or O_2 [*Duddleston et al.*, 2002]. The overall process has been termed "acetate-decoupling" by *Hines et al.* [2001] to denote the separation of acetic acid from its more typical fate of becoming a substrate for methanogenesis during the final stages of anaerobic decay in freshwater environments.

Acetate dissimilation yielding CH_4 may be absent in ombrotrophic bogs because of poor dissociation of acetic acid at low pH [*Fukuzaki et al.*, 1990; *Lansdown et al.*, 1992]. Alternatively, another plausible explanation is that acidophilic or acid-tolerant methanogens capable of acetate fermentation simply are absent in bogs (or may not exist at all). Recent profiling of microbial populations in peatlands using molecular biology techniques has shown that acidic bogs overwhelmingly host primarily H_2-oxidizing methanogens [*Horn et al.*, 2003; *Sizova et al.*, 2003; *Berestovskaya et al.*, 2005; *Juottonen et al.*, 2005; *Galand et al.*, 2005; *Metje and Frenzel*, 2005; *Rooney-Varga et al.*, 2007], which notably is a similar finding to work conducted nearly two decades ago by *Williams and Crawford* [1983, 1984, 1985] on Minnesota peatlands via conventional microbiological approaches. To date, the most acid-tolerant methanogen isolated from an ombrotrophic bog is an H_2-oxidizer belonging to the order Methanomicrobiales [*Bräuer et al.*, 2006]. Under in situ conditions, mixed populations of methanogens in peatlands have a broader and lower pH tolerance than this isolate, suggesting that other factors or associations of microorganisms afford additional means to cope with acidic conditions. Regardless of the lower pH limit for methanogenesis in peat bogs, the key point is that acetic acid is not an important substrate for CH_4 in ombrotrophic mires. Indeed, the addition of acetate to incubations of peat from acidic bogs has been shown to inhibit methanogenesis, whereas supplements of H_2 or glucose have a stimulatory effect [*Williams and Crawford*, 1984; *Horn et al.*, 2003]. Noteworthy is that *Kotsyurbenko et al.* [2004] measured a pathway proportion of 2:1 for acetate fermentation and CO_2/H_2 methanogenesis in Bakchar Bog, a site which possesses the same distribution of pore water $\delta^{13}C(CH_4)$ values as the fens in Figure 1. In a subsequent study, increasing the acidity of Bakchar Bog peat by an order of magnitude to pH 3.8 in incubations resulted in a predominance of H_2/CO_2 methanogenesis and a shift in the dominant methanogen community to Methanobacteriaceae, accompanied by production of CH_4 having $\delta^{13}C$ values of −91 to −74‰ [*Kotsyurbenko et al.*, 2007].

Thus the inverted distribution of $\delta^{13}C(CH_4)$ values in ombrotrophic bogs compared to groundwater-influenced peatlands shown in Figure 1 reasonably can be attributed to a predominance of CO_2/H_2 methanogenesis in the former. Knowledge of these patterns of $\delta^{13}C(CH_4)$ values and the underlying causes for their origins in rain-fed and groundwater-influenced peatlands are important because CH_4 is produced in both types of wetland primarily at shallow depths immediately below the ambient water table level [*Daulat and Clymo*, 1998]. The $\delta^{13}C$ values of CH_4 in the zone of peak methanogenesis differs between the different classes of peatlands as summarized in Table 2. Shallow dissolved CH_4 in fens (−54.8 ± 3.5‰; $n = 129$) is notably more ^{13}C-enriched than in ombrotrophic bogs (−72.6 ± 7.8‰; $n = 78$), perhaps even more than the values suggest in Table 2 because inclusion of only blanket and raised bog sites decreases the mean $\delta^{13}C(CH_4)$ value for ombrotrophic peatlands by a further ~5‰. The $\delta^{13}C$ values for CH_4 from Point Pelee Marsh have been excluded from the fen average because *Typha*-dominated wetlands do not cover significant areas at northern latitudes.

The average $\delta^{13}C$ values are much more similar for shallow CH_4 in fens (−54.8 ± 3.5‰, $n = 129$) and bogs (−58.0 ± 5.2‰, $n = 30$) that possess the same subsurface pattern of $\delta^{13}C(CH_4)$ values, and it would be difficult to distinguish pore water CH_4 from these two types of peatlands on the basis of their $\delta^{13}C$ compositions. The contrast in mean $\delta^{13}C$ values is more significant for CH_4 from >50 cm depth in these peatlands (fens = −56.9 ± 5.2‰, $n = 85$ versus bog = −64.5 ± 5.0‰, $n = 33$) and the difference to ombrotrophic bogs more notable still (−70.7 ± 5.4‰, $n = 91$). As noted previously, the compiled data set is not comprehensive, but the similarity of the $\delta^{13}C(CH_4)$ values between broadly similar types of peatland provides a measure of confidence for extrapolating the mean values to other minerotrophic and ombrotrophic wetlands.

Finally, the stable carbon isotope data reported by *Bellisario et al.* [1999] for four peatlands (Collapse Bog, Poor Fen, Intermediate Fen, and Rich Fen) from a wetland complex located near Thompson, Manitoba, Canada were included

Table 2. Summary of $\delta^{13}C(CH_4)$ Values From Shallow and Deep Peat

Peatland	Set	$\delta^{13}C(CH_4)^a$ ‰ 0 to 50 cm	$\delta^{13}C(CH_4)^a$ ‰ >50 cm
Fens			
Crymlyn Bog	1	−51.5 (1.0, 17)	−50.7 (0.5, 9)
	2	−52.0 (0.9, 13)	−51.8 (0.6, 7)
	3	−51.0 (1.5, 14)	−51.5 (0.4, 9)
	4	−52.5 (1.3, 14)	−53.4 (0.5, 9)
Gors Lwyd	1	−60.1 (2.2, 16)	−58.7 (0.6, 8)
	2	−56.2 (0.7, 14)	−57.4 (0.8, 8)
	3	−55.6 (0.8, 8)	−57.9 (0.7, 8)
	4	−55.7 (0.5, 12)	−56.9 (0.9, 8)
Alaskan Transect (alkaline)	1	−57.5 (2.1, 5)	
	2	−56.9 (3.9, 5)	
Alberta Fen	1	−57.1 (3.8, 4)	
	2	−57.6 (4.6, 4)	
Upper Red Lake Fen	1	−57.8	−64.7 (3.2, 9)
	2	−59.4 (0.6, 2)	−64.4 (3.4, 10)
Point Pelee Marsh	1	−61.7 (1.1, 4)	−68.4 (3.5, 9)
	2		−70.1 (6.1, 4)
	Avg:	**−54.8 (3.5, 129)[b]**	**−56.9 (5.2, 85)[b]**
Bogs 1[c]			
Sifton Bog	1	−55.0 (3.1, 3)	−60.2 (1.2, 3)
	2	−55.8 (2.2, 3)	−60.6 (0.9, 2)
	3	−58.4 (1.3, 3)	−63.0 (1.9, 3)
	4	−56.5	−61.6 (2.8, 5)
	5	−56.0 (3.7, 2)	−62.7 (1.1, 6)
Bog 3850		−59.3 (2.8, 6)	−69.0 (0.9, 3)
Bakchar Bog		−52.4 (7.2, 5)	−64.3 (5.2, 7)
Bog S4		−63.5 (2.9, 7)	−73.9 (2.3, 4)
	Avg:	**−58.0 (5.2, 30)**	**−64.5 (5.0, 33)**
Bogs 2[d]			
Turnagain Bog		−70.1	−69.7 (0.7, 3)
Alaskan Transect (acidic)	1	−55.4 (1.0, 8)	
	2	−63.4 (5.3, 6)	
Upper Red Lake Bog	1	−63.1 (5.1, 2)	−65.8 (0.5, 12)
	2	−67.7	−66.7 (0.6, 12)
Rainy River Bog	1		−66.0 (7.0, 3)
	2		
Blaen Fign	1	−71.6 (2.9, 7)	
	2	−71.0 (1.5, 13)	
Kings Lake Bog	1	−74.8 (4.7, 4)	−68.4 (1.2, 3)
	2	−76.4 (3.0, 5)	−68.9 (0.9, 2)
	3	−79.4 (3.9, 7)	−70.7
Cors Caron	1	−79.0 (1.3, 11)	−76.5 (0.7, 9)
	2	−77.3 (3.1, 14)	−73.5 (0.6, 9)
	3	−79.3 (1.7, 6)	−75.8 (1.1, 8)
Ellergower Moss	1		−77.8 (3.3, 5)
	2		−77.3 (3.8, 4)
	3		−75.0 (1.6, 5)
	4		−73.0 (0.8, 4)
Ellergower Moss		−70.2	−63.4 (3.0, 11)
	Avg:	**−72.6 (7.8, 86)**	**−70.7 (5.4, 91)**

[a]Mean (SD, n); SD and n not listed when $n = 1$; where $n = 2$ half the range of the $\delta^{13}C$ values is used instead of the SD.

[b]Excludes Point Pelee Marsh data.

[c]Bogs that have a pore water distribution of $\delta^{13}C$ values similar to fens.

[d]Bogs in which $\delta^{13}C$ values are most negative at shallow depths, becoming more positive with greater depth into the catotelm.

in Figure 1 despite lacking a detailed depth distribution (i.e., CH_4 is from either 15 or 50 cm depth) because the data set is unique: it appears to be the sole example of a collection of $\delta^{13}C(CH_4)$ values from a contiguous peatland succession. The relationship between pore water $\delta^{13}C(CH_4)$ values versus trophic level and pH is striking with an ~30‰ ^{13}C-enrichment occurring along the gradient. Data for the Rich Fen deviate from the trend and the slightly more negative $\delta^{13}C(CH_4)$ values in the highly alkaline fen may result from the greater prevalence of ericaceous shrubs versus graminoids (i.e., compared to the Intermediate Fen). Lesser quantities of root exudates may be released from the shrubs, but it is likely that enhanced loss of $^{12}CH_4$ through graminoid aerenchyma contributes to the more positive $\delta^{13}C(CH_4)$ values in the Intermediate Fen [*Chanton*, 2005]. Regardless, the absolute $\delta^{13}C$ values and ^{13}C-enrichment trend in CH_4 from nutrient-poor, acidic bog to increasingly minerotrophic fens within one landscape agrees well with the patterns of $\delta^{13}C(CH_4)$ values present in the diverse collection of peatland data in Figure 1.

3.2. CH_4 Flux

The quantity of CH_4 dissolved in pore water at shallow depths at any point in time normally is less than within deeper peat horizons; however, the residence time of shallow CH_4 is much shorter because CH_4 formed in this highly productive zone is exported continuously to the atmosphere or oxidized, in whole or in part, by methanotrophs in the rhizosphere or at the water-air interface. For example, *Lombardi et al.* [1997] showed that the residence time of CH_4 at shallow depths is measured in days to weeks during the growing season. The modern ^{14}C-age of CH_4 emitted to the atmosphere from both bogs and fens demonstrates that shallow methanogenesis based upon recent plant litter or exudates is the primary source of persistent diffusive flux that accounts for much of the CH_4 emissions annually from northern peatlands [*Chanton et al.*, 1995]. Thus, given the key depths for methanogenesis and the distributions of pore water $\delta^{13}C(CH_4)$ values shown in Figure 1, CH_4 emissions from fens predictably should be ^{13}C-enriched relative to flux from bogs as reported by *Kelley et al.* [1992], *Chanton et al.* [1995, 2000], and *Hornibrook and Bowes* [2007].

Stable carbon isotope data for CH_4 flux and subsurface CH_4 are available only for a small subset of the peatlands listed in Table 1. In general, there remains a scarcity of published measurements of CH_4 flux collected in situ from peatlands, which is surprising given the proliferation during the past decade of commercial CH_4 preconcentration devices for rapid (~30 min) online stable isotope analysis of low concentrations of CH_4 in small volume gas samples.

The average $\delta^{13}C$ values ($\pm 1\sigma$ error bars) for dissolved CH_4 in the 0- to 50-cm depth interval of the peatlands for which $\delta^{13}C(CH_4)$ emission data are available are plotted against $\delta^{13}C$ values of CH_4 flux in Figure 2. The range of $\delta^{13}C$ values for CH_4 emissions is ~40‰ compared to ~30‰ for pore water CH_4. The latter is smaller because of mean values being employed from the 0 to 50 cm depth interval (i.e., if

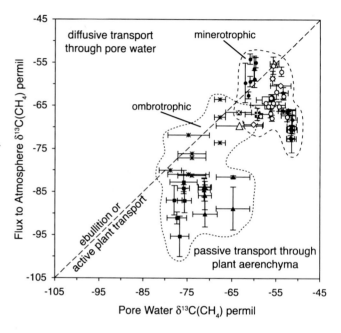

Figure 2. $\delta^{13}C$ values of CH_4 emitted to the atmosphere versus CH_4 dissolved in pore water of peatlands. The pore water $\delta^{13}C(CH_4)$ values are the average of available data for the depth interval 0 to 50 cm for each peatland. The dashed diagonal line represents a 1:1 relationship between $\delta^{13}C$ values of CH_4 flux and subsurface CH_4, corresponding to transport of CH_4 to the atmosphere without partitioning of $^{13}CH_4$ and $^{12}CH_4$ (i.e., via ebullition or bulk flow by pressurized ventilation in aquatic plants). Data that plot below the 1:1 line exhibit kinetic isotope effects associated with preferential transport of $^{12}CH_4$ by passive diffusion through aerenchyma of vascular plants, which yields CH_4 flux having $\delta^{13}C$ values ~5 to 20‰ more negative than the pore water CH_4 pool [*Chanton*, 2005]. The $\delta^{13}C$ values plotting above the 1:1 line result from bacterial oxidation of CH_4 during diffusive transport through water or air-saturated pore spaces in peat soil. Data legend: Kings Lake Bog (asterisk) [*Lansdown et al.*, 1992; *Lansdown*, 1992]; Cors Caron (filled squares), Crymlyn Bog (filled diamonds), Gors Lwyd (open diamonds) [*Hornibrook and Bowes*, 2007]; Blaen Fign (filled triangles) [*Bowes and Hornibrook*, 2006; *Hornibrook and Bowes*, 2007]; Alaskan Tundra acidic (open triangles) and alkaline (open squares), Bog S4 (filled square with white X) and Bog 3850 (filled square with white cross) [*Quay et al.*, 1988; *Lansdown*, 1992]; Alberta fen natural (open circles) and vegetation removed (filled circles) [*Popp et al.*, 1999].

data from >50 cm were included the range of $\delta^{13}C$ values, or pore water CH_4 would be comparable or greater than that for CH_4 emissions). The stable carbon isotope compositions of emissions and pore water from minerotrophic and ombrotrophic peatlands occupy distinctly separate areas of Figure 2. Rain-fed sites emit CH_4 that is highly ^{13}C-depleted similar to the CH_4 in the subsurface of those peatlands. As expected, CH_4 flux from fens has more positive $\delta^{13}C$ values reflecting the more ^{13}C-enriched CH_4 in the subsurface of the minerotrophic mires.

Although stable carbon isotope data from minerotrophic and ombrotrophic peatlands cluster separately in Figure 2, the majority of $\delta^{13}C(CH_4)$ pairs plot beneath the 1:1 diagonal line, which indicates that CH_4 flux from both types of peatland generally is ^{13}C-depleted relative to the pore water CH_4 pool. The negative offset between $\delta^{13}C$ values of CH_4 emissions and CH_4 dissolved in pore water is a common feature of wetlands hosting an abundance of graminoids and results from the greater translational velocity of $^{12}CH_4$ versus $^{13}CH_4$ during transit by passive diffusion through lacunae or aerenchymatous tissue [e.g., *Popp et al.*, 1999; *Chasar et al.*, 2000; *Chanton et al.*, 2002]. The commonly observed ^{13}C-depletion of ~15 to 20‰ agrees well with the theoretical magnitude of C-isotope fractionation estimated from the reduced masses of $^{12}CH_4$ and $^{13}CH_4$ diffusion in air [*Chanton*, 2005]. Thus, CH_4 in ombrotrophic bogs that already has a highly negative $\delta^{13}C$ value can become even further ^{13}C-depleted during transport to the troposphere. For example, the two acidic rain-fed Welsh peatlands Blaen Fign and Cors Caron emit CH_4 that has $\delta^{13}C$ values, respectively, of –90 to –82‰ [*Bowes and Hornibrook*, 2006] and –95 to –83‰ [*Hornibrook and Bowes*, 2007].

Further information about gas transport mechanisms can be gleaned from the position of data in Figure 2. Ebullition is nonfractionating with respect to isotopologues of CH_4 [*Chanton*, 2005], resulting in the $\delta^{13}C$ values of subsurface CH_4 being preserved in flux and $\delta^{13}C(CH_4)$ pairs plotting near or on the 1:1 line. Similarly, active ventilation of the rhizosphere of plants such as *Typha* spp. results in bulk flow of gas through porous vascular tissue, and consequently, the $\delta^{13}C$ values of pore water CH_4 are not altered significantly during transport. A few data plot above the 1:1 line for either fens or bogs, indicating that diffusion of CH_4 through pore water contributes little to the total diffusive flux of CH_4 from these peatlands. During passive diffusion through pore water, methanotrophic bacteria preferential consume $^{12}CH_4$ causing ^{13}C enrichment of residual CH_4 [*Barker and Fritz*, 1981]. The methanotrophic barrier in peatlands tends to be highly efficient, and CH_4 oxidation potential typically exceeds rates of pore water CH_4 diffusion and localized CH_4 production [*Whalen and Reeburgh*, 2000; *Hornibrook et al.*,

2008] resulting in little or no residual CH_4 surviving for export to the atmosphere.

Noteworthy in Figure 2 are the data from the Alberta Fen reported by *Popp et al.* [1999] which show the effects of removing surface vegetation on the $\delta^{13}C$ values of CH_4 emissions. In vegetated control plots, $\Delta_{CH_4}(^{13}C)$ between emissions and subsurface CH_4 averaged approximately –9‰ (i.e., the open circles plotting beneath the 1:1 line in Figure 2). Removal of surface vegetation eliminated the passive diffusion transport isotope effect and resulted in partial oxidation of CH_4 escaping to the atmosphere leading to a positive shift of ~12‰ in $\Delta_{CH_4}(^{13}C)$ (i.e., to +3‰; the filled circles plotting above the 1:1 line in Figure 2).

While the occurrence of CH_4 flux via pore water diffusion can be discounted in the majority of peatlands in Figure 2 based upon the absence of ^{13}C enrichment in the CH_4 emitted, the importance of ebullition as a transport mechanism is more difficult to judge because of its sporadic nature. Emission samples collected using flux chambers are almost exclusively of some form of diffusive flux (i.e., plant or pore water) because, typically, there is a linear increase of CH_4 concentration in the headspace of the chambers, and if there is not, then data commonly are discarded because of concerns that an artificial ebullition event has been caused by the act of sampling. A related issue is that methods employed for collecting emission samples for analysis of $^{13}C/^{12}C$ ratios in CH_4 in some early studies, may have unintentionally caused ebullition or bulk flow through plants. Use of modern CH_4 preconcentration systems requires the collection of only small volumes of gas (~100 cm^3) from the headspace of flux chambers, an amount that is typically <1% of the chamber volume and can be collected without the use of a pump if the flasks are pre-evacuated. In the past, $\delta^{13}C$ analysis of CH_4 by conventional dual-inlet stable isotope ratio mass spectrometry necessitated collection of much larger volumes of gas (several liters or more), which commonly was withdrawn from flux chambers using an electric pump. Creation of even a small negative pressure in the headspace of a flux chamber during sampling can initiate ebullition or cause bulk flow of gas through plant aerenchyma. Neither transport process partitions $^{13}CH_4$ and $^{12}CH_4$ any measurable amount [*Chanton*, 2005], which may account for the larger proportion of $\delta^{13}C(CH_4)$ values from some of the older literature plotting in the vicinity of the 1:1 line in Figure 2.

The $\delta^{13}C$ values and rates of CH_4 flux for 13 peatlands listed in Table 1 are shown in Figure 3. Also shown is the linear regression curve reported by *Bellisario et al.* [1999] for CH_4 flux and pore water $\delta^{13}C(CH_4)$ values (data not shown) from the peatland complex situated near Thompson, Manitoba, Canada, and the regression curve for CH_4 flux and $\delta^{13}C$ values of CH_4 emissions (data shown) for four Welsh

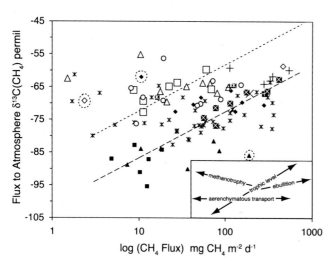

Figure 3. $\delta^{13}C$ values of CH_4 flux versus logarithm of CH_4 emission rate. The fine dashed line is the regression equation $y = 14.18$ $\log(x) - 86.6‰$ ($r^2 = 0.81$) reported by *Bellisario et al.* [1999] determined from pore water $\delta^{13}C(CH_4)$ values and CH_4 flux rates (data not shown) from the Collapse Bog, Poor Fen, Intermediate Fen, and Rich Fen, Manitoba, Canada. The coarse dashed line is the regression equation $y = 6.13 \ln(x) - 100.8‰$ ($r^2 = 0.63$) reported by *Hornibrook and Bowes* [2007] for the data from Cors Caron, Blaen Fign, Gors Lwyd, and Crymlyn Bog, UK (minus the three circled data points). The offset of ~14 ‰ between these two lines corresponds approximately to the amount of $^{13}CH_4$ and $^{12}CH_4$ partitioning that typically occurs during transport of pore water CH_4 to the atmosphere by passive diffusion through plant aerenchyma. The inset diagram shows the effects of peatland trophic level (first-order effect) and transport processes (second order effects) on the $\delta^{13}C$ values and rates of CH_4 flux. Data legend: Kings Lake Bog (asterisk) [*Lansdown et al.*, 1992; *Lansdown*, 1992]; Cors Caron (filled squares), Crymlyn Bog (filled diamonds), Gors Lwyd (open diamonds) [*Hornibrook and Bowes*, 2007]; Blaen Fign (filled triangles) [*Bowes and Hornibrook*, 2006; *Hornibrook and Bowes*, 2007]; Alaskan Tundra acidic (open triangles) and alkaline (open squares), Bena (X), Bog S2 (open circles), Bog S3 (filled square with white asterisk), Bog S4 (filled square with white X), Bog 3850 (filled square with white cross) and Junction Fen (cross) [*Quay et al.*, 1988; *Kelley et al.*, 1992; *Lansdown*, 1992].

peatlands reported by *Hornibrook and Bowes* [2007]. The curves have remarkably similar slopes despite being determined from data collected in disparate peatlands. The difference in *y*-intercept of ~14‰ for the two curves is consistent with the $\delta^{13}C$ data from *Bellisario et al.* [1999] being from subsurface CH_4 and the $\delta^{13}C$ values from *Hornibrook and Bowes* [2007] being from CH_4 emissions (i.e., it corresponds approximately to the isotope effect associated with passive diffusion of CH_4 through plant aerenchyma). Noteworthy as well is how the $\delta^{13}C$ values and rates for CH_4 flux from Bog S4, Minnesota, United States of America (filled squares with

white Xs) plot tightly along the emission curve from *Hornibrook and Bowes* [2007].

There is considerable scatter in the data shown in Figure 3, and no rigorous relationship can be drawn overall at present between the $\delta^{13}C$ composition of CH_4 flux and rates of CH_4 emission from northern peatlands. It appears there is a general trend of negative $\delta^{13}C$ values being associated with low CH_4 flux rates from ombrotrophic bogs, which is consistent with the correlation between CH_4 emissions rates and net primary productivity reported by *Whiting and Chanton* [1993]. The relationship between $\delta^{13}C$ values and rates of CH_4 flux shown in Figure 3 is impacted by differences in the density of vascular plant cover on certain peatlands, which can significantly influence CH_4 emissions rates but has little effect on $\delta^{13}C$ values of CH_4 flux [e.g., *Bowes and Hornibrook*, 2006]. Induced ebullition events will cause a positive shift in $\delta^{13}C$ values but not necessarily the perceived rate of CH_4 flux because the two measurements commonly are made independently (i.e., the flask sample for $\delta^{13}C(CH_4)$ analysis is filled by pumping after smaller samples have been collected from the chamber in syringes for determining CH_4 flux rate). A generalized inset diagram in Figure 3 shows how $\delta^{13}C$ values and rates of CH_4 flux should change due to differences in trophic status (first-order control) and transport or alteration processes (second-order control). The slope of the curves for methanotrophy and ebullition will depend upon the amount of CH_4 oxidized or emitted, respectively, by the two processes in relation to the background rate of CH_4 diffusion through plant aerenchyma.

There are several examples in Figure 3 where more positive $\delta^{13}C(CH_4)$ values are associated with either high or low rates of CH_4 flux, but there are very few instances in which highly negative $\delta^{13}C$ values are linked to high CH_4 emission rates. In general, the ombrotrophic peatlands in Figure 3 that produce and emit highly ^{13}C-depleted CH_4 are characterized by low CH_4 diffusive flux rates [*Hornibrook and Bowes*, 2007].

The mean $\delta^{13}C$ values for CH_4 flux calculated from the data compilation for peatlands listed in Table 1 are: fens $= -64.8 \pm 4.0‰$, range -72.6 to $-55.6‰$ ($n = 38$) and bogs $= -74.9 \pm 9.8‰$, range -95.3 to $-55.2‰$ ($n = 42$) (data shown in Figures 2 and 3). These are simple means of the available data and the $\delta^{13}C$ values are not weighed with emission strength. The average values for fens agree well between this study and summary values from both *Chanton et al.* [2000] (range -67 to $-64‰$) and *Hornibrook and Bowes* [2007] (average $-67.2 \pm 4.5\%$, range -72.6 to $-60.3‰$; $n = 13$). It is acknowledged that data from both those studies comprise a significant portion of the current data compilation, and consequently, this is not an independent comparison. The agreement predictably is poorer for bogs: *Chanton et al.* [2000] (range -70 to

−68‰) and *Hornibrook and Bowes* [2007] (average −86.6 ± 3.9‰; range −95.3 to −81.5‰; *n*=13) because of the two types of subsurface distributions of $\delta^{13}C(CH_4)$ values in bogs (Figure 1). *Hornibrook and Bowes* [2007] reported data only for two ombrotrophic peatlands, which result in the emission $\delta^{13}C(CH_4)$ data being skewed to more negative values.

The data presented in Figures 1 to 3 suggest that fens and bogs should be assigned independent $\delta^{13}C(CH_4)$ values in mass balance models of the global CH_4 cycle. The division is straightforward regarding minerotrophic peatlands versus acidic ombrotrophic peatlands (i.e., raised bogs and blanket bogs) because the latter possess highly ^{13}C-depleted CH_4 and a trend of pore water $\delta^{13}C(CH_4)$ values that is inverted relative to fens and, most probably, all other types of wetland. The distinction is less clear when attempting to subdivide further the category of bog. As shown in Figure 1, other acidic peatlands considered to be bogs produce and emit CH_4 having $\delta^{13}C$ values that are similar to fens. The situation is complicated further by carbon flow pathways differing significantly in peatlands that otherwise appear to be very similar. For example, Sifton Bog is a kettle hole bog that formed by gradual infilling of a small lake. Surface water in the floating mat has a low pH (3.8) and contains vegetation similar to Kings Lake Bog, which also formed by lakefill (Table 1), yet dissolved CH_4 at shallow depths in the two peatlands has $\delta^{13}C$ values that differ by ~30‰ and exhibit the opposite trend with depth (Figure 1). Clearly, there must be other factors at play in controlling pathways of carbon degradation in these peatlands.

The database is too small at present to explore further ways to subdivide bogs on the basis of their $\delta^{13}C(CH_4)$ values. Moreover, the data listed in Table 1 were collected in many instances for different purposes and by a range of methods. A pattern appears to be emerging for discerning the controls on $\delta^{13}C(CH_4)$ values, but ideally, a parameter such as vegetation assemblage, which is sensitive to nutrient status and pH, and which critically can be quantified by remote sensing, needs to be characterized systematically in relation to $\delta^{13}C(CH_4)$ values of pore water and flux for a range of peatlands in a manner similar to the analysis of *Bubier* [1995] for CH_4 flux, hydrology, and water chemistry. Recent advances in gas analysis instrumentation should greatly reduce the level of effort required to conduct such a task. Cavity ring-down spectrometers presently enable CH_4 flux rates to be measured in real time in either flux chambers or via eddy covariance towers. The lower detection limit of field-portable cavity ring-down spectrometers capable of measuring $\delta^{13}C(CH_4)$ values (rather than CH_4 concentration) is presently too high (~25 to 125 ppmv) to enable in situ measurement of the stable carbon isotope composition of CH_4 flux; however, commercial instruments capable of analyzing $\delta^{13}C$

(and δ^2H) values of CH_4 at low ppmv levels arc likely to be forthcoming as a result of ongoing development work for other applications [e.g., *Webster*, 2005; *Mungas and Dreyer*, 2006; *Onstott et al.*, 2006].

4. SUMMARY AND FUTURE WORK

Methane produced in ombrotrophic peatlands has a $\delta^{13}C$ composition that is significantly more negative than that of CH_4 formed in fens and other groundwater-influenced wetlands. Water acidity appears to be a critical factor determining which biochemical pathway of methanogenesis operates during the final stages of terminal carbon mineralization. Methanogenesis in nutrient-poor rain-fed peatlands is largely dominated by archaea that utilize CO_2/H_2, and for reasons that are not fully understood, acetic acid accumulates in such peatlands as an end product of anaerobic decay [*Hines et al.*, 2008]. Graminoid plants are typically present in abundance in fens and contribute in several ways to increasing $\delta^{13}C$ values of subsurface CH_4: (1) exudates from their roots provide labile substrates that promote a prevalence of the acetate fermentation pathway, (2) transport of O_2 to their roots enhances rates of methanotrophy causing localized ^{13}C enrichment of CH_4, and (3) preferential export of $^{12}CH_4$ via aerenchyma further increases the $\delta^{13}C$ value of residual CH_4 in the pore water pool. The latter two processes are sufficiently pervasive in fens that a pristine or unaltered $\delta^{13}C$ value for pore water CH_4 may not exist in the acrotelm and shallow catotelm (i.e., within the depth range of the rhizosphere) [*Chanton et al.*, 2002].

Differences in organic matter degradation in ombrotrophic and minerotrophic peatlands are reflected in the $\delta^{13}C$ values of CH_4 emissions. Methane fluxes from both bogs and fens are ^{13}C-depleted relative to pore water CH_4 because of isotope effects that occur during gas transport through plant aerenchyma. The absence of ^{13}C enrichment in CH_4 flux relative to pore water CH_4 in most peatlands suggests that little CH_4 escapes to the atmosphere as a result of diffusion through pore water (i.e., there is no evidence of a methanotrophy-associated kinetic isotope effect).

The difference in $\delta^{13}C$ composition of CH_4 flux is ~10‰ between ombrotrophic bogs and minerotrophic fens, with the former being more ^{13}C-depleted. Bogs are diverse though in terms of their hydrology and water chemistry, which complicates extrapolation of the observed difference in $\delta^{13}C$ values to all wetlands classified as "bog" because the designation encompasses a wide variety of peatlands that are nutrient-poor and *Sphagnum*-dominated. There is sufficient evidence to recommend that blanket bogs and raised bogs be attributed lower $\delta^{13}C$ values (approximately −74‰) for CH_4 flux compared to fens (approximately −64‰) in mass balance budgets. Further work is needed to establish the origin of fen-like distributions of $\delta^{13}C(CH_4)$ values in what appear to be nutrient-poor bogs. The issue may be one of wetland classification, but it could also result from subtle pH-mediated changes in the pore water chemistry of acetic acid.

The $\delta^{13}C$ values presented here were collected from studies that focused almost exclusively on the measurement of steady state CH_4 flux (i.e., diffusion through pore water or plant aerenchyma). Ebullition is capable of releasing large quantities of CH_4 in brief periods of time and can account for a significant proportion of the annual flux of CH_4 from a peatland [Glaser et al., 2004; Tokida et al., 2007; Comas et al., 2008]. The $\delta^{13}C$ value of CH_4 emitted in such events will be similar to that of pore water CH_4 because of the absence of a kinetic isotope effect during rapid gas transport. High temporal resolution measurement of CH_4 flux and $\delta^{13}C$ values in concert should be possible in the near future via cavity ring-down spectrometry. The potential of such instruments in this capacity has been demonstrated recently in the detection of a major late season CH_4 pulse from a high Arctic peatland during the onset of soil freezing [Mastepanov et al., 2008]. High resolution measurements of $\delta^{13}C(CH_4)$ values in parallel with CH_4 emission rates have the potential to confirm conclusively mechanisms of gas transport and to link initiation of abrupt flux events to internal and external factors. The advent of widespread use of cavity ring-down spectrometers should address not only gaps in our temporal knowledge of $\delta^{13}C(CH_4)$ values but also deficiencies in spatial coverage, which are notable for vast areas of peatland in Siberia and Scandinavia.

Another unknown at this juncture are changes in $\delta^{13}C$ values of CH_4 emitted from peatlands on short and long timescales, and how stable the relationship is between the $\delta^{13}C$ composition of CH_4 flux and CH_4 within the soil of peatlands. Both types of information are required for accurate interpretation of $\delta^{13}C$ values of atmospheric CH_4 from monitoring networks and de-convolution of ice core records of tropospheric $\delta^{13}C(CH_4)$ values.

This review has focused exclusively on $\delta^{13}C$ values of CH_4 in peatlands, in part, because of the lack of δ^2H data for CH_4 in both pore water and emissions. To date, there has not been a rigorous examination of $\delta^2H(CH_4)$ values in the context of kinetic isotope effects associated with transport mechanisms or potential differences in microbial methanogenic pathways between ombrotrophic and minerotrophic peatlands. Similar to $\delta^{13}C(CH_4)$ values, large changes in $\delta^2H(CH_4)$ values occur with depth in peatlands [Hornibrook et al., 1997; Chanton et al., 2006] and differences almost certainly must exist in CH_4 released by ebullition versus diffusion-based transport processes. The stable hydrogen isotope composition of CH_4 should have potential for tracing emission sources along

back trajectories of air masses at high latitudes because δ^2H values of CH_4 in peatlands change systematically with latitude as a result of differences in $\delta^2H(H_2O)$ values [Waldron et al., 1999a, 1999b; Chanton et al., 2006].

Acknowledgments. I am grateful to John Lansdown for permission to use unpublished data from his PhD dissertation and to Stephen Sebestyn and Nancy Dise for providing information about peatlands in the Marcel Experimental Forest, Minnesota, USA. Jeff Chanton and Mark Hines are thanked for insightful discussion about this topic at the 2007 PeatNet meeting in Wageningen, Netherlands. This manuscript was improved by comments and suggestions from reviewers Yit Arn Teh, Jeff Chanton, and Xavier Comas.

REFERENCES

Allan, W., M. R. Manning, K. R. Lassey, D. C. Lowe, and A. J. Gomez (2001), Modeling the variation of $\delta^{13}C$ in atmospheric methane: Phase ellipses and the kinetic isotope effect, *Global Biogeochem. Cycles*, *15*, 467–481.

Aravena, R., B. G. Warner, D. J. Charman, L. R. Belyea, S. P. Mathur, and H. Dinel (1993), Carbon isotopic composition of deep carbon gases in an ombrogenous peatland, northwestern Ontario, Canada, *Radiocarbon*, *35*, 271–276.

Avery, G. B., Jr., R. D. Shannon, J. R. White, C. S. Martens, and M. J. Alperin (1999), Effect of seasonal changes in the pathways of methanogenesis on the $\delta^{13}C$ values of pore water methane in a Michigan peatland, *Global Biogeochem. Cycles*, *13*, 475–484.

Barker, J. F., and P. Fritz (1981), Carbon isotope fractionation during microbial methane oxidation, *Nature*, *293*, 289–291.

Bellisario, L. M., J. L. Bubier, T. R. Moore, and J. P. Chanton (1999), Controls on CH_4 emissions from a northern peatland, *Global Biogeochem. Cycles*, *13*, 81–91.

Berestovskaya, Y. Y., et al. (2005), The processes of methane production and oxidation in the soils of the Russian Arctic tundra, *Microbiology*, *74*, 221–229.

Bergamaschi, P., M. Bräunlich, T. Marik, and C. A. M. Brenninkmeijer (2000), Measurements of the carbon and hydrogen isotopes of atmospheric methane at Izaña, Tenerife: Seasonal cycles and synoptic-scale variations, *J. Geophys. Res.*, *105*, 14,531–14,546.

Botz, R., et al. (1996), Carbon isotope fractionation during bacterial methanogenesis by CO_2 reduction, *Org. Geochem.*, *25*, 255–262.

Bowes, H. L., and E. R. C. Hornibrook (2006), Emission of highly ^{13}C-depleted methane from an upland blanket mire, *Geophys. Res. Lett. 33*, L04401, doi:10.1029/2005GL025209.

Bräuer, S. L., et al. (2006), Isolation of a novel acidiphilic methanogen from an acidic peat bog, *Nature*, *442*, 192–194.

Bubier, J. L. (1995), The relationship of vegetation to methane emission and hydro-chemical gradients in northern peatlands, *J. Ecol.*, *83*, 403–420.

Chanton, J. P. (2005), The effect of gas transport on the isotope signature of methane in wetlands, *Org. Geochem.*, *36*, 753–768.

Chanton, J., P. Crill, K. Bartlett, and C. Martens (1989), Amazon capims (floating grassmats): A source of ^{13}C enriched methane to the troposphere, *Geophys. Res. Lett.*, *16*, 799–802.

Chanton, J. P., et al. (1995), Radiocarbon evidence for the substrates supporting methane formation within northern Minnesota peatlands, *Geochim. Cosmochim. Acta*, *59*, 3663–3668.

Chanton, J. P., C. M. Rutkowski, C. C. Schwartz, D. E. Ward, and L. Boring (2000), Factors influencing the stable carbon isotopic signature of methane from combustion and biomass burning, *J. Geophys. Res.*, *105*, 1867–1877.

Chanton, J. P., et al. (2002), Diel variation in lacunal CH_4 and CO_2 concentration and $\delta^{13}C$ in *Phragmites australis*, *Biogeochemistry*, *59*, 287–301.

Chanton, J. P., D. Fields, and M. E. Hines (2006), Controls on the hydrogen isotopic composition of biogenic methane from high-latitude terrestrial wetlands, *J. Geophys. Res.*, *111*, G04004, doi:10.1029/2005JG000134.

Chasar, L. S., et al. (2000), Methane concentration and stable isotope distribution as evidence of rhizospheric processes: Comparison of a fen and bog in the Glacial Lake Agassiz Peatland complex, *Ann. Bot.*, *86*, 655–663.

Clymo, R. S., and C. L. Bryant (2008), Diffusion and mass flow of dissolved carbon dioxide, methane, and dissolved organic carbon in a 7-m deep raised peat bog, *Geochim. Cosmochim. Acta*, *72*, 2048–2066.

Comas, X., L. Slater, and A. Reeve (2008), Seasonal geophysical monitoring of biogenic gases in a northern peatland: Implications for temporal and spatial variability in free phase gas production rates, *J. Geophys. Res.*, *113*, G01012, doi:10.1029/2007JG000575.

Conrad, R. (2005), Quantification of methanogenic pathways using stable carbon isotopic signatures: a review and a proposal, *Org. Geochem.*, *36*, 739–752.

Conrad, R., et al. (1986), Thermodynamics of H_2-consuming and H_2-producing metabolic reactions in diverse methanogenic environments under in situ conditions, *FEMS Microbiol. Ecol.*, *38*, 353–360.

Crill, P. M., K. B. Bartlett, R. C. Harriss, E. Gorham, E. S. Verry, D. I. Sebacher, L. Madzar, and W. Sanner (1988), Methane flux from Minnesota peatlands, *Global Biogeochem. Cycles*, *2*, 371–384.

Daulat, W. E., and R. S. Clymo (1998), Effects of temperature and water table on the efflux of methane from peatland surface cores, *Atmos. Environ.*, *32*, 3207–3218.

Dise, N. B. (1993), Methane emission from Minnesota peatlands: Spatial and seasonal variability, *Global Biogeochem. Cycles*, *7*, 123–142.

Duddleston, K. N., M. A. Kinney, R. P. Kiene, and M. E. Hines (2002), Anaerobic microbial biogeochemistry in a northern bog: Acetate as a dominant metabolic end product, *Global Biogeochem. Cycles*, *16*(4), 1063, doi: 10.1029/2001GB001402.

Ferretti, D. F., et al. (2005), Unexpected changes to the global methane budget over the past 2000 years, *Science*, *309*, 1714–1717.

Ferretti, D. F., et al. (2007), Stable isotopes provide revised global limits of aerobic methane emissions from plants, *Atmosp. Chem. Phys.*, *7*, 237–241.

Francey, R. J., M. R. Manning, C. E. Allison, S. A. Coram, D. M. Etheridge, R. L. Langenfelds, D. C. Lowe, and L. P. Steele (1999), A history of $\delta^{13}C$ in atmospheric CH_4 from the Cape Grim air archive and Antarctic firn air, *J. Geophys. Res.*, *104*, 23,631–23,643.

Fukuzaki, S., et al. (1990), Kinetics of the methanogenic fermentation of acetate, *Appl. Environ. Microbiol.*, *56*, 3158–3163.

Galand, P. E., et al. (2005), Pathways for methanogenesis and diversity of methanogenic archaea in three boreal peatland ecosystems, *Appl. Environ. Microbiol.*, *71*, 2195–2198.

Gerard, G., and J. Chanton (1993), Quantification of methane oxidation in the rhizosphere of emergent aquatic macrophytes—Defining upper limits, *Biogeochemistry*, *23*, 79–97.

Glaser, P. H., J. P. Chanton, P. Morin, D. O. Rosenberry, D. I. Siegel, O. Ruud, L. I. Chasar, and A. S. Reeve (2004), Surface deformations as indicators of deep ebullition fluxes in a large northern peatland, *Global Biogeochem. Cycles*, *18*, GB1003, doi:10.1029/2003GB002069.

Hein, R., P. J. Crutzen, and M. Heimann (1997), An inverse modeling approach to investigate the global atmospheric methane cycle, *Global Biogeochem. Cycles*, *11*, 43–76.

Hines, M. E., K. N. Duddleston, and R. P. Kiene (2001), Carbon flow to acetate and C_1 compounds in northern wetlands, *Geophys. Res. Lett.*, *28*, 4251–4254.

Hines, M. E., K. N. Duddleston, J. N. Rooney-Varga, D. Fields, and J. P. Chanton (2008), Uncoupling of acetate degradation from methane formation in Alaskan wetlands: Connections to vegetation distribution, *Global Biogeochem. Cycles*, *22*, GB2017, doi:10.1029/2006GB002903.

Horn, M. A., et al. (2003), Hydrogenotrophic methanogenesis by moderately acid-tolerant methanogens of a methane-emitting acidic peat, *Appl. Environ. Microbiol.*, *69*, 74–83.

Hornibrook, E. R. C., and H. L. Bowes (2007), Trophic status impacts both the magnitude and stable carbon isotope composition of methane flux from peatlands, *Geophys. Res. Lett.*, *34*, L21401, doi:10.1029/2007GL031231.

Hornibrook, E. R. C., et al. (1997), Spatial distribution of microbial methane production pathways in temperate zone wetland soils: Stable carbon and hydrogen isotope evidence, *Geochim. Cosmochim. Acta*, *61*, 745–753.

Hornibrook, E. R. C., et al. (1998), Reply to comment by S. Waldron, A. Fallick, and A. Hall on "'Spatial distribution of microbial methane production pathways in temperate zone wetland soils: Stable carbon and hydrogen isotope evidence", *Geochim. Cosmochim. Acta*, *62*, 373–375.

Hornibrook, E. R. C., et al. (2000a), Carbon-isotope ratios and carbon, nitrogen and sulfur abundances in flora and soil organic matter from a temperate-zone bog and marsh, *Geochem. J.*, *34*, 237–245.

Hornibrook, E. R. C., et al. (2000b), Evolution of stable carbon isotope compositions for methane and carbon dioxide in freshwater wetlands and other anaerobic environments, *Geochim. Cosmochim. Acta*, *64*, 1013–1027.

Hornibrook, E. R. C., et al. (2000c), Factors influencing stable-isotope ratios in CH_4 and CO_2 within subenvironments of freshwater wetlands: Implications for δ-signatures of emissions, *Isotopes Environ. Health Stud.*, *36*, 151–176.

Hornibrook, E. R. C., et al. (2008), Methanotrophy potential versus methane supply by pore water diffusion in peatlands, *Biogeosci. Discuss.*, 5, 2607–2643.

Houweling, S., G. R. van der Werf, K. Klein Goldewijk, T. Röckmann, and I. Aben (2008), Early anthropogenic CH_4 emissions and the variation of CH_4 and $^{13}CH_4$ over the last millennium, *Global Biogeochem. Cycles*, 22, GB1002, doi:10.1029/2007GB002961.

Juottonen, H., et al. (2005), Methanogen communities and *Bacteria* along an ecohydrological gradient in a northern raised bog complex, *Environ. Microbiol.*, 7, 1547–1557.

Kaplan, J. O. (2002), Wetlands at the Last Glacial Maximum: Distribution and methane emissions, *Geophys. Res. Lett.*, 29(6), 1079, doi:10.1029/2001GL013366.

Kelley, C. A., et al. (1992), Temporal variations in the stable carbon isotopic composition of methane emitted from Minnesota peatlands, *Global Biogeochem. Cycles*, 6, 263–269.

Kotsyurbenko, O. R., et al. (2004), Acetoclastic and hydrogenotrophic methane production and methanogenic populations in an acidic West-Siberian peat bog, *Environ. Microbiol.*, 6, 1159–1173.

Kotsyurbenko, O. R., et al. (2007), Shift from acetoclastic to H_2-dependent methanogenes is in a West Siberian peat bog at low pH values and isolation of an acidophilic *Methanobacetium* strain, *Appl. Environ. Microbiol.*, 73, 2344–2348.

Lansdown, J. M. (1992), The carbon and hydrogen stable isotope composition of methane released from natural wetlands and ruminants, Doctoral Dissertation thesis, 225 pp., University of Washington.

Lansdown, J. M., et al. (1992), CH_4 production via CO_2 reduction in a temperate bog: A source of ^{13}C-depleted CH_4, *Geochim. Cosmochim. Acta*, 56, 3493–3503.

Lassey, K. R., D. C. Lowe, and M. R. Manning (2000), The trend in atmospheric methane $\delta^{13}C$ and implications for isotopic constraints on the global methane budget, *Global Biogeochem. Cycles*, 14, 41–49.

Lassey, K. R., et al. (2007), Centennial evolution of the atmospheric methane budget: what do the carbon isotopes tell us?, *Atmos. Chem. Phys.*, 7, 2119–2139.

Lombardi, J. E., et al. (1997), Investigation of the methyl fluoride technique for determining rhizospheric methane oxidation, *Biogeochemistry*, 36, 153–172.

Lowe, D. C., et al. (1999), Shipboard determinations of the distribution of ^{13}C in atmospheric methane in the Pacific, *J. Geophys. Res.*, 104, 26,125–26,135.

MacDonald, G. M., et al. (2006), Rapid early development of circumarctic peatlands and atmospheric CH_4 and CO_2 variations, *Science*, 314, 285–288.

Mastepanov, M., et al. (2008), Large tundra methane burst during onset of freezing, *Nature*, 456, 628–630.

Metje, M., and P. Frenzel (2005), Effect of temperature on anaerobic ethanol oxidation and methanogenesis in acidic peat from a northern wetland, *Appl. Environ. Microbiol.*, 71, 8191–8200.

Mikaloff Fletcher, S. E., P. P. Tans, L. M. Bruhwiler, J. B. Miller, and M. Heimann (2004), CH_4 sources estimated from atmospheric observations of CH_4 and its $^{13}C/^{12}C$ isotopic ratios: 2. Inverse modeling of CH_4 fluxes from geographical regions, *Global Biogeochem. Cycles*, 18, GB4005, doi:10.1029/2004GB002224.

Miller, J. B. (2004), The carbon isotopic composition of atmospheric methane and its constraint on the global methane budget, in *Stable Isotopes and Biosphere-Atmosphere Interactions: Processes and Biological Controls*, edited by L. B. Flanagan, et al., pp. 288–310, Elsevier, London.

Miller, J. B., K. A. Mack, R. Dissly, J. W. C. White, E. J. Dlugokencky, and P. P. Tans (2002), Development of analytical methods and measurements of $^{13}C/^{12}C$ in atmospheric CH_4 from the NOAA Climate Monitoring and Diagnostics Laboratory Global Air Sampling Network, *J. Geophys. Res.*, 107(D13), 4178, doi:10.1029/2001JD000630.

Miyajima, T., et al. (1997), Anaerobic mineralization of indigenous organic matters and methanogenesis in tropical wetland soils, *Geochim. Cosmochim. Acta*, 61, 3739–3751.

Mungas, G. S., and Dreyer, C. (2006), Pulsed cavity ringdown laser absorption spectroscopy in a hollow waveguide, *IEEE*, 13 pp.

Onstott, T. C., D. McGown, J. Kessler, B. Sherwood Lollar, K. K. Lehmann, and S. M. Clifford (2006), Martian CH_4: Sources, flux, and detection, *Astrobiology*, 6(2), 377–395.

Penning, H., et al. (2005), Variation of carbon isotope fractionation in hydrogenotrophic methanogenic microbial cultures and environmental samples at different energy status, *Global Change Biol.*, 11, 2103–2113.

Popp, T. J., J. P. Chanton, G. J. Whiting, and N. Grant (1999), Methane stable isotope distribution at a *Carex* dominated fen in north central Alberta, *Global Biogeochem. Cycles*, 13, 1063–1077.

Prater, J. L., J. P. Chanton, and G. J. Whiting (2007), Variation in methane production pathways associated with permafrost decomposition in collapse scar bogs of Alberta, Canada, *Global Biogeochem. Cycles*, 21, GB4004, doi:10.1029/2006GB002866.

Quay, P., J. Stutsman, D. Wilbur, A. Snover, E. Dlugokencky, and T. Brown (1999), The isotopic composition of atmospheric methane, *Global Biogeochem. Cycles*, 13, 445–461.

Quay, P. D., S. L. King, J. M. Lansdown, and D. O. Wilbur (1988), Isotopic composition of methane released from wetlands: Implications for the increase in atmospheric methane, *Global Biogeochem. Cycles*, 2, 385–397.

Rooney-Varga, J. N., et al. (2007), Links between archaeal community structure, vegetation type and methanogenic pathway in Alaskan peatlands, *FEMS Microbiol. Ecol.*, 60, 240–251.

Schaefer, H., and M. J. Whiticar (2008), Potential glacial-interglacial changes in stable carbon isotope ratios of methane sources and sink fractionation, *Global Biogeochem. Cycles*, 22, GB1001, doi:10.1029/2006GB002889.

Schaefer, H., et al. (2006), Ice record of $\delta^{13}C$ for atmospheric CH_4 across the Younger Dryas-Preboreal Transition, *Science*, 313, 1109–1112.

Sizova, M. V., et al. (2003), Isolation and characterization of oligotrophic acido-tolerant methanogenic consortia from a sphagnum peat bog, *FEMS Microbiol. Ecol.*, 45, 301–315.

Smith, L. C., et al. (2004), Siberian peatlands a net carbon sink and global methane source since the early Holocene, *Science*, 303, 353–356.

Sowers, T. (2006), Late quaternary atmospheric CH_4 isotope record suggests marine clathrates are stable, *Science, 311*, 838–840.

Stevens, C. M. (1993), Isotopic abundances in the atmosphere and sources, in *Atmospheric Methane: Sources, Sinks, and Role in Global Change*, edited by M. A. K. Khalil, pp. 62–88, Springer, New York.

Stevens, C. M., and A. Engelkemeir (1988), Stable carbon isotopic composition of methane from some natural and anthropogenic sources, *J. Geophys. Res., 93*, 725–733.

Tokida, T., T. Miyazaki, M. Mizoguchi, O. Nagata, F. Takakai, A. Kagemoto, and R. Hatano (2007), Falling atmospheric pressure as a trigger for methane ebullition from peatland, *Global Biogeochem. Cycles, 21,* GB2003, doi:10.1029/2006GB002790.

Valentine, D. L., et al. (2004), Carbon and hydrogen isotope fractionation by moderately thermophilic methanogens, *Geochim. Cosmochim. Acta, 68*, 1571–1590.

Wahlen, M. (1994), Carbon dioxide, carbon monoxide and methane in the atmosphere: Abundance and isotopic composition, in *Stable Isotopes in Ecology and Environmental Science*, edited by K. Lajtha and R. H. Michener, pp. 93–113, Blackwell, London.

Waldron, S., et al. (1998), Comment on "Spatial distribution of microbial methane production pathways in temperate zone wetland soils: Stable carbon and hydrogen isotope evidence" by E. R. C. Hornibrook, F. J. Longstaffe, and W. S. Fyfe, *Geochim. Cosmochim. Acta, 62*, 369–372.

Waldron, S., A. J. Hall, and A. E. Fallick (1999a), Enigmatic stable isotope dynamics of deep peat methane, *Global Biogeochem. Cycles, 13*, 93–100.

Waldron, S., et al. (1999b), The global influence of the hydrogen isotope composition of water on that of bacteriogenic methane from shallow freshwater environments, *Geochim. Cosmochim. Acta, 63*, 2237–2245.

Webster, C. R. (2005), Measuring methane and its isotopes [12]CH_4, [13]CH_4, and CH_3D on the surface of Mars with in situ laser spectroscopy, *Appl. Opt., 44*, 1226–1235.

Whalen, S. C., and W. S. Reeburgh (2000), Methane oxidation, production, and emission at contrasting sites in a boreal bog, *Geomicrobiol. J., 17*, 237–251.

Whiticar, M., and H. Schaefer (2007), Constraining past global tropospheric methane budgets with carbon and hydrogen isotope ratios in ice, *Philos. Trans. A Math. Phys. Eng. Sci., 365*, 1793–1828.

Whiticar, M. J. (1993), Stable isotopes and global budgets, in *Atmospheric Methane: Sources, Sinks, and Role in Global Change*, edited by M. A. K. Khalil, pp. 138–167, Springer, New York.

Whiticar, M. J. (1999), Carbon and hydrogen isotope systematics of bacterial formation and oxidation of methane, *Chem. Geol., 161*, 291–314.

Whiticar, M. J., et al. (1986), Biogenic methane formation in marine and freshwater environments: CO_2 reduction *vs.* acetate fermentation—Isotope evidence, *Geochim. Cosmochim. Acta, 50*, 693–709.

Whiting, G. J., and J. P. Chanton (1993), Primary production control of methane emission from wetlands, *Nature, 364*, 794–795.

Williams, R. T., and R. L. Crawford (1983), Microbial diversity of Minnesota peatlands, *Microb. Ecol., 9*, 201–214.

Williams, R. T., and R. L. Crawford (1984), Methane production in Minnesota peatlands, *Appl. Environ. Microbiol., 47*, 1266–1271.

Williams, R. T., and R. L. Crawford (1985), Methanogenic bacteria, including an acid-tolerant strain, from peatlands, *Appl. Environ. Microbiol., 50*, 1542–1544.

E. R. C. Hornibrook, Bristol Biogeochemistry Research Centre, Department of Earth Sciences, University of Bristol, Wills Memorial Building, Queens Road, Bristol BS8 1RJ, UK. (ed. hornibrook@bristol.ac.uk)

Laboratory Investigations of Methane Buildup in, and Release From, Shallow Peats

Mikhail Mastepanov and Torben R. Christensen

GeoBiosphere Science Centre, Physical Geography and Ecosystems Analysis, Lund University
Lund, Sweden

Peatlands are significant sources of the important greenhouse gas methane (CH_4). Here we explore the design and uses of experimental systems developed for the determination of continuous fluxes of CO_2 and CH_4 in closed ecosystem monoliths including the capture of $^{14}CO_2$ and $^{14}CH_4$ following radiolabeling. In this chapter, we detail the experimental design developed in our laboratory that may be used to study a range of different processes controlling formation, oxidation, and emission of methane in shallow peat soils under controlled conditions. These include methanogenic production versus methanotrophic oxidation; relative contribution of different methanogenic pathways; the importance of rhizospheric processes including root exudation for methanogenic substrate provision; the time lag between photosynthetic carbon assimilation and CH_4 emission; and finally the relative proportion of different emission pathways: plant mediated transport, diffusion, and ebullition. The results that may be produced using the range of experimental measurement systems presented have implications for better understanding wetland ecosystem/atmosphere interactions, including possible feedback effects on climate change. Much attention has been devoted to field-based ecosystem-atmosphere flux measurement campaigns investigating the relationship between peatland net ecosystem productivity and CH_4 emission. The experimental laboratory-based systems presented here represent useful compliments to these field studies. Under controlled conditions, the complex relationship between plant composition, productivity, and CH_4 emission may be studied. Here we show examples of how this complex relationship may change considerably in nature depending on plant species composition as this varies both within and between ecosystems.

1. INTRODUCTION

The composition of the atmosphere is critically dependent on microbial activities in soils and sediments as well as those of the terrestrial plant cover. While natural wetlands in the northern hemisphere constitute only about 3% of the total land surface area [*Mathews and Fung*, 1987], they represent an enormous deposit of carbon estimated at about one third of the global total [*Gorham*, 1991; *Zimov et al.*, 2006]. Gradual plant decomposition and microbial decay processes lead to slow but continuous deposition of peat which, over a millennium, may increase in depth by a meter or more [*MacDonald et al.*, 2006]. Anaerobic subsurface microbial fermentation and methanogenesis culminate in gaseous exchange of end products from land to atmosphere. Globally,

Carbon Cycling in Northern Peatlands
Geophysical Monograph Series 184
Copyright 2009 by the American Geophysical Union.
10.1029/2008GM000820

these processes are estimated to produce methane (CH_4) emissions from the world's soils in the order of 150 to 250 Tg CH_4 a^{-1} [*Prather et al.*, 2001], with a quarter to a third of the total emitted from peat soils of high latitudes [*Walter et al.*, 2001a, 2001b]. Current global circulation models predict the strongest warming in northern continental areas where most of the northern wetlands are located [*IPCC*, 2007]. Interannual variability in wetland emissions [*Bousquet et al.*, 2006] and the pronounced temperature sensitivity of CH_4 emissions from peat at many different sites, often giving Q_{10} values greater than 3 (Q_{10} is defined as the relative change in the rate of a process over a ten-degree C increase. In this case meaning a threefold increase, see *Svensson*, 1984) indicates a significant potential for positive feedback on climate change.

As peatland exchanges of CO_2 and CH_4 hold a significant importance for the atmospheric concentrations of these greenhouse gases, it is crucial that we have the best tools for investigating the dynamics of the production and transformation of CO_2 and CH_4 in peatland environments. One such tool is laboratory investigation under controlled environment conditions. This chapter summarizes a range of methods used in our laboratory as an approach to the study of entire peatland ecosystems functioning using mesocosm or "monoliths" under controlled conditions.

A mixture of new and well-established techniques is presented with an admitted bias toward those applied in our laboratory. The chapter does not aim to provide a full review of these types of techniques and their application in peatland CH_4 emission studies. Rather, it aims to communicate details and applications of selected relevant techniques for the study under controlled environment conditions of peatland ecosystem dynamics in relation to CH_4 emissions.

2. SHALLOW PEAT MONOLITHS

Laboratory investigation of CH_4 buildup and emission can be conducted on different scales. On the traditional microbiological scale, cultivation of pure lines of microorganisms is possible [e.g., *Castro et al.*, 2004]. In biological studies of peat, often peat samples are taken. Important information about plant functioning on the ecosystem scale can be obtained with controlled planted vegetation. However, the smallest object that maintains the main ecosystem properties is a live plant-peat monolith. Surprisingly, a peat block of relatively small size [e.g., 30-cm diameter and 30- to 40-cm depth, *Benstead and Lloyd*, 1994; 15-cm diameter and 30-cm deep, *Thomas et al.*, 1998; 30-cm diameter and 40-cm deep, *Daulat and Clymo*, 1998; 7-cm diameter and 35-cm deep, *Sheppard and Lloyd*, 2002; 25 × 25-cm and 40-cm deep, *Christensen et al.*, 2003] with all plants growing naturally is large enough to sustain the main natural pro-

cesses involved in ecosystem element cycling, including the processes leading to CH_4 production, oxidation, and emission to the atmosphere. Monoliths incubated in the laboratory at a natural temperature, light regime, and ambient CO_2 concentration generally show levels of net ecosystem exchange (NEE) and gross primary productivity (GPP) and CH_4 emissions equivalent to those found under natural conditions in the field [*Ström et al.*, 2005].

2.1. Sampling Technique

The technique of monolith sampling should be as easy as possible, as reproducible as possible, and as nondestructive as possible. In the literature, there are several examples of ways to cut out monoliths and transfer them to incubation [e.g., *Billings et al.*, 1982; *Benstead and Lloyd*, 1994; *Daulat and Clymo*, 1998]. Figure 1 shows a simplified schematic of the approaches we have taken in our laboratory. One method is to use a shell with a detachable bottom. The shell without its bottom can be cut directly into the peat whereupon the monolith is cut off below the shell and the bottom placed underneath (Figure 1a). The monolith can then be lifted out, and the connection between shell bottom and walls further insulated. The advantage of this method over a separate sampler (see below) is that it imposes less disturbance to the monolith, as the shell is pushed through only once. A potentially major disadvantage of the detachable bottom is nevertheless the difficulty of obtaining an entirely waterproof seal between the bottom and walls, and monoliths in such shells have a tendency to leak.

The alternative is to use a separate sampler of the same size as the shell, but without a bottom, for removing the monolith and transferring it to a solid waterproof shell (Figure 1b). This method causes greater disturbance of the sides of the monolith, first when the sampler is pushed through and, second, when the monolith is pushed out from the sampler into the shell. However, the solid waterproof shell is more convenient and becomes easy to transport and manipulate once the monolith has been transferred to it. In general, the first method can be recommended for dry monoliths, while the second is preferable for monoliths with a high water level.

Before sampling the monoliths, it is important to prepare any holes and connector ports on the shell that may be required later (e.g., for soil water sampling, water level monitoring). Drilling holes in a shell with a monolith in place causes serious disturbance of the microcosm.

2.2. Incubation and Maintenance

Once the monolith has been obtained, the first thing to address is the water table level. During sampling, a large

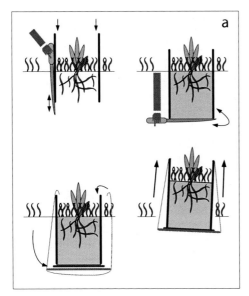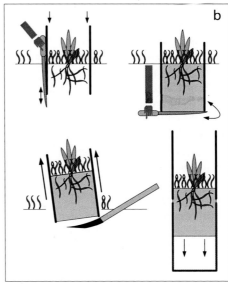

Figure 1. Monolith sampling techniques: (a) shell with detachable bottom and (b) sampler and solid shell.

fraction of pore water usually drains out, and water should therefore be replenished to the original level as soon as the monolith has been placed into a proper shell (or a detachable bottom attached). For this purpose, water from the hole from which the monolith originated can be used, since it has the necessary water chemistry. After a monolith has been obtained and the water content compensated, it should be transported to a laboratory and placed under stable ambient conditions. Monoliths obtained during the first half of the growing season can be incubated under close to natural conditions (light cycle and intensity, temperature, etc.). However, even in this case, the monolith sampling itself may cause stressful conditions for the plants. Consequently, since the ecosystem may require up to a month to adapt to new conditions, the natural seasonal cycle may become perturbed. It is therefore advantageous to obtain monoliths late in the season and first incubate them under dormant conditions at low temperature and in the dark. This preincubation allows the ecosystem to recover from the disturbance; damaged roots are gradually decaying, Eh is restoring, etc.

After one or more months of such preincubation, the monoliths may be brought to controlled environment conditions where they are carried through springtime conditions, gradually increasing temperature and natural light regime. Following this procedure, the monoliths may be considered as approximating a natural seasonal cycle. When the monolith ecosystem is incubated under ombrotrophic conditions, it needs only regular addition of distilled water to compensate for evaporation losses, natural (or chosen) CO_2 concentra-

tion, and a natural (or chosen) temperature and light regime. The hydrology of the site of monolith origin is important. If this origin is from minerotrophic conditions, these represent a challenge to mimic in the laboratory. Under the ombrotrophic conditions of concern here, distilled water is considered appropriate to compensate for transpirational losses, as these largely leaves the mineral contents of the peat water behind.

An important aspect of the whole ecosystem functioning captured in a monolith is the gas exchange of carbon (mainly CO_2 and CH_4) and nitrogen (mainly N_2O) carrying gasses with the atmosphere. For this purpose, the monoliths should be covered by transparent chambers as soon as they have been transferred to controlled environmental conditions. While the chambers may be removable (i.e., placed on a water seal), or more or less permanently fixed, the entire chamber plus shell system should be closed, so that gas exchange can be measured, while control may be kept of key gas concentrations, such as CO_2 as close to natural conditions as possible. Once the monolith has been sealed in its chamber, an air circulation system should be activated. In the simplest case, this should provide regulated air inflow, regulated air outflow, and air mixing in the chamber. Air inflow and outflow should be equal, and the pressure in the chamber kept as close to ambient as possible. Air mixing in the chamber can be provided by high in/outflow, or additionally by an internal fan.

When the monolith is incubated in the dark, there are no specific incoming gas requirements, as in the absence of photosynthesis an artificial atmospheric CO_2 concentration does not appreciably affect ecosystem processes. Ambient air with atmospheric or higher (laboratory air) CO_2 concentrations will suffice. However, when the monolith is photosynthesizing, CO_2 concentration plays a key role in the ecosystem production followed by all other ecological parameters. In our systems, we have used different strategies for setting the CO_2 concentration in a flow-through chamber.

1. The incoming concentration is kept constant and close to ambient (370–380 ppmV CO_2). In this case the concentration inside the chamber under light conditions is lower because of photosynthetic fixation. Highly productive ecosystems may be CO_2 limited in this case.

2. The incoming concentration is kept constant and artificially higher than ambient to prevent the concentration in the chamber under light conditions from falling below a certain threshold. In this case, plants are not CO_2 limited during light periods, while the concentration may become artificially high during dark periods.

3. Incoming CO_2 concentration is dynamically regulated to keep the internal concentration always close to ambient. This is the most natural mode for the microcosm, but also the most complicated technically.

When more than one monolith is incubated with flow-through chambers, the CO_2 supply system may become complicated, and some form of compromise solution may be required.

Water loss from evaporation can be an issue with isolated microcosms. Wetland monoliths always provide close to 100% relative humidity in the headspace, and if the incoming air is drier, water loss can therefore be significant. Due to this problem, the incoming air should be humidified before entering the chamber. Heating of the soil surface under light conditions keeps the air temperature in a chamber somewhat higher than the temperature of the incoming air, and even at 100% relative humidity, it can therefore cause removal of an additional amount of water vapor. The water table level in the monoliths should therefore be monitored and regulated, e.g., by spraying distilled water into the headspace.

An easy and efficient way to establish controlled gas flow through the chamber is to install a pump with flow regulator after the chamber in the gas line (Figure 2c) and a pump providing the inflow mixture (Figure 2a) with a little overflow before the chamber (Figure 2b). Several monoliths being incubated at the same time can share one pushing pump (Figure 2a), but each should have its own pulling pump with flow regulator (Figure 2c). To maintain the incoming gas mixture close to 100% relative humidity at the temperature of the monolith headspace, a heat exchanger and humidifier can be used, but the simplest solution may be to have the

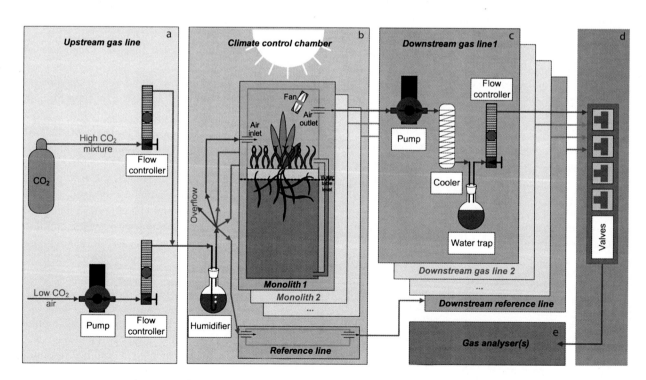

Figure 2. Experimental setup for parallel studies of several monoliths: (a) upstream gas line, (b) climate control chamber, (c) downstream gas lines, (d) valve system, and (e) gas analyzers.

incoming gas percolate through a large water bottle (Figure 2b). The gas mixture after a monolith may become over-saturated by water vapor, which can cause condensation in the tubes, on the flow controller, and in other parts of the gas line. Flow controllers of any type are sensitive to water droplets. A malfunctioning flow controller can cause disturbance not only in the measurements, but also in the carbon exchange of the microcosm. To prevent this problem, the humidity of outgoing air should be reduced, e.g., by means of a cooler with water trap (Figure 2c). Since the condensation is pressure dependent, the gas line design should take into account how high- and low-pressure zones are being altered. A good option (Figure 2c) is the outgoing pump positioned immediately after the monolith which permits some under-pressure in the chamber headspace. This helps to detect possible leaks in the chamber; in this case, high-CO_2 laboratory air is leaking into the chamber and can be easily detected by a CO_2 analyzer (see below). During experiments with ^{14}C labeling (see below), leakage of ambient air into a chamber is preferable to the opposite, also for safety reasons. A high-pressure zone is then formed between the pump and the flow regulator and can be used for efficient removal of water vapor, making this a good location for the cooler and the water trap (Figure 2c). For optimal functioning of the flow regulator, the pressure after the regulator should be kept stable and close to ambient. This should be considered when the gas line is being extended by analytical instrumentation (Figure 2e, see below). Finally, the necessary CO_2 concentration in the incoming gas mixture (see above) can be obtained by adding a subflow of high-CO_2 mixture to the main flow of low-CO_2 air (Figure 2a).

Incubating a number of monoliths of the same size at the same conditions is very important to provide experiments with replicates. The setup shown in Figure 2 was successfully used for instantaneous experiments with up to nine monoliths (three groups of three replicates at a time).

3. GAS EXCHANGE DYNAMICS

3.1. Instruments and Methods

The isolated monolith system with flow-through headspace chamber is a very good object for gas exchange studies. Virtually any gas compound with an atmospheric lifetime longer than a few minutes can be examined. Here we will focus on the main carbon gases CO_2 and CH_4 acknowledging that many minor C-carrying gases, such as CO and volatile organic compounds, may also play key roles for the ecosystem atmosphere under certain conditions. To monitor the gas exchange of CO_2 and CH_4 determination of their concentration in incoming (C_{in}) and outgoing (C_{out})

air is necessary. The principal formula for an emission (net exchange when positive means a source to the atmosphere) in long-time terms (e.g., in daily scale) is

$$E = (C_{out} - C_{in}) \times F, \qquad (1)$$

where F represents flow rate through the chamber. The "long-time" limitation above applies when the integration time for the emission is definitely longer than the exchange time for the chamber volume. In this case, we can assume that, e.g., all the CH_4 emitted during a certain time exits from the chamber and is registered as a difference of concentrations. However, if we are interested in detailed emission dynamics, air dilution in the chamber headspace should be considered. Each brief fluctuation in emission (e.g., bubble release) is followed by a smooth change in outgoing air concentration as it returns to steady state, which is why rapid flux dynamics cannot be directly observed in the outgoing concentration. This can nevertheless be derived from reverse modeling, as shown in section 3.4. Even if the experimental setup consists of only one monolith, the necessity of monitoring both incoming and outgoing concentrations imposes time constraints on a given analyzer shifting between two gas lines, sample and reference (Figures 2b–2d). The reference line should preferably be as similar as possible to the sample line, having the same water traps and flow controllers, although they may not be compulsory.

For time sharing of analyzer(s) between the two gas lines, a system of electropneumatic valves can be used (Figure 2d). The simplest setup for two channels (sample and reference) consists of two three-way valves, directing gas flow from one channel to an analyzer, while the other goes to an exhaust, and vice versa. However, as noted above, variations in aerodynamic resistance (an analyzer or an exhaust) after a flow controller can cause unstable gas flow through the chamber and thus disturb the ecosystem. More sustainable is a system based on a main flow through the chamber (in practical terms, in the order of 1 L min^{-1}) going to an exhaust with subflows going to analytical instruments. In this case, each instrument (or each chain of nondestructive instruments) should have its own pump and flow regulator. Since valves, switching hardware and software, etc. are required even for a single monolith setup, it is more efficient in practice to design setups with several monoliths (depending on the space available in a controlled environment cabinet), which are incubated and measured simultaneously. All chambers will then have constant gas flow-through, while the subflow to the analyzer(s) is switched between them, including the reference line.

3.2. CO_2 Exchange

Even if we are mainly focused on CH_4 fluxes, CO_2 exchange (may be called NEE in the closed microecosystem) provides important information on the ecosystem status, and as it is often related to CH_4 emissions, it should therefore generally be monitored [*Whiting and Chanton*, 1993; *Christensen et al.*, 2000]. A nondestructive infrared CO_2 analyzer, such as a LI-COR or PP Systems instrument, can be connected to the described system. Those instruments can provide CO_2 concentration data in close to real-time resolution in the order of one measurement per second, which is quite sufficient for such inert system. In practical terms, instrumentation is much cheaper and more robust for CO_2 than for CH_4, and it is generally quite easy to complement a system designed for CH_4 studies with an appropriate CO_2 analyzer. CO_2 dynamics in the chamber headspace of a monolith with fixed incoming CO_2 concentration usually involves alternating low steady state daytime concentrations with high steady state nighttime concentrations. Important ecosystem parameters, such as NEE, GPP, and net ecosystem respiration, can be derived from these values.

3.3. CH_4 Fluxes

The principle of CH_4 flux monitoring in the isolated monolith microcosm is the same as for CO_2. An instrument capable of monitoring the concentration at natural levels in close to real-time is needed, and this instrument should be switched between gas subflows from a number of monoliths and the reference line. Continuous gas flow through a chamber volume provides a smooth outgoing concentration picture. This may be considered a shortcoming of the system, as it is the perturbation of the chamber atmosphere that is used for the actual quantification of the ecosystem-atmosphere exchange. However, an advantage of this is that the concentration can be measured less frequently and still provide a good estimation of the CH_4 balance. For example, the Innova 1412 multigas monitor providing one CH_4 measurement about every 2 min was successfully used for a system with three monoliths plus one reference line [*Christensen et al.*, 2003; *Ström et al.*, 2005]. An instrument with more frequent readings provides an opportunity to monitor more monoliths and shorten the measurement cycle (e.g., *Lund et al.*, 2008: using DLT-100, LGR with six monoliths and 10-min cycles). Net CH_4 emission, which can be directly monitored in the described setup, is a product of at least three different groups of processes: CH_4 generation in anaerobic conditions, vertical CH_4 transport, and its partial oxidation under aerobic conditions. Subsurface CH_4 storage, in dissolved form or in entrapped bubbles, is a significant but not infinite pool in the isolated monolith. When the ecosystem of a monolith has been stabilized under summer conditions as defined in the controlled environment, this pool appears to stay constant, allowing the emission to be admitted as a difference between CH_4 production and oxidation. Unfortunately, simple concentration monitoring cannot resolve these processes (other approaches are described in sections 5 and 6). However, it can provide a very accurate net emission value and some interesting details of its dynamics.

Firstly, isolated monoliths with a flow-through headspace chamber offer a unique opportunity for direct measurement of CH_4 fluxes for a theoretically unlimited time. There is no field method with such capabilities: the closed chamber method does measure direct emission, but only over minutes after which the chamber should be open; micrometeorological methods can provide an estimate of emissions from a certain area, but without a direct measurement. Open-top chambers can be installed for a long period, but again do not provide direct measurements. Flow-through chambers in situ can, in essence, provide direct emission measurements from the fixed area for long time periods, but their artificial effect at the ecosystem level can be quite high. Even minor differences between ambient and internal air pressure can cause a fraction of CH_4 formed under the chamber to leak into the surrounding area or vice versa. Isolated monoliths under controlled laboratory conditions are indeed highly artificial in many respects, but under these artificial conditions, all CH_4 exchange can be directly measured for months. In this case, there is no escape for CH_4 to be uncounted, neither in time, nor in space. If the CH_4 concentration in the incoming air of the monolith is stable, its concentration dynamics in the outgoing air can have two different patterns. One is a more or less stable concentration, the product of steady state between the constant emission and constant flushing by incoming air. This constant emission can be recognized as a result of molecular diffusion and vascular transport. The other pattern has distinctive peaks with rapid increases in CH_4 concentration followed by slow descent to a stable level. This second pattern is a clear indication of ebullition [*Christensen et al.*, 2003; *Ström et al.*, 2005], worthy of consideration.

3.4. Ebullition

CH_4, produced in the anaerobic horizons inside a monolith as well as in a natural wetland, can find different pathways to the atmosphere. One is molecular diffusion in an aqueous solution. This is not very efficient and is known to provide about 2% of the total emission [*Christensen et al.*, 2002, 2003]. Another pathway is so-called vascular transport: enhanced diffusion through plant tissues [*Christensen*, 1993; *Morrissey et al.*, 1993; *Schimel*, 1995]. This process can be

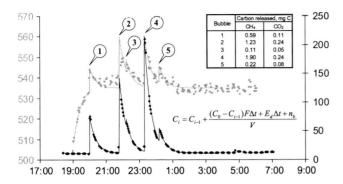

Bubble	Carbon released, mg C	
	CH_4	CO_2
1	0.59	0.11
2	1.23	0.24
3	0.11	0.05
4	1.90	0.24
5	0.22	0.08

$$C_i = C_{i-1} + \frac{(C_0 - C_{i-1})F\Delta t + E_d \Delta t + n_b}{V}$$

Figure 3. An example of the detailed dynamics of CO_2 (light circles, left scale, ppmV) and CH_4 (dark diamonds, right scale, ppmV) concentrations in a system with high ebullition. Solid lines: concentrations modeled by the equation, where E_d is constant and n_b differs from zero in five points, shown in the table.

very efficient, and sometimes, all the CH_4 emission appears to emanate from vascular transport [*Schimel*, 1995]. Both molecular emission and vascular transport are characterized by very stable CH_4 release which, in the case of an isolated monolith, provides stable steady state concentrations in outgoing air. The two types of emission cannot be distinguished by concentration monitoring and are usually grouped together as steady emission (SE). Ebullition is a very different way of CH_4 transport from anaerobic horizons. Since CH_4 is poorly soluble in water [e.g., *Wilhelm et al.*, 1977], a large fraction of subsurface CH_4 is present in so-called entrapped bubbles, gas bubbles which remain below the water table level [see *Rosenberry et al.*, 2006]. Under certain circumstances, these bubbles can move upward and be released at a given moment in a gas phase above the water table [*Chasar et al.*, 2000; *Ström et al.*, 2005]. For high-latitude lake environments, these types of emissions have been shown to be of global significance [*Walter et al.*, 2006]. Since active diffusion in gas is much higher than in liquid, a gas bubble released under a closed chamber with high air circulation disseminates its content throughout the headspace almost instantly. This is followed by a significant spike in CH_4 concentration in the outgoing air. If the bubbles are not very frequent, the concentration gradually decreases due to dilution by incoming air. Such an event can be clearly seen in the concentration dynamics (Figure 3). As each particular gas bubble contains not only CH_4, but also a high concentration of CO_2, the synchronous peak is visible in the CO_2 data.

With sufficient resolution in the concentration data, the actual instant emissions can be back calculated. In the flow-through system with both SE and ebullition, the CH_4 (and CO_2) balance in the headspace appears as:

$$n_i = n_{i-1} + n_0 + n_d + n_b - n_1, \quad (2)$$

where n_i represents the amount of source component (e.g., in mg C) in the headspace at timepoint i, while n_{i-1} is the amount at the time of the previous concentration measurement; n_0 is the amount of source component which entered with the incoming air during the period between measurements; n_1 is the amount of source component discharged with the outgoing air during the period between measurements; and n_d and n_b are the amounts of source component emitted by diffusion and bubble transport, respectively. To convert amounts to concentrations,

$$C_i V = C_{i-1}V + C_0 F\Delta t + E_d \Delta t + n_b - C_{i-1}F\Delta t, \quad (3)$$

so

$$C_i = C_{i-1} + \frac{(C_0 - C_{i-1})F\Delta t + E_d \Delta t + n_b}{V}, \quad (4)$$

where C_i and C_{i-1} are the concentrations of source component in the headspace at timepoints i and $i-1$, (mg C/mL), C_0 is the concentration in incoming air, V is the chamber headspace volume (mL), F is the flow rate (mL/min), E_d is the diffusional emission (mg C/min), and Δt is the time interval from $i-1$ to i. If we assume E_d to be constant over a certain time period and bubble release to be instantaneous ($n_b > 0$ only for single moments), the equation can be used to fit E_d and n_b as shown in Figure 3.

When bubbles are large and infrequent, this analysis can be used to calculate the amount of CH_4 (as well as CO_2) in each individual bubble. From their mixing ratio, it is also possible to estimate the depth from which each bubble emerged, using CH_4 and CO_2 dissolved concentration profiles and their Henry's constant ratio at peat incubation temperature.

Unfortunately, the bubbles are often frequent, and the picture is too complex to allow individual bubbles to be distinguished. In such case, the alternate method of differentiation can be used based on frequency analysis of instant flux values (see Appendix A in *Ström et al.*, 2005). These values have a normal distribution around the mean when SE is stable, and ebullition is negligible. At moments of gas bubble release, the instant flux values shift away from the normal distribution curve and may form subsequent frequency peaks. However, even in this case, the first peak may be used to estimate SE.

4. PEAT GAS PROFILE

Isolated plant-peat monoliths can be studied not only from the standpoint of surface fluxes, but can also be used for investigations in the subsurface CH_4 pool. As under natural conditions, CH_4 and other gases are present in the peat profile of isolated monoliths in two main pools: dissolved in peat water and as gas inside entrapped gas bubbles. These two pools are closely interlinked by mass exchange and equilibration between dissolved concentration and partial pressure in the gas phase (Henry's law). However, these pools are very unequal in terms of CH_4. The solubility of CH_4 is quite low, and under normal conditions, a given gas bubble volume contains about 30 times more CH_4 than the equivalent volume of solution. Molecular diffusion in the solution is fairly slow, and as long as the peat water itself is immobile, the spread of dissolved CH_4 is limited. By contrast, molecular diffusion in the gas phase is very rapid, and if entrapped bubbles interconnect to form chains or sometime even gas layers, CH_4 concentrations can rapidly equalize. However, in submerged peat layers, most of the liquid phase is usually connected into a single body, while individual entrapped bubbles are distributed within it. In this case, horizontal movement of bubble CH_4 is possible only through the liquid phase, while vertical movement also may occur when the entire bubble moves upward through the peat structure. The dissolved and gaseous pools react differently to changes in temperature and pressure. Emission of CH_4 from the dissolved pool to the atmosphere most likely takes place by diffusion, while CH_4 from entrapped bubbles will be released by ebullition.

4.1. Sampling Methods

To monitor CH_4 concentrations in the monolith subsurface, two groups of methods can be used: direct sampling, where a small volume of peat water or gas is sampled and transferred to the analytical setup, and membrane diffusion sampling, where the source component (CH_4, CO_2, etc.) diffuses through a membrane into a carrier substance which is then analyzed.

The space resolution achieved by some of these methods can be up to one centimeter [see *Mastepanov and Christensen*, 2008], and practical feasibility is mainly limited by the number of manual or membrane probes which can be inserted into a monolith. However, theoretical limitations also exist: both direct sampling and membrane diffusion disturbs natural concentrations in an area around the probe tip (in the order of a 5- to 10-mm circumference, depending on the method). Probes placed closer can therefore affect each other.

While it may be interesting to place several probes at the same depth to determine the characteristics of the artificial effect of monolith walls, for practical reasons, monolith profiles are usually measured in one dimension only with the probes placed one below the other along the axis of the monolith.

When the experiment with monoliths includes analyses of subsurface components, the location of ports for direct sampling should be planned in advance. Before the monolith is sampled, the shell should be prepared, all necessary holes made, and some type of fast and watertight connections provided. The simplest port is a small round hole in the shell which is covered by gas impermeable tape while the monolith is being taken. The tape is then removed and the hole quickly closed by a rubber plug with a sampling tube. The sampling tube may, for example, be of stainless steel. The internal diameter should be at least 1 mm. The end of this tube may be flattened and sharpened to provide easier intrusion into peat and prevent blockage of the tube aperture. At 5–10 mm from the tip, the sampling tube should have a side hole through which the samples will be drawn. The outer end of the sampling tube should be kept locked between samplings. Probe sampling can be done by a syringe: first, a portion of peat water somewhat larger than the internal volume of the sampling tube should be withdrawn to flush the tube. The next portion is the actual sample, its volume depending on the analytical methods and instruments used.

For peat water CH_4 concentration analysis, a small amount of liquid sample is injected into a vial filled with N_2 or any other gas mixture with a known concentration of CH_4. The vial is then shaken and the gas sample analyzed by gas chromatography [e.g., *McAuliffe*, 1971]. The direct sampling provides the exact amount of the source component clearly defined in time and space (the moment of sampling and a sampled volume around the sampling point). However, it cannot provide any high resolution either in time (because frequent sampling affects the object) or in space (depends of sampling volume necessary for analyses). For a further discussion, see *Mastepanov and Christensen* [2008].

The simplest membrane diffusion probe is a sampling tube, covered by a membrane which is impermeable to liquid water but permeable to gas molecules. A thinner tube is placed inside to supply carrier gas to the membrane (Figure 4). CH_4, diffusing through the membrane, is carried out with the carrier gas flow and can be analyzed by any appropriate gas analyzer. There are several modifications of this method (for a review, see *Mastepanov and Christensen*, 2008), some of which can provide almost real-time monitoring of dissolved gas concentrations.

4.2. Dissolved Gases and Entrapped Bubbles

Both direct sampling and membrane probe sampling work well for solutions, but their application is problematic in mixed media such as peat water with many entrapped bubbles.

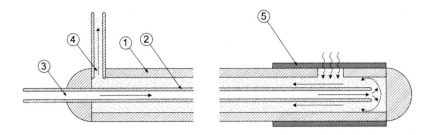

Figure 4. Example of a membrane probe construction. 1, outer stainless steel tube; 2, inner stainless steel tube; 3, gas inlet; 4, gas outlet; 5, silicone membrane. Adapted from *Mastepanov and Christensen* [2008], with permission from Elsevier.

Even a small gas bubble with a volume of 3.5% of the sample withdrawn together with liquid by manual sampling will more than double the amount of CH_4 in the probe compared to a pure liquid sample (considering 1:30 solubility). It is therefore best to sample only pure liquid or pure gas phase, although this may be unfeasible if there is a high content of entrapped bubbles. Membrane diffusion methods also have certain problems in mixed media [e.g., *Krämer and Conrad*, 1993; *Rothfuss and Conrad*, 1994; *Beckmann and Lloyd*, 2001]. Diffusion often depends on the immediate surroundings of the membrane. The amplitude of the signal obtained in pure liquid may be raised several times by an adjacent gas bubble. For such situations, a recently developed bimembrane probe [*Mastepanov and Christensen*, 2008] can be used instead of conventional monomembrane probes, making signals from probes in contact with liquid and the equilibrium gas phase almost identical. Another complication is that under natural conditions, not all gas bubbles are in equilibrium with the surrounding solution. This can depend on the pressure within the individual bubble and on its history (when the bubble was formed, at what depth, how fast it was ascending, *Romanovicz et al.*, 1995). Finally, even if the concentrations in the liquid and gas phase can somehow be estimated, estimating the amount of entrapped bubbles (gas content) through the profile is essentially a challenge. Gas content entrapped within peat soils can range from 0 to 19% per volume [*Rosenberry et al.*, 2006]. In the monoliths, this range can be different because limitation in the size of the monoliths excludes bubble-confining layers [*Romanowicz et al.*, 1995].

One approach to quantify the entrapped gas content, applicable for peat monoliths, as well as for in situ measurements, is using soil moisture probes (time domain reflectometry type, e.g., *Kellner et al.*, 2005, or capacitance type, e.g. *Tokida et al.*, 2005), which can provide an estimate of gas content within a submerged zone. Dielectric permittivity [*Comas and Slater*, 2007] and resistivity [*Slater et al.*,

2007] measurements have also proved very useful to investigate gas dynamics in peat monoliths. Certain galvanic and electromagnetic methods for detecting gas bubbles are under development. Applying magnetic resonance imaging [*Glaser et al.*, 2006] or X-ray tomography [*Kettridge and Binley*, 2008] for peat structure monitoring are expensive, but promising methods. There are also simpler options, such a visual estimation of the gas volume per volume of manually sampled medium. Monitoring the exact weight of the monolith together with the water table level can provide an estimate of peat density and thus of the gas content. A destructive way of determining subsurface CH_4 storage in the monolith profile is by fast freezing and dividing it into slices and grinding in the closed gas volume. A problem with this approach is that the freezing itself can alter the native peat structure.

5. MANIPULATION EXPERIMENTS

For studies of isolated peat-plant monoliths, the monitoring of surface fluxes and subsurface concentrations under close to natural conditions is indeed very important in order to examine to what extent, if at all, the isolated ecosystem has changed from its in situ status. However, manipulation experiments, where specific factors are artificially changed in order to track the ecosystem's response to this event, are certainly more interesting. Unlike natural ecosystems where many factors usually change together, laboratory-controlled conditions offer an important opportunity to determine the response to a single-factor event.

5.1. Temperature Manipulations

Technically, the easiest factor to manipulate is temperature. Depending on the type of controlled environment, the entire monolith can be incubated at the same temperature, which can be changed gradually or stepwise, or the monolith can have a temperature gradient between soil and surface.

This gradient can then be altered, and if the controlled environment setup allows for a diurnal variation in temperature, the scale of difference between day and nigh may also be manipulated.

Temperature manipulations can be divided into fast (when the target is the reaction of a balanced ecosystem to the temperature change) and slow (when the target is a new steady state at different temperature). Slow temperature response experiments are complicated by the fact that even under stable laboratory conditions, the monolith's ecosystem has its own seasonal cycle. Driven probably by a biological clock in the plants, the monolith gradually moves from summer to autumn even if the temperature and light regime are constant at the controlled summer levels. It is therefore important to be aware of where the monoliths are in their natural seasonal cycles in particular when conducting temperature manipulations.

5.2. Water Table Manipulations

The water table level is also a well-known factor affecting CH_4 emission [e.g., *Moore and Knowles*, 1989]. It divides the monolith into aerobic and anaerobic zones and influences both CH_4 production and oxidation. Fast changes in the water table can also cause changes in the amount of trapped subsurface CH_4, thus causing significant short-term changes in emissions [*Romanovicz et al.*, 1995]. Short-term drainage with associated degassing can be used to study methanogenic activity in the monolith. Opening a hole in the lower part of the monolith and draining its gravitational water (5–10 min) causes release of entrapped gas bubbles into the headspace. The water is then degassed by flushing with N_2 (for 10 min) and returned to the monolith. When conducted fast (e.g., 15–20 min, as specified above), the procedure is found out to have no detrimental effect on methanogenic activity [*Mastepanov*, 2004]. After refilling, the initial dissolved CH_4 concentration is close to zero across the entire profile but then starts to increase within the methanogenic layers and disseminate and re-establish a steady state profile. This process may take a few days. Such manipulation is indeed highly artificial for the entire microcosm, first, because it equilibrates all the dissolved components, such as DOCs, within the peat. It can nevertheless be used to study potential methanogenic activity in the previously stable ecosystem. A continued change of the water table level causes redistribution of aerobic and anaerobic zones and adaptation of microbial communities to the new state.

6. THE ^{14}C IMPULSE LABELING

When manipulation experiments are used to study ecosystem processes by artificial changes in their controlling parameters, the impulse labeling method makes it possible to study processes that are ongoing under stable conditions. Adding a small portion of ^{14}C isotope in one or another molecular form (see below) at a certain stage of the ecosystem's carbon turnover provides an opportunity to track its further fate, which follows the natural course of carbon turnover in this ecosystem. The technique may vary depending on the compound used for labeling. Below, we will describe two approaches to impulse ^{14}C labeling using ^{14}C–CO_2 and ^{14}C acetate.

6.1. The ^{14}C–CO_2 Labeling

^{14}C–CO_2 can be injected into the monolith headspace with incoming air under light conditions. Labeled CO_2 is fixed by plants as well as usual $^{12}CO_2$. Fixation of atmospheric CO_2 is the first step in the ecosystem's carbon turnover, and in theory, the labeled CO_2 should therefore undergo virtually all natural ecosystem processes. Labeling can be done by connecting a vial of sodium [^{14}C]bicarbonate mixed with an acid to the chamber inlet gas line. First, the CO_2 concentration in the monolith headspace is artificially increased two to three times. The ventilation is then looped via a nondestructive IR analyzer (air from the chamber is pumped through the analyzer and then back to the chamber). Next the vial containing sodium [^{14}C]bicarbonate in acid solution is connected to the gas line loop, and air is bubbled through it for 1–2 min to ensure that all ^{14}C enters the gas phase. The vial is then disconnected, but the gas circulation loop is maintained for 60–90 min, while the CO_2 concentration in the headspace decreases due to photosynthesis; however, it should not fall down to the level when the plants become CO_2 limited, as it was set artificially high in the beginning. Headspace air samples are taken during labeling, one immediately after the injection and a few during the closed loop circulation. These samples are analyzed for ^{14}C–CO_2 to monitor its consumption; the last sample being taken just before the monolith is connected back to the flow-through ventilation system. Monitoring the decrease in total CO_2 at the same time allows resolving the true photosynthetic fixation rate of ^{14}C–CO_2 and ^{12}C–CO_2 over the labeling period.

The setup to monitor ^{14}C–CO_2 and ^{14}C–CH_4 emission must be ready before the actual labeling and activated as soon as the labeled monolith is returned to a flow-through regime. An example of such setup is shown in Figure 5 [see also *Christensen et al.*, 2003; *Ström et al.*, 2003, 2005]. From the main gas outflow of the monoliths (see section 2.2 and Figure 2), a subflow of about 100 mL/min is taken by a separate pump with flow regulator. This subflow is passed through traps with alkali solution (e.g., 0.1 M NaOH), then through a furnace and a second set of alkali traps. In the first trap, all

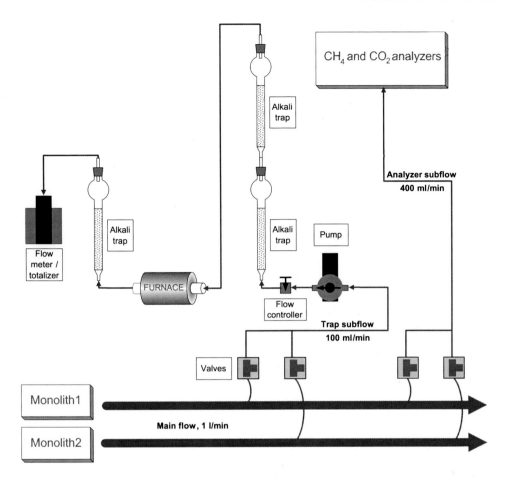

Figure 5. Setup for ^{14}C–CO_2 and ^{14}C–CH_4 emissions analysis.

the CO_2, including ^{14}C–CO_2 is fixed; in the furnace (CuO catalyzer, 850°C), CH_4 is oxidized to CO_2, this CO_2 being fixed in the second trap. If the system is correctly regulated, ^{14}C from CO_2 will be accumulated in the first trap and ^{14}C from CH_4 in the second. During the initial days after labeling, ^{14}C–CO_2 emission may be two to three orders of magnitude higher then ^{14}C–CH_4. In this situation, even <1% of CO_2 escaping through the first trap can find its way to the CH_4 trap and be mistaken for CH_4. To prevent this, two CO_2 traps can be used consecutively. The ratio between ^{14}C fixed in the first and second CO_2 traps shows the efficiency of the trapping. Based on this ratio, the final amount of CO_2 escaping through the second trap can be estimated as well and is usually found to be negligible.

The efficiency of CH_4 oxidation in the furnace should also be checked prior to labeling. The traps should be regularly refilled and ^{14}C determined in aliquots by the scintillation technique. Since ^{14}C–CO_2 emitted to the chamber headspace can be re-fixed by plants, it is recommended that traps be changed at least twice daily, once during light hours and once during darkness. The ^{14}C–CO_2 emitted in the dark will show the true emission, while the difference between light and dark emissions represents refixation. For the accurate calculation of ^{14}C emissions, all flow rates and volumes should be monitored as accurately as possible: the main flow rate through the monolith headspace chamber, subflow to the traps (or the exact amount of gas passed through the traps), volume of alkali solution in each trap, and the volume of aliquots counted for ^{14}C. If more than one monolith is labeled at the same time, each should have a separate trap system; alternatively, one trap system can be switched between two monoliths, e.g., on a daily basis. For example, six labeled monoliths can share three trap systems [*Lund et al.*, 2008]. ^{14}C can also be determined in subsurface liquid samples in the form of DOC, dissolved CO_2, and dissolved CH_4. To provide all necessary analyses using only 1 mL of liquid sample, a fairly complex routine has to be designed, such as the one shown in Figure 6.

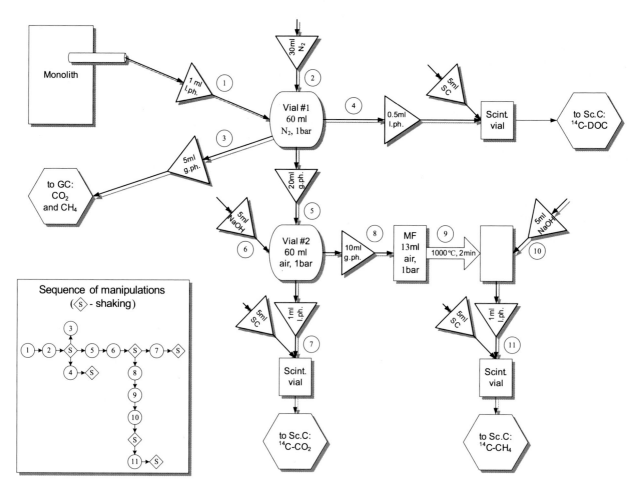

Figure 6. Scheme for complex analysis of subsurface samples. l.ph., liquid phase; g.ph., gas phase; GC, gas chromatograph; Sc.C, scintillation counter; MF, microfurnace; SC, scintillation cocktail.

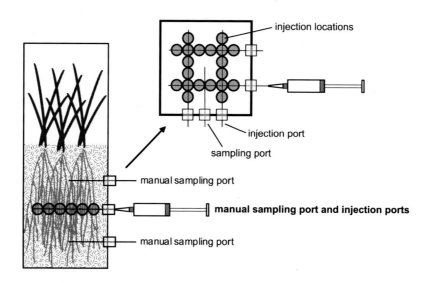

Figure 7. Example of ^{14}C-acetate labeling setup. Adapted from *Ström et al.* [2003], copyright 2003 John Wiley and Sons, Inc.

6.2. The ^{14}C-Acetate Labeling

To document the presence of acetoclastic methanogenesis and more closely study the second half of the CH_4 chain, ^{14}C-acetate impulse labeling can be used. Labeled acetate should be injected into the monolith's peat water with care taken not to alter the total natural acetate concentration. The labeled carbon should then pass through the natural processes, which can be examined by following the ^{14}C in DOC, CO_2 and CH_4 [*Ström et al.*, 2003, 2005]. Before labeling, the natural acetate concentrations in the subsurface profile are monitored and the appropriate depth and concentration for labeling determined. The solution of ^{14}C-acetate is prepared according the natural pH and flushed with N_2 to remove the dissolved oxygen [*Ström et al.*, 2005]. During labeling, the solution is distributed over a specific horizon, for example, as shown in Figure 7. The setup for ^{14}C monitoring is generally the same as described above for ^{14}C–CO_2 labeling.

7. CONCLUSION

This chapter has described a range of methods for detailed controlled environment studies of processes affecting net emissions of CH_4 from peatland environments. These elements include quantification of (1) methanogenic production versus methanotrophic oxidation, (2) relative contribution of different methanogenic pathways, (3) the importance of rhizospheric processes including root exudation for methanogenic substrate provision, (4) the time lag between photosynthetic carbon assimilation and CH_4 emission, and (5) the relative proportion of different emission pathways: plant mediated transport, diffusion, and ebullition.

These applications are all valuable in their own right but even more so when they are complementing field-based atmospheric flux measurements quantifying landscape scale emissions of CH_4 from peatlands. In such integrated studies, a comprehensive understanding of ecosystem-atmosphere exchanges may be obtained ranging from the detailed process-oriented understanding obtained through the type of approaches presented here to the landscape scale flux measurements, which are now widely applied in peatland ecosystems.

Acknowledgments. Lena Ström, Angela Stewart, Magnus Lund, Albert Kölbener, Anna Ekberg, Nicolai Panikov are all acknowledged for contributing to the laboratory studies over the past decade that we have summarized the methods used in here. Numerous funding agencies have supported these studies including the European Union (FP4, FP5), the Swedish Research Councils, the Swedish Research Council for Environment, Agricultural Sciences and Spatial Planning (FORMAS), and the Crafoord Foundation.

REFERENCES

Beckmann, M., and D. Lloyd (2001), Mass spectrometric monitoring of gases (CO_2, CH_4, O_2) in a mesotrophic peat core from Kopparas Mire, Sweden, *Global Change Biol.*, 7(2), 171–180.

Benstead, J., and D. Lloyd (1994), Direct mass spectrometric measurement of gases in peat cores, *FEMS Microbiol. Ecol.*, 13(3), 233–240.

Billings, W. D., J. O. Luken, D. A. Mortensen, and K. M. Peterson (1982), Arctic tundra: A source or sink for atmospheric carbon dioxide in a changing environment, *Oecologia*, 53, 7–11.

Bousquet, P., et al. (2006), Contribution of anthropogenic and natural sources to atmospheric methane variability, *Nature*, 443, 439–443.

Castro, H., A. Ogram, and K. R. Reddy (2004), Phylogenetic characterization of methanogenic assemblages in eutrophic and oligotrophic areas of the Florida Everglades. *Appl. Environ. Microbiol.*, 70, 6559–6568.

Chasar, L. S., J. P. Chanton, P. H. Glaser, and D. I. Siegel (2000), Methane concentration and stable isotope distribution as evidence of rhizospheric processes: Comparison of a fen and bog in the glacial lake Agassiz peatland complex, *Ann. Bot. (London)*, 86, 655–663.

Christensen, T. R. (1993), Methane emission from Arctic tundra, *Biogeochemistry*, 21, 117–139.

Christensen, T .R., T. Friborg, M. Sommerkorn, J. Kaplan, L. Illeris, H. Soegaard, C. Nordstroem, and S. Jonasson (2000), Trace gas exchange in a high-Arctic valley 1: Variations in CO_2 and CH_4 flux between tundra vegetation types, *Global Biogeochem. Cycles*, 14, 701–713.

Christensen, T. R., D. Lloyd, B. Svensson, P. J. Martikainen, R. Harding, H. Óskarsson, T. Friborg, H. Soegaard, and N. Panikov (2002), Biogenic controls on trace gas fluxes in northern wetlands, *Global Change Newsl.*, 51, 9–15.

Christensen, T. R., N. Panikov, M. Mastepanov, A. Joabsson, A. Stewart, M. Öquist, M. Sommerkorn, S. Reynauld, and B. Svensson (2003), Biotic control on CO_2 and CH_4 exchange in wetlands—A closed environment study, *Biogeochemistry*, 64, 337–354.

Comas, X., and L. Slater (2007), Evolution of biogenic gases in peat blocks inferred from noninvasive dielectric permittivity measurements, *Water Resour. Res.*, 43, W05424, doi:10.1029/2006WR005562.

Daulat, W. E., and R. S. Clymo (1998), Effects of temperature and watertable on the efflux of methane from peatland surface cores, *Atmos. Environ.*, 32(19), 3207–3218.

Glaser, P. H., P. J. Morin, N. Tsekos, D. I. Siegel, D. O. Rosenberry, J .P. Chanton, and A. S. Reeve (2006), The transport of free-phase methane in hydraulically confined and unconfined peat strata analyzed by magnetic resonance imaging (MRI), *Eos Trans.*, AGU, 87(52), Fall Meet. Suppl., Abstract B31D-05.

Gorham, E. (1991), Northern peatlands: Role in the carbon cycle and probable responses to climatic warming, *Ecol. Appl.*, 1, 182–195.

IPCC (2007), *Climate Change 2007: The Physical Science Basis*, edited by S. Solomon, et al., Cambridge Univ. Press, Cambridge, U. K.

Kellner, E., J. M. Waddington, and J. S. Price (2005), Dynamics of biogenic gas bubbles in peat: Potential effects on water storage and peat deformation, *Water Resour. Res.*, *41*, W08417, doi:10.1029/2004WR003732.

Kettridge, N., and A. Binley (2008), X-ray computed tomography of peat soils: Measuring gas content and peat structure, *Hydrol. Processes*, *22*(25), 4827–4837.

Kramer, H., and R. Conrad (1993), Measurement of dissolved H_2 concentrations in methanogenic environments with a gas diffusion probe, *FEMS Microbiol. Ecol.*, *12*(3), 149–158.

Lund, M., M. Mastepanov, T. R. Christensen, and L. Ström (2008), Nutrient availability and carbon cycling in a subarctic wetland— A pulse labeling experiment, *Eos Trans., AGU*, *89*(53), Fall Meet. Suppl., Abstract B13A-0411.

MacDonald, G. M., D. W. Beilman, K. V. Kremenetski, Y. W. Sheng, L. C. Smith, and A. A. Velichko (2006), Rapid early development of circumarctic peatlands and atmospheric CH_4 and CO_2 variations, *Science*, *314*, 285–288.

Mastepanov, M. (2004), Kinetics of gas exchange in *Sphagnum* bog profile (in Russian), Ph.D. dissertation, 174 pp., Moscow State Univ., Moscow.

Mastepanov, M., and T. R. Christensen (2008), Bimembrane diffusion probe for continuous recording of dissolved and entrapped bubble gas concentrations in peat, *Soil Biol. Biochem.*, *40* (12), 2992–3003

Matthews, E., and I. Fung (1987), Methane emission from natural wetlands: Global distribution, area, and environmental characteristics of sources, *Global Biogeochem. Cycles*, *1*, 61–86.

McAuliffe, C. (1971), GC determination of solutes by multiple phase equilibration, *Chem. Technol.*, *1*, 46–51.

Moore, T. R., and R. Knowles (1989), The influence of water table levels on methane and carbon dioxide emissions from peatland soils, *Can. J. Soil Sci.*, *69*, 33–38.

Morrissey, L. A., D. B. Zobel, and G. P. Livingston (1993), Significance of stomatal controls on methane release from Carex-dominated wetlands, *Chemosphere*, *26*, 339–355.

Prather, M., et al. (2001), Atmospheric chemistry and greenhouse gases, in *Climate Change 2001: The Scientific Basis*, edited by J. T. Houghton et al., pp. 239–287, Cambridge Univ. Press, Cambridge, U. K.

Romanowicz, E. A., D. I. Siegel, J. P. Chanton, and P. H. Glaser (1995), Temporal variations in dissolved methane deep in the Lake Agassiz Peatlands, Minnesota, *Global Biogeochem. Cycles*, *9*, 197–212.

Rosenberry, D. O., P. H. Glaser, and D. I. Siegel (2006) The hydrology of northern peatlands as affected by biogenic gas: Current developments and research needs, *Hydrol. Processes*, *20*, 3601–3610.

Rothfuss, F., and R. Conrad (1994), Development of a gas diffusion probe for the determination of methane concentrations and diffusion characteristics in flooded paddy soil, *FEMS Microbiol. Ecol.*, *14*(4), 307–318.

Schimel, J. P. (1995), Plant transport and methane production as controls on methane flux from arctic wet meadow tundra, *Biogeochemistry*, *28*, 183–200.

Sheppard, S. K., and D. Lloyd (2002), Direct mass spectrometric measurement of gases in soil monoliths, *J. Microbiol. Methods*, *50*(2), 175–188.

Slater, L., X. Comas, D. Ntarlagiannis, and M. R. Moulik (2007), Resistivity-based monitoring of biogenic gasses in peat soils, *Water Resour. Res.*, *43*, W10430, doi:10.1029/2007WR006090.

Ström, L., A. Ekberg, M. Mastepanov, and T. R. Christensen (2003), The effect of vascular plants on carbon turnover and methane emissions from a tundra wetland, *Global Change Biol.*, *9*, 1185–1192.

Ström, L., M. Mastepanov, and T. R. Christensen (2005), Species-specific effects of vascular plants on carbon turnover and methane emissions from wetlands, *Biogeochemistry*, *75*(1) 65–82.

Svensson, B. H. (1984), Different temperature optima for methane formation when enrichments from acid peat are supplemented with acetate or hydrogen, *Appl. Environ. Microbiol.*, *48*, 389–394.

Thomas, K. L., J. Benstead, S. H. Lloyd, and D. Lloyd (1998), Diurnal oscillations of gas production and effluxes (CO_2 and CH_4) in cores from a peat bog, *Biol. Rhythm Res.*, *29*(3), 247–259.

Tokida, T., T. Miyazaki, M. Mizoguchi, and K. Seki (2005), In situ accumulation of methane bubbles in a natural wetland soil, *Eur. J. Soil Sci.*, *56*(3), 389–396.

Walter, B. P., M. Heimann, and E. Matthews (2001a), Modeling modern methane emissions from natural wetlands: 1. Model description and results, *J. Geophys. Res.*, *106*(D24), 34,189–34,206.

Walter, B. P., M. Heimann, and E. Matthews (2001b), Modeling modern methane emissions from natural wetlands 2. Interannual variations 1982–1993, *J. Geophys. Res.*, *106*(D24), 34,207–34,219.

Walter, K. M., S. A. Zimov, J. P. Chanton, D. Verbyla, and F. S. Chapin III (2006), Methane bubbling from Siberian thaw lakes as a positive feedback to climate warming, *Nature*, *443*, 71–75.

Whiting, G. J., and J. P. Chanton (1993), Primary production control of methane emission from wetlands, *Nature*, *364*, 794–795.

Wilhelm, E., R. Battino, and R. J. Wilcock (1977), Low-pressure solubility of gases in liquid water, *Chem. Rev.*, *77*(2), 219–262.

Zimov, S. A., E. A. G. Schuur, and F. S. Chapin III (2006), Permafrost and the global carbon budget, *Science*, *312*, 1612–1613.

T. R. Christensen and M. Mastepanov, GeoBiosphere Science Centre, Physical Geography and Ecosystems Analysis, Lund University, Sölvegatan 12, SE-22362 Lund, Sweden. (Mikhail. Mastepanov@nateko.lu.se)

Physical Controls on Ebullition Losses of Methane From Peatlands

Takeshi Tokida,[1,2] Tsuyoshi Miyazaki,[3] and Masaru Mizoguchi[4]

Recent studies indicate that direct escaping of CH_4-containing gas bubbles, i.e., ebullition, plays a considerable role in determining the total CH_4 emission from peatlands into the atmosphere. Although methane is a biological product, a large bubble-storage capacity of peat leads to a partial decoupling between the production and release of methane, allowing for physical factors to act as a trigger for the ebullition. Buoyancy-induced ebullition can be controlled by (1) atmospheric pressure, (2) peat temperature, and (3) water table level. Falling atmospheric pressure exerts a dominant role in determining the timing of ebullition in some peatlands. Rapid rise in water table position by rain would result in the suppression of the bubble volume, and hence, halting ebullition. Diurnal temperature modulation might affect ebullition; however, its significance is expected to depend heavily on the position of water table, thermal characteristics of the peat, and the depth distribution of the CH_4-containing bubbles. Wind-induced surface turbulence also gives rise to ebullition as demonstrated by eddy covariance studies. Another type of ebullition includes a release of entrapped CH_4 accumulated during winter at spring-thaw period, but its significance is largely unknown. Further technical development is necessary to examine recently suggested massive CH_4 ebullition ($>g\ CH_4\ m^{-2}\ h^{-1}$) in terms of surface flux monitoring. Future research also needs to address subsurface behavior of the bubbles in relation to physical characteristics of peat and other transport modes.

[1]Agro-Meteorology Division, National Institute for Agro-Environmental Sciences, Tsukuba, Japan.

[2]Japan Society for the Promotion of Science, Tokyo, Japan.

[3]Department of Biological and Environmental Engineering, Graduate School of Agricultural and Life Sciences, University of Tokyo, Tokyo, Japan.

[4]Interfaculty Initiative in Information Studies, University of Tokyo, Tokyo, Japan.

Carbon Cycling in Northern Peatlands
Geophysical Monograph Series 184
Copyright 2009 by the American Geophysical Union.
10.1029/2008GM000805

1. INTRODUCTION

Ebullition, i.e., episodic bubbling, is attracting growing attention as a significant transport mechanism of methane (CH_4) from peatland into the atmosphere. In spite of apparent water saturation, recent research has revealed that tremendous gas bubbles are trapped in peat matrix underneath the water table level [e.g., *Tokida et al.*, 2005a]. The bubble CH_4 inventory is generally greater than the dissolved inventory due to the low solubility of CH_4 [e.g., *Tokida et al.*, 2005a; *Strack and Waddington*, 2008]. Compared to the other transport modes, i.e., diffusive transport of dissolved-form CH_4 [e.g., *Chasar et al.*, 2000] and release of CH_4 through vascular plants [e.g., *Shannon et al.*, 1996; *Bellisario et al.*, 1999; *Frenzel and Karofeld*, 2000], the rate of ebullition might exhibit much greater spatiotemporal variation as it occurs episodically both in space and time. However, at present, most estimates of CH_4 emissions are based upon infrequent (commonly weekly to monthly intervals) and spatially discontinuous observations which might be insufficient to capture episodic ebullitions. Knowledge of environmental factors that regulate ebullition would be useful to design a better observation strategy as well as to improve process-based models that can simulate the CH_4 cycle under varying environmental conditions.

To date, many researchers have examined variations in CH_4 emission from biochemical aspects, presumably because CH_4 is a biological product in wetland ecosystems. In particular, temperature [e.g., *Dise et al.*, 1993], water table position [e.g., *Waddington et al.*, 1996], and substrate availability [e.g., *Whiting and Chanton*, 1993; *Chanton et al.*, 1995; *Christensen et al.*, 2003a] have attracted great attention because they can serve to limit net CH_4 production. These studies greatly improve our knowledge related to CH_4 cycle in peatland. Nevertheless, substantial variation in CH_4 emission rates persists that cannot be explained from a biochemical viewpoint alone [e.g., *Daulat and Clymo*, 1998]. Given that the biochemical factors are unlikely to explain short-term variation in emission rates that ebullition might bring about, environmental controls on ebullition must be understood also from physical perspectives.

In this section, we review and discuss the recent findings on gas-phase (free-phase) CH_4 storage in and release from peatlands with a particular emphasis on environmental control on ebullition losses from a physical perspective. Issues and concerns related to these subjects will also be discussed to clarify potential future research needs.

2. IMPORTANCE OF GAS-PHASE CH_4 STORAGE IN RELATION TO PHYSICAL CONTROL ON EBULLITION

Physical control of ebullition can take place if decoupling occurs between net CH_4 production and the CH_4 emission rate. Presuming that produced CH_4 evolves to the atmosphere instantaneously, only factors that affect net CH_4 production should be a relevant determinant for ebullition. Conversely, if storage of CH_4 in peat sediment serves as an important buffer, release of CH_4 might not have a close link with production and might be controlled by changing environmental parameters, especially in short time scales [*Fechner-Levy and Hemond*, 1996].

Abundance of CH_4 storage in peatlands depends not only on CH_4 concentration, but also on the volume of the gas phase (bubbles). Presuming that there is a unit volume of peat where gas phase as well as liquid and solid phase occurs,

methane might exist both in the gas and liquid phases. Under equilibrium conditions, the fraction of gas-form storage to the total storage (gas form + solution form) can be expressed as a function of the volumetric ratio of air to water content (x) as shown in equation (1).

$$r = \frac{x}{x + \beta}. \qquad (1)$$

In that equation, the Ostwald coefficient, β, is defined as the inverse of dimensionless Henry's constant from equation (3) and is thus equivalent to the ratio of the liquid-phase concentration to the gas-phase concentration. Figure 1 clearly illustrates that the fraction of the gas form CH_4 increases dramatically with increased x; for instance, at 20°C, more than 60% of CH_4 exists in the gas phase, even if the volume of the gas phase is only one twentieth that of the water phase (i.e., $x = 0.05$). This presents a sharp contrast to CO_2, which is a rather soluble gas species (Figure 1). Recent field investigations suggested the volumetric gas content to be approximately 10% (see Table 1 in *Rosenberry et al.*, 2006), suggesting the volumetric ratio of air content to water content (x) as >0.12 (given the solid content to be approximately 10%). Consequently, from Figure 1, the main inventory of CH_4 can be seen to originate in gaseous form, not the dissolved form. The dominance of bubble CH_4 storage itself underscores the potential importance of bubbling release for the CH_4 transport mechanisms.

Turnover of CH_4 in peatlands can be a good indicator to evaluate the degree of decoupling between the production and the release, providing information to evaluate whether

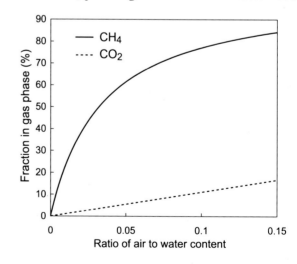

Figure 1. Fraction of gas-phase storage to the total (gas form + solution form) as a function of the volumetric ratio of air to water content at 20°C, as calculated from equation (1). We assumed that the bubble behaves as an ideal gas and that equilibrium between the gas phase and solution phase is established according to Henry's law.

or not physical factors can control the events of ebullition. *Tokida et al.* [2005a] have estimated the abundance of CH_4 down to 1 m depth as approximately 20–34 g CH_4 m^{-2} in a Japanese bog with a gaseous contribution of approximately 60–70%. Meanwhile, the rate of steady CH_4 emission (excluding episodic releases) was approximately 1 mg CH_4 m^{-2} h^{-1}, even during high summer [*Tokida et al.*, 2007a]. Therefore, the average CH_4 turnover is calculated as being on the order of several hundred days or even years if ebullient flushes are not considered. *Strack and Waddington* [2008] estimated a mean residence time of CH_4 at a poor fen in Québec as 28–120 days for which calculation ebullition fluxes were examined. Accordingly, physical environmental factors are likely to serve as important controllers of ebullitions because they modulate in time scales that are much shorter than such a long residence time.

3. AN APPROACH FOR ASSESSING EFFECTS OF CHANGING TEMPERATURE, WATER TABLE, AND ATMOSPHERIC PRESSURE ON THE GAS VOLUME AND BUOYANCY

Buoyancy is the primary driving force for bubbles to move upward through the peat for the eventual ebullition. From this standpoint, we specifically examine the environmental factors which affect buoyancy forces. Presuming that a bubble in peat exists below the water table level, the upward buoyancy force of the bubble might balance with forces that work collectively downward at the bubble-peat boundary when the bubble is trapped. The bubble can migrate upward when the balance fails, i.e., when the upward buoyancy surpasses the downward force. For each bubble, the buoyancy force is related directly to its volume. For that reason, it seems reasonable to infer that a certain threshold volume exists, beyond which the buoyancy moves the bubble upward [*Baird et al.*, 2004].

The trapped bubble size might depend not only on biological gas production but also on physicochemical parameters, such as gas solubility, pressure regime, and temperature. In peatlands, the peat temperature might vary according to diurnal and seasonal cycles. Fluctuations in the water table level and atmospheric pressure can also affect the volume through a changing pressure regime. Therefore, these three parameters, peat temperature, water table position, and atmospheric pressure, are candidates as physically controlling factors of the bubble size or buoyancy and eventual ebullition.

Below is a method for the quantitative evaluation of the effects of changing variables on the bubble volume. It is noteworthy that, in the following calculation, "gas phase" and "bubbles" are used in a similar sense: trapped bubbles in peat are considered together as if they formed a single gas phase. We acknowledge that simulating the actual behavior of an individual gas bubble requires a more complex assumption. Nevertheless, considering the lack of knowledge related to the detailed geometry of the trapped bubbles, our simplification might be an educated guess that markedly eases calculation.

Given that the ideal gas law is valid for the gas-phase gas in peat beneath the water table level, we can simply write the following expression,

$$P_i V_g = n_{g,i} RT, \qquad (2)$$

where P_i is the partial pressure of gas species i, V_g is the volume of the gas, $n_{g,i}$ represents the moles of gas species i in the gas, R is the universal gas constant, and T is the absolute temperature. Because the bubbles occur under the water table position, not only bubble-form gases but also dissolved-form gases must exist. Simply, a similar equation may be valid for the dissolved-form gas.

$$\frac{P_i}{H_i} V_w = n_{w,i} RT. \qquad (3)$$

Therein, H_i is the dimensionless Henry's constant, defined as the ratio of the gas-phase concentration to the liquid-phase concentration of gas species i. In addition, V_w is the volume of the water surrounding the bubble in a unit space, and $n_{w,i}$ is the moles of the dissolved-state gas. Actually, H_i itself might have a considerable dependence on temperature. Regarding the formulation of pressure, we can assume that the sum of internal partial pressures equilibrates with the pressure that acts on the bubble, which in turn, is equal to the sum of atmospheric pressure (P_a) and hydrostatic pressure (P_h).

$$\sum P_i = P_a + P_h. \qquad (4)$$

In other words, atmospheric pressure and the water table position can affect the bubble volume through controlling the total pressure of the bubbles.

Given the lack of production and consumption of any gas species (we here specifically address physical processes), we might regard that the total moles of the gas species i ($C_i \overset{\text{def}}{=} n_{g,i} + n_{w,i}$) do not change with time (here, transport of the gas is not examined). Consequently, from equations (2) and (3), the following equation might be obtained.

$$\frac{P_i V_g}{RT} + \frac{P_i V_w}{H_i RT} = C_i = \text{const.} \qquad (5)$$

Equation (5) suggests that once a set of C_i (C_i should be measured for each gas species) is obtained under a specific condition, we can estimate change in V_g or new bubble size under varying thermal and pressure conditions, as either of the following:

$$\frac{P_i' V_g'}{RT'} + \frac{P_i' V_w'}{H_i' RT'} = C_i \qquad (6)$$

or

$$P_i' = C_i RT' \left(V_g' + \frac{V_w'}{H_i'} \right)^{-1}. \qquad (7)$$

Therein, prime marks signify new environmental conditions. Relative change in the water volume (dV_w) is likely to be small. Therefore, V_w can be treated as constant ($V_w = V_w'$) for simplicity. Solving the simultaneous equation (7) under the constraint of total pressure [equation (4)] will provide a new bubble size (V_g') and partial pressure of each gas species (P_i').

Under specific conditions, relations between bubble volume and temperature or pressures can be shown more explicitly. If temperature is fixed, a "pressure-volume" relationship can be ascertained by differentiation of equation (5) with respect to pressure, yielding,

$$\frac{dV_g}{dP_i} = -\frac{V_g}{P_i} - \frac{V_w}{P_i H_i}, \qquad (8)$$

which is identical to equation 2a of *Fechner-Levy and Hemond* [1996] and equation 1 of *Tokida et al.* [2005b]. Regarding the "temperature-volume" relationship, differentiation of equation (5) with respect to temperature yields the equation below.

$$\frac{dV_g}{dT} = \left(\frac{V_g}{T} + \frac{V_w}{H_i T} \right) + \frac{V_w}{H_i^2} \frac{dH_i}{dT}. \qquad (9)$$

It is identical to equation 6b of *Fechner-Levy and Hemond* [1996] and equation 2 of *Kellner et al.* [2006], but it uses different notation. It is noteworthy that equation (9) assumes that P_i is not a function of T. This assumption is valid for a single gas species. However, if a mixture of gas species is considered, such might not be the case. The solubility (H_i) and its temperature dependency might vary among the gas species. Therefore, P_i might change with T as repartitioning of gases occur between the dissolved and the gas phases. An exact solution is obtainable by solving equation (6) or (7) directly, although they are implicitly defined.

4. PHYSICAL FACTORS CONTROLLING EBULLITION

In this subsection, we review and discuss previous investigations of the effects of physical parameters on ebullition. Considering the importance of the buoyancy as a driving force of the ebullition as described above, we specifically examine atmospheric pressure (section 4.1), temperature (section 4.2), and water table elevation (section 4.3). Ebullition of other types (as independent of buoyancy-driven ebullition) is described later (sections 4.4 and 4.5).

4.1. Atmospheric Pressure

Fechner-Levy and Hemond [1996] provided a conceptual discussion on the role of atmospheric pressure by using equation (8), suggesting that rising atmospheric pressure (10 hPa d^{-1}) is expected to have a capacity to arresting bubble volume growth by microbial production, thereby halting ebullition. Conversely, periods of decreasing atmospheric pressure are expected to engender enhanced bubble growth and thereby trigger ebullition.

Observational evidence of the importance of atmospheric pressure has been reported from both laboratory and field studies. *Tokida et al.* [2005b] conducted a laboratory experiment where the intact peat column was subjected to natural air pressure fluctuations, while the temperature and water table were controlled and kept constant. Results show that both moderate and steep drops in atmospheric pressure can trigger CH$_4$ ebullitions from an intact peat column (Figure 2) and that the amount of released bubbles was identical to the increased gas volume caused by decreased pressure [*Tokida et al.*, 2005b]. Unlike the laboratory experiments, peat temperature and water table also change with time under field conditions. They can also affect the volume of the bubbles. In a 90-h intensive field campaign [*Tokida et al.*, 2007a], events of ebullition were found during or shortly after falling air pressure phases (Figure 3a). Further analysis of the effects of changing variables on the trapped bubble volume in peat was done by solving simultaneous equation (7) (refer to *Tokida et al.*, 2007a, for details). Results showed a close link between the ebullition and the increased gas volume caused by reduction in the air pressure with no correlations with any other parameters (Figure 3b), thereby confirming the idea that fluctuations in the atmospheric pressure played a dominant role in determining the timing and magnitude of the ebullition events.

Other flux measurements, both in the field [*Shurpali et al.*, 1993] and laboratory [*Moore and Dalva*, 1993], also revealed episodic emissions with concurrent reductions in the atmospheric pressure. The importance of atmospheric pressure is also evidenced by several independent studies, all of which quantified change in gas-phase storage using different methodologies [*Rosenberry et al.*, 2003; *Glaser et al.*, 2004; *Strack et al.*, 2005; *Comas et al.*, 2008]. Irrespective of the method, these studies observed decrease in the gas-phase storage associated with drops in atmospheric pressure, clearly indicating the occurrence of ebullition with falling barometric pressure. Recently, *Sachs et al.* [2008] demon-

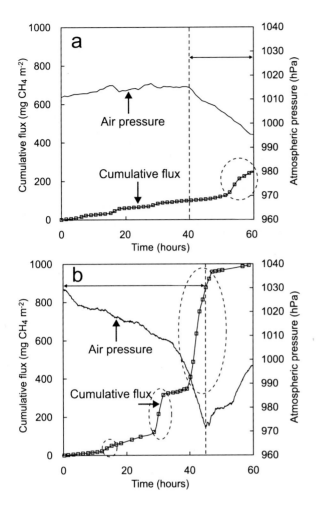

Figure 2. Cumulative methane flux is shown with a change in atmospheric pressure during two independent measurement periods (adopted from Figure 2 of *Tokida et al.* [2005b]). Circles indicate a rapid increase in cumulative emissions and events of CH_4 ebullition. Periods of falling pressure are indicated by horizontal arrows: (a) moderate, and (b) steep pressure drops. Open squares present the timing of flux measurement by a closed chamber.

strated that inclusion of the air-pressure term markedly improves the performance of a model (proposed by *Wille et al.* [2008]) against an entire-season flux record measured using eddy covariance method at a Siberian polygonal tundra ecosystem (see also section 4.4).

4.2. Temperature

Temperature has long been recognized as an important determinant for microbial CH_4 production [e.g., *Dunfield et al.*, 1993]. However, its importance has rarely been investigated from a physical viewpoint. Here we try to evaluate the effect of temperature using conceptual and computational analyses.

Effects of rising temperature on the change in gas volume (ΔV_g) can be estimated by solving equation (7), as portrayed in Figure 4. For comparison, a pressure-induced effect is also presented. The following values are assumed and given to equation (7) for illustrative purposes: 45%, CH_4; 12%, CO_2; 43%, N; for the composition of the gas bubbles [*Tokida et al.*, 2005b]; volumetric ratio of V_g and V_w to be 10 and 80%, respectively; and $P_a + P_h = 106.4$ kPa $= 1.05$ atm. Hypothetical calculations were made to isolate the direct effect of temperature (T) on V_g (denoted as T alone in Figure 4) as well as the effect of temperature dependency of Henry's constant (H_i; H alone) from the combined effect ($T + H$) by assuming that either H_i or T is kept constant. Figure 4 quantitatively portrays the potential importance of temperature on ΔV_g. Temperature elevation by 1°C has an almost identical effect

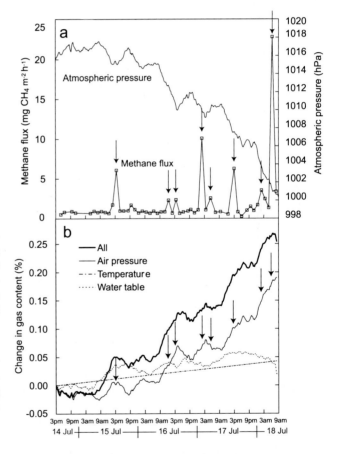

Figure 3. Time series of CH_4 flux and atmospheric pressure (a) and effects of changing atmospheric pressure, peat temperature, water table position, and all variables on the change in volumetric gas content (ΔV_g) during the corresponding time (b). The vertical arrows indicate the episodic fluxes, which are significantly greater ($P < 0.05$, by one-tailed t test) than the other emission rates (results of plot B in Figures 3 and 4 of *Tokida et al.* [2007a]).

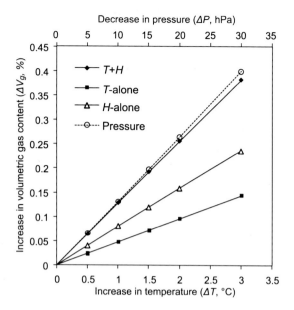

Figure 4. Effect of temperature elevation (bottom axis) in comparison with the effect of decreased pressure (top axis) on the increase in gas content (ΔV_g) computed from equation (7). Temperature dependence of Henry's constant (based on the solubility data of *Wilhelm et al.* [1977], *Battino* [1982], *Clever and Young* [1987]) is taken into consideration in cases of $T + H$ and H alone, which does not account for actual change in temperature (T) itself.

(in case of $T + H$) to that of a 10-hPa decrease in atmospheric pressure or a 10-cm drop in water table position (see the next part), which is likely to be sufficient to induce bubbling events. The temperature dependency of H_i (H alone) has a stronger effect than the direct effect of increased temperature (T alone). This indicates that degassing from aqueous to gas phase is the major cause of increased gas volume by temperature elevation, as also reported by *Fechner-Levy and Hemond* [1996].

In the actual field condition, peat soil is subjected to both diurnal and annual temperature fluctuations. Effects of these two cycles should be evaluated separately because they have different frequency and amplitude. The annual temperature cycle affects the bubble volume when considering seasonal-scale time period. Because of its deeper damping depth, the amplitude can reach deep anoxic layers, where a large amount of CH_4 might be trapped [e.g., *Brown et al.*, 1989]. However, because ebullition is a short-lived phenomenon, other mechanisms having a shorter time scale variation can take over the control of ebullition. In the intensive field study of *Tokida et al.* [2007a], where peat temperature increased following the annual cycle, atmospheric pressure exerted a dominant effect in determining the timing of ebullition, and temperature effect was of minor importance (Figure 3b).

Diurnal temperature variation might have greater influence on ebullition because it occurs more rapidly than the annual cycles. Reported diurnal variation in CH_4 emissions [e.g., *Suyker et al.*, 1996] might invoke potential correlation between the triggering of ebullition by diel temperature fluctuation. However, the amplitude of diel temperature fluctuation converges rapidly with peat depth because of very low thermal conductivity of the unsaturated moss surface, if present, in addition to the large heat capacity of subwater-table layers [*van der Molen and Wijmstra*, 1994]. Therefore, diurnal temperature variation below the water table position can become pronounced only where the water table is close to the peat surface. For that reason, we suggest that the effect of diurnal temperature change on ebullition, if any, is expected to depend heavily on the position of water table, thermal characteristics of the peat, and the depth distribution of CH_4-containing bubbles along the peat profile.

It is worth mentioning that when considering the amount of cumulative ebullition, temperature as a "biological" control should become important because it is natural to think that time-integrated CH_4 emission exhibits a good correlation with cumulative production, which, in part, is controlled by the mean temperature of the corresponding period [*Worthy et al.*, 2000; *Treat et al.*, 2007]. In other words, from a *biological* standpoint, longer-term mean temperature and its intraseasonal and interseasonal variations have large impacts on the cumulative emissions, meanwhile from a *physical* viewpoint, short-time variation is important as it determines the timing of each ebullition.

4.3. Water Table Elevation

The water table can affect ebullition by increasing or decreasing hydrostatic pressure in peat pore water. The amplitude of the water table oscillation might be comparable to that of atmospheric pressure: 1 hPa corresponds to ~1 cm of water. For that reason, it is reasonable to presume that drops in water table can be a trigger for ebullition in the same way as the air pressure [*Moore et al.*, 1990]. Meanwhile, the temporal pattern of water table position can differ considerably from that of atmospheric pressure. In general, lowering of water table takes a longer period during which atmospheric pressure can fluctuate several times [*Tokida et al.*, 2007a; *Sachs et al.*, 2008]. Consequently, although ebullitions are often recorded during a period of falling water table [*Moore and Dalva*, 1993; *Shurpali et al.*, 1993; *Glaser et al.*, 2004], drops in barometric pressure are likely to act as a direct trigger for the ebullition in most cases (Figure 3b) [*Tokida et al.*, 2007a; *Sachs et al.*, 2008]. On the other hand, suppression of ebullition because of increased hydrostatic pressure is likely during the rising water table phase because it occurs rapidly with rain events. *Frolking and Crill* [1994] observed a decrease in CH_4 flux for several days following rainfall.

It is noteworthy that changes in water table can engender other processes that are independent of the buoyancy-induced ebullition. First, a falling water table might cause release of CH_4 because trapped bubbles would be directly exposed to the atmosphere. Secondly, change in the water table will decisively affect the activity of methanogens (CH_4 production) and methanotrophs (CH_4 consumption) because it determines redox boundary in the peat column. Interestingly, "physical" versus "biological" functions of the water table might have opposite effects on CH_4 emissions: for instance, when the water table drops, enhancement of ebullitive CH_4 release is expected from a physical viewpoint, whereas suppression is likely from biological perspective because of enhanced CH_4 oxidation and decreased production [*Moore and Dalva*, 1993].

4.4. Surface Turbulence by Wind

Physical disturbance of the surface of peatlands can also trigger the release of bubbles that are unable to escape by their buoyancy alone. Wind-induced sediment disturbance is a known mechanism triggering ebullition in vegetation-free, open-water environments, such as reservoirs and lakes [*Keller and Stallard*, 1994; *Joyce and Jewell*, 2003]. Reports of a series of eddy covariance studies illustrate that this process has considerable importance also in peatland for controlling CH_4 ebullition. *Fan et al.* [1992] observed the wind speed dependence of CH_4 flux from subarctic tundra in Alaska if the footprint contained small lakes and ponds. In a Swedish peatland, a significant positive relationship between hourly mean of friction velocity and CH_4 flux was found during a summer period (during 5 days in late June and early July) although no correlation was apparent during the thawing period, suggesting that wind-induced surface turbulence can play an important role if surface roughness is high [*Friborg et al.*, 1997]. *Hargreaves et al.* [2001] also observed a close relation between CH_4 flux and momentum flux for periods up to a day in August in a Finnish appa mire with concurrent visual observation of bubbling at hummock/carpet and hummock/pool interfaces. *Wille et al.* [2008] found the friction velocity to be a second important factor for CH_4 flux during a snow-free period of 2003–2004 in a Siberian polygonal tundra. They speculated that wind-induced ebullition might have occurred especially during a period of high winds. *Sachs et al.* [2008] found that friction velocity was the single best parameter accounting 57% of variation in CH_4 flux observed at the same site of *Wille et al.* [2008], but in a different year.

All the examples noted above involve open-water areas, and we are not aware of studies reporting wind-induced CH_4 ebullition in peatlands having no open-water surfaces. Some studies have shown that wind turbulence can enhance trace gas effluxes from unsaturated mineral soils [*Arneth et al.*, 1998; *Takle et al.*, 2004]; however, it remains uncertain whether or not high wind affects rate of ebullition in wetlands where unsaturated layer occurs. Further research is needed to identify the specific hydrological or vegetative condition that is sensitive to wind-induced ebullition.

4.5. Melting Frozen Layer at Spring Thaw

Another type of ebullition was observed during a period of spring thaw [*Windsor et al.*, 1992; *Friborg et al.*, 1997; *Hargreaves et al.*, 2001; *Rinne et al.*, 2007; *Tokida et al.*, 2007b; *Comas et al.*, 2008]. This emission is attributable to the release of stored CH_4 in and beneath the surface frozen layer. Visual observations of bubbling from cracks of the thawing ice layer have been made [*Hargreaves et al.*, 2001; *Tokida et al.*, 2007b; *Comas et al.*, 2008]. To date, a few reported studies have estimated the contribution of spring-thaw CH_4 pulse to the annual total emission: 11% in a Kaamanen appa mire [*Hargreaves et al.*, 2001], and less than 3% in a minerotrophic fen at Siikaneva [*Rinne et al.*, 2007].

5. FUTURE RESEARCH NEEDS

Recent advancements related to gas-phase CH_4 dynamics in peatland have raised new questions that deserve future research effort. Some potential research directions that are relevant to physical processes of ebullition are described below.

5.1. Methodological Development of a Better Quantification of Ebullition Loss From Peatland

Accurate quantification of surface ebullition flux is a fundamental basis for identifying relevant environmental factors. Recent intensive field studies have revealed that the conventional closed chamber method can capture episodic ebullitions if sampling is made with sufficient frequency, i.e., at least every several hours [*Tokida et al.*, 2007a]. *Christensen et al.*, [2003b] obtained continuous CH_4 emission records by a steady-state chamber method from peat monoliths in a laboratory environment. Both studies observed spiky peaks which superimposed on the "steady emission." However, the problem is how one should set and justify criteria needed to separate ebullitions out from the other transport components. One method is to compare subjectively selected "episodic" releases with the other emissions by statistical analysis [*Windsor et al.*, 1989; *Tokida et al.*, 2007a]. Another option is to define the "steady emission" visually at a fixed flux rate [*Christensen et al.*, 2003]. Because any choice of method will be arbitrary, a simpler approach may be preferred. The most important thing is that one should ensure the full transparency of the method adopted [*Christensen et al.*, 2003]. Depending on the topography, the "inverted funnel method" is applicable in areas such as pools and inundated hollows [*Strack and Waddington*, 2008]; it is often used in open-water ecosystems [*Chanton and Whiting*, 1995], such as coastal marine basin [*Martens and Klump*, 1980], and lakes and ponds [*Weyhenmenyer*, 1999; *Casper et al.*, 2000]. This method may have some advantages over the chamber method because it can directly quantify ebullition without any assumptions. However, both the chamber and inverted funnel method are generally labor-intensive and spatially limited. Thereby, other methods are necessary for long-term and larger-scale flux monitoring. In addition, the possibility exists that the chamber method suffers from underestimation of wind-induced ebullition (see section 4.4) because enclosure by the chamber inherently eliminates surface turbulence [*Sachs et al.*, 2008].

Use of eddy covariance methods can overcome some issues described above because this method provides a noninvasive, spatially integrated, and near-continuous flux record. However, it remains unclear whether the eddy covariance method can capture all episodic ebullition or not because this method requires a stationary concentration of target gas over the averaging time [*Foken and Wichura*, 1996]. The CH_4 concentration in the air might be unstable and might not meet this time requirement if ebullition occurs episodically. Furthermore, this method might be unsuitable if bubbling release occurs patchily through discrete channels because horizontal homogeneity is another requirement for this method. Several studies have estimated the size of ebullition by investigating time course gas-form CH_4 storage accumulated in peats. They reported that several tens of grams of CH_4 per square meter (values exceeding annual total reported either by chamber or eddy covariance methods) are lost within several days or even hours [*Rosenberry et al.*, 2003; *Glaser et al.*, 2004; *Comas et al.*, 2008]. Further development in flux monitoring technique is necessary to assess such a massive CH_4 loss from surface observations. We suggest the micrometeorological mass balance method [*Harper et al.*, 1999; *Desjardins et al.*, 2004] as a promising approach because it requires no spatial heterogeneity or temporal stationarity.

5.2. Subsurface Behavior of Bubbles in Relation to Physical Characteristics of Peat and Other Transport Modes

A certain threshold bubble volume is believed to exist, beyond which the bubble can move upward through the peat by its buoyancy force [e.g., *Baird et al.*, 2004]; this threshold concept has been supported by several studies [e.g., *Tokida et al.*, 2005b]. Nevertheless, little is known about the exact value of the threshold itself or its potential variation among varying types of peat. Further investigations are necessary to relate the threshold value to, e.g., some mechanical characteristics of the peat medium. It is noteworthy that, because previous investigations have treated the entrapped bubbles as if they formed a single gas phase, it remains unknown how the bulk threshold gas content relates to the buoyancy force of the individual gas bubbles. Recent application of X-ray computed tomography scanning to an intact peat core has provided detailed information related to the size and configuration of trapped bubbles and fabric structure of peat [*Kettridge and Binley*, 2008]. Such an approach might improve our understanding on each bubble movement within the peat and ensuing ebullition.

Physical behavior of the bubbles within the peat must be investigated in relation to other transport processes. *Amos and Mayer* [2006] conducted sand column experiments showing a rapid vertical transport of CH_4 at rates several times faster than the more soluble tracers. Similar experiments using peat would support quantitative evaluation of bubble-form transfer through peat with simultaneous estimation of the diffusive transport. Studies of the potential effect of vegetation on the ebullition also deserve much attention because some vascular plants might act as a conduit of the trapped bubbles into the atmosphere. We speculate that there might be a trade-off relation between ebullition and plant-mediated transport.

Acknowledgments. We thank Tim Moore, Xavier Comas, and one anonymous reviewer for their thorough examination and constrictive comments for improving the earlier version of this manuscript.

REFERENCES

Amos, R. T., and K. U. Mayer (2006), Investigating ebullition in a sand column using dissolved gas analysis and reactive transport modeling, *Environ. Sci. Technol.*, *40*, 5361–5367.

Arneth, A., F. M. Kelliher, S. T. Gower, N. A. Scott, J. N. Byers, and T. M. McSeven (1998), Environmental variables regulating soil carbon dioxide efflux following clear-cutting of a *Pinus radiata* D. Don plantation, *J. Geophys. Res.*, *103*, 5695–5705.

Baird, A. J., C. W. Beckwith, S. Waldron, and J. M. Waddington (2004), Ebullition of methane-containing gas bubbles from near-surface *Sphagnum* peat, *Geophys. Res. Lett.*, *31*, L21505, doi:10.1029/2004GL021157.

Battino, R. (Ed.) (1982), *Nitrogen and Air, IUPAC Solubility Data Ser.*, vol. 10, Elsevier, New York.

Bellisario, L. M., J. L. Bubier, T. R. Moore, and J. P. Chanton (1999), Controls on CH_4 emissions from a northern peatland, *Global Biogeochem. Cycles*, *13*, 81–91.

Brown, A., S. P. Mathur, and D. J. Kushner (1989), An ombrotrophic bog as a methane reservoir, *Global Biogeochem. Cycles*, *3*, 205–213.

Casper, P., S. C. Maberly, G. H. Hall, and B. J. Finlay (2000), Fluxes of methane and carbon dioxide from a small productive lake to the atmosphere, *Biogeochemistry*, *49*, 1–19.

Chanton, J. P., and G. J. Whiting (1995), Trace gas exchange in freshwater and coastal marine environments: Ebullition and transport by plants, in *Biogenic Trace Gases: Measuring Emissions From Soil and Water*, edited by P. A.Matson and R. C. Harriss, pp. 98–125, Blackwell, Oxford, U. K.

Chanton, J. P., J. E. Bauer, P. A. Glaser, D. I. Siegel, C. A. Kelley, S. C. Tyler, E. H. Romanowicz, and A. Lazrus (1995), Radiocarbon evidence for the substrates supporting methane formation within northern Minnesota peatlands, *Geochim. Cosmochim. Acta*, *59*, 3663–3668.

Chasar, L. S., J. P. Chanton, P. H. Glaser, and D. I. Siegel (2000), Methane concentration and stable isotope distribution as evidence of rhizospheric processes: Comparison of a fen and bog in the Glacial Lake Agassiz Peatland complex, *Ann. Bot.*, *86*, 655–663.

Christensen, T. R., A. Ekberg, L. Ström, M. Mastepanov, N. Panikov, M. Öquist, B. H. Svensson, H. Nykänen, P. J. Martikainen, and H. Oskarsson (2003a), Factors controlling large scale variations in methane emissions from wetlands, *Geophys. Res. Lett.*, *30*(7), 1414, doi:10.1029/2002GL016848.

Christensen, T. R., N. Panikov, M. Mastepanov, A. Joabsson, A. Stewart, M. Öquist, M. Sommerkorn, S. Reynaud, and B. Svensson (2003b), Biotic controls on CO_2 and CH_4 exchange in wetlands—A closed environment study, *Biogeochemistry*, *64*, 337–354.

Clever, H. L., and C. L. Young (Eds.) (1987), *Methane, IUPAC Solubility Data Ser.*, vol. 27/28, Elsevier, New York.

Comas, X., L. Slater, and A. Reeve (2008), Seasonal geophysical monitoring of biogenic gases in a northern peatland: Implications for temporal and spatial variability in free phase gas production rates, *J. Geophys. Res.*, *113*, G01012, doi:10.1029/2007JG000575.

Daulat, W. E., and R. S. Clymo (1998), Effects of temperature and water table on the efflux of methane from peatland surface cores, *Atmos. Environ.*, *32*, 3207–3218.

Desjardins, R. L., O. T. Denmead, L. Harper, M. McBain, D. Masse, and S. Kaharabata (2004), Evaluation of a micrometeorological mass balance method employing an open-path laser for measuring methane emissions, *Atmos. Environ.*, *38*, 6855–6866.

Dise, N. B., E. Gorham, and E. S. Verry (1993), Environmental factors controlling methane emissions from peatlands in Northern Minnesota, *J. Geophys. Res.*, *98*, 10,583–10,594.

Dunfield, P., R. Knowles, R. Dumont, and T. R. Moore (1993), Methane production and consumption in temperate and subarctic peat soils: Response to temperature and pH, *Soil Biol. Biochem.*, *25*, 321–326.

Fan, S. M., S. C. Wofsy, P. S. Bakwin, D. J. Jacob, S. M. Anderson, P. L. Kebabian, J. B. McManus, C. E. Kolb, and D. R. Fitzjarrald (1992), Micrometeorological measurements of CH_4 and CO_2 exchange between the atmosphere and subarctic tundra, *J. Geophys. Res.*, *97*, 16,627–16,643.

Fechner-Levy, E. J., and H. F. Hemond (1996), Trapped methane volume and potential effects on methane ebullition in a northern peatland, *Limnol. Oceanogr.*, *41*, 1375–1383.

Foken, T., and B. Wichura (1996), Tools for quality assessment of surface-based flux measurements, *Agric. For. Meteorol.*, *78*, 83–105.

Frenzel, P., and E. Karofeld (2000), CH_4 emission from a hollow-ridge complex in a raised bog: The role of CH_4 production and oxidation, *Biogeochemistry*, *51*, 91–112.

Friborg, T., T. R. Christensen, and H. Søgaard (1997), Rapid response of greenhouse gas emission to early spring thaw in a subarctic mire as shown by micrometeorological techniques, *Geophys. Res. Lett.*, *24*, 3061–3066.

Frolking, S., and P. Crill (1994), Climate controls on temporal variability of methane flux from a poor fen in southeastern New Hampshire: Measurement and modeling, *Global Biogeochem. Cycles*, *8*, 385–397.

Glaser, P. H., J. P. Chanton, P. Morin, D. O. Rosenberry, D. I. Siegel, O. Ruud, L. I. Chasar, and A. S. Reeve (2004), Surface deformations as indicators of deep ebullition fluxes in a large northern peatland, *Global Biogeochem. Cycles*, *18*, GB1003, doi:10.1029/2003GB002069.

Hargreaves, K. J., D. Fowler, C. E. R. Pitcairn, and M. Aurela (2001), Annual methane emission from Finnish mires estimated from eddy covariance campaign measurements, *Theor. Appl. Climatol.*, *70*, 203–213, doi:10.1007/s007040170015.

Harper, L. A., O. T. Denmead, J. R. Freney, and F. M. Byers (1999), Direct measurements of methane emissions from grazing and feedlot cattle, *J. Anim. Sci.*, *77*, 1392–1401.

Joyce, J., and P. W. Jewell (2003), Physical controls on methane ebullition from reservoirs and lakes, *Environ. Eng. Geosci.*, *9*, 167–178.

Keller, M., and R. F. Stallard (1994), Methane emission by bubbling from Gatun Lake, Panama, *J. Geophys. Res.*, *99*, 8307–8319.

Kellner, E., A. J. Baird, M. Oosterwoud, K. Harrison, and J. M. Waddington (2006), Effect of temperature and atmospheric pressure on methane (CH_4) ebullition from near surface peats, *Geophys. Res. Lett.*, *33*, L18405, doi:10.1029/2006GL027509.

Kettridge, N., and A. Binley (2008), X-ray computed tomography of peat soils: Measuring gas content and peat structure, *Hydrol. Processes*, *22*, 4827–4837, doi:10.1002/hyp.7097.

Martens, C. S., and J. V. Klump (1980), Biogeochemical cycling in an organic-rich coastal marine basin I. Methane sediment-water exchange processes. *Geochim. Cosmochim. Acta*, *44*, 471–490.

Moore, T. R., and M. Dalva (1993), The influence of temperature and water-table position on carbon-dioxide and methane emissions from laboratory columns of peatland soils, *J. Soil Sci.*, *44*, 651–664.

Moore, T. R., N. Roulet, and R. Knowles (1990), Spatial and temporal variations of methane flux from subarctic/northern boreal fens, *Global Biogeochem. Cycles*, *4*, 29–46.

Rinne, J., T. Riutta, M. Pihlatie, M. Aurela, S. Haapanala, J.-P. Tuovinen, E.-S. Tuittila, and T. Vesala (2007), Annual cycle of methane emission from a boreal fen measured by the eddy covariance technique, *Tellus, Ser. B*, *59*, doi:10.1111/j.1600-0889.2007.00261.x.

Rosenberry, D. O., P. H. Glaser, D. I. Siegel, and E. P. Weeks (2003), Use of hydraulic head to estimate volumetric gas content and ebullition flux in northern peatlands, *Water Resour. Res.*, *39*(3), 1066, doi:10.1029/2002WR001377.

Rosenberry, D. O., P. H. Glaser, and D. I. Siegel (2006), The hydrology of northern peatland as affected by biogenic gas: Current developments and research needs, *Hydrol. Processes*, *20*, 3601–3610, doi:10.1002/hyp.6377.

Sachs, T., C. Wille, J. Boike, and L. Kutzbach (2008), Environmental controls on ecosystem-scale CH_4 emission from polygonal tundra in the Lena River Delta, Siberia, *J. Geophys. Res.*, *113*, G00A03, doi:10.1029/2007JG000505.

Shannon, R. D., J. R. White, J. E. Lawson, and B. S. Gilmour (1996), Methane efflux from emergent vegetation in peatlands, *J. Ecol.*, *84*, 239–246.

Shurpali, N. J., S. B. Verma, R. J. Clement, and D. P. Billesbach (1993), Seasonal distribution of methane flux in a Minnesota peatland measured by eddy correlation, *J. Geophys. Res.*, *98*, 20,649–20,655.

Strack, M., and J. M. Waddington (2008), Spatiotemporal variability in peatland subsurface methane dynamics, *J. Geophys. Res.*, *113*, G02010, doi:10.1029/2007JG000472.

Strack, M., E. Kellner, and J. M. Waddington (2005), Dynamics of biogenic gas bubbles in peat and their effects on peatland biogeochemistry, *Global Biogeochem. Cycles*, *19*, GB1003, doi:10.1029/2004GB002330.

Suyker, A. E., S. B. Verma, R. J. Clement, and D. P. Billesbach (1996), Methane flux in a boreal fen: Season-long measurement by eddy correlation, *J. Geophys. Res.*, *101*, 28,637–28,648.

Takle, E. S., W. J. Massman, J. R. Brandle, R. A. Schmidt, X. Zhou, I. V. Litvina, R. Garcia, G. Doyle, and C. W. Rice (2004), Influence of high-frequency ambient pressure pumping on carbon dioxide efflux from soil, *Agric. For. Meteorol.*, *124*, 193–206.

Tokida, T., T. Miyazaki, M. Mizoguchi, and K. Seki (2005a), In situ accumulation of methane bubbles in a natural wetland soil, *Eur. J. Soil Sci.*, *56*, 389–395, doi:10.1111/j.1365-2389.2004.00674.x.

Tokida, T., T. Miyazaki, and M. Mizoguchi (2005b), Ebullition of methane from peat with falling atmospheric pressure, *Geophys. Res. Lett.*, *32*, L13823, doi:10.1029/2005GL022949.

Tokida, T., T. Miyazaki, M. Mizoguchi, O. Nagata, F. Takakai, A. Kagemoto, and R. Hatano (2007a), Falling atmospheric pressure as a trigger for methane ebullition from peatland, *Global Biogeochem. Cycles*, *21*, GB2003, doi:10.1029/2006GB002790.

Tokida, T., M. Mizoguchi, T. Miyazaki, A. Kagemoto, O. Nagata, and R. Hatano (2007b), Episodic release of methane bubbles from peatland during spring thaw, *Chemosphere*, *70*, 165–171.

Treat, C. C., J. L. Bubier, R. K. Varner, and P. M. Crill (2007), Timescale dependence of environmental and plant-mediated controls on CH_4 flux in a temperate fen, *J. Geophys. Res.*, *112*, G01014, doi:10.1029/2006JG000210.

van der Molen, P. C., and T. A. Wijmstra (1994), The thermal regime of hummock-hollow complexes on Clara bog, Co. Offaly, *Proc. R. Irish Acad. B*, *94*, 209–221.

Waddington, J. M., N. T. Roulet, and R. V. Swanson (1996), Water table control of CH_4 emission enhancement by vascular plants in boreal peatlands, *J. Geophys. Res.*, *101*, 22,775–22,785.

Weyhenmeyer, C. E. (1999), Methane emissions from beaver ponds: Rates, patterns, and transport mechanisms, *Global Biogeochem. Cycles*, *13*, 1079–1090.

Whiting, G. J., and J. P. Chanton (1993), Primary production control of methane emission from wetlands, *Nature*, *364*, 794–795.

Wilhelm, E., R. Battino, and R. J. Wilcock (1977), Low-pressure solubility of gases in liquid water, *Chem. Rev.*, *52*, 219–262.

Wille, C., L. Kutzbach, T. Sachs, D. Wagner, and E.-M. Pfeiffer (2008), Methane emission from Siberian arctic polygonal tundra: Eddy covariance measurements and modeling, *Global Change Biol.*, *14*, 1395–1408, doi:10.1111/j.1365-2486.2008.01586.x.

Windsor, J., T. R. Moore, and N. T. Roulet (1992), Episodic fluxes of methane from subarctic fens, *Can. J. Soil Sci.*, *72*, 441–452.

Worthy, D. E. J., I. Levin, F. Hopper, M. K. Ernst, and N. B. A Trivett (2000), Evidence for a link between climate and northern wetland methane emissions, *J. Geophys. Res.*, *105*, 4031–4038.

T. Miyazaki, Department of Biological and Environmental Engineering, Graduate School of Agricultural and Life Sciences, University of Tokyo, 1-1-1 Yayoi, Bunkyo-ku, Tokyo, 113-8657, Japan. (amiyat@soil.en.a.u-tokyo.ac.jp)

M. Mizoguchi, Interfaculty Initiative in Information Studies, University of Tokyo, 1-1-1 Yayoi, Bunkyo-ku, Tokyo, 113-8657, Japan. (amizo@mail.ecc.u-tokyo.ac.jp)

T. Tokida, Agro-Meteorology Division, National institute for Agro-Environmental Sciences 3-1-3 Kannondai, Tsukuba, 305-8604, Japan. (tokida@affrc.go.jp)

Dissolved Organic Carbon Production and Transport in Canadian Peatlands

Tim R. Moore

Department of Geography and Global Environmental and Climate Change Centre, McGill University
Montréal, Québec, Canada

It is clear that dissolved organic carbon (DOC) plays a significant role in northern peatlands, through its participation in the overall C cycle, in internal cycling and export and in participating in chemical transformations. I examine the evidence on the rates of DOC production and their controls in northern peatlands and on the processes leading to retention in the soil, such as biodegradability and sorption. Transport of DOC within and export from peatlands is a function of net production rates and hydrologic pathways, and export ranges from <5 to 40 g DOC m^{-2} a^{-1}, depending on the proportion of catchment that is peatland and climate and runoff. DOC dynamics are changed through drainage and flooding of peatlands, both of which lead to increased DOC export. DOC is important through its link with the N cycle (dissolved organic nitrogen being the dominant form in most peatlands), its combination with metals, such as mercury, and its capacity as both a reducing and an oxidizing substance in peatlands. There needs to be a more coherent assessment of DOC chemistry and a better integration of DOC into the developing models of C cycling in peatlands.

1. INTRODUCTION

Dissolved organic matter (DOM), of which dissolved organic carbon (DOC) is a part, is perhaps the most visible attribute of peatlands: waters draining peatlands and catchments containing peatlands have a brown color, indicating both high DOC concentrations and DOM with light-absorbing (chromophoric) attributes. DOC is significant in peatlands in several ways. First, it represents part of the C released from decomposing peat and plant tissues, and thus is part of the overall C cycle, along with carbon dioxide (CO_2) and methane (CH_4). Second, not only is DOC produced in decomposing materials, it can also be used as a substrate for microbial activity which involves the further production of DOC. Third, DOC plays an important role in chemical processes within peatlands, such as redox reactions and production and transport of soluble mercury.

There are two major sets of drivers that control production and transport of DOC. One is a function of the biological and physical processes that release DOC and the retention processes such as microbial utilization and sorption. The second is the hydrologic pathways in which water moves and retention times: where systems are fast moving, DOC may be exported rapidly, whereas in slow moving systems, retention times are long, and the DOC becomes more susceptible to transformations and retention within the peatland, resulting in DOC being a less important part of C cycling. In this context, it should be noted that many upland, blanket bogs in Europe, and maritime Canada have relatively steep hydrologic

Carbon Cycling in Northern Peatlands
Geophysical Monograph Series 184
10.1029/2008GM000816

gradients, whereas many of the large peatlands in central Canada have very gentle gradients: in the 320,000-km^2 Hudson Bay Lowlands, for example, the overall gradient is 0.00005 to 0.00001 [*Glooschenko et al.*, 1994], and at the domed, 28-km^2 Mer Bleue peatland, the gradient is 0.0008 [*Fraser et al.*, 2001]. These are low-energy systems with generally slow water movement and long residence times, which affect the role of DOC in C cycling.

2. CONTROLS ON DOC PRODUCTION

DOC production in peatlands is usually measured from the increase in DOC concentration within peat pore water, though this increase (or decrease) represents the balance between release of DOC to the pore water and its consumption, by biological, chemical, or physical processes. The release of DOC can be associated with desorption of organic C from soil organic matter, with the release of DOC from the decomposition of peat and plant tissues by soil organisms, or through the exudation of organic C from plant roots [e.g., *Fenner et al.*, 2004; *Trinder et al.*, 2008].

Several attempts have been made to identify some of the controls on the rate at which DOC is produced from peat

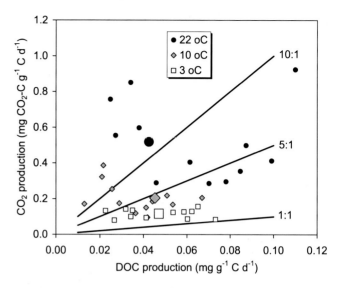

Figure 2. Relative production of DOC and CO_2 over the first 24 days of the incubation of 12 bog and fen peat samples as a function of temperature (3°, 10°, and 22°C). Large symbols represent the mean DOC and CO_2 production at each temperature (adapted from *Moore et al.* [2008]).

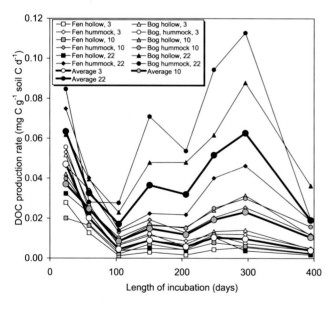

Figure 1. Production of DOC during a 395-day long incubation of 12 peat samples collected from hummock and hollow locations in a fen and a bog, at 3°, 10°, and 22°C (adapted from *Moore et al.* [2008]). The production rates represent the DOC found in leached solutions, over the period since the last leaching and expressed per gram of C in the sample. Large symbols and bold lines represent the average of the four peat types at each temperature (adapted from *Moore et al.* [2008]).

samples. *Moore et al.* [2008] incubated a wide range of Canadian forest soil samples, including four samples of peat, over 1 year under aerobic conditions at temperatures of 3°, 10°, and 22°C and measured rates of DOC leached from the soil as well as the amount of CO_2 produced. The rates of DOC production were high initially, probably through microbial activities, then fell and rose in the middle stages of the incubation (Figure 1), following the pattern also seen by *Neff and Hooper* [2002]. The temperature sensitivity on production rates of DOC and CO_2 during the first 24 days of incubation show that although both rates were higher at warmer temperature, the sensitivity of DOC production to temperature is less than that of CO_2, which results in larger CO_2-C/DOC production ratios as temperatures increase (Figure 2). This may explain, in part, why DOC concentrations are high in northern peatlands. The temperature sensitivity of DOC desorption is weak, with Q_{10} values of between 1.1 and 1.4, which may contribute to the low Q_{10} values for DOC production [*Kaiser et al.*, 2001]. Warmer temperatures increase the rate at which organic C is released from organic matter, but also the biodegradation of DOC into CO_2 [*Bengtson and Bengtsson*, 2007]. The CO_2-C/DOC production ratio would also increase if the water residence time is long, allowing DOC to be recycled and released as CO_2 (Figure 3). Over the range of forest organic soil samples incubated, *Moore et al.* [2008] were unable to find any strong soil chemical pre-

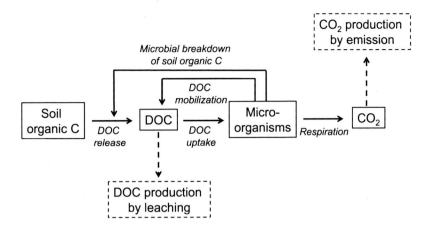

Figure 3. A conceptual model of DOC production in soils and the relationship to microbial cycling and CO_2 emission (adapted from *Bengtson and Bengtsson* [2007]). The dashed line boxes are the usual measurements in DOC production studies.

dictor of DOC production rates, with lignin content (negatively) being the strongest.

In an earlier experiment, *Moore and Dalva* [2001] incubated four peat samples representing different degrees of decomposition and observed higher DOC production rates in less decomposed samples (*Sphagnum* > fibric > hemic > sapric). They also observed a weak sensitivity of DOC production to temperature (average Q_{10} value of 1.6) and CO_2-C/DOC production ratios of 10 to 100:1. Incubation of the samples for 60 days under anaerobic conditions at 22°C resulted in only minor decreases in DOC production compared to aerobic incubation This was attributed to a slowing of DOC utilization for CO_2 production, as there were major decrease in CO_2 production in the samples and resultant decreases in the CO_2-C/DOC production ratios to <1:1.

Although these laboratory incubation experiments give some indication of the rates of DOC and CO_2 production in peat soils, and the controls on these production rates, the results need to be applied to field conditions with caution. The rates of DOC are much higher than the observed rates of DOC export from peatlands (see calculation in *Moore and Dalva* 2001), and retention times in laboratory incubations may be longer than those experienced in the field.

There have been few attempts to measure the biodegradability of DOC in peat pore water, though this has been common in upland soils. *Wickland et al.* [2007] recently reported that the leachates of *Sphagnum* and feather mosses were rapidly biodegraded, with 90% of the DOC mineralized under laboratory conditions, in contrast to 20% for *Eriophorum* and 10% for *Picea* needles. Thus, there is the strange fact that *Sphagnum* has very slow rates of decomposition in northern peatlands, with exponential k mass loss values ranging from

0.03 to 0.08 year^{-1} [*Moore and Basiliko*, 2006], yet its DOC leachate is rapidly mineralized.

Sorption of DOC by soil organic matter has been given little attention, apart from the Florida Everglades *Qualls and Richardson* [2003], but in attempting to model DOC transport in Swedish catchments, *Yurova et al.* [2008] have recently suggested that DOC sorption may play an important role, beyond biological production. Adsorption of DOC within the peat profile may be associated with Fe leached in from the surrounding upland mineral soils in the reduced form and then being oxidized at the peat water table [*Moore*, 1988]. There is also a suggestion that the generally small DOC concentrations in peatlands fed by Ca- and Mg-rich groundwaters may be associated with DOC precipitation by Ca and Mg [*Moore*, 1987]. When DOC-rich peat pore water percolates into the underlying mineral soil, large amounts (4 to 10 kg C m^{-2}) of DOC can be retained by adsorption on Fe and Al minerals, particularly if the peatland has developed through paludification onto upland soils rich in Fe and Al, such as podzols [*Moore and Turunen*, 2004; *Turunen and Moore*, 2003].

At the Mer Bleue peatland, *Blodau et al.* [2007a] used detailed pore water sampling to estimate C turnover rates within the saturated zone of peat profile and thus the partitioning of C released into DOC, CO_2, and CH_4 under field conditions. In this 'dry' bog they found that DOC concentrations ranged from 30 to 100 mg L^{-1}, with larger values during the summer. They estimated DOC production rates fell in the range of 1200 to –1200 (i.e., consumption) ng cm^{-3} d^{-1}, which was larger than the rates of CO_2 or CH_4 production, the former being 60 to 240 ng CO_2-C cm^{-3} d^{-1}, and the latter 24 to 60 ng CH_4-C cm^{-3} d^{-1}. Integrating the measurements over the

Table 1. Net Production and Stock of DOC, CO_2, and CH_4 in the Profile of an Ombrotrophic Bog Site, Mer Bleue[a]

Form	Net Production (mg C m^{-2} d^{-1})	Stock (g m^{-2})
DOC	102	29.64
CO_2	12.48	5.64
CH_4	−0.06[b]	0.66

[a]Adapted from *Blodau et al.* [2007a].
[b]Gross production of CH_4 estimated to be 2.28 mg C m^{-2} d^{-1}.

70-cm profile, the importance of DOC as both stocks and as production rates is clear (Table 1), and provides some support of the laboratory incubations.

3. DOC TRANSPORT WITHIN AND EXPORT FROM CATCHMENTS

Transport of DOC within peatlands will be dependent on rates of net DOC production and hydrologic pathways. Peatlands exhibit a rapid decline in hydraulic conductivity with depth: at the Mer Bleue peatland, *Fraser et al.* [2001] observed a decrease of 1×10^{-3} m s^{-1} in the top 20 cm to 1×10^{-5} m s^{-1} at 40 cm and $<1 \times 10^{-6}$ m s^{-1} below 50 cm. Thus, the major zone of transport of DOC in peatlands is generally in the saturated upper part of the profile, and in peatlands in the Experimental Lakes Area, northwestern Ontario, *Moore et al.* [2003] observed maximum summer DOC transport in the upper 50 cm, except where there was hydrologic convergence or increased hydraulic conductivity, such as in gyttja sediments near pools. As the water table drops, transport is reduced, creating strong relationships between DOC transport from peatlands and the water table within the peatland [*Fraser et al.*, 2001].

During the spring snowmelt, the surface layers of the peatland may be frozen, so that part of the melt passes rapidly across the peat surface, with smaller DOC concentrations. Thus, there are often strong seasonal variations in the concentration of DOC draining northern peatlands, dependent on hydrologic pathways and the rates of DOC production. This is illustrated by measurements of DOC concentration over 4 years in a beaver pond and the main outlet at Mer Bleue, with low concentrations (20–30 mg L^{-1}) during the snowmelt, rising to 40–80 mg L^{-1} during the summer and falling during the autumn (Figure 4), with a negative relationship between DOC concentration and runoff [*Fraser et al.*, 2001]. Although DOC concentrations are always small in peak runoff, there is considerable variation in concentrations at low discharges, probably related to seasonal patterns and antecedent conditions. Increase in runoff (associated with changes in precipitation and evapotranspiration) lead

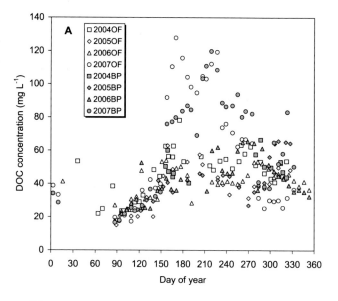

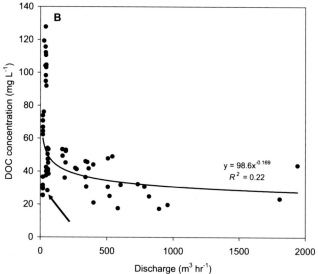

Figure 4. (a) Variation in DOC concentration in a beaver pond (BP) and the main outflow (OF) at the Mer Bleue peatland, 2004–2007; (b) relationship between DOC concentration and discharge at the main outflow, 2006 and 2007. Many of the low-flow, low-DOC points (see arrow) occurred in late autumn 2007 when cold conditions may have hampered DOC production.

to increases in DOC export, as illustrated by the 6-year Mer Bleue record [*Roulet et al.*, 2007] and at sites in the Dorset watersheds in Ontario [e.g., *Dillon and Molot*, 1997].

Export from peatland catchments varies considerably, with small values (<5 g DOC m^{-2} a^{-1}) in environments with low runoff rates (300 mm a^{-1}), such as the interior of North America [e.g., *Moore*, 2003; *Moore et al.*, 2003; *Urban et al.*,

1989] and under cold conditions. DOC export increases as runoff increases, such as at Mer Bleue and eastern Canadian peatlands [*Roulet et al.*, 2007; *Creed et al.*, 2008], where DOC export reaches 5 to 15 g m^{-2} a^{-1}, and in maritime locations where runoff is large (>1000 mm a^{-1}), DOC export is >25 g m^{-2} a^{-1} [e.g., *Moore and Jackson*, 1989].

For many years, it has been known that the export of DOC from catchments is generally related to the proportion of peatland or wetland in the catchment. The reason for this is that hydrologic pathways in peatland wetlands rarely encounter mineral soils upon which DOC adsorption occurs, and the movement in the saturated, upper layers of wetland soils is fast, reducing the opportunity for biodegradation of DOC. Several studies have tested this relationship, with mixed results, some of which are presented in Figure 5 for several regions of Canada, Finland, and Siberia. While the relationship holds in most cases, the coefficient of determination in linear regression of DOC concentration on peatland/wetland proportion is modest (generally 0.3–0.6). This indicates that other C cycling processes are important, such as position in the catchment of the peatland/wetland and the presence of "cryptic" wetlands, which are not measured [*Creed et al.*, 2003]. The poor relationship developed for Siberia (Frey, personal communication, 2003) is probably because this contains catchments in a wide variety of settings, including permafrost. In addition, several studies in northern areas have reported poor relationships between DOC export or concentration and catchment peatland area [e.g., *Frost et al.*, 2006; *Anderson and Nyborg*, 2008].

4. EFFECT OF DISTURBANCE ON DOC

Peatlands, as with other ecosystems, are increasingly affected by anthropogenic disturbances, which may alter the DOC. Although decreased atmospheric sulfate deposition has been implicated in increased DOC export from peatlands in Europe [e.g., *Clark et al.*, 2006; *Evans et al.*, 2006], there is little evidence for this in North America [*Monteith et al.*, 2007]. This may arise from atmospheric sulfate deposition rates in peatland areas in North America never achieving the magnitude of that in Europe.

Climate change, through its influence on precipitation and temperature, and thus evapotranspiration rates, water table position and hydrologic pathways, and vegetation productivity, may influence DOC in peatlands [e.g., *Fenner et al.*, 2007], though natural variability is large and may obscure any response. The most profound change in DOC export is likely to occur in northern landscapes with permafrost, which is undergoing melting in response to warmer temperatures. In the Thompson area of northern Manitoba, *Moore* [2003] observed high DOC concentrations in peatland collapse scars, created by the thawing of spruce-moss palsa. In Siberia, *Frey and Smith* [2005] suggest that thawing of permafrost in peatlands will release large amounts of DOC to rivers and then the Arctic Ocean. On the other hand, *Walvoord and Striegl* [2007] noted that permafrost thaw in the Yukon River basin may decrease DOC export, through deeper flow paths and DOC sorption by mineral soils, emphasizing the importance of flow paths in DOC transport.

Lowered water tables, either through modest drainage or dramatic change for horticultural peat moss production, may affect DOC concentrations and export. In southern Quebec, *Strack et al.* [2008] experimentally lowered the water table by about 20 cm in a poor fen peatland and noted an immediate export of DOC followed by higher DOC concentrations in the pore water of the drained peatland, compared to the control, which they ascribed to elevated net DOC production, associated with higher plant productivity and larger water table fluctuations after drainage. Drainage of bogs in eastern Quebec and New Brunswick increases runoff, and DOC export and vegetative restoration of peatlands leads to increased DOC concentrations in peat pore water [e.g., *Glatzel et al.*, 2003] and in drainage ditches [*Waddington et al.*, 2007].

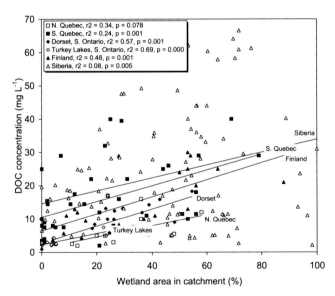

Figure 5. Relationship between DOC concentration in streams containing varying amounts of wetland. Methods of sample collection and calculation of DOC vary, so relationships within each region are important, rather than among regions. Sources: N. Quebec [*Koprivnjak and Moore*, 1992], S. Quebec [*Eckhardt and Moore*, 1990], Dorset [*Creed et al.*, 2008], Turkey Lakes [*Creed et al.*, 2008], Finland [*Kortelainen et al.*, 2006], Siberia [*Frey et al.*, 2007].

In northern regions, peatlands have been flooded during the creation of hydroelectric reservoirs. *Moore et al.* [2003] measured changes in DOC concentration and export from a small peatland catchment that was experimentally flooded to a depth of 1.2 m, in the Experimental Lakes Area of northwestern Ontario. They observed an increase in DOC concentrations in the flooded peat pore water and the central pond, ascribed to the decomposition of vegetation biomass killed during the flood, and increased migration of DOC through the creation of floating peat islands and changes in hydrologic pathways. In addition, the chemistry of the DOC changed, with an increase in the proportion of the hydrophilic neutral fraction after flooding. Thus, large areas of peatland flooded in northern catchments under reservoirs may increase DOC export, at least until the fresh plant biomass has been leached and decomposed.

5. SIGNIFICANCE OF DOC

The first significance of DOC within peatlands is in terms of the overall C budget. This budget includes atmospheric exchange as CO_2 (the balance of photosynthesis and respiration) and DOC in wet and dry deposition, the export of C as CH_4 and volatile organic carbon to the atmosphere, and hydrologic export of particulate carbon, DOC, CH_4, and inorganic C. The residual of inputs less outputs is the accumulation of C within the peatland, as peat. Although the long-term C accumulation rates have been established frequently and generally fall in the range 15–30 g C m^{-2} a^{-1} [e.g., *Vitt et al.*, 2000; *Turunen et al.*, 2001], there are fewer data on the contribution of the various forms to the overall C budget. At the Mer Bleue peatland in eastern Canada, over 6 years of measurements, *Roulet et al.* [2007] estimated that the 6-year mean CO_2-C exchange, CH_4-C exchange, and net DOC loss were 40.2, 3.7, and 14.9 g m^{-2} a^{-1}, resulting in a net C accumulation of 21.5 g m^{-2} a^{-1}. There was, however, considerable interannual variability in each component, with coefficients of variation of 20 to 100%. Thus, in some peatlands, C loss as DOC can be as large as the C accumulation rate.

A second is in the form of nitrogen (N) in the pore water. In some upland soil systems, most of the dissolved N is in the form of ammonium or nitrate, whereas in many peatlands, the majority is in the form of dissolved organic nitrogen (DON), which is intimately tied to the DOC and participates in biodegradation and sorption processes. As an illustration, the dissolved N concentrations in pore water at the Mer Bleue peatland are quite large (1.8–2.4 mg L^{-1} in the top 1 m) and are dominated by the DON fraction, which forms 80% in the surface layers and <50% beneath a depth of 1 m (Table 2). The DOC/DON ratio falls between 26:1 and 45:1, similar to the ratio in the peat at these depths. Thus,

Table 2. Mean Annual Concentration of DOC, DON, NH_4^+-N, and NO_3^--N and Percentage of Total Dissolved Nitrogen (DON, ΣDON, NH_4^+-N, and NO_3^--N) as DON in Pore Water at Different Depths in the Mer Bleue Peatland[a]

Depth (m)	DOC (mg L^{-1})	DON (mg L^{-1})	DOC/ DON	NH_4^+-N (mg L^{-1})	NO_3^--N (mg L^{-1})	DON/ TDN (%)
0.25	68	1.5	45	0.3	0.03	82
0.50	58	1.5	39	0.7	0.11	65
0.75	56	1.4	40	1.0	0.21	54
1	50	1.3	38	1.5	0.27	42
2	46	1.2	38	3.3	0.16	26
4.75	34	1.3	26	7.9	0.07	14

[a]Data from *Rattle* [2006].

movement of dissolved N in the upper layers of peatlands is strongly dependent on the movement of DOC. Given the low concentrations of NH_4 and NO_3 in pore water in the rooting zone, there is also the possibility that the plant uptake of N may be in organic forms, which has been shown to occur in upland forests and tundra [e.g., *Lipson and Näsholm*, 2001].

Peatlands export Hg, primarily attached to DOC [e.g., *Driscoll et al.*, 1995]. Peatlands have also been shown to be sites of mercury (Hg) methylation and thus a precursor to high concentrations of Hg in the food chain. Although the mechanisms controlling rates of Hg methylation in peatlands are still unclear, and methylation occurs in biogeochemical "hot spots" with high spatial and temporal variability, it appears that DOC and the lability of DOC to methylating organisms may be important, with DOC from upland soils stimulating the methylation [*Mitchell et al.*, 2008a, 2008b].

Finally, DOC has the capacity to be engaged in redox reactions through electron acceptance and donation, particularly by the quinone moieties present in polyphenols [*Cory and McKnight*, 2005]. In the Mer Bleue peatland, *Heitmann et al.* [2007] determined electron acceptance and donation capacities of peat pore water DOC of 0.2 to 6.1 and 0.0 to 1.4 meq g^{-1} C, respectively. Thus, DOC may lead to the oxidation of reduced organic substrates by anaerobic heterotrophic respiration, with less methanogenesis. Secondly, the cycling of S can be affected strongly by DOC leading to the anaerobic cycling between oxidized and reduced S pools [*Blodau et al.*, 2007b].

REFERENCES

Andersson, J.-O., and L. Nyberg (2008), Spatial variation of wetlands and flux of dissolved organic carbon in boreal headwater streams, *Hydrol. Processes*, 22, 1965–1975.

Bengtson, P., and G. Bengtsson (2007), Rapid turnover of DOC in temperate forests accounts for increased CO_2 production at elevated temperatures, *Ecol. Lett.*, 10, 783–790.

Blodau, C., N. T. Roulet, T. Heitmann, H. Stewart, J. Beer, P. Lafleur, and T. R. Moore (2007a), Belowground C turnover in a temperate ombrotrophic bog, *Global Biogeochem. Cycles*, *21*, GB1021, doi:10.1029/2005GB002659.

Blodau, C., B. Mayer, S. Peiffer, and T. R. Moore (2007b), Support for an anaerobic sulfur cycle in two Canadian peatland soils, *J. Geophys. Res.*, *112*, G02004, doi:10.1029/2006JG000364.

Clark, J. A. M., P. J. Chapman, A. L. Heathwaite, and J. K. Adamson (2006), Suppression of dissolved organic carbon by sulfate induced acidification during simulated droughts, *Environ. Sci. Technol.*, *40*, 1776–1783.

Cory, R. M., and D. M. McKnight (2005), Fluorescence spectroscopy reveals ubiquitous presence of oxidized and reduced quinines in dissolved organic matter, *Environ. Sci. Technol.*, *39*, 8142–8149.

Creed, I. F., S. E. Sanford, F. D. Beall, L. A. Molot, and P. J. Dillon (2003), Cryptic wetlands: Integrating hidden wetlands in regression models of export of dissolved organic carbon from forested landscapes, *Hydrol. Processes*, *17*, 3629–3648.

Creed, I. F., F. D. Beall, T. A. Clair, P. J. Dillon, and R. H. Hesslein (2008), Predicting export of dissolved organic carbon from forested catchments in glaciated landscapes with shallow soils, *Global Biogeochem. Cycles.*, *22*, GB4024, doi:10.1029/2008GB003294.

Dillon, P. J., and L. A. Molot (1997), Effect of landscape form on export of dissolved organic carbon, iron, and phosphorus from forested stream catchments, *Water Resour. Res.*, *33*, 2591–2600.

Driscoll, C. T., V. Blette, C. Yan, C. L. Schofield, R. Munson, and J. Holsapple (1995), The role of dissolved organic carbon in the chemistry and bioavailability of mercury in remote Adirondack lakes, *Water Air Soil Pollut.*, *80*, 499–508.

Eckhardt, B. W., and T. R. Moore (1990), Controls on dissolved organic carbon concentrations in streams, southern Quebec, *Can. J. Fish. Aquat. Sci.*, *47*, 1537–1544.

Evans, C. D., P. J. Chapman, J. M. Clark, D. T. Monteith, and M. S. Cresser (2006), Alternative explanations for rising dissolved organic carbon export from organic soils, *Global Change Biol.*, *12*, 2044–2053.

Fenner, N., N. Ostle, C. Freeman, D. Sleep, and B. Reynolds (2004), Peatland carbon afflux partitioning reveals that *Sphagnum* photosynthate contributes to the DOC pool, *Plant Soil*, *259*, 345–354.

Fenner, N., C. Freeman, M. A. Lock, H. Harmens, B. Reynolds, and T. Sparks (2007), Interactions between elevated CO_2 and warming could amplify DOC exports from peatland catchments, *Environ. Sci. Technol.*, *41*, 3146–3152.

Fraser, C. J. D., N. T. Roulet, and T. R. Moore (2001), Hydrology and dissolved organic carbon biogeochemistry in an ombrotrophic bog, *Hydrol. Processes*, *15*, 3151–3166.

Frey, K. E., and L. C. Smith (2005), Amplified carbon release from vast west Siberian peatlands by 2100, *Geophys. Res. Lett.*, *32*, L09401, doi:10.1029/2004GL022025.

Frey, K. E., D. I. Siegel, and L. C. Smith (2007), Geochemistry of west Siberian streams and their potential response to permafrost degradation, *Water Resour. Res.*, *43*, W03406, doi:10.1029/2006WR004902.

Frost, P. C., J. H. Larson, C. A. Johnston, K. C. Young, P. A. Maurice, G. A. Lamberti, and S. D. Bridgham (2006), Landscape predictors of stream dissolved organic matter concentration and physicochemistry in a Lake Superior river watershed, *Aquat. Sci.*, *68*, 40–51.

Glatzel, S., K. Kalbitz, M. Dalva, and T. Moore (2003), Dissolved organic matter properties and their relationship to carbon dioxide efflux from restored peat bogs, *Geoderma*, *113*, 397–411.

Glooschenko, W. A., N. T. Roulet, L. A. Barrie, H. I. Schiff, and H. G. McAdie (1994), The Northern Wetlands Study (NOWES): An overview, *J. Geophys. Res.*, *99*, 1423–1428.

Heitmann, T., T. Goldhammer, J. Beer, and C. Blodau (2007), Electron transfer of dissolved organic matter and its potential significance for anaerobic respiration in a northern bog, *Global Change Biol.*, *13*, 1771–1785.

Kaiser, K., M. Kaupenjohann, and W. Zech (2001), Sorption of dissolved organic carbon in soils: Effects of soil sample storage, soil-to-solution ratio, and temperature, *Geoderma*, *99*, 317–328.

Koprivnjak, J.-F., and T. R. Moore (1992), Sources, sinks and fluxes of dissolved organic carbon in subarctic fen catchments, *Arctic Alpine Res.*, *24*, 204–210.

Kortelainen, P, T. Mattsson, L. Finer, L, M. Ahtiainen, S. Saukkonen, and T. Sallantaus (2006), Controls on the export of C, N, P and Fe from undisturbed boreal catchments, Finland, *Aquat. Sci.*, *68*, 453–468.

Lipson, D. A., and T. Näsholm (2001), The unexpected versatility of plants: Organic nitrogen use and availability in terrestrial ecosystems, *Oecologia*, *128*, 305–316.

Mitchell, C. P. J., B. A. Branfireun, and R. K. Kolka (2008a), Spatial characteristics of net methylmercury production hot spots in peatlands, *Environ. Sci. Technol.*, *42*, 1010–1016.

Mitchell, C. P. J., B. A. Branfireun, and R. K. Kolka (2008b), Assessing sulfate and carbon controls on net methylmercury production in peatlands: An in situ mesocosm approach, *Appl. Geochem.*, *23*, 503–518.

Monteith, D. T., et al. (2007), Dissolved organic carbon trends resulting from changes in atmospheric deposition chemistry, *Nature*, *450*, 537–541.

Moore, T. R. (1987), Patterns of dissolved organic matter in subarctic peatlands, eastern Canada, *Earth Surf. Processes Landforms*, *12*, 387–397.

Moore, T. R. (1988), Dissolved iron and organic matter in northern peatlands, *Soil Sci.*, *145*, 70–76.

Moore, T. R. (2003), Dissolved organic carbon in a northern boreal landscape, *Global Biogeochem. Cycles*, *17*(4), 1109, doi:10.1029/2003GB002050.

Moore, T. R., and N. Basiliko (2006), Decomposition, in *Boreal Peatland Ecosystems*, Ecological Studies, vol. 188, edited by R. K. Wieder and D. H. Vitt, pp. 126–143, Springer-Verlag.

Moore, T. R., and M. Dalva (2001), Some controls on the release of dissolved organic carbon by plant tissues and soils, *Soil Sci.*, *166*, 38–47.

Moore, T. R., and R. J. Jackson (1989), Dynamics of dissolved organic carbon in forested and disturbed catchments, Westland, New Zealand: 2. Larry River, *Water Resour. Res.*, *25*, 1331–1339.

Moore, T. R., and J. Turunen (2004), Subsoil accumulation of carbon beneath forest and peat in Michigan, *Soil Sci. Soc. Am. J.*, *68*, 690–696.

Moore, T. R., L. Matos, and N. T. Roulet (2003), Dynamics and chemistry of dissolved organic carbon in Precambrian Shield catchments and an impounded wetland, *Can. J. Fish. Aquat. Sci.*, *60*, 612–623.

Moore, T. R., D. Paré, and R. Boutin (2008), Production of dissolved organic carbon in Canadian forest soils, *Ecosystems*, *11*, 740–751.

Neff, J. C., and D. U. Hooper (2002), Vegetation and climate controls on potential CO_2, DOC and DON production in northern latitude soils, *Global Change Biol.*, *8*, 872–884.

Qualls, R. G., and C. J. Richardson (2003), Factors controlling concentration, export and decomposition of dissolved organic nutrients in the Everglades of Florida, *Biogeochemistry*, *62*, 197–229.

Rattle, J. (2006), Dissolved nitrogen dynamics in an ombrotrophic bog, M.Sc. thesis, McGill Univ., Montreal, Quebec, Canada.

Roulet N. T., P. M. Lafleur, P. J. H. Richard, T. R. Moore, E. R. Humphreys, and J. Bubier (2007), Comparison of a six year contemporary carbon balance and the carbon accumulation for the last 3,000 years for a northern peatland, *Global Change Biol.*, *13*, 397–411.

Strack, M., J. M. Waddington, R. A. Bourbonnière, E. L. Buckton, K. Shaw, P. Whittington, and J. S. Price (2008), Effect of water table drawdown on peatland dissolved organic carbon export and dynamics, *Hydrol. Processes*, *22*, 3373–3385.

Trinder, C. J., R. E. R Artz, and D. Johnson (2008), Contribution of plant photosynthate to soil respiration and dissolved organic carbon in a naturally recolonising cutover peatland, *Soil Biol. Biochem.*, *40*, 1622–1628.

Turunen, J., and T. R. Moore (2003), Controls on C accumulation in the mineral subsoil beneath peat in Lakkasuo mire, central Finland, *Eur. J. Soil Sci.*, *54*, 279–286.

Turunen, J., T. Tahvanainen, K. Tolonen, and A. Pitkänen (2001), Carbon accumulation in west Siberian mires, Russia, *Global Biogeochem. Cycles*, *15*, 285–296.

Vitt, D. H., L. A. Halsey, I. E. Bauer, and C. Campbell (2000), Spatial and temporal trends in carbon storage of peatlands of continental western Canada through the Holocene, *Can. J. Earth Sci.*, *37*, 683–693.

Urban, N. R., S. E. Bayley, and S. J. Eisenreich (1989), Export of dissolved organic carbon and acidity from peatlands, *Water Resour. Res.*, *25*, 1619–1628.

Waddington, J. M., K. Tóth, and R. Bourbonniere (2007), Dissolved organic carbon export from a cutover and restored peatland, *Hydrol. Processes*, *22*, 2215–2224.

Walvoord, M. A., and R. G. Striegl (2007), Increased groundwater to stream discharge from permafrost thawing in the Yukon River basin: Potential impacts on lateral export of carbon and nitrogen, *Geophys. Res. Lett.*, *34*, L12402, doi:10.1029/2007GL030216.

Wickland, K. P., J. C. Neff, and G. R. Aiken (2007), Dissolved organic carbon in Alaskan boreal forest: Sources, chemical characteristics, and biodegradability, *Ecosystems*, *10*, 1323–1340.

Yurova, A., A. Sirin, I. Buffam, K. Bishop, and H. Laudon (2008), Modeling the dissolved organic carbon output from a boreal mire using the convection-dispersion equation: Importance of representing sorption, *Water Resour. Res.*, *44*, W07411 doi:10.1029/2007WR006523.

T. R. Moore, Department of Geography and Global Environmental and Climate Change Centre, McGill University, 805 Sherbrooke Street W., Montréal, Québec, Canada H3A 2K6. (tim.moore@mcgill.ca)

Hydrological Controls on Dissolved Organic Carbon Production and Release From UK Peatlands

Nathalie Fenner and Chris Freeman

School of Biological Sciences, Bangor University, Bangor, UK

Fred Worrall

Department of Earth Sciences, University of Durham, Durham, UK

Long-term increases in dissolved organic carbon (DOC) release from peatlands to British aquatic ecosystems are widely acknowledged, and are now confirmed to occur in a wide variety of boreal and subboreal settings. Depth to water table is probably the single most important hydrological factor governing that DOC generation and will modulate the response of the system to other environmental factors (such as warming and rising atmospheric carbon dioxide) in a changing climate. Many workers have attempted to attribute the rising DOC trend to a single "universal" driving variable. However, two fundamental problems prevent this: (1) universal theories, i.e., climate change theories that can account for rising trends in diverse catchment types, seem insufficient to account for the large observed increases, and (2) regional theories cannot account for the trend in all catchment types. Here it is suggested that multiple and possibly different drivers can modify DOC exports at four stages, namely, production, diffusion, solubility, and transport, with hydrology undoubtedly having a direct or indirect role in all the potential drivers considered here. These mechanisms, and the interactions between them, need to be more fully understood if we are to predict the response of the United Kingdom and global peatland carbon stores to environmental changes. Moreover, if we are to attempt to ameliorate rising DOC, we will need to fully appreciate the implications of restoration of drained peatlands and land management practices, to ensure that carbon losses are reduced on various temporal scales. These are research topics that remain in their infancy.

1. INTRODUCTION

Long-term increases in stream water dissolved organic carbon (DOC) concentrations for a range of boreal and subboreal settings are now widely reported. Increases have been observed in North America [*Driscoll et al.*, 2003; *Stoddard et al.*, 2003], in Central Europe [*Hejzlar et al.*, 2003] and across the northern subboreal and boreal zones [*Monteith et al.*, 2007]. For the United Kingdom, *Worrall and Burt* [2007a] have shown that out of 315 records, 68% showed a significant increase in stream DOC concentrations over timescales of between 9 and 42 years (catchment areas ranging between 400 and 2120 km^2). *Worrall and Burt* [2007a] have shown widespread increases in DOC flux from peat-covered

Carbon Cycling in Northern Peatlands
Geophysical Monograph Series 184
10.1029/2008GM000823

catchments, but nevertheless have identified significant decreases in DOC flux from all the peatlands of southwestern England (Figure 1).

Increasing DOC concentrations are of concern for several reasons. First, the removal of DOC from water is a major treatment cost for water companies, and the incomplete removal of DOC results in the following problems: (1) water of low aesthetic quality, (2) increased threat of biological contamination, as DOC consumes free residual chlorine used to protect water in transit, and (3) formation of trihalomethanes, which are potential carcinogens and whose concentration in drinking water is limited by law in the United Kingdom [*Hsu et al.*, 2001]. Second, high DOC concentrations in rivers are particularly associated with catchments where there is extensive peat cover [*Aitkenhead et al.*, 1999]. Therefore, with respect to climate change, increased DOC concentrations in rivers may be indicative of a translocation of terrestrial carbon reserves, especially from peat soils [*Limpens et al.*, 2008]. If much of this carbon is respired [*Cole and Caraco*, 2001], then it represents a positive feedback to rising atmospheric CO_2.

In this chapter, DOC production in, and loss from peatlands will be considered focusing on hydrological controls. How these mechanisms are affected by changing environmental factors (warming, elevated atmospheric CO_2, drought, atmospheric deposition, discharge, land use, and management), which form the main hypothesized drivers of the rising DOC trend, will then be discussed. Management practices to reduce DOC losses from peatlands will also be considered.

2. DOC EXPORT

Export of DOC from a catchment (also referred to as transport or flux) is the product of concentration and discharge. In a review of annual DOC flux from a range of catchments, *Hope et al.* [1997] document values up to 10.3 mg C km^{-2} a^{-1} (River Halladale, Sutherland, Scotland). *Dawson et al.* [2004] report DOC exports for a range of small, upland peat catchments, with values ranging from 8.3 to 26.2 mg C km^{-2} a^{-1} in catchments between 0.41 and 46.3 km^2, with the higher exports being reported in the smaller catchments. These studies, however, do not allow for in-stream loss. *Worrall et al.* [2006a] have studied a peat-covered catchment and by allowing for in-stream losses and/or gains, calculated DOC export from peat soils to be between 10 and 25 mg C km^{-2} a^{-1}. Going further, for a single drain (a zero-order channel cut into peat but not entering the substrate) in peat, *Gibson et al.* [2009] have found exports as high as 84 mg C km^{-2} a^{-1}.

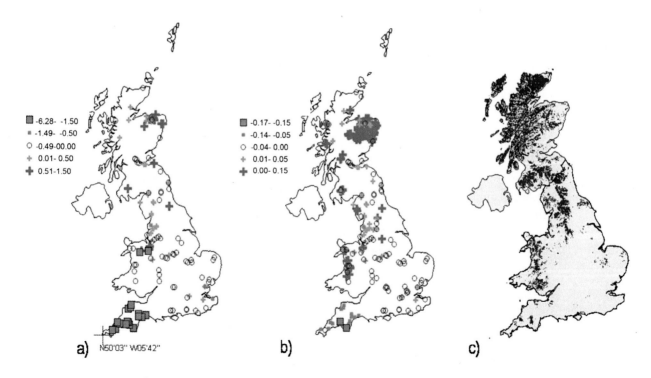

Figure 1. The change in DOC concentration: (a) percentage change and (b) absolute trend for all 315 records from this study and that of *Worrall et al.* [2004a], in comparison to (c) the distribution of organic-rich soils [*Milne and Brown*, 1997]. Adapted from *Worrall and Burt* [2007a].

When considering DOC exports, it is helpful to take into account two stages: DOC production in the peat followed by transport to the drainage network. The transport stage is controlled by discharge, so hydrology influences short-term fluctuations in riverine DOC export. However, long-term changes in discharge, unless accompanied by changes in DOC production, do not generate a sustained trend in DOC flux [*Evans et al.*, 2002]. An increase in DOC production within peat does not necessarily result in increased river transport if, for example, consumption is also increased.

3. HYDROLOGICAL CONTROLS UPON DOC PRODUCTION

DOC production within peat can occur via aerobic or anaerobic organic matter decomposition pathways. The former tends to favor a higher proportion of mineralization to CO_2; the latter tends to favor DOC end products and CH_4. However, it has generally been considered that extra DOC production is an aerobic process occurring above the water table in the acrotelm and that this releases carbon from the matrix allowing mobilization upon rehydration [e.g., *Fenner et al.*, 2005b; *Freeman et al.*, 2001b]. Increased water tables (i.e., reduced depth to the water table) have received less attention. This is probably because increasing the wetness of peatlands is considered unlikely to promote decomposition (therefore DOC release), given that slow decomposition rates are primarily attributed to waterlogging [*Freeman et al.*, 1998]. Water table depth is the single most important factor governing DOC production and loss, since it is the single most important factor governing wetland properties in general [*Ponnamperuma*, 1972]. It will modulate the response of the system to other environmental factors (such as warming) in a changing climate, but in turn, will also be affected by them.

Existing models of DOC release from peat (and forest soils) have tended to be based either on production or on solubility. For example, *Michalzik et al.* [2003] model production of DOC and carbon turnover, with the DOC being included as part of the turnover of soil organic carbon and the soluble fraction partitioned into mobile water. Similarly, *Futter et al.* [2007] use the Integrated Nitrogen Catchment Model-Carbon (INCA-C) model, which includes DOC production and makes this available for transport. *Worrall et al.* [2005a] produced a model of DOC runoff that used the same principle with a production rate of DOC based upon a first-order process. *Worrall et al.* [2004b] have modeled DOC flux from peaty catchments and included both changes in DOC production rate with changes in air temperature and with changes in water table depth (due to evaporation); it is assumed that although temperature controls the rate of pro-

duction, the water table depth controls the volume of source soil in which production can occur.

4. TREND GENERATION MECHANISMS

4.1. Warming

Temperature has a key role in controlling the rates of decomposition in peat [e.g., *Scanlon and Moore*, 2000; *Fenner et al.*, 2005a], including DOC production and consumption [*Freeman et al.*, 2001a; *Fenner et al.*, 2005a, 2007a]. *Freeman et al.* [2001a] suggested that increasing air temperature could be a driver of the rising DOC concentrations. They subjected "waterlogged" peat soil to a thermal gradient of $2°$–$20°C$ and found that phenol oxidase activity (responsible for phenolic compound cycling) was greater at higher temperatures, despite the known constraints of waterlogging (lack of O_2) on this enzyme. An increase of $10°C$ led to a 36% increase in activity ($Q_{10} = 1.36$), accompanied by an equivalent increase in DOC release ($Q_{10} = 1.33$), and an even greater increase in release of phenolic compounds within this pool ($Q_{10} = 1.72$) from the peat matrix. This selective enrichment with phenolics is potentially important because of the inhibitory character of these materials, which can impair the heterotrophic metabolism (consumption) of the remaining DOC, allowing even more DOC to accumulate in the receiving waters [*Freeman et al.*, 2001a; *Fenner et al.*, 2005a]. However, some studies have found that peat is not sufficiently thermally responsive for warming to account for *all* of the increase in DOC observed [*Freeman et al.*, 2001a, 2004; *Pastor et al.*, 2003]. In contrast, subsequent experiments on "aerobic" peat suggest Q_{10} values can be much more sensitive to warming ($Q_{10} = 3.66$) [*Clark et al.*, 2005], confirming that the height of the water table will indeed be critical in governing any response to changing temperatures. A similar scale of increase has been predicted by *Cole et al.* [2002] based upon increased enchytraeid worm activity with increased temperature. *Evans et al.* [2006] proposed a dual driver of increased DOC levels, namely, warming with changes in deposition chemistry, which can account for much of the DOC trend in 11 Acid Waters Monitoring Network lakes in the United Kingdom.

In addition to the direct effects above, warming will also induce indirect effects on DOC production, both in the short and the long term, through hydrological mechanisms, namely, increased evapotranspiration leading to lowered water tables. The modeling approach by *Worrall et al.* [2004b], which includes changes in DOC production rate with changes in air temperature (and the consequently increased evaporation), predicted a 10–20% increase in DOC concentration over a 30-year period from 1970.

In the short term, lowered water tables due to warming might lead to increased decomposition of the peat matrix [*Freeman et al.*, 2001b; *Fenner et al.*, 2005b], i.e., aerobic DOC production. In the longer term, warming and/or lowered water tables will promote vascular plant growth over *Sphagnum* spp. [*Toet et al.*, 2006; *Fenner et al.*, 2007a], leading to a number of biogeochemical changes relevant to DOC production: (1) An increase in vascular plant cover at the expense of *Sphagnum* spp. can stimulate decomposition by increasing new carbon inputs, as labile exudates, into the rhizosphere [*Freeman et al.*, 2004; *Fenner et al.*, 2007a, 2007b], (2) Increased litter inputs, both in terms of magnitude and lability [*Painter*, 1983; *Rasmussen et al.*, 1995; *Verhoeven and Toth*, 1995], could increase carbon turnover and losses as DOC [*Fenner et al.*, 2007a], and (3) Enhanced vascular plant growth can promote evapotranspiration, leading to even more aerobic decomposition and positive feedback to lowered water tables [*Fenner et al.*, 2007a, 2007b]. *Evans et al.* [2007] have used ^{14}C dating to show that DOC lost from a peat profile is young, i.e., from turnover of relatively fresh material, rather than humified peat, consistent with the increased plant contributions above.

4.2. Elevated CO_2

Rising CO_2 levels have also been proposed as a driver of increased DOC production by stimulating vascular plant exudates and litter inputs [*Freeman et al.*, 2004; *Fenner et al.*, 2007a, 2007b]. This would also be expected to increase the proportion of new carbon in the DOC pool in line with ^{14}C evidence [*Evans et al.*, 2007] and again has indirect effects on hydrological processes. Aerobic decomposition is promoted via reduced saturation [*Fenner et al.*, 2007b] as well as increased oxygen release into the root zone [*Visser et al.*, 2000]. Indeed, the response was described as a "drought-like" effect [*Fenner et al.*, 2007b] and could exacerbate DOC losses due to drought events. The mechanisms for this are discussed in detail below (part D).

4.3. Elevated CO_2 Plus Warming

Information of the interactive effects of elevated CO_2 plus warming on DOC production and export is sparse, but DOC release increased synergistically (i.e., significantly more than additively) in a study by *Fenner et al.* [2007a] on northern peat. This was attributed to increased vascular plant inputs coupled with reduced microbial consumption of that carbon. Again, the changed vegetation induced a marked shift in hydrological properties, lowering the water table and increasing aerobic production (see below).

4.4. Increased Drought Frequency

The frequency of severe drought within the UK uplands is known to be increasing [*Worrall et al.*, 2006b], and such events have been associated with rising DOC trends. There are several lines of evidence to support this proposal. First, there are step changes in DOC flux from several UK catchments that coincide with the aftermath of the most severe droughts during the periods of available records [*Worrall and Burt*, 2007b]. Second, there is a change in the impulsive relationship between flow and DOC concentration after a severe drought, i.e., the relationships between pulses of flow and those of DOC changes. Furthermore, these changes persist even through more minor droughts [*Worrall and Burt*, 2004]. Third, soil respiration and DOC production become decoupled after a severe drought, implying a change in the DOC production mechanism, which persists for years [*Worrall et al.*, 2005b]. Fourth, *Worrall et al.* [2005a] have shown that inclusion of kinetically limited DOC production, proportional to the severity of summer drought, greatly improves the modeling of DOC fluxes and solves the problem of poor fit between DOC flux records and linearly trending drivers (such as increased air temperature or atmospheric CO_2).

There are several possible theories that could explain the influence of severe drought upon DOC production and export including (1) biogeochemical mechanisms and (2) changes in flow path.

4.4.1. Biogeochemical mechanisms. Two biogeochemical mechanisms have been proposed to explain any severe drought effect. *Freeman et al.* [2001b] have proposed an enzymic-latch mechanism, whereby hydrolase enzymes, the primary agents of decomposition, become active as the water table falls because inhibitory phenolic materials are removed by extracellular phenol oxidase. These hydrolases are not repressed again after the water table recovers until phenolics re-accumulate and so trigger indirect additional production. Similarly, *Fenner et al.* [2005b] found an increased diversity of bacteria possessing the capacity to catabolize phenolics intracellularly under simulated drought, confirming the increased decomposition potential. In the longer term, the lowered water table and increased nutrient cycling as a result of accelerated decomposition will favor vascular plant growth over *Sphagnum* spp., potentially stimulating more rapid carbon turnover [*Fenner et al.*, 2007a, 2007b], with positive feedback to hydrological changes (as discussed above). However, an alternative drought mechanism was proposed by *Clark et al.* [2005]. A lowering of the water table in peat leads to the oxidation of sulfide minerals to SO_4. The increase in SO_4 concentration suppresses the mobility of DOC; when the drought ends, this suppression is released, and DOC concentrations rise.

Is there direct evidence for biogeochemical effects of drought upon DOC flux? *Worrall et al.* [2008b] compared a long-term record of DOC flux from a peat-covered catchment with drought measures, but were unable to find any significant links between magnitude of flux and any drought severity measure. Further, *Worrall and Burt* [2009a] examined DOC and river flow records from 97 sites across Great Britain and again found no significant additional production for the two most severe droughts experienced within Great Britain over the last 100 years. But given a lack of direct correlative evidence for droughts' influence upon DOC production and export, how do we explain the supporting evidence already cited?

First, step changes were observed in the DOC flux record for several catchments containing peat soils after severe droughts [*Worrall et al.*, 2003]. Such changes can now be viewed, not so much as the result of the drought itself but as a result of the change in runoff volumes after a severe drought. DOC flux is directly related to the amount of water flowing through peat soils and so will increase from periods of very low flow during droughts to periods of high or average flow after droughts, i.e., there is a step change in river flow that causes a step change in DOC flux to occur. Second, the inclusion of kinetically limited DOC production, proportional to the severity of summer drought, greatly improved the fit of DOC flux models. This kinetically limited reservoir of additional DOC was shown to have a time constant of about a decade [*Worrall et al.*, 2005a]. However, if there is no additional DOC production, then this kinetically limited reservoir is simply reflecting the return period of severe droughts and periods of high river discharge. Third, decoupling between soil respiration of CO_2 and DOC production is only evidence for a drought effect if it is expected that they would be coupled under nondrought conditions. The link between the two suggests that soil CO_2 and DOC are produced in the same or closely related steps and that this production is what normally limits the supply of DOC [*Worrall et al.*, 2005b]. However, if DOC production is not directly linked to turnover of organic matter but limited by another process, such as adsorption, solubility, or diffusion (see later), then there is no necessary link between DOC loss and soil respiration.

4.4.2. Changes in flow pathways. The occurrence of severe drought could also increase DOC runoff by altering runoff pathways within the peat after the drought. Is there evidence that physical changes occur in the peat? *Evans et al.* [1999] have shown that extreme drying out of peat during droughts has an immediate effect on peat hydrology and runoff characteristics, but could only speculate on the longer term, inter-

annual hydrological consequences. *Holden and Burt* [2002a, 2002b] showed, in laboratory and field studies of peat, that the short-term consequences of drought are to increase infiltration and flow at depth, and to decrease surface runoff, but do not affect mixing processes. However, the scale of changes observed in records of DOC concentration is years rather than months. *Worrall et al.* [2006c] used a multivariate analysis of long-term records of stream and soil water chemistry but produced no evidence for substantive new flow paths being generated during a drought, though there appeared to be a persistent change in the runoff chemistry due to a hydrophobic effect in the peat. Alternatively, *Worrall et al.* [2007a] have examined the change in runoff initiation probability and showed that it returned to a predrought state within months of the cessation of the severe drought. Therefore, studies have found only hydrophobic effects to be persistent beyond the year of the severe drought, but even that effect is not sufficient to explain the multiannual increases in DOC concentration or flux.

4.4.3. Remaining lines of evidence. Several lines of evidence for a drought effect still remain though, with two proposed mechanisms relying on drastic changes in water table and/or moisture content [*Freeman et al.*, 2001b; *Clark et al.*, 2005]. Both mechanisms would imply a compositional change in DOC leaving peat catchments. For the period 1993–2000 (including the 1 in 33 year drought in 1995), *Worrall et al.* [2006a] examined the multiannual records of specific absorbance of DOC from an upland, peat-covered catchment in Northern England and found no long-term trend was apparent. On the other hand, blocking of peat drains has been viewed as an analogue for the recovery from severe drought, and this significantly increased the specific absorbance of DOC until the end of a 9-month study [*Worrall et al.*, 2007b]. Similarly, *Wallage et al.* [2006] examined blocked peat drains and also found a difference in the composition of DOC in blocked drains compared to unblocked drains. *Toberman et al.* [2008] found a phenoloxidase-mediated increase in phenolic DOC mobilization following impeded peat drainage. *Worrall and Burt* [2009b] have examined a daily record of coagulant ratio data for 7 years from a peat-covered catchment and found a significant change in DOC composition that was independent of flow, temperature, pH, or alkalinity changes, and that suggested long-term decreasing specific absorbance. This record started in 1999, and there was no data before or after a severe drought.

Some of these additional production mechanisms may be small relative to other production mechanisms or short-lived relative to the records presented here. *Worrall et al.* [2008c] showed that the sulfide oxidation mechanism proposed by

Clark et al. [2005] gave increased DOC production for a maximum of only 2 months after a severe drought. In addition, records used to test for a drought effect are not normally on first-order streams, while that is where drought mechanisms are developed, i.e., there may be a decoupling between responses. However, rising trends are observed both at the tidal limit and in headwaters of catchments. Further, a detailed study of the DOC flux from a headwater catchment showed no correlation with measures of drought severity [*Worrall et al.*, 2008b]. This evidence implies that the proposed mechanisms of augmenting DOC production during severe droughts are limited in magnitude.

4.5. Increased Solubility

The amount of DOC that leaves a peat soil may not only be controlled by changes in production (be that due to warming or drought), but rather by its own solubility, which is also dependent on temperature [*Fenner et al.*, 2005a], pH, and the ionic strength of the peat pore water [*Lofts et al.*, 2001; *Lumsdon et al.*, 2005]. Changes in the solubility of DOC have also been invoked to explain observed trends in DOC concentration.

Changes in atmospheric deposition of both acidity and SO_4 could lead to declines in soil pore water ionic strength and increased pH, leading to enhanced DOC solubility and thus increased runoff. In the UK uplands, streams and lakes are showing initial indications of recovering from acidification [*Monteith and Evans*, 2000], while DOC concentrations are increasing. Other components of deposition could also have an effect on DOC concentrations. *Evans et al.* [2005, 2006] suggested that changes in SO_4 deposition and not pH are leading to increased release of DOC. Changes in SO_4 deposition could lead to DOC solubility changes either by changing pH or soil solution ionic strength, or both. *Evans et al.* [2006] have gone further and suggested that decreases in sea-salt deposition could also lead to changes in soil solution pH and ionic strength, with a concomitant influence upon DOC concentrations. Conversely, increased soil solution SO_4 has also been associated with suppression of DOC [*Clark et al.*, 2005] and so decreasing atmospheric sulfate inputs could also increase DOC release.

However, there are problems with proposing that components of atmospheric deposition control the observed increases in DOC release. (1) *Lumsdon et al.* [2005] have shown that it is not necessary to invoke seasonal changes in deposition to explain seasonal cycles in DOC within organic-rich soils; changes in DOC solubility can simply be a matter of the seasonal cycle in air temperature. (2) *Evans et al.* [2005] have shown that decreases in SO_4 deposition did not begin until the 1980s in the United Kingdom, but increases in DOC flux and concentration started at least as early as the 1960s, when SO_4 deposition was still increasing. Correlations between declines in atmospheric deposition and increases in DOC concentrations may therefore be apparent only if records of certain lengths are considered. (3) Decreases in DOC have been observed in areas where SO_4 deposition has been observed also to decrease [*Worrall and Burt*, 2007a]. (4) Experimental and field evidence is equivocal. For example, *Worrall et al.* [2008c] showed that impulses of SO_4 or acid deposition did not cause a significant increase in DOC concentration (in pore or runoff water) for a peat-covered catchment. Thus, it seems that these deposition mechanisms do not offer a universal explanation, but may be important regional drivers in certain catchments. Similarly, nitrogen deposition could be important in governing carbon losses from peat [*Bragazza et al.*, 2006], but cannot account for increasing DOC trends in pristine areas.

If the solubility of DOC is a control upon its release, then it is not only the position of the solubility equilibrium that governs the loss of DOC from a peat, but also the amount of water that goes through the soil. *Worrall et al.* [2008b] and *Worrall and Burt* [2007b] have shown that the most important factor explaining long-term trends in DOC flux is the water yield. In itself, this is not surprising; what is surprising is that this relationship shows no exhaustion effects, i.e., no leveling off of DOC flux was found as water yield increased. The lack of exhaustion effect suggests that loss of DOC from peat soils is never supply-limited. It is noteworthy that extracellular enzymes could be considered as "solubilizing enzymes" with an essentially infinite substrate (i.e., peat) [cf. *Fenner et al.*, 2005a; *Freeman et al.*, 2001a].

Worrall et al. [2008a] examined 170 DOC chemohydrographs in a peat drain in order to assess the controls on the amount and concentration of DOC in each rainfall/storm event, in terms of the nature of the event and the antecedent conditions (including the time between events). The hypothesis was that if production was limiting DOC runoff, then the time between events would be significant, but if not, and solubility was limiting, then the size of the event would be the most important. The results showed that event size was the dominant control, and the rate of DOC release was fast, i.e., on timescales resolvable within the study (8 h). On timescales of up to 2 days, the time between events was significant, but DOC release seemed to be limited by diffusion not by production. Only on timescales of greater than 2 days between rainfall events did production, in the sense of biologically mediated turnover of organic matter to generate DOC, become important. These event-based observations led to a conceptual model of DOC production in peat that combines production, diffusion, and sorption/solubility (Figure 2).

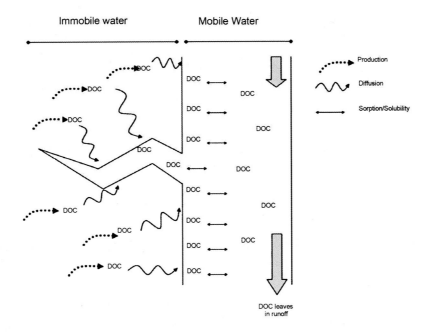

Figure 2. Schematic diagram of the conceptual model proposed by this study combining immobile and mobile water with production, diffusion, and sorption/solubility. Adapted from *Worrall et al.* [2008a].

4.6. Changes in Discharge

Tranvik and Jansson [2002] have suggested that observed increases in DOC concentration could result from changes in hydrology, i.e., a change in discharge should result in a change in concentration. At the same time as DOC fluxes have been observed to increase in the United Kingdom, river discharges in the Scotland have been increasing [*Werrity*, 2002], but increases are small and not widespread. Alternatively, it may not be changes in the amount of discharge but rather its nature, for example, if flow paths at the soil or catchment scale shift (due to changes in soil moisture, etc.), new, richer sources of DOC may be accessed. However, *Worrall et al.* [2003] showed that for two long-term time series of increasing DOC concentration, the DOC increase occurred equally for both the base flow and storm runoff. Given the discussion above regarding solubility control, the major hydrological influence upon DOC loss would seem to be its quantity and its timing. This however, does suggest a new and alternative explanation of DOC trends, namely, that it is not so much an increase in total flow (as this has already been described as insufficient), but rather changes in intensity. *Huntington* [2005] has described global evidence for the intensification of the water cycle with climate change. In the case of DOC, an increase in concentration could result from increased number of storms and a decrease in the time between storms.

4.7. Changes in Land Use and Management

Changes in peatland management can have dramatic effects upon hydrology and water table position and thus have been linked to observed changes in DOC concentrations. A multitude of management practices have been implemented in the UK uplands during the period for which DOC concentrations have been observed to increase, and this itself suggests that no one management intervention could have caused such a widespread phenomenon. However, management interventions could have had a local effect. In the United Kingdom, the following are worth considering relative to DOC release: (1) peat drainage and the blocking of those drains, (2) changes in forestation, (3) occurrence of fire (managed or wildfire), and (4) vegetation changes other than trees.

4.7.1. Drainage. A common land management technique in UK peatlands is the use of open drainage channels. Peat drainage has been common in many European countries. In the United Kingdom, it has been estimated that 1.5 million ha of the country's 2.9 million ha of peat has been drained [*Milne and Brown*, 1997; *Stewart and Lance*, 1991]. Typically, peat drains are spaced 20 m apart, but drain spacing can be as little as 7 m. The drains are trapezoidal in section, typically 50-cm deep, 90-cm broad at the top, and 40-cm broad at the base. Draining peatlands to lower the water table is done to improve grazing, game production, or to allow forestry. However, *Stewart and Lance* [1991] have shown that there is little evidence for any of these claims. The consequences of peat drainage for the environment have been recently reviewed by *Holden et al.* [2004].

Drains may lower water tables and cause additional carbon production due to increasing the depth to the water table [cf. *Freeman et al.*, 2001b], and *Mitchell and McDonald* [1995] have shown that DOC concentration in runoff is correlated with peat drain density. *Worrall et al.* [2007c] have combined a model for the DOC flux from peat catchments [*Worrall et al.*, 2005a] with a description of the behavior of peat drains [*Braekke*, 1984] to predict the consequences for catchment DOC fluxes of blocking those drains. The modeling suggests that drain blocking could lead to a decrease in DOC flux, although the amount of decrease was critically dependent upon the spacing between drains. *Wallage et al.* [2006] have observed a decrease in DOC concentration and water color (absorbance at 400 nm) between blocked and unblocked peat drains, but their study contained no preblocking data. *Worrall et al.* [2007b] examined a series of peat drains before and after blocking and showed a statistically significant difference upon drain blocking, but showed an increase in DOC concentration and water color within blocked drains. Neither study considered the flux of DOC from the drains. *Gibson et al.* [2009] considered DOC flux and found a significant 39% decline in DOC flux, but most of the change was accounted for by changes in water yield, i.e., water retention by the blocked drain caused the decrease in flux. As a consequence, only a 1% decline in DOC concentration was observed upon drain blocking. However, a simple consideration of water balance shows that a 39% decline in the water flowing through a drain cannot be compensated for by increased evaporation, i.e., there must be increased water flow elsewhere and around the drain block, and the effectiveness of drain blocking at scales larger than that of an individual drain is limited.

4.7.2. Afforestation. Afforestation has been the main cause of the net loss of UK moorland habitat over the past century. Nine percent of upland UK peatland has been afforested [*Cannell et al.*, 1993], and in Scotland, 25% of Caithness and Sutherland peatlands have been affected by afforestation [*Ratcliffe and Oswald*, 1988]. Narrowly spaced drainage ditches (ribbon plough furrows) are commonly dug across moorland areas before forests are planted. Fertilizer is also often applied. These drains lower the water table and result in associated subsidence of the peat surface due to compression and shrinkage [*Anderson et al.*, 2000]. The peat tends to further dry out after canopy closure, with increased interception and transpiration causing a much greater lowering in the water table than drainage alone, further encouraging surface

subsidence [*Shotbolt et al.*, 1998] and increasing hydraulic conductivity in the upper layers, often with large-scale cracking of the peat. Felling of the trees causes the water table to rise, but the water table tends to fluctuate much more than in intact moorlands because of changes to soil structure and enhanced hydraulic conductivity. The impact of afforestation is not restricted to the planted area alone. Drying and shrinkage of organic soils can occur at some distance away from the forest, depending on local topography and drainage. Runoff is also affected by afforestation. Streamflow tends to increase in both total and in peakedness with increased low-flow levels in the first years following drainage (perhaps 20 years), followed by decreases in water yield as the forest matures. Water quality may also change downstream as afforested moorland streams often become more acidic with higher concentrations of aluminum. While carbon is taken up by tree biomass as the forest grows, there may be severe depletion of the soil carbon store through enhanced decomposition of the organic soil [*Cannell et al.*, 1993]. As for consequences for DOC, then the change and increase in runoff would be expected to increase DOC losses, and the harvesting of trees on peat soils has been shown to increase DOC concentration over several years after harvesting [*Neal et al.*, 1998].

4.7.3. Fire. In the United Kingdom, vegetation on peat soils is commonly burned in order to improve grazing (especially for sheep) and habitat for game species (especially red grouse). In England, it has been estimated that 40% of moorland has received some burn management [*Thomas et al.*, 2004, quoted in *DEFRA*, 2005]. There have been very few direct studies of the fate of any form of carbon under managed burns. *Garnett et al.* [2000] examined peat accumulation of carbon under three treatments (grazed/unburnt, grazed/burnt, and ungrazed/unburnt), and recalculating their data shows that the mean difference between burnt and unburnt treatments is 2.48 kg m^{-2} (not 2.3 as reported); this gives a mean effect of burning of 55 mg C km^{-2} a^{-1} (not 73 mg C km^{-2} a^{-1} as reported). In a recent review of the consequences of heather and grassland burning, including that on peat [*DEFRA*, 2005; *Tucker*, 2005], there were no studies reporting upon individual components of the carbon cycle (including DOC) and only very few studies which examined the consequences of burning for hydrology or water quality. *Worrall et al.* [2007d] have shown that there are significant differences in water table depth between different burn frequencies, with the lowest water tables being found on sites where there has been no burning for 40 years, and the vegetation is mature. Significantly lower DOC concentrations were observed in soil water beneath more frequently burnt sites. This latter study was limited to only 7 months over one summer, but an extended study lasting 33 months, and including a prescribed burn, has shown no significant difference in DOC concentrations in either soil. However, it could be that any observed differences are not a consequence of burnt soil or burn products but just a change in vegetation.

Information for accidental and catastrophic, as opposed to managed and prescribed, burns is scarce with respect to DOC. Such fires, though, have been linked to increased peat erosion [*Mackay and Tallis*, 1996], and burning in other settings has been associated with the development of water repellency that limits infiltration (e.g., for Californian wildlands) [*DeBano*, 2000].

4.7.4. Grazing. Grazing on uplands in the United Kingdom can be by cattle or deer but is predominantly by sheep. At 11 sites in the Scottish Highlands, where sheep had been excluded for up to 25 years, dwarf shrubs (*Calluna vulgaris*, *Erica* spp., and *Vaccinium myrillus*) came to dominate over grasses, with one site showing the invasion of birch woodland [*Hope et al.*, 1998]. A range of studies have associated increased sheep numbers with changes in runoff which may change fluvial carbon losess [e.g., *Langlands and Bennett*, 1973; *Evans*, 1990]. *Meyles et al.* [2006] argued that more intense grazing causes conditions which promote increased delivery of soil water to rapid flow paths. As with the effects of burning, the number of studies into the effects on soil and water quality are limited. *Sansom* [1990] noted that in the north Derwent catchment (Lake District, Cumbria, UK), sheep numbers had doubled between 1944 and 1975 to 24,000, and in that time, annual water yield had increased by 25%. The effect of grazing on other carbon uptake and release pathways is little reported. *Garnett et al.* [2000] showed no significant difference in peat accumulation beneath grazed and ungrazed plots. In contrast, *Worrall et al.* [2007d] have reported significant changes. However, the authors know of no studies that consider the interaction of management upon DOC pathways specifically.

5. MANAGEMENT OPTIONS TO REDUCE DOC LOSSES FROM PEATLANDS

Given the current concern over carbon capital in peatlands and the effects of DOC on water quality, management options to reduce DOC losses would be valuable [e.g., *Fenner et al.*, 2001, 2005b]. A potential method of controlling carbon (including DOC) release would be to raise the water table. While this action would favor CH_4 flux [*Huttunen et al.*, 2003], it may be offset by decreases in aerobic decomposition of the peat matrix and therefore CO_2 and DOC release, assuming that carbon uptake pathways, especially primary

productivity, were unaffected, although, in considering the latter, it would be expected that the growth of *Sphagnum* would be promoted eventually, and these species retain carbon for far longer than vascular plants [*Fenner et al.*, 2004, 2007a, 2007b] while also inhibiting decomposition processes [*Painter*, 1983; *Rasmussen et al.*, 1995; *Verhoeven and Toth*, 1995]. Thus, raising the water table could directly reduce DOC production by limiting the aerobic zone and promoting carbon storage, although this could be offset in the short term, after water table restoration, by any enzymic-latch or sulfide oxidation mechanisms (initiated during periods of low water table).

On the other hand, changing the depth to the water table could increase the flux of DOC as runoff changes. Indeed, experimentally rewetting a previously (naturally) well-drained peatland produced large increases in DOC concentration [*Fenner et al.*, 2001, 2005b]. This has been attributed to both a simple flush effect [*Hughes et al.*, 1998] and also repressed microbial and enzymic processing of DOC, allowing it to accumulate in pore waters [*Fenner et al.*, 2001, 2005b; *Freeman et al.*, 1998] characterized by high molecular weight fractions, phenolic compounds, and iron [*Fenner et al.*, 2001, 2005a]. However, the length of such a response is unknown, and the causes require more research. It may be though, that rewetted subcatchments could be left "off-line" for water collection, until they recover to normal DOC levels. Indeed, simple turnout sequences have long been considered in relation to potable water quality [*Fenner et al.*, 2001; A. T. McDonald et al., Discoloured runoff in the Yorkshire Pennines, paper presented at Second National Hydrological Symposium, Institute of Hydrology, Wallingford, U. K., 1989].

6. CONCLUSIONS

The loss of DOC from peat soil is controlled by an interplay of production (versus consumption), diffusion, solubility, and transport (Figure 2, Table 1). The differentiation of these four mechanisms will aid our understanding of DOC trend generation, both in the United Kingdom and elsewhere, because multiple and possibly different drivers can act on each of them. Furthermore, each driver has multiple levels of regulation with, for example, warming potentially inducing increased evaporation, which, in turn, leads to lowered water tables (Table 1).

Two fundamental problems apparently remain with attributing rising DOC trends to any one driver: (1) universal theories, i.e., climate change theories that can account for rising trends in diverse catchment types, seem insufficient to account for the large observed increases, and (2) regional theories cannot account for the trend in all catchment types. More research is needed in the case of universal drivers

Table 1. Effects of Universal and Regional Drivers on DOC Production, Diffusion, Solubility, and Transport[a]

	Biological Production	Diffusion	Solubility	Transport	Secondary Effects
			Universal Drivers		
Warming	+	+	+	− short term	increased aerobic production ("drought-like")
Elevated CO_2	+				increased aerobic production ("drought-like")
Warming + eCO_2	+				increased aerobic production ("drought-like")
Drought	+			− short term	physical/flow path changes/SO_4 release/ lowered pH/vegetation changes
Increased discharge			+	+	
Flow path changes			+	+	
Intensity of hydrological cycle	+		+	+	
			Regional Drivers		
Changes in atmospheric deposition (pH and ionic strength)			+		
Drainage	+			− short term	increased aerobic production/vegetation changes
Afforestation	+			+ short term − long term	increased aerobic production ("drought-like")
Fire	+			−	increased aerobic production ("drought-like")/ vegetation changes
Grazing				+	vegetation changes

[a]Plus symbol (+) indicates increase; minus symbol (−) indicates decrease.

though, such as the interactive effect of environmental factors. Similarly, combined regional and universal drivers are likely to be important in determining the net change in DOC flux over time, possibly even explaining decreasing trends in some regions. However, with all the potential universal drivers of the rising trend considered here, induced changes in hydrology have a direct or indirect role (Table 1). Similarly, all but one hypothesized regional driver (SO_4 deposition) induced indirect hydrological changes, and original hydrological status will moderate the effect of all drivers. Thus, these mechanisms need to be more fully understood, if we are to predict the response of the United Kingdom and global peatland carbon stores to environmental changes. This includes restoration of drained peatlands and land management practices, which need careful monitoring to determine whether carbon losses are reduced on various temporal scales. Concern over widespread increasing DOC has often focused upon understanding DOC production; however, transport of DOC might be more easily controlled to limit its impact upon water resources.

Acknowledgments. The authors would like to acknowledge The Leverhulme Trust, Royal Society, and NERC.

REFERENCES

Aitkenhead, J. A., D. Hope, and M. F. Billet (1999), The relationship between dissolved organic carbon in stream water and soil organic carbon pools at different spatial scales, *Hydrol. Processes*, *13*, 1289–1302.

Anderson, A. R., D. Ray, and D. G. Pyatt (2000), Physical and hydrological impacts of blanket bog afforestation at Bad a' Cheo, Caithness: The first 5 years, *Forestry*, *73*, 467–478.

Braekke, F. H. (1984), Water table levels at different drainage intensities on deep peat in Northern Norway, *For. Ecol. Manage.*, *5*, 169–192.

Bragazza, et al. (2006), Atmospheric nitrogen deposition promotes carbon loss from peat bogs, *Proc. Natl. Acad. Sci. U. S. A.*, *103*, 19,386–19,389.

Cannell, M. G. R., R. C. Dewar, and D. G. Pyatt (1993), Conifer plantations on drained peatlands in Britain: A net gain or loss of carbon?, *Forestry*, *66*, 353–369.

Clark, J. M., P. J. Chapman, J. K. Adamson, and S. J. Lane (2005), Influence of drought-induced acidification on the mobility of dissolved organic carbon in peat soils, *Global Change Biol.*, *11*, 791–809.

Cole, J. J., and N. F. Caraco (2001), Carbon in catchments: Connecting terrestrial carbon losses with aquatic metabolism, *Mar. Freshwater Res.*, *52*, 101–110.

Cole, L., R. D. Bardgett, P. Ineson, and J. K. Adamson (2002), Relationships between enchytraeid worms (Oligochaeta), climate change, and the release of dissolved organic carbon from blanket peat in northern England, *Soil Biol. Biochem.*, *34*, 599–607.

Dawson, J. J. C., M. F. Billet, D. Hope, S. M. Palmer, and C. M. Deacon (2004), Sources and sinks of aquatic carbon in a peatland stream continuum, *Biogeochemistry*, *70*, 71–92.

DeBano, L. F. (2000), The role of fire and soil heating on water repellency in wildland environments: A review, *J. Hydrol.*, *231–232*, 195–206.

DEFRA (2005), *Review of the Heather and Grass Etc. (Burning) Regulations 1986 and the Heather and Grass Burning Code 1994 in England*, Dep. of Environ., Food and Rural Affairs, London.

Driscoll, C. T., K. M. Driscoll, K. M. Roy, and M. I. Mitchell (2003), Chemical response of lakes in the Adirondack region of New York to declines in acidic deposition, *Environ. Sci. Technol.*, *37*, 2036–2042.

Evans, C. D., C. Freeman, D. T. Monteith, B. Reynolds, and N. Fenner (2002), Terrestrial export of organic carbon, *Nature*, *415*, 861–862.

Evans, C. D., D. T. Montieth, and D. M. Cooper (2005), Long-term increases in surface water dissolved organic carbon: Observations, possible causes and environmental impacts, *Environ. Pollut.*, *137*, 55–71.

Evans, C. D., P. J. Chapman, J. M. Clark, D. T. Monteith, and M. S. Cresser (2006), Alternative explanations for rising dissolved organic carbon export from organic soils, *Global Change Biol.*, *12*, 2044–2053.

Evans, C. D., C. Freeman, L. G. Cork, D. N. Thomas, B. Reynolds, M. F. Billett, M. H. Garnett, and D. Norris (2007), Evidence against recent climate-induced destabilisation of soil carbon from ^{14}C analysis of riverine dissolved organic matter, *Geophys. Res. Lett.*, 34, L07407, doi:10.1029/2007GL029431.

Evans, M. G., T. P. Burt, J. Holden, and J. K. Adamson (1999), Runoff generation and water table fluctuations in blanket peat: Evidence from UK data spanning the dry summer of 1995, *J. Hydrol.*, *221*, 141–160.

Evans, R. (1990), Water erosion in British Farmers fields—Some causes, impacts, predictions. *Prog. Phys. Geogr.*, *14*, 199–219.

Fenner, N., C. Freeman, S. Hughes, and B. Reynolds (2001), Molecular weight spectra of dissolved organic carbon in a rewetted Welsh peatland and possible implications for water quality, *Soil Use Manage.*, *17*, 106–112.

Fenner, N., N. Ostle, C. Freeman, D. Sleep, and B. Reynolds (2004), Peatland carbon efflux partitioning reveals that *Sphagnum* photosynthate contributes to the DOC pool, *Plant Soil*, *259*, 345–354.

Fenner, N., C. Freeman, and B. Reynolds (2005a), Observations of a seasonally shifting thermal optimum in peatland carbon-cycling processes; implications for the global carbon cycle and soil enzyme methodologies, *Soil Biol. Biochem.*, *37*, 1814–1821.

Fenner, N., C. Freeman, and B. Reynolds (2005b), Hydrological effects on the diversity of phenolic degrading bacteria in a peatland: Implications for carbon cycling, *Soil Biol. Biochem.*, *37*, 1277–1287.

Fenner, N., C. Freeman, M. A. Lock, H. Harmens, and T. Sparks (2007a), Interactions between elevated CO_2 and warming could amplify DOC exports from peatland catchments, *Environ. Sci. Technol.*, *41*, 3146–3152.

Fenner, N., N. J. Ostle, N. McNamara, T. Sparks, and C. Freeman (2007b), Elevated CO_2 effects on peatland plant community carbon dynamics and DOC production, *Ecosystems*, *10*, 635–647.

Freeman, C., G. B. Nevison, S. Hughes, B. Reynolds, and J. Hudson (1998), Enzymic involvement in the biogeochemical responses of a Welsh peatland to a rainfall enhancement manipulation. *Biol. Fertil. Soils*, *27*, 173–178.

Freeman, C., C. D. Evans, D. T. Monteith, B. Reynolds, and N. Fenner (2001a), Export of organic carbon from peat soils, *Nature*, *412*, 785.

Freeman, C., N. Ostle, and H. Kang (2001b), An enzymic "latch" on a global carbon store—A shortage of oxygen locks up carbon in peatlands by restraining a single enzyme, *Nature*, *409*, 149.

Freeman, C., N. Fenner, N. J. Ostle, H. Kang, D. J. Dowrick, B. Reynolds, M. A. Lock, D. Sleep, and J. A. Hudson (2004), Export of dissolved organic carbon from peatlands under elevated carbon dioxide levels, *Nature*, *430*, 195–198.

Futter, M. N., D. Butterfield, B. J. Cosby, P. J. Dillon, A. J. Wade, and P. G. Whitehead (2007), Modeling the mechanisms that control in-stream dissolved organic carbon dynamics in upland and forested catchments, *Water Resour. Res.*, *43*, W02424, doi:10.1029/2006WR004960.

Garnett, M. H., P. Ineson, and A. C. Stevenson (2000), Effects of burning and grazing on carbon sequestration in a Pennine blanket bog, UK, *Holocene*, *10*, 729–736.

Gibson, H. S., F. Worrall, T. P. Burt, and J. K. Adamson (2009), DOC budgets of drained peat catchments—Implications for DOC production in peat soils, *J. Hydrol.*, in press.

Hejzlar, J., M. Dubrovsky, J. Buchtele, and M. Ruzicka (2003), The apparent and potential effects of climate change on the inferred concentration of dissolved organic matter in a temperate stream (the Malse River, South Bohemia), *Sci. Total Environ.*, *310*, 143–152.

Holden, J., and T. P. Burt (2002a), Laboratory experiments on drought and runoff in blanket peat, *Eur. J. Soil Sci.*, *53*(4), 675.

Holden, J., and T. P. Burt (2002b), Infiltration, runoff and sediment production in blanket peat catchments; implications of field rainfall simulation experiments, *Hydrol. Processes*, *16*, 2537–2557.

Holden, J., P. J. Chapman, and J. C. Labadz (2004), Artificial drainage of peatlands: Hydrological and hydrochemical process and wetland restoration, *Prog. Phys. Geogr.*, *28*, 95–123.

Hope, D., M. F. Billett, R. Milne, and T. A. W. Brown (1997), Exports of organic carbon in British rivers, *Hydrol. Processes*, *11*, 325–344.

Hope, D., N. Picozzi, D. C. Catt, and R. Moss (1998), Effects of reducing sheep grazing in the Scottish Highlands, *J. Range Manage.*, *49*, 301–310.

Hsu, C. H., W. L. Jeng, R. M. Chang, L. C. Chien, and B. C. Han (2001), Estimation of potential lifetime cancer risks for trihalomethanes from consuming chlorinated drinking water in Taiwan, *Environ. Res.*, *85*(2), 77–82.

Hughes, S., B. Reynolds, S. A. Brittain, J. A. Hudson, and C. Freeman (1998), Temporal trends in bromide release following rewetting of a naturally drained gully mire, *Soil Use Manage.*, *14*, 248–250.

Huntington, T. G. (2005), Evidence for intensification of the global water cycle: Review and synthesis, *J. Hydrol.*, *319*, 83–95

Huttunen, J. T., H. Nykanen, J. Turunen, and P. J. Martikainen (2003), Methane emissions from natural peatlands in the northern boreal zone in Finland, Fennoscandia, *Atmos. Environ.*, *37*, 147–151.

Langlands, J. P., and I. L. Bennett (1973), Stocking intensity and pastoral production. I. Changes in the soil and vegetation of a sown pasture grazed by sheep at different stocking rates, *J. Agric. Sci.*, *81*, 193–194.

Limpens, J., F. Berendse, C. Blodau, J. G. Canadell, C. Freeman, J. Holden, N. Roulet, H. Rydin, and G. Schaepman-Strub (2008), Peatlands and the carbon cycle: From local processes to global implications—A synthesis, *Biogeosciences*, *5*, 1475–1491.

Lofts, S., B. M. Smith, E. Tipping, and C. Woof (2001), Modelling the solid-solution partitioning of organic matter in European forest soils, *Eur. J. Soil Sci.*, *52*, 215–226.

Lumsden, D. G., M. I. Stutter, R. J. Cooper, and J. R. Manson (2005), Model assessment of biogeochemical controls on dissolved organic carbon partitioning in an acid organic soil, *Environ. Sci. Technol.*, *39*, 8057–8063.

Mackay, A. W., and J. H. Tallis (1996), Summit type blanket mire erosion in the Forest of Bowland, Lancashire, UK: Predisposing factors and implications for conservation, *Biol. Conserv.*, *76*, 31–44.

Meyles, E. W., A. G. Williams, J. L. Ternan, J. M. Anderson, and J. F. Dowd (2006), The influences of grazing on vegetation, soil properties and stream discharge in a small Dartmoor catchment, southwest England, UK, *Earth Surf. Processes Landforms*, *31*, 622–631.

Michalzik, B., E. Tipping, J. Mulder, J. F. G. Lancho, E. Matzner, C. L. Bryant, N. Clarke, S. Lofts, and M. A. V. Esteban (2003), Modelling the production and transport of dissolved organic carbon in forest soils, *Biogeochemistry*, *66*, 241–264.

Milne, R., and T. A. Brown (1997), Carbon in the vegetation and soils of Great Britiain, *J. Environ. Manage.*, *49*, 413–433.

Mitchell, G., and A. T. McDonald (1995), Catchment characterization as a tool for upland water quality management, *J. Environ. Manage.*, *44*, 83–95.

Monteith, D. T., and C. D. Evans (2000), *The UK Acid Waters Monitoring Network: 10 Year Report*, ENSIS, London.

Monteith, D. T., et al. (2007), Dissolved organic carbon trends resulting from changes in atmospheric deposition chemistry, *Nature*, *450*(7169), 537–541.

Neal, C., B. Reynolds, J. Wilkinson, T. Hill, M. Neal, S. Hill, and M. Harrow (1998), The impacts of conifer harvesting on runoff water quality: A regional survey for Wales, *Hydrol. Earth Syst. Sci.*, *2*, 323–344.

Painter, T. J. (1983), Residues of D-lyxo-5-hexosulopyranuronic acid in *Sphagnum* holocellulose, and their role in cross-linking, *Carbohydr. Res.*, *124*, C18–C21.

Pastor, J., J. Solin, S. D. Bridgham, K. Updegraff, C. Harth, P. Weishampel, and B. Dewey (2003), Global warming and the export of dissolved organic carbon from boreal peatlands, *Oikos*, *100*, 380–386.

Ponnamperuma, F. M. (1972), The chemistry of submerged soils, *Adv. Agron.*, *24*, 29–96.

Rasmussen, S., C. Wolff, and H. Rudolph (1995), Compartmentalization of phenolic constituents in *Sphagnum*, *Phytochemistry*, *38*, 35–39.

Ratcliffe, D. A., and P. H. Oswald (1988), *The Flow Country*, Nat. Conserv. Counc., Peterborough, U. K.

Sansom, A. (1990), Upland vegetation management, the impacts of overstocking, *Water Sci. Technol.*, *39*, 83–92.

Scanlon, D., and T. Moore (2000), Carbon dioxide production from peatland soil profiles: The influence of temperature, oxic/anoxic conditions and substrate, *Soil Sci.*, *165*, 153–160.

Shotbolt, L., A. R. Anderson, and J. Townend (1998), Changes in blanket bog adjourning forest plots at Bad a' Cheo, Rumster Forest, Caithness, *Forestry*, *71*, 311–324.

Stewart, A. J. A., and A. N. Lance (1991), Effects of moor-draining on the hydrology and vegetation of northern Pennine blanket bog, *J. Appl. Ecol.*, *28*, 1105–1117.

Stoddard, J. L., J. S. Karl, F. A. Devinev, D. R. DeWalle, C. T. Driscoll, A. T. Herlihy, J. H. Kellogg, P. S. Murdoc, J. R. Webb, and K. E. Webster (2003), Response of surface water chemistry to the Clean Air Act Amendments of 1990, *Rep. EPA 620/R-03/001*, U.S. Environ. Prot. Agency, Research Triangle, N. C.

Thomas, G., A. R. Yallop, J. Thacker, T. Brewer, and C. Sannier (2004), A history of burning as a management tool in the English uplands, draft report to English Nature, Cranfield Univ., Silsoe, U. K.

Toberman, H., C. Freeman, R. R. E. Artz, C. D. Evans, and N. Fenner (2008), Impeded drainage stimulates extracellular phenol oxidase activity in riparian peat cores, *Soil Use Manage.*, *24*, 357–365.

Toet, S., J. H. C. Cornelissen, R. Aerts, R. S. P. Van Logtestijn, M. De Beus, and R. Stoevelaar (2006), Moss responses to elevated CO_2 and variation in hydrology in a temperate lowland peatland, *Plant Ecol.*, *182*, 27–40.

Tranvik, L. J., and M. Jansson (2002), Terrestrial export of organic carbon, *Nature*, *415*, 861–862.

Tucker, G. (2005), Review of the impacts of heather and grassland burning in the uplands on soils, hydrology and biodiversity, *Engl. Nat. Res. Rep. 550*, Engl. Nat., Peterborough, U. K.

Verhoeven, J. T. A., and W. M. Liefveld (1997), The ecological significance of organochemical compounds in *Sphagnum*, *Acta Bot. Neerlandica*, *46*, 117–130.

Visser, E. J. W., T. D. Colmer, C. W. P. M. Blom, and L. A. C. J. Voesenek (2000), Changes in growth, porosity, and radial oxygen loss from adventitious roots of selected mono-and dicotyledonous wetland species with contrasting types of aerenchyma, *Plant Cell Environ.*, *23*, 1237–1245.

Wallage, Z. E., J. Holden, and A. T. McDonald (2006), Drain blocking: An effective treatment for reducing dissolved organic carbon loss and water discolouration in a drained peatland, *Sci. Total Environ.*, *367*, 811–821.

Werrity, A. (2002), Living with uncertainty: Climate change, river flows and water resource management in Scotland, *Sci. Total Environ.*, *294*, 29–40.

Worrall, F., and T. P. Burt (2004), Time series analysis of long term river DOC records, *Hydrol. Processes*, *18*, 893–911.

Worrall, F., and T. P. Burt (2007a), Trends in DOC concentration in Great Britain, *J. Hydrol.*, *346*, 81–92.

Worrall, F., and T. P. Burt (2007b), Flux of dissolved organic carbon from U.K. rivers, *Global Biogeochem. Cycles*, *21*, GB1013, doi:10.1029/2006GB002709.

Worrall, F., and T. P. Burt (2009a), The effect of severe drought on the dissolved organic carbon (DOC) concentration and flux from British rivers, *J. Hydrol.*, in press.

Worrall, F., and T. P. Burt (2009b), Changes in DOC treatability: Indications of molecular changes in DOC trends, *J. Hydrol.*, *366*, 1–8.

Worrall, F., T. P. Burt, and R. Shedden (2003), Long terms records of riverine carbon flux, *Biogeochemistry*, *64*, 165–178.

Worrall, F., et al. (2004a), Trends in dissolved organic carbon in UK rivers and lakes, *Biogeochemistry*, *70*(3), 369–402.

Worrall, F., T. P. Burt, and J. K. Adamson (2004b), Can climate change explain increases in DOC flux from upland peat catchments?, *Sci. Total Environ.*, *326*, 95–112.

Worrall, F., T. P. Burt, and J. K. Adamson (2005a), Predicting the future DOC flux form upland peat catchments, *J. Hydrol.*, *300*, 126–139.

Worrall, F., T. P. Burt, and J. K. Adamson (2005b), Fluxes of dissolved carbon dioxide and inorganic carbon from an upland peat catchment: Implications for soil respiration, *Biogeochemistry*, *73*, 515–539.

Worrall, F., T. P. Burt, and J. K. Adamson (2006a), The rate of and controls upon DOC loss in a peat catchment, *J. Hydrol.*, *321*, 311–325.

Worrall, F., T. P. Burt, and J. K. Adamson (2006b), Trends in drought frequency—The fate of Northern Peatlands, *Clim. Change*, *76*, 339–359.

Worrall, F., T. P. Burt, and J. K. Adamson (2006c), Long-term changes in hydrological pathways in an upland peat catchment—Recovery from severe drought?, *J. Hydrol.*, *321*, 5–20.

Worrall, F., T. P. Burt, and J. K. Adamson (2007a), Change in runoff initiation probability over a severe drought—Implications for peat stability, *J. Hydrol.*, *345*, 16–26.

Worrall, F., A. Armstrong, and J. Holden (2007b), Short-term impact of peat drain-blocking on water colour, dissolved organic carbon concentration, and water table depth, *J. Hydrol.*, *337*, 315–325.

Worrall, F., H. S. Gibson, and T. P. Burt (2007c), The consequences of drainage and drain-blocking on the release of DOC from peat catchments: Is drain-blocking a solution to the DOC problem?, *J. Hydrol.*, *337*, 315–325.

Worrall, F., A. Armstrong, and J. K. Adamson (2007d), The effect of burning and sheepgrazing on water table depth and soil water quality in a blanket bog, *J. Hydrol.*, *339*, 1–14.

Worrall, F., H. S. Gibson, and T. P. Burt (2008a), Production vs. solubility in controlling runoff of DOC from peat soils—The use of an event analysis, *J. Hydrol.*, *358*, 84–95.

Worrall, F., T. P. Burt, and J. K. Adamson (2008b), Long-term records of DOC flux from peat-covered catchments—Evidence for a drought effect?, *Hydrol. Processes*, *22*(16), 3181–3193.

Worrall, F., T. P. Burt, and J. K. Adamson (2008c), Linking pulses of atmospheric deposition to DOC release in an upland peat-covered catchment, *Global Biogeochem. Cycles*, *22*, GB4014, doi:10.1029/2008GB003177 .

N. Fenner and C. Freeman, School of Biological Sciences, Bangor University, Deiniol Road, Bangor LL57 2UW, UK. (n.fenner@bangor.ac.uk)

F. Worrall, Department of Earth Sciences, University of Durham, Science Laboratories, South Road, Durham DH1 3LE, UK.

The Role of Natural Soil Pipes in Water and Carbon Transfer in and From Peatlands

J. Holden, R. P. Smart, P. J. Chapman, and A. J. Baird

School of Geography, University of Leeds, Leeds, UK

M. F. Billett

Centre for Ecology and Hydrology Edinburgh, Edinburgh, UK

Natural piping has been reported in peatlands around the world. This chapter reviews the role of natural pipes in peatland hydrology and carbon fluxes. There is a growing body of evidence to suggest that pipes are important hydrological agents in peatlands typically delivering over 10% of streamflow. Deep and shallow pipes respond rapidly to rainfall inputs demonstrating strong connectivity with the peat surface. While ephemeral pipes respond quickly to rainfall events, they also appear to obtain water from deep peat layers and underlying mineral strata. This mix of different sources of water results in highly variable concentrations of dissolved organic carbon and dissolved carbon dioxide and methane within pipe water. Early results from an intensive monitoring study in a blanket peatland in northern England suggest that pipe flows may account for around half of the dissolved organic carbon that is delivered to the stream network. Episodic pulses of particulate organic carbon from pipe outlets are common during storm events. Work that is underway to understand more about the sources of carbon being released from pipe networks is outlined, and several areas of further research are highlighted including examination of the role of natural pipes in bog pool hydrology and carbon cycling.

1. INTRODUCTION

Recent research on carbon cycling within peatlands has focused on the relationships between soils, water table position, temperature, plants, microbes, and carbon flux [e.g., *Billett et al.*, 2006a; *Cole et al.*, 2002; *McNeil and Waddington*, 2003; e.g., *Strack et al.*, 2008; *Worrall et al.*, 2006]. There is

Carbon Cycling in Northern Peatlands
Geophysical Monograph Series 184
Copyright 2009 by the American Geophysical Union.
10.1029/2008GM000804

very little work examining the role that water "movement" through peatlands plays in the retention and release of particulate, dissolved, and gaseous forms of carbon. Recent strong interest in waterborne carbon exports from peatlands has focused mainly on concentrations and fluxes of dissolved organic carbon (DOC) [*Andersson and Nyberg*, 2008; *Billett et al.*, 2006a; *Dawson et al.*, 2002]. We know relatively little about what controls the transport of DOC, particulate organic carbon (POC), and dissolved gaseous forms of carbon within peatlands [e.g., *Billett and Moore*, 2008] and the hydrological processes leading to their delivery to rivers.

Natural soil pipes consist of connected natural conduits, often many centimeters in diameter, which transport water, sediment, and solutes through soil systems. These pipes can often be several hundred meters in length and typically form branching subsurface networks. While some pipe networks form at the interface of soil horizons, other networks may occur at a variety of depths within the soil profile and are therefore potentially able to connect shallow and deep sources of water, sediments, and solutes. This may be of great importance in a number of environments such as in ombrotrophic peatlands, where deep sediments and solutes (including dissolved gases) were thought to be largely disconnected from the stream network [Holden and Burt, 2003].

Natural soil pipes have been reported on every continent, except Antarctica, and in a broad range of environments and soil types including, and most commonly, in tropical forest soils [Baillie, 1975; Chappell and Sherlock, 2005; Sayer et al., 2006], collapsible loess soils [Zhu, 2003], boreal forests [Roberge and Plamondon, 1987], subarctic hillslopes [Carey and Woo, 2000], steep, temperate, humid hillslopes [Terajima et al., 2000; Uchida et al., 1999, 2005], and dispersive semiarid soils, where severe gully erosion has often resulted from pipe development [Bryan and Jones, 1997; Crouch et al., 1986; Gutierrez et al., 1997]. However, one of the most commonly reported environments where soil piping has been reported is peatlands. For example, natural soil pipes have been reported in the peatlands of Scandanavia, New Zealand, Tasmania, Indonesia, Canada, Siberia, Ireland, and the United Kingdom [e.g., Gunn, 2000; Holden, 2005a; Holden and Burt, 2002; Holden et al., 2004; Jones, 1981; Jones et al., 1997; Norrstrom and Jacks, 1996; Price, 1992; Rapson et al., 2006; Thorp and Glanville, 2003; Markov and Khoroshev, 1988]. Despite this, we still know relatively little about their hydrological role in peatlands or how they affect carbon cycling in different environmental settings. This chapter provides an overview of recent work on peatland piping processes and presents early results from a new study to examine the link between natural pipes and water and carbon transfers within and from peatlands. Since most of the published and ongoing work on peat piping has focused on blanket peat (histosols) and shallow peats (histic podzols), these soils provide the focus for this chapter in text.

2. PEAT PIPE FORMATION AND DISTRIBUTION

Published work on natural pipes in peatlands has most frequently reported their occurrence in blanket peats [Holden, 2005a; Jones, 1981; Jones et al., 1997; McCaig, 1983; Price, 1992]. However, they have also been reported in many other peatland types including low gradient peatlands such as the James Bay Lowlands of Canada [Woo and DiCenzo, 1988], raised bogs [Ingram, 1983], various peatlands in Germany [Dittrich, 1952; Egglesmann, 1960; Rudolf and Firbas, 1927],

and gully-head fens in New Zealand [Rapson et al., 2006]. Glaser [1998] noted that "pools can suddenly drain when they are breached by the pipe systems which honeycomb many bog deposits" and presented a photograph of such a pipe on a bog in Quebec. The above demonstrates that piping is a widespread phenomenon occurring in peatlands across the world. Despite this, most research into peat pipe hydrology has taken place in blanket peatlands or histic podzols.

Peatlands are conducive to piping because of the combined effect of a plentiful supply of water and sharp transitions in hydraulic conductivity both laterally and vertically within the peat profile [Rosa and Larocque, 2008]. While pipe formation instigated by faunal burrows (e.g., small mammals, earthworms, and crustaceans) is common in many environments [e.g., Onda and Itakura, 1997; Wilson and Smart, 1984], this has not been reported in peatlands where the environmental conditions are generally too harsh for such fauna.

When water flows through a porous medium there is drag force. When this drag force becomes sufficient to entrain material at an open stream bank, then a subsurface conduit can be formed which erodes backward from the outlet. This process is sometimes referred to as sapping or spring sapping. Such a sapping pipe formation process is different from pipe formation caused by root channels or desiccation cracks, although the subsequent enlargement processes will be the same [Bryan and Jones, 1997]. Enlargement of desiccation cracks or root channels is reported to be the main causative factor of pipe formation in peat and histic podzols [Gilman and Newson, 1980; Holden, 2005b; Jones, 1982], but overgrowth by the peat mass of an existing small stream channel or natural seepage line [Anderson and Burt, 1982; Thorp and Glanville, 2003] is also common. Indeed, there are many peatland streams that run through roofed sections of peat.

Pipes are most commonly observed at their outlets on peatland stream banks (Figure 1a). Occasionally, during storm events, water can be seen emerging onto the peat surface indicating the location of a pipe which is surcharging and discharging [Jones, 1982] (Figure 1b), while more often, pipes can be observed where their roof has collapsed to form a sinkhole in the peat (Figure 1c). Behind the pipe outlet, pipes can form complex branched networks, and the diameter and shape of the pipes can be quite different from that observed at their outlet. In the histic podzols of Plynlimon in mid-Wales, which until recently dominated the world's literature on soil piping, the pipe networks tend to exist at the interface between the peat and the underlying mineral substrate [Jones, 1994; Jones and Crane, 1984]. However, in deep peats, pipes form complex undulating networks, which can occur throughout the peat profile and extend for

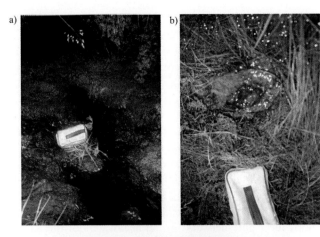

Figure 1. Natural peatland pipe outlets (a) flowing pipe on a streambank, (b) pipe overflowing onto the peat surface during a storm, (c) collapsed pipe roof. The camera case used for scale is 13 cm × 7 cm.

several hundred meters [*Holden and Burt*, 2002; *Holden et al.*, 2002]. Indeed, *Chapman* [1993] reported peat pipes that were so large in the deep peats of the Peak District of northern England that cavers explored them, often following them underground for over 50 m. The collapse of roofs of large pipes appears to be a common cause of gully development in peats, and this is very prominent in the most eroded peats of the Peak District in northern England. At the head of many gullies in Ireland, Scotland, and northern England natural pipes are often found, and while piping can lead to gully development, gullies may also steepen the hydraulic gradient and lead to exposed peat subject to desiccation which favors pipe development. Ultimately, piping may lead to stream network development, and at the head of many streams in mid-Wales, pipes are commonly found.

Ground-penetrating radar (GPR) has been used for peatland stratigraphic mapping for the last three decades [*Lapen et al.*, 1996; *Warner et al.*, 1990], and penetration depths and resolution are good despite the expected high attenuation due to high water contents. This is because the attenuation is reduced by the low electrical conductivity of peat pore water (in ombrogenous peats). *Holden et al.* [2002] have shown GPR can be used to map natural pipes, larger than around 10 cm in diameter, in peatlands. The hydrological connectivity of these large pipes can also be remotely mapped using simple tracers (such as a weak sodium chloride solution) that can be detected by the GPR [*Holden*, 2004]. Such work has revealed that, where pipe networks appear to meet on their longitudinal paths across peatlands, they are sometimes connected. They also sometimes pass each other at different depths in the peat profile and are hydrologically disconnected. GPR provides a technique that allows for a good comparison of the occurrence of piping between peat catchments. GPR has major advantages over techniques that rely on observation at streambanks and sinkholes, dye tracing, or listening for gurgling water beneath the peat surface [*Jones*, 1981]. The drawback of GPR is that it cannot detect small diameter pipes (<10 cm) which may be hydrologically important.

Through GPR survey, *Holden* [2005a] has recently shown piping to be ubiquitous in UK blanket peat catchments. One hundred and sixty UK blanket peat catchments were surveyed, with pipes found in all with a mean frequency of 69 km^{-1} of surveyed transect. Aspect did not influence pipe frequency in blanket peats, except through a weak role in catchments with annual precipitation <1500 mm. In these drier catchments, southwesterly facing peat slopes tended to have more frequent piping indicative of desiccation cracking on sun-facing slopes as a driver of pipe formation. Topographic position was an important control on the frequency of pipes found per unit length of GPR transect. However, while piping has been reported for mineral soils to be more likely to occur as slope gradient increases [*Gutierrez et al.*, 1997; *Jones*, 1981], this was not the case for the blanket peats surveyed by GPR. Blanket peat topslopes had significantly greater pipe frequencies than footslopes, with midslopes having the lowest pipe frequency. As shown in Figure 2, this pattern is similar to surface erosion features found in many peatlands [*Bower*, 1961] and indicates that, while there may be fewer pipes on steeper midslopes, they may still transmit cumulative flow from more branching networks upslope. *Holden* [2005a, 2005b] proposed that the nature of the preexisting topography (and its associated drainage conditions), on which blanket peatlands have developed, promotes differential buildup of peat deposits that influences later pipe development. The differential buildup of peat occurs because of the development of small pools in surface depressions and larger bog pool systems on hilltops and hilltoes, which are colonized by a mosaic of plants with specialist positions within the local microtopography.

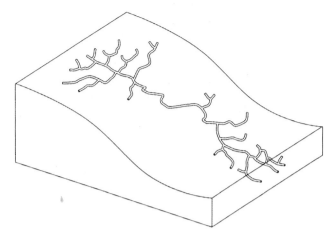

Figure 2. Schematic of a typical branching pipe network on a blanket peatland.

The remains of these plants are then incorporated into the peat as it thickens or as the pool is recolonized by lawn-hummock assemblages [*Glaser*, 1998], resulting in a peat of variable properties throughout its profile. This means that the peat on hilltops and hilltoes is inherently more susceptible to preferential flow and piping than on midslopes where hummock-pool formation is lacking, and a more uniform peat structure has developed [*Holden*, 2005a]. Pipes have commonly been reported that link bog pools [*Holden and Burt*, 2002], and indeed, many drained bog pools have been found to have pipes on their walls or floor [*Rapson et al.*, 2006; *Glaser*, 1998].

The UK GPR survey of blanket peats has also shown that peatland management can have a strong influence on pipe distributions. Drained (ditched) peatlands were found to have a significantly greater frequency of pipes (over twice as many) than intact peatlands, and there was a significant relationship between age of ditches and the frequency of pipes, with the number of pipes doubling 35 years after drain cutting [*Holden*, 2006]. *Calluna vulgaris* L. growth has been shown to be a causative factor in macropore development, and piping was more common where *Calluna* was present compared to sedges and mosses [*Holden*, 2005b]. *Calluna*, which grows better on drier upland peats or mineral-organic soils, is encouraged on many upland UK peats due its perceived role in supporting game bird populations.

3. PIPE FLOW

In a small (4 ha) headwater catchment of the River Wye in mid-Wales, the contribution of peak pipe flow from the major pipe outlet in the catchment to maximum stream discharge during rainfall events ranged from 3.3 to 32.2%, with a mean of 10% [*Chapman*, 1994]. The smallest contributions tended to occur in long duration, low intensity rainfall events, while the largest contributions were associated with short, high intensity events. *Jones* [1990], working in the nearby Maesnant catchment, also observed that the contribution of pipe water to streamflow decreased when the catchment was very wet and/or the rainstorm very heavy, even though the absolute quantity of pipe flow continued to increase.

Pipe flow appears to be controlled by three major factors; total storm rainfall, soil moisture antecedent conditions, and rainfall intensity [*Chapman*, 1994; *Holden and Burt*, 2002; *Muscutt et al.*, 1990]. The relationship between total event runoff, for a pipe network, in the headwaters of the River Wye, and the corresponding storm rainfall is shown in Figure 3. The relationship is approximately linear, although a wide scatter exists. This scatter can be related to soil moisture antecedent conditions. An antecedent runoff index (ARI) of the hydrological conditions in the catchment prior to the rainfall event was derived by *Chapman et al.* [1997]. Similar indices have been used elsewhere [e.g., *Foster*, 1979] to represent the diminishing impact over time of preceding rainfall events. The index takes the form;

$$\mathrm{ARI} = \sum_{i=1}^{n} k^{i-1} R_{t-i}$$

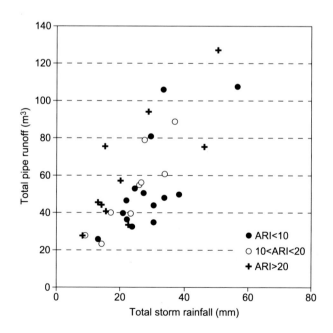

Figure 3. Total event rainfall versus total event runoff from a pipe outlet in the Upper Wye catchment. ARI, antecedent runoff index.

where t is the day on which the event occurred, R_{t-i} is the total runoff recorded on day $t - i$, and k is a coefficient between 0 and 1. The effect on present runoff of past rainfall can be set either by changing n or by changing k. Using $k = 0.5$, as is used here, means that runoff conditions over the preceding 7 days are represented; thereafter, events have little effect on the value of the ARI. Events with low short-term ARI (<10; dry antecedent conditions) tend to produce less runoff from the pipe than events with high ARI (>20) from a similar total storm rainfall. The effect of antecedent conditions on pipe flow is clearly illustrated by the consecutive events of 11, 12, and 13 March 1992 as shown in Table 1. The increase in pipe runoff generated per millimeter of precipitation in each consecutive event is controlled by the wetter antecedent conditions in the catchment as indicated by the ARI.

Around 10% of streamflow was found to move through pipe networks in Little Dodgen Pot Sike, a deep blanket peat catchment of 0.44 km^2 in northern England [*Holden and Burt*, 2002] with discharges over 4.6 L s^{-1} from single pipes being reported. The proportion of stream water moving though pipes varied during storms and with antecedent conditions. Maximum contributions (typically over 30%) were most commonly observed for intermediate-sized storm events. *Jones and Crane* [1984] reported that 49% of streamflow was produced by soil pipes in histic podzols in mid-Wales. The pipes also transmit water to the stream from an area on the hillslope of 10 to 20 times greater than would be the case if all stormwater were drained via surface and near surface flow [*Jones*, 1997]. This shows the potential of pipes to transport carbon from more remote areas of the peatland, which may appear disconnected from the drainage network.

Figure 4 shows the typical response of a pipe monitored during a 30-day period during 1999 in Little Dodgen Pot Sike. The pipe was monitored at three points along its course. The first (site a) was adjacent to an area of bog pools where the pipe was first evident because the roof had collapsed. It was then gauged 70 m downslope (mean gradient 0.6°) at the next natural opening (a roof collapse) (site b) and then a further 95 m downslope at its outlet at the head of a vegetated gully (site c). At all three sites, flow was gauged using V notch weirs and floats attached to potentiometers to gauge level. The response was broadly similar at all three points in the pipe network, although the hydrograph characteristics at sites b and c most closely matched each other. While discharge increased along the pipe from site b to site c, discharge at site b was lower than upslope at site a. *Newson and Harrison* [1978] reported significant losses from pipe flow during experiments using artificially pumped water in natural ephemeral pipes in shallow peats in Wales and proposed that this situation was normal. However, *Jones* [1982] suggested that pipes in histic podzols gained more pore water via groundwater seepage than they lost so that discharge continued to increase downslope. Blocked sections of pipe, which cause water in upstream parts of the pipe to spill out to the surface or to overflow to other connected pipes may explain the situation in Little Dodgen Pot Sike. In fact, 10 m upstream of site b another pipe was found discharging vertically upward into the monitored pipe. These findings are indicative of a complex interconnected network of pipes in peatland systems.

Peat pipe flow hydrographs tend to be "flashy" in nature. This suggests that most pipe flow results from rapid transmission of water from the surface through cracks and pipe openings into pipe networks. Nevertheless, data from Little Dodgen Pot Sike [*Holden and Burt*, 2002] showed that the stream hydrograph from the catchment was actually more flashy (narrower spike) than the hydrograph from the pipes, although streamflow and pipe flow peaks (when separate pipe flows were summed) tended to occur, on average, at approximately the same time. Thus, it appears that pipes rapidly transmit new surface water but also tend to act as conduits for slower drainage from the peat as the water table recedes through the acrotelm once rainfall has stopped.

Figure 5 provides some example hydrographs along with DOC and POC concentrations for pipes in the Cottage Hill Sike catchment on the Moor House National Nature Reserve in northern England. This newly instrumented catchment is in blanket peat, typically 2 to 7 m thick, with natural piping. Earlier work on the site has focused on stream DOC fluxes and water table elevation [*Clark et al.*, 2007; *Evans et al.*, 1999]. A total of 88 pipe outlets have so far been identified within the 20-ha catchment, with outlets ranging in depth from those open at the surface to those over 2 m deep and from 1 to 30 cm in diameter. A peak flow of 10.8 L s^{-1} has been reported during the first 7 months of monitoring, and the eight continuously gauged pipes alone have accounted for 2.9% of streamflow to date. Of the eight gauged pipes, three flow continuously, and five are ephemeral (Table 2); one of these is thought to be an overflow pipe from a continuously flowing pipe network nearby.

Table 1. Rainfall-Pipe Runoff Relationship During Three Consecutive Events in a 4-ha Headwater Catchment of the River Wye, Mid-Wales

Day	Total Rain (mm)	Total Pipe Runoff (m^3)	Runoff per Millimeter Rain (m^3)	ARI
11.3.92	27.36	78.8	2.88	10.02
12.3.92	28.62	93.9	3.28	20.11
13.3.92	15.12	59	3.90	28.61

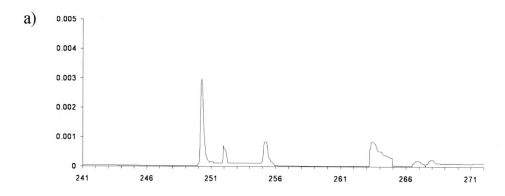

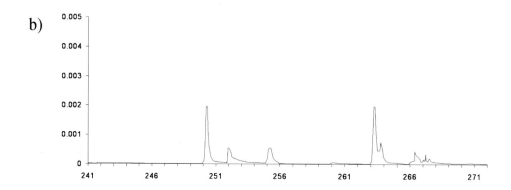

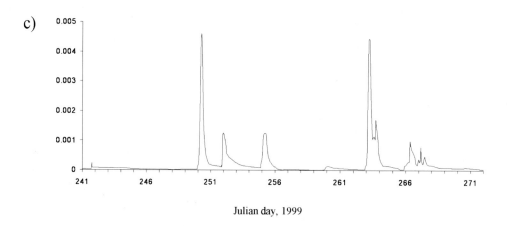

Julian day, 1999

Figure 4. Discharge, cumecs, at points within the same pipe (a) at an opening near bog pools, (b) 70 m downslope from Figure 4a, and (c) 165 m downslope from Figure 4a.

Both perennial and ephemeral pipes respond rapidly to rainfall events as do both deep and shallow pipes. This suggests that deep pipes are well connected hydrologically with the peat surface. However, not all pipes show the same hydrological behavior. Some pipes only flow during high-flow events suggesting that they are "overflow" components of other pipe networks; such pipes have also been reported for peats in Wales [*Jones and Crane*, 1984] and Canada [*Woo and DiCenzo*, 1988]. Other pipes produce hydrographs with a greater proportion of "base flow." This suggests that both new and old sources of water (and potentially carbon) are being released by these pipe systems.

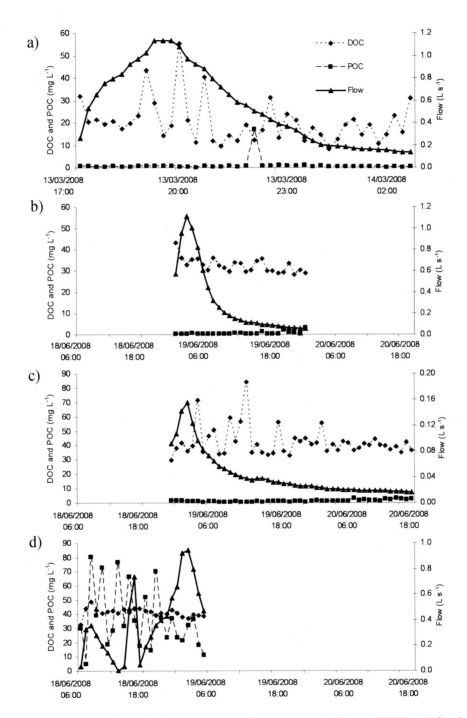

Figure 5. Example pipe flow, DOC, and POC concentrations for storm events in Cottage Hill Sike; (a) for pipe 5 starting on 13 March 2008 (10 h); (b) for pipe 5; (c) for pipe 32; (d) for pipe 39. Figures 5b–5d are for the same storm starting on 18 June 2008 although samples were collected by the autosamplers for different durations for the storm at each pipe. Table 2 describes the physical characteristics of each pipe outlet.

Table 2. Physical Characteristics of Flow Gauged Pipes Within Cottage Hill Sike[a]

	Site							
	P1	P3[b]	P5[b]	P6	P9	P32[b]	P35[b]	P39[b]
Pipe outlet diameter, cm	17	3	30	6.5	1	20.5	10	10
Pipe roof depth from peat surface, cm	47	75	25	60	100	100	30	160
Height above stream, cm	115	59	>100	>100	>100	45	75	10
Flow characteristics	E	E	C	E	E	C	C	E
Maximum recorded flow	0.015	0.015	10.8	0.012	0.013	0.29	U	2.01
Minimum recorded flow	0	0	0.006	0	0	0.005	U	0

[a]Abbreviations are E, ephemeral; C, continuous flow, all flows measured in L s^{-1}; and U, unknown.
[b]Pipes also have autosamplers installed.

It is not clear whether peatland pipes (1) decouple overland flow from near surface matrix flow by allowing water to by-pass large parts of the peat mass and/or (2) couple deep and shallow pore water (water in the peat matrix) with overland flow. Peatland pipes may alter the proportions of overland flow and flow through the peat matrix (throughflow), thus affecting water residence time, evaporation, and the rainfall to runoff ratio. The new instrumentation at Cottage Hill Sike seeks to examine how pipes interact with other peatland hydrological processes through their role as drainage channels (escape routes) and as sources of water from different parts of the peatland. Such assessment will aid our process understanding of the dual role of pipes in water movement and carbon cycling.

Recent advances in the study of subsurface water flow in peatlands, such as the North American and United Kingdom work on hydraulic conductivity determination and the effect of biogenic gas bubbles on hydrological and hydrochemical processes [*Baird and Waldron*, 2003; *Beckwith and Baird*, 2001; *Beckwith et al.*, 2003] illustrates that peats are often not always saturated below the water table. Biogenic gas bubbles contain no oxygen, but it is possible that pipe networks allow entry of (relatively) oxygen-rich air deep into the peat, thus enhancing rates of peat decomposition close to pipe walls. *Woo and DiCenzo* [1988] found that peat pipes could act as zones of local water table drawdown during high water table periods, while during drier periods, they could act as a source of water for the surrounding peat, demonstrating the complex effect that peat pipes might have on local hydrology and carbon production and export.

4. PIPE CARBON EXPORT

Most research on particulate carbon loss from peatlands focuses on streambank or surface erosion [*Evans and Warburton*, 2005; *Tallis*, 1995; *Warburton*, 2003], and there is very little research on subsurface particulate erosion. Pipes appear to be components of peatlands around the world, and yet, there are no data on how important pipes might be for peatland sediment or carbon export.

The relationship between pipe frequency and the age of artificial peat drainage networks identified by *Holden* [2006] for UK blanket peats, suggests that contemporary pipe erosion actively follows ditch cutting. Using GPR data and measurements of pipe outlet sizes and organic matter content measurements of the peat at each site, he estimated that the rate of pipe erosion increased over time as the pipe network branched and advanced following desiccation caused by ditches. Particulate carbon loss from subsurface pipes was greatest where drains were oldest. For slopes where the drainage system was 40 years old, data suggested there would be an extra 5.8×10^3 kg C km^{-2} a^{-1} exported from subsurface pipe erosion alone over that 40-year period, compared to that from an undrained slope. This value would be in addition to any surface erosion related to ditch channel incision or other surface processes. However, these data were estimates, and there were no direct measurements of POC.

Until now, there have been no direct measurements of POC loss from peat pipes in intact or disturbed peats. It is possible that, unlike disturbed peats, pipes in intact peatlands remain relatively stable and produce little POC. However, one pipe that was monitored by *Holden and Burt* [2002] suddenly ceased flowing during their monitoring period and did not resume. This is likely to have occurred because the pipe became blocked by sediment, or a collapse occurred somewhere in the network, suggesting that pipe networks can behave in a dynamic and nonlinear way. Such phenomena have also been reported in loess pipes [*Zhu*, 1997] and a cambisol in Japan [*Uchida et al.*, 1999]. Data from Plynlimon, Wales, suggested that areas of the peat catchments which seemed to contain more pipes also yielded more sediment to the stream network [*Jones*, 2004], and it was estimated that the pipes produced 25% of stream sediment in the Maesnant basin [*Jones*, 1990]. Nevertheless, most of this sediment appeared to be mineral sediment because the pipes at Plynlimon occurred on the mineral substrate at the base of a shallow peat layer.

In the Cottage Hill Sike catchment, five of the eight continuously gauged pipes have been fitted with automatic pump samplers so that stormflow DOC and POC can be monitored. The monitored pipe characteristics are given in Table 2. During storms, samples are collected at intervals varying from 15 to 60 min. Samples are then analyzed in the laboratory for POC, DOC, and other parameters while there is in situ monitoring of flow, temperature, pH, and conductivity. Figure 5 shows, for the first time, POC production during storm events for pipes in deep peat. The figure shows data from two storms for one pipe (Figures 5a and 5b) and for the same storm for three different pipes (Figures 5b, 5c, and 5d). There are marked differences in response both between storms for a single pipe and between pipes for the same storm. There was a tendency for episodic flushes of POC during the storms, with peak concentrations of 99 mg C L^{-1} observed, which is indicative of occasional pipe wall failures within the pipe. For example, Figure 5a shows fairly uniform POC concentrations during the pipe flow storm hydrograph except for one sudden spike in POC. These spikes in POC are commonly seen on all the pipes being monitored. Figure 5d shows a different pipe (from that in Figures 5a, 5b, and 5c) with generally greater and more fluctuating concentrations of POC during storms.

POC export from the sampled pipes over the intermediate-sized storm events studied so far have ranged from 3.75 to 234 g from single pipes over a 24-h period, representing between 1.5 to 21% of organic carbon export from the pipes. Our early data suggest that peat pipes may be major sources of POC for the stream network and that they provide episodic delivery of POC to the stream system. To date, we only have data from one complete storm event when both stream and pipe POC and DOC flux were monitored from base flow through peak back to base flow again. During this storm, the monitored pipes produced 7.9% of POC when compared to the overall catchment POC yield. If we consider that 88 pipe outlets have so far been detected, and our monitored pipes are considered to be representative of those, then it is estimated that the pipes produced more than twice as much POC as was lost from the stream outlet.

Despite the fact that only a few studies have investigated the hydrochemistry of pipe flow, it is possible to make a number of generalizations. Pipe water is commonly acidic, high in DOC, and low in base cation concentrations reflecting the chemistry of the organic soil through which they flow [*Chapman et al.*, 1993, 1997; *Cryer*, 1980; *Jones*, 2004; *Muscutt et al.*, 1990]. In mid-Wales, *Chapman* [1994] observed that DOC concentrations at an ephemeral pipe outlet ranged between 12 and 38 mg L^{-1} and displayed no consist-

ent temporal variations through storms. Table 3 shows that DOC concentrations increased in successive summer storms to peak in the event of 24 July 1991. Concentrations in the autumn and March 1992 events were considerably lower than those observed in the summer events. Hence, greater variation in DOC concentrations was observed between rather than during events. *Chapman* [1994] also noted that DOC concentrations in pipe water were considerably higher than those in soil solutions from the peat horizon of the soil (maximum of 20 mg C L^{-1} in August). In the nearby Maesnant catchment, *Jones* [2004] reported considerably lower concentrations of DOC than *Chapman* [1994] in perennial pipe outlets (mean = 4.08 mg L^{-1}), heads of perennial pipes (mean = 1.5 mg L^{-1}), and ephemeral pipe outlets (2.2 and 15.6 mg L^{-1} for two different outlets). Given the uncertainties governing the generation of pipe flow, the processes which control pipe water chemistry still remain elusive and apparently highly variable in both space and time.

Figures 5a–5d show the export of DOC from pipes during storms in Cottage Hill Sike. The response of DOC to storm events varies between events for the same pipe (compare Figures 5a and 5b) and between pipes for the same event (compare Figures 5b and 5c). For pipe 5, DOC concentrations fluctuate widely (10–55 mg L^{-1}) during the event of 13 March 2008 displaying no clear relationship with pipe flow. In contrast, during the storm of 19 June 2008 DOC concentrations displayed considerably less variation (30–40 mg L^{-1}). During the storm of 19 June 2008, DOC concentrations for pipes 5 and 18 showed similar patterns, displaying no relationship with pipe flow and little variation throughout the storm. In contrast, concentrations fluctuated widely (30–80 mg L^{-1}) at the outlet of pipe 32, particularly following the peak in pipe flow (Figure 5c). Thus, the response of pipe water DOC to storms is very variable. These pipe flow DOC concentration

Table 3. The Mean and Range in Concentration of DOC at a Pipe Outlet in the Headwaters of the River Wye, Mid-Wales[a]

Storm	DOC (mg L^{-1})	Number of Samples
10 Jun 1991	21 (20–22)	3
3 Jul 1991	29 (28–30)	3
24 Jul 1991	35 (31–38)	7
31 Oct 1991 to 1 Nov 1991	16 (14–18)	8
10–1 Mar 1992	13.5 (12–16)	7

[a]Range is shown in parentheses.

data are higher than those reported by *Chapman* [1994] and *Jones* [2004] for pipes in shallow peats in mid-Wales but, as observed by *Chapman* [1994], display no consistent relationship with pipe flow. One might have expected a smoother flushing or dilution effect to be seen, but these data are the first such high-resolution data from peat pipes during storms. One possible explanation for the very spiky response in pipe water DOC concentrations observed in Figures 5a and 5c is that the pipe network contains hundreds of small sumps (or U-shaped bends), as the pipes undulate within the peat mass. Some of the sumps within the pipe network may contain water, which during low flow periods have accumulated high concentrations of DOC produced following oxidation around pipe walls or supplied from the draining acrotelm. Following rainfall, water can begin to move rapidly into the pipes via macropores and surface openings and flush through the pipe network. Different parts of the pipe network will fill at different rates depending on the macropore network and hydraulic conductivity of the peat in their vicinity. After the beginning of a storm, we may expect an increasing proportion of the pipe network to become an active contributing area to the pipe outlet. However, DOC appears to fluctuate well into the flow recession, and so it may be that there are different water sources contributing to the pipe flows at different points in time which accounts for the variation. The differential response of different parts of the pipe network to rainfall means that there will be mixtures of DOC-rich peat-derived soil water and DOC-poor recent rainwater that vary over time. DOC export from the monitored pipes accounted for 3.0% of stream DOC export during one storm for which we have complete data from all eight monitored pipes at Cottage Hill Sike. Scaling this up (as with POC) suggests that pipe DOC export accounted for 52.8% of the catchment DOC export during that single storm.

Of course, the amount of DOC entering pipes will depend, in part, on the source of the water. Therefore, the sources of pipe water need to be determined if we are to understand carbon production processes in peatland pipe systems. As part of the monitoring program at Cottage Hill Sike, we will measure the isotopic composition of POC and DOC produced from the pipes to assess their source and age. This will help us locate the origin of these forms of C within the peat mass and so map the subsurface "micro-carbon catchments" of the pipes.

Conflicting isotopic and solute data from upland soils (not deep peats) suggest that pipes transport both old and young water [*Hyett*, 1990; *Sklash et al.*, 1996]. *Billett et al.* [2007] found that DOC released from the stream system of four UK peatland catchments was young (modern-202 years BP), while CO_2 being released from the stream water surface was much older (modern-1449 years BP) suggesting (in associa-

tion with the $\delta^{13}C$ values) that the sources of fluvial DOC and CO_2 were in most cases different.

As CO_2 concentrations in the soil atmosphere of peats may be 10 to 100 times higher than concentrations in the aboveground atmosphere, we expect concentrations in pipe water to be significantly higher than those in stream water. Pipe water supersaturated with CO_2 and CH_4 will degas rapidly when experiencing lower ambient pressures once it leaves the pipe system. In peatland systems, this outgassing in the drainage network potentially contributes a significant loss of terrestrially derived C to the atmosphere [*Hope et al.*, 2004]. Pipe outgassing has never been measured, but it is known that peatland headwater streams are important zones of degassing [*Dawson et al.*, 2002], and there are parts of peatland catchments which act as hotspots of degassing [*Billett and Moore*, 2008]. In a recent study within the Cottage Hill Sike catchment, gullies were found to be significant degassing hotspots of CO_2 and CH_4 [*McNamara et al.*, 2008]. These hotspots may be associated with pipe systems which could provide a focus for gas escape from deep peats.

Table 4 provides data on the CO_2 and CH_4 concentrations in the pipe waters of Cottage Hill Sike measured to date. The first 7 months of pipe and streamwater CO_2 and CH_4 data from Cottage Hill Sike show large spatial variation in dissolved gas concentrations within the catchment. Concentrations range from 0.22 to 22.36 mg L^{-1} of CO_2-C and 0.0006 to 12.97 mg L^{-1} of CH_4-C in pipe waters and 0.22 to 5.73 mg L^{-1} of CO_2-C and 0.0008 to 0.11 mg L^{-1} of CH_4-C in streamwater at the catchment outlet. However, there are also differences in concentration found between continuously flowing pipes and ephemeral pipes. Generally, the greatest dissolved gas concentrations are found in waters from continuously flowing pipes. This suggests that continuously flowing pipes are fed by water from deeper within the peat and possibly from the mineral substrate. Hence, pipes can provide conduits for the flow of water from deep within peat deposits to surface waters such as streams. This hypothesis is corroborated by pH data. pH values from continuously flowing pipes at Cottage Hill Sike are between 0.5 and 1.0 pH units higher than those found in ephemeral pipes. For example, the pH values from continuously flowing pipe P32, the pipe associated with the highest CO_2 and CH_4 concentrations, range from 3.53 to 5.90 while those from ephemeral pipe P3 range from 3.53 to 4.29. The same pattern is found in conductivity values from the continuous monitoring of both ephemeral and continuously flowing pipes. The mean conductivity of waters from P32 is 0.128 mS cm^{-1}, whereas that from the ephemeral P3 is 0.014 mS cm^{-1}. These data suggest that a portion of water flowing from P32 is supplied from a source richer in base cations under low-flow conditions, while during storms, the majority of the supply is from recent precipi-

Table 4. Variation in Dissolved Carbon Dioxide and Methane in Waters From Cottage Hill Sike (CHS) Over 7 Months From January to July 2008 Based on Biweekly Samples[a]

	Site								
	CHS	P1	P3	P5	P6	P9	P32	P35	P39
Mean CO_2	2.63	0.83	1.94	2.31	1.04	1.22	9.36	3.34	2.20
Range	0.22–5.73	0.56–1.07	0.83–3.71	0.34–4.22	0.23–1.97	0.47–2.12	0.22–22.36	1.35–6.99	1.24–3.63
σ	1.54	0.20	1.04	1.13	0.57	0.69	7.98	1.91	1.03
Mean CH_4	0.025	0.009	0.005	0.012	0.008	0.014	4.877	0.389	0.081
Range	0.0008–	0.0009–	0.002–	0.0006–	0.001–	0.003–	0.0008–	0.004–	0.029–
	0.11	0.39	0.007	0.042	0.019	0.036	12.99	1.33	0.21
σ	0.029	0.015	0.002	0.013	0.006	0.011	5.226	0.529	0.089

[a]Sites labeled with P are pipe outlets, and CHS is the flume installed at the catchment outlet. CO_2 measured as CO_2–C (mg L^{-1}), CH_4 measured as CH_4–C (mg L^{-1}), and σ, standard deviation. See Table 2 for details of the physical characteristics of each pipe.

tation. The pipe therefore, provides a link between deep and shallow stores of water within the peat profile. In association with the isotopic measurements of DOC and POC in Cottage Hill Sike, we plan to use a new direct method for trapping CO_2 using molecular sieves [*Billett et al.*, 2006b] to both source and age CO_2 within the pipe network.

5. GEOCHEMICAL INTERACTIONS WITHIN PIPE NETWORKS

As pipe networks undulate throughout the peat profile, different parts of a single network may be entirely surrounded by peat, entirely surrounded by the mineral substrate, or show a combination of the two. This means that there is potential for connectivity between different sources of water and changes in water chemistry along these pathways that would otherwise not be assumed to occur in deep peats. Concentrations of DOC and free CO_2 are known to be affected by both biotic and abiotic processes within streams. However, little information is available regarding changes in their concentrations within pipe networks. *Chapman et al.* [1993] observed a decline in DOC along a pipe network within shallow peat soils in mid-Wales, with differences in concentrations between the sampled point furthest upslope from the outlet and the outlet itself being largest during the summer events when concentrations were higher than in the autumn (Table 5). They suggested that this decline in DOC was probably due to adsorption onto clay particles exposed in the mineral soil on the floor of the pipe network. As the cumulative area of mineral soil exposed to the pipe water increases downslope, the length of the pipe pathway may determine the magnitude of the decrease in DOC concentration along the pipe network. However, *Chapman et al.* [1993] did not carry out any experiments to determine if this was the process accounting for the decline in DOC. Experiments have been carried out in small headwater streams to determine the

relative importance of biotic and abiotic mechanisms in the retention of DOC within stream channels. *McDowell* [1985] suggested that abiotic adsorption by iron and aluminum oxides was responsible for the rapid removal of DOC, whereas *Dahm* [1981] found that microorganisms were more effective in the removal and degradation of stream DOC. In reality, a combination of biotic and abiotic processes probably accounts for the removal of DOC from headwater streams. However, little information regarding the processes and factors controlling the concentration, chemical composition, and reactivity of DOC in pipes in peatlands exists, and this is an interesting area for further research.

There may also be further geochemical interactions where pipes and bog pools are connected. In Little Dodgen Pot Sike, *Holden and Burt* [2002] found that all four areas that contained bog pools were intercepted by pipes. There is also some evidence that pools in continental peatlands have pipes associated with them [e.g., *Glaser*, 1998]. Pools are

Table 5. The Concentration of DOC (mg L^{-1}) at Sampling Points Within a Pipe Network in the Headwaters of the River Wye at Various Times During Storm Events[a]

Storm	Time	P1 (−85 m)	P2 (−75 m)	P3 (−50 m)	P4 (−20 m)	Pipe Outlet
3 Jul 1991	08:30	36	32	31	–	28
	09:20	36	33	32	34	29
	10:50	36	37	32	33	30
24 Jul 1991	11:50	41	41	31	27	35
	13:15	41	41	31	29	31
	16:00	38	–	31	–	35
31 Oct 1991	12:30	15	16	14	14	16
	15:00	17	15	17	14	17
1 Nov 1991	09:30	17	–	15	–	16
10 Mar 1992	11:10	13	12	12	12	12
11 Mar 1992	16:25	14	14	13	13	13

[a]Number in parentheses is distance from the pipe outlet.

hotspots of carbon release (both CH_4 and CO_2) [*Waddington and Roulet*, 2000]. Given their low productivity, much of the carbon lost from pools must be derived from areas around the pool or via deepening of the pool [*Waddington and Roulet*, 2000]. If the former, we can envisage pools as having their own mini-catchments in terms of carbon. Where pools receive water from pipes, the pipes might play an important role in delivering POC, DOC, and dissolved gases to the pool system. Dissolved gases arriving in the pool may be directly lost to the atmosphere via diffusion through the pool water column. In addition, POC and DOC may be mineralized, so producing more CO_2 and CH_4 which could be lost to the atmosphere via ebullition and diffusion. That pools may act in this way is a hypothesis, but it would be interesting to know what role pipes have in pool carbon dynamics.

6. NEXT STEPS

A growing body of evidence is emerging, at least for some types of peatland, that natural pipe flow is an important component of the system. There is now a significant amount of research on blanket peat pipe hydrology and an ongoing program of research into blanket peat pipe carbon export (Cottage Hill Sike) including isotopic measurements of POC, DOC, and CO_2 in pipe water. Data presented in this chapter in text from Cottage Hill Sike are from the early stages of the research program and indicative of the work being performed. There are also sets of experiments planned to examine water flow and oxidation processes in the vicinity of pipe walls. However, further work is needed in other types of peatland, including continental peatlands (especially raised bogs). It may be that for most types of peatland, pipes are indeed present, but they have not been examined and their role in carbon cycling not considered. Work is also needed to understand whether pipes provide an additional source of water and carbon to bog pool systems, and to gather information on whether environmental change is leading to changes in the rate of pipe formation and pipe-mediated carbon export in peatlands across the world.

Acknowledgments. The collection of new data from Cottage Hill Sike was funded by the UK Natural Environment Research Council (NERC) grant NE/E003168/1. Cottage Hill Sike is a NERC Centre for Ecology and Hydrology (CEH) Carbon Catchment and is part of the Moor House Environmental Change Network site. We gratefully acknowledge the technical assistance of David Ashley and Kirsty Dyson.

REFERENCES

Anderson, M. G., and T. P. Burt (1982), Throughflow and pipe monitoring in the humid temperate environment, in *Badland Geomorphology and Piping*, edited by R. B. Bryan, and A. Yair, pp. 337–354, Geo Books, Norwich, U. K.

Andersson, J. O., and L. Nyberg (2008), Spatial variation of wetlands and flux of dissolved organic carbon in boreal headwater streams, *Hydrol. Processes*, *22*, 1965–1975.

Baillie, I. C. (1975), Piping as an erosion process in the uplands of Sarawak, *J. Trop. Geogr.*, *41*, 9–15.

Baird, A. J., and S. Waldron (2003), Shallow horizontal groundwater flow in peatlands is reduced by bacteriogenic gas production, *Geophys. Res. Lett.*, *30*(20), 2043, doi:10.1029/2003GL018233.

Beckwith, C. W., and A. J. Baird (2001), Effect of biogenic gas bubbles on water flow through poorly decomposed blanket peat, *Water Resour. Res.*, *37*, 551–558.

Beckwith, C. W., et al. (2003), Anisotropy and depth-related heterogeneity of hydraulic conductivity in a bog peat. I: Laboratory measurements, *Hydrol. Processes*, *17*, 89–101.

Billett, M. F., and T. R. Moore (2008), Supersaturation and evasion of CO_2 and CH_4 in surface waters at Mer Bleue peatland, Canada, *Hydrol. Processes*, *22*, 2044–2054.

Billett, M. F., C. M. Deacon, S. M. Palmer, J. J. C. Dawson, and D. Hope (2006a), Connecting organic carbon in stream water and soils in a peatland catchment, *J. Geophys. Res.*, *111*, G02010, doi:10.1029/2005JG000065.

Billett, M. F., et al. (2006b), A direct method to measure 14CO2 lost by evasion from surface waters, *Radiocarbon*, *48*, 61–68.

Billett, M. F., M. H. Garnett, and F. Harvey (2007), UK peatland streams release old carbon dioxide to the atmosphere and young dissolved organic carbon to rivers, *Geophys. Res. Lett.*, *34*, L23401, doi:10.1029/2007GL031797.

Bower, M. M. (1961), Distribution of erosion in blanket peat bogs in the Pennines, *Trans. Inst. Br. Geogr.*, *29*, 17–30.

Bryan, R. B., and J. A. A. Jones (1997), The significance of soil piping processes: inventory and prospect, *Geomorphology*, *20*, 209–218.

Carey, S. K., and M. K. Woo (2000), The role of soil pipes as a slope runoff mechanism, Subarctic Yukon, Canada, *J. Hydrol.*, *233*, 206–222.

Chapman, P. (1993), *Caves and Cave Life*, Harper Collins.

Chapman, P. J. (1994), Hydrogeochemical processes influencing episodic stream water chemistry in a headwater catchment, Plynlimon, mid-Wales, PhD thesis, Imperial College, Univ. of London, London.

Chapman, P. J., et al. (1993), Hydrochemical changes along stormflow pathways in a small moorland headwater catchment in Mid-Wales, UK, *J. Hydrol.*, *151*, 241–265.

Chapman, P. J., et al. (1997), Sources and controls of calcium and magnesium in storm runoff: role of groundwater and ion exchange reactions along water flowpaths, *Hydrol. Earth Syst. Sci.*, *1*, 671–685.

Chappell, N. A., and M. D. Sherlock (2005), Contrasting flow pathways within tropical forest slopes of Ultisol soils, *Earth Surf. Processes Landforms*, *30*, 735–753.

Clark, J. M., et al. (2007), Export of dissolved organic carbon from an upland peatland during storm events: Implications for flux estimates, *J. Hydrol.*, *347*, 438–447.

Cole, L., et al. (2002), Relationships between enchytraeid worms (Oligochaeta), climate change, and the release of dissolved organic carbon from blanket peat in northern England, *Soil Biol. Biochem.*, *34*, 599–607.

Crouch, R. J., et al. (1986), Tunnel formation processes in the Riverina Area of N.S.W., Australia, *Earth Surf. Processes Landforms*, *11*, 157–168.

Cryer, R. (1980), The chemistry quality of some pipeflow waters in upland mid-Wales and its implications, *Cambria*, *6*, 28–46.

Dahm, C. N. (1981), Pathways and mechanisms for removal of dissolved organic carbon from leaf leachate in streams, *Can. J. Fish. Aquat. Sci.*, *38*, 68–76.

Dawson, J. J. C., et al. (2002), A comparison of particulate, dissolved and gaseous carbon in two contrasting upland streams in the UK, *J. Hydrol.*, *257*, 226–246.

Dittrich, J. (1952), Zur natürlichen Entwasserung der Moore, *Wasser Boden*, *4*, 286–288.

Egglesmann, R. (1960), Uber den unterirdischen Abfluss aus Mooren, *Wasserwirtschaft*, *50*, 149–154.

Evans, M., and J. Warburton (2005), Sediment budget for an eroding peat-moorland catchment in northern England, *Earth Surf. Processes Landforms*, *30*, 557–577.

Evans, M. G., et al. (1999), Runoff generation and water table fluctuations in blanket peat: evidence from UK data spanning the dry summer of 1995, *J. Hydrol.*, *221*, 141–160.

Foster, I. D. L. (1979), Intra-catchment variability in solute response, an East Devon example, *Earth Surf. Processes*, *4*, 381–394.

Gilman, K., and M. D. Newson (1980), *Soil Pipes and Pipeflow: A Hydrological Study in Upland Wales*, Geo Books, Norwich, U. K.

Glaser, P. H. (1998), The distribution and origin of mire pools, in *Patterned Mires and Mire Pools: Origin and Development, Flora and Fauna*, edited by V. Standen, J. H. Tallis, and R. Meade, pp. 4–25, Durham Univ., Durham, U. K.

Gunn, J. (2000), Introduction, in *The Geomorphology of Cuilcagh Mountain, Ireland: A Field Guide for the British Geomorpholical Research Group Spring Field Meeting, May 2000*, edited by J. Gunn, pp. 1–3, Limestone Research Group, Univ. of Huddersfield.

Gutierrez, M., et al. (1997), Quantitative study of piping processes in badland areas of Ebro basin, NE Spain, *Geomorphology*, *20*, 121–134.

Holden, J. (2004), Hydrological connectivity of soil pipes determined by ground-penetrating radar tracer detection, *Earth Surf. Processes Landforms*, *29*, 437–442.

Holden, J. (2005a), Controls of soil pipe frequency in upland blanket peat, *J. Geophys. Res.*, *110*, F01002, doi:10.1029/2004JF000143.

Holden, J. (2005b), Piping and woody plants in peatlands: Cause or effect? *Water Resour. Res.*, *41*, W06009, doi:10.1029/2004WR003909.

Holden, J. (2006), Sediment and particulate carbon removal by pipe erosion increase over time in blanket peatlands as a consequence of land drainage, *J. Geophys. Res.*, *111*, F02010, doi:10.1029/2005JF000386.

Holden, J., and T. P. Burt (2002), Piping and pipeflow in a deep peat catchment, *Catena*, *48*, 163–199.

Holden, J., and T. P. Burt (2003), Hydrological studies on blanket peat: The significance of the acrotelm-catotelm model, *J. Ecol.*, *91*, 86–102.

Holden, J., et al. (2002), Application of ground-penetrating radar to the identification of subsurface piping in blanket peat, *Earth Surf. Processes Landforms*, *27*, 235–249.

Holden, J., et al. (2004), Artificial drainage of peatlands: Hydrological and hydrochemical process and wetland restoration, *Progr. Phys. Geogr.*, *28*, 95–123.

Hope, D., et al. (2004), Variations in dissolved CO_2 and CH_4 in a first-order stream and catchment: An investigation of soil-stream linkages, *Hydrol. Processes*, *18*, 3255–3275.

Hyett, G. A. (1990), The effect of accelerated throughflow on the water yield chemistry under polluted rainfall, unpublished PhD thesis, Univ. of Wales, Aberystwyth.

Ingram, H. A. P. (1983), Hydrology, in *Ecosystems of the World 4A, Mires: Swamp, Bog, Fen and Moor*, edited by A. J. P. Gore, pp. 67–158, Elsevier, Oxford, U. K.

Jones, J. A. A. (1981), *The Nature of Soil Piping: A Review of Research*, Geo Books, Norwich, U. K.

Jones, J. A. A. (1982), Experimental studies of pipe hydrology, in *Badland Geomorphology and Piping*, edited by R. B. Bryan and A. Yair, pp. 355–371, Geobooks, Norwich, U. K.

Jones, J. A. A. (1990), Piping effects in humid lands, in *Groundwater Geomorphology: The Role of Subsurface Water in Earth-Surface Processes and Landforms*, edited by C. G. Higgins and D. R. Coates, pp. 111–137, Geol. Soc. of Am., Boulder, Colo.

Jones, J. A. A. (1994), Subsurface flow and subsurface erosion, in *Process and Form in Geomorphology*, edited by D. R. Stoddart, pp. 74–120, Routledge, London.

Jones, J. A. A. (1997), Pipeflow contributing areas and runoff response, *Hydrol. Processes*, *11*, 35–41.

Jones, J. A. A. (2004), Implications of natural soil piping for basin management in upland Britain, *Land Degrad. Dev.*, *15*, 325–349.

Jones, J. A. A., and F. G. Crane (1984), Pipeflow and pipe erosion in the Maesnant experimental catchment, in *Catchment Experiments in Fluvial Geomorphology*, edited by T. P. Burt and D. E. Walling, pp. 55–72, Geo Books, Norwich, U. K.

Jones, J. A. A., et al. (1997), Factors controlling the distribution of piping in Britain: A reconnaissance, *Geomorphology*, *20*, 289–306.

Lapen, D. R., et al. (1996), Using ground-penetrating radar to delineate subsurface features along a wetland catena, *Soil Sci. Soc. Am. J.*, *60*, 923–931.

Markov, V. D., and P. I. Khoroshev (1988), Contemporary estimation of the USSR peat reserves, in *Proceedings of the 8th International Peat Congress, Lenningrad 1988*, pp. 72–77, Int. Peat Soc., Jyväskylä, Finland.

McCaig, M. (1983), Contributions to storm quickflow in a small headwater catchment—The role of natural pipes and soil macropores, *Earth Surf. Processes Landforms*, *8*, 239–252.

McDowell, W. H. (1985), Kinetics and mechanisms of dissolved organic carbon retention in a headwater stream, *Biogeochemistry*, *1*, 329–352.

McNamara, N. P., T. Plant, S. Oakley, S. Ward, C. Wood, and N. Ostle (2008), Gully hotspot contribution to landscape methane

(CH_4) and carbon dioxide (CO_2) fluxes in a northern peatland, *Sci. Total Environ.*, *404*, 354–360.

McNeil, P., and J. M. Waddington (2003), Moisture controls on Sphagnum growth and CO_2 exchange on a cutover bog, *J. Appl. Ecol.*, *40*, 354–367.

Muscutt, A. D., et al. (1990), Stormflow hydrochemistry of a small Welsh upland catchment, *J. Hydrol.*, *116*, 239–249.

Newson, M. D., and J. G. Harrison (1978), Channel studies in the Plynlimon experimental catchments, Institute of Hydrology Report No. 47, 67 pp., Wallingford.

Norrström, A. C., and G. Jacks (1996), Water pathways and chemistry at the groundwater/surface water interface to Lake Skjervatjern, Norway, *Water Resour. Res.*, *32*, 2221–2229.

Onda, Y., and N. Itakura (1997), An experimental study in the burrowing activity of river crabs on subsurface water movement and piping erosion, *Geomorphology*, *20*, 279–288.

Price, J. S. (1992), Blanket Bog in Newfoundland 2. Hydrological Processes, *J. Hydrol.*, *135*, 103–119.

Rapson, G. L., et al. (2006), Subalpine gully-head ribbon fens of the Lammerlaw and Lammermoor Ranges, Otago, New Zealand, *N. Z. J. Bot.*, *44*, 351–375.

Roberge, J., and A. P. Plamondon (1987), Snowmelt runoff pathways in a boreal forest hillslope, the role of pipe throughflow, *J. Hydrol.*, *95*, 39–54.

Rosa, E., and M. Larocque (2008), Investigating peat hydrological properties using field and laboratory methods: Application to the Lanoraie peatland complex (southern Quebec, Canada), *Hydrol. Processes*, *22*, 1866–1875.

Rudolf, K., and F. Firbas (1927), Die Moore des Riesengebirges, *Beih. Bot. Zentbl.*, *43*, 69–144.

Sayer, A. M., et al. (2006), Pipeflow suspended sediment dynamics and their contribution to stream sediment budgets in small rainforest catchments, Sabah, Malaysia, *Forest Ecol. Manage.*, *224*, 119–130.

Sklash, M. G., et al. (1996), Isotope studies of pipeflow at Plynlimon, Wales, UK, *Hydrol. Processes*, *10*, 921–944.

Strack, M., et al. (2008), Effect of water table drawdown on peatland dissolved organic carbon export and dynamics, *Hydrol. Processes*, *22*, 3373–3385, doi:10.1002/hyp.6931.

Tallis, J. H. (1995), Climate and erosion signals in British blanket peats: The significance of *Racomitrium lanuginosum* remains, *J. Ecol.*, *83*, 1021–1030.

Terajima, T., et al. (2000), Morphology, structure and flow phases in soil pipes developing in forested hillslopes underlain by a Quaternary sand-gravel formation, Hokkaido, northern main island in Japan, *Hydrol. Processes*, *14*, 713–726.

Thorp, M., and P. Glanville (2003), Mid-Holocene sub-blanket peat alluvia and sediment sources in the upper Liffet Valley, Co. Wicklow, Ireland, *Earth Surf. Processes Landforms*, *28*, 1013–1024.

Uchida, T., et al. (1999), Runoff characteristics of pipeflow and effects of pipeflow on rainfall-runoff phenomena in a mountainous watershed, *J. Hydrol.*, *222*, 18–36.

Uchida, T., et al. (2005), The role of lateral pipe flow in hillslope runoff response: An intercomparison of non-linear hillslope response, *J. Hydrol.*, *311*, 117–133.

Waddington, M. J., and N. T. Roulet (2000), Carbon balance of a boreal patterned peatland, *Global Change Biol.*, *6*, 87–97.

Warburton, J. (2003), Wind-splash erosion of bare peat on UK upland moorlands, *Catena*, *52*, 191–207.

Warner, B. G., et al. (1990), An application of ground penetrating radar to peat stratigraphy of Ellice Swamp, Southwestern Ontario, *Can. J. Earth Sci.*, *27*, 932–938.

Wilson, C. M., and P. L. Smart (1984), Pipes and pipe flow processes in an upland catchment, *Catena*, *11*, 145–158.

Woo, M.-K., and P. DiCenzo (1988), Pipe flow in James Bay coastal wetlands, *Can. J. Earth Sci.*, *25*, 625–629.

Worrall, F., et al. (2006), The rate of and controls upon DOC loss in a peat catchment, *J. Hydrol.*, *231*, 311–325.

Zhu, T. X. (1997), Deep-seated, complex tunnel systems—A hydrological study in a semi-arid catchment, Loess plateau, China, *Geomorphology*, *20*, 255–267.

Zhu, T. X. (2003), Tunnel development over a 12 year period in a semi-arid catchment of the Loess Plateau, China, *Earth Surf. Processes Landforms*, *28*, 507–525.

A. J. Baird, P. J. Chapman, J. Holden, and R. P. Smart, School of Geography, University of Leeds, Leeds LS2 9JT, UK. (j.holden@leeds.ac.uk)

M. F. Billett, Centre for Ecology and Hydrology Edinburgh, Bush Estate, Penicuik, Midlothian EH26 0QB, UK.

Improving Conceptual Models of Water and Carbon Transfer Through Peat

Jeffrey M. McKenzie

Earth and Planetary Sciences, McGill University, Montreal, Quebec, Canada

Donald I. Siegel

Earth Sciences, Syracuse University, Syracuse, New York, USA

Donald O. Rosenberry

U.S. Geological Survey, Lakewood, Colorado, USA

Northern peatlands store 500×10^{15} g of organic carbon and are very sensitive to climate change. There is a strong conceptual model of sources, sinks, and pathways of carbon within peatlands, but challenges remain both in understanding the hydrogeology and the linkages between carbon cycling and peat pore water flow. In this chapter, research findings from the glacial Lake Agassiz peatlands are used to develop a conceptual framework for peatland hydrogeology and identify four challenges related to northern peatlands yet to be addressed: (1) develop a better understanding of the extent and net impact of climate-driven groundwater flushing in peatlands; (2) quantify the complexities of heterogeneity on pore water flow and, in particular, reconcile contradictions between peatland hydrogeologic interpretations and isotopic data; (3) understand the hydrogeologic implications of free-phase methane production, entrapment, and release in peatlands; and (4) quantify the impact of arctic and subarctic warming on peatland hydrogeology and its linkage to carbon cycling.

1. INTRODUCTION

Peatlands [landscapes accumulating peat soils >40 cm thickness; *National Wetland Working Group*, 1997] constitute a major global carbon repository. Northern peatlands store 500×10^{15} g of organic carbon [*Gorham*, 1991], equivalent to ~100 years of current fossil-fuel combustion

Carbon Cycling in Northern Peatlands
Geophysical Monograph Series 184
10.1029/2008GM000821

[*Moore et al.*, 1998]. The interactions of peatlands with the Earth's atmosphere are dynamic and complex; on an annual basis, peatlands are a source of methane to the atmosphere, but on a longer timescale, they serve as net repositories of carbon dioxide [*Roulet et al.*, 1994]. Although many studies quantify carbon cycling in peatlands [e.g., *Siegel et al.*, 2001; *Rivers et al.*, 1998], how they will respond to climate change remains uncertain [e.g., *Waddington et al.*, 1998] because conceptual models of these systems are usually based on specific case studies and are difficult to generalize to all northern peatlands.

Peatlands were originally assumed to have effectively stagnant pore water due to highly decomposed layers of peat with very low permeability at depth [e.g., *Boelter and Verry*, 1977; *Ivanov*, 1975; *Verry and Boelter*, 1978]. These results led to a view that the exchange of atmospheric carbon with peatlands is essentially only through the peat surface [*Clymo*, 1984]. However, *Ingram* [1982] suggested that horizontal flow throughout a peat profile is a vital component of peatland hydrogeology and can partially explain a bog's size and morphology.

Current research shows that the movement of water and carbon through peatlands is variable, as there are many spatial and temporal factors that can control the movement of water and solutes. First is peat accumulation. Bogs and fens accumulate carbon and grow vertically and horizontally. Vertical peatland accumulation rates in North American peatlands range from 18 to over 100 cm per 1000 years [*Glaser et al.*, 2004a; *Gorham et al.*, 2003]. Second is permeability and porosity. The permeability and porosity of peat soils are temporally and spatially variable and dependant on numerous factors, including peat substrates, compaction, and decomposition. Changes in the pore water solute concentrations also can affect permeability by interacting with organic acid functional groups and dilating larger pores [*Ours et al.*, 1997]. *Price* [1997] found considerable hysteretic changes in permeability of near-surface peat driven by seasonal water table lowering and peat compaction. Third is macroporosity. Peatland soils can have extensive networks of macropores that control the movement of water. Networks of macropores within the peat soils (e.g., decayed roots) create preferential pathways by which water is rapidly transmitted [*Beven and Germann*, 1982] and create a dual porosity peat matrix. Additionally, macropores in the upper, aerobic layer (acrotelm) control the runoff of surface water [*Holden et al.*, 2001], and can be enhanced by desiccation [*Strack et al.*, 2008].

This chapter examines the connections between hydrogeology and carbon cycling in peatland systems and presents four future challenges for research in large northern peatlands. Discussion in this chapter is based on observations over the past 20 years at the patterned glacial Lake Agassiz peatlands (GLAP) in northern Minnesota, which has sufficient size and hydrologic variability to serve as a possible template for circumboreal peatlands, in general. The discussion herein is primarily focused on processes at depth within the peat profile and largely excludes research questions related to processes in the very top few tens of centimeters of the peat profile [e.g., *Belyea and Baird*, 2006].

2. GLACIAL LAKE AGASSIZ PEATLANDS

GLAP are a 7000-km^2 expanse of sub-boreal patterned peatlands in Northern Minnesota and Southern Manitoba (Figure 1). The patterned ecosystems at GLAP alternate between bogs that have total dissolved solids (TDS) less than 15 mg L^{-1} and pH less than 4.0, and fens with surface-water TDS greater than 50 mg L^{-1} and pH greater than 6.0 [*Glaser et al.*, 1981]. The raised bogs of the GLAP are dominated by *Picea mariana*, *Carex oligosperm*, and ericaceous shrubs with a continuous mat of *Sphagnum*, whereas fens are dominated by sedges, such as *Carex lasiocarpa* and *Rhynchospora alba* and various *Amblystigeaceae* mosses [*Glaser et al.*, 1981; *Heinselman*, 1970]. The peat thickness in the GLAP ranges from 2 to >4 m, reflecting mainly less than 5000 years of accumulation [*Glaser et al.*, 1997; *Janssen*, 1968]. The research summarized in this chapter is primarily from two sites (Figure 1), the Red Lake Bog which has an area of 151.4 km^2 and a basal depth of 341–345 cm, and the Lost River Bog which has an area of 16.3 km^2 and a basal depth of 320–325 cm [*Glaser et al.*, 1997]. The Red Lake bog is referred to as the Red Lake II bog in some publications [e.g., *Glaser et al.*, 1997, *Hogan et al.*, 2000].

The regional climate is more arid to the west and colder to the north. Average precipitation declines from 63.5 cm in the eastern part of GLAP to 55.8 cm in the west, whereas evapotranspiration increases from east to west [*Glaser et al.*, 1997]. The entire region is subject to extreme multiyear droughts [*Glaser et al.*, 1997]. The annual mean temperature ranges from 2.2° to 2.7°C (NCDC/NOAA Waskish Meteorological Station data from 1997 to 2000), and the GLAP is typically covered by at least 15 cm of snow 70 to 100 d a^{-1}. The peatlands straddle a north-south climate divide along which average annual evapotranspiration equals average annual precipitation. Evapotranspiration exceeds precipitation to the west and is less than precipitation to the east of the line [*Siegel et al.*, 1995]. Consequently, the peatland is highly sensitive to changes in regional climate.

Research over more than 20 years has shown that the evolution of the GLAP is closely controlled by local climate variability within the study area and regional-scale hydrogeologic systems. The importance of regional groundwater that discharges at the base of these peatlands, and which comprises a major component of their annual water budget, cannot be overstated. This upwelling is confirmed by (1) upward vertical hydraulic gradients at the base of many peat profiles [e.g., *Rosenberry et al.*, 2003], (2) elevated dissolved-ion species in pore water near the base of peat profiles [*Siegel et al.*, 1995], and (3) regional hydrogeologic models that elucidate the connection between regional groundwater flow systems and peatlands [*Siegel*, 1993].

The peatland hydrogeology within the GLAP, while moving relatively small amounts of water in terms of total flux, is dynamic. As in most groundwater systems, the flow of water through the GLAP is primarily horizontal, but there is also complex variability in vertical flow. At raised ombrotrophic

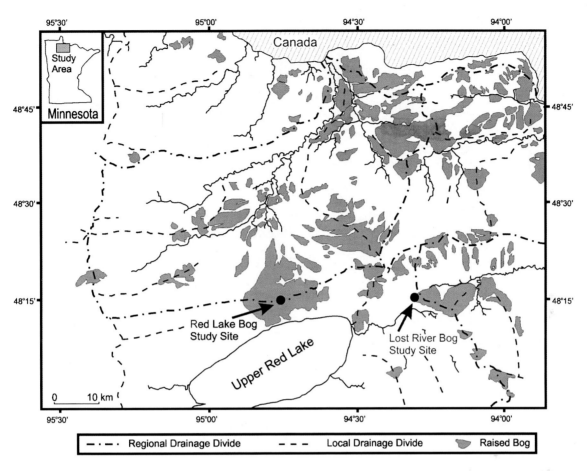

Figure 1. Map of the glacial Lake Agassiz peatlands (GLAP) with the three study sites indicated [*McKenzie et al.*, 2006].

bogs, such as Lost River, there is flow downward from the water table of the bog and upward flow from the base of the peat profile. These two vertical flow systems meet at an intermediate depth within the peat profile that varies over time (Figure 2). These groundwater flow systems connect to fens and water-track landforms originating on the bog flanks. Here, bog vegetation is succeeded by fen plants at what appear to be seepage faces. The fens are primarily dominated by the upward and lateral flow of minerotrophic groundwater [*Siegel and Glaser*, 1987]. Numerical modeling simulations by *Reeve et al.* [2000] show hydrologic interactions among local to intermediate scale groundwater flow systems beneath these peatlands and demonstrate the importance of the permeability of these underlying sediments.

Within this GLAP hydrogeologic framework, the direction and velocity of pore water flow in peatlands varies over seasonal, annual, and longer timescales. The interaction of meteorological events (such as drought) with regional

groundwater flow systems creates a dynamic hydrogeologic flow system [*Siegel and Glaser*, 1987; *Reeve et al.*, 2006].

For example, flow reversals, when the vertical direction of pore water movement changes seasonally, have been documented at GLAP as well as in many peatland morphologies and locations [*Devito et al.*, 1997; *McKenzie et al.*, 2002; *Siegel et al.*, 1995]. At GLAP, detailed studies in the Lost River bog site show that groundwater flows upward into large, raised bogs during droughts, but in temperate years, groundwater moves predominately downward [*Glaser*, 1992; *Romanowicz et al.*, 1993; *Siegel*, 1983; *Siegel and Glaser*, 1987]. During long droughts, almost the entire peat column can be flushed by upwelling mineral water, and during long wet periods, surface water can penetrate almost to the mineral soil because of vertical recharge downward under developing water table mounds [*Siegel et al.*, 1995]. In general, the degree to which advective flow within raised bog peat occurs is related to the kind of subsurface mineral

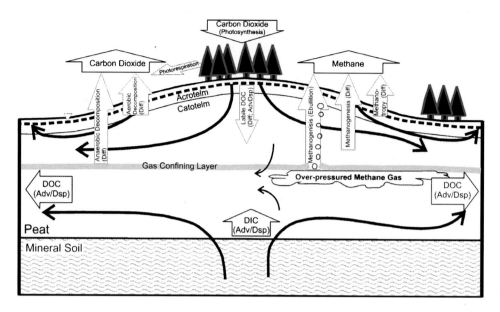

Figure 2. Conceptual model of the hydrogeology and carbon cycling of a representative peatland in the glacial Lake Agassiz peatlands. The dashed line represents the water table, and the line arrows indicate idealized groundwater flow paths. Image adapted from *Siegel et al.* [1995] and *Glaser et al.* [2004b]. Abbreviations are DOC, dissolved organic carbon; DIC, dissolved inorganic carbon; Diff, diffusion; and Adv/Dsp, advection/dispersion.

soil beneath them. *Reeve et al.* [2000] used heuristic numerical modeling experiments to suggest that where bogs are underlain by permeable mineral soil, such as is in parts of the GLAP, their water table mounds can efficiently drive downward vertical flow. However, where bogs evolved over clayey low-permeable soils, such as in the Hudson Bay Lowlands, vertical flow in inhibited.

Advection may also be affected by hydraulic conductivity that varies temporally in response to changing volumes of biogenic gas trapped beneath semielastic, semipermeable layers of woody debris in the peat [*Rosenberry et al.*, 2003]. Trapped gas can occlude pores in the peat and locally reduce hydraulic conductivity, preventing vertical flow of water in a finite volume of the peat beneath the trapped gas. Sudden releases of trapped gas could result in near instantaneous changes in direction of flow within the peat in response to rapidly changing pressure fields that occur as gas is released, either to a shallower depth within the peat column or to the atmosphere.

3. CARBON CYCLING

The simplest view of peatlands is a two-layer system where the acrotelm is responsible for the majority of organic matter decomposition (90% of total organic carbon decomposition), and the lower, anaerobic layer (catotelm) stores the bulk of undecomposed carbon and has very slow decom-

position rates [*Clymo*, 1984]. Active and alive vegetation at the surface of the peat column sequesters carbon dioxide from the atmosphere, photorespires carbon dioxide back to the atmosphere, and produces organic litter. This organic litter is either oxidized in the oxic zone above the water table or is buried and stored below the water table [*Clymo*, 1984]. *Blodau* [2002] and *Limpens et al.* [2008] provide thorough reviews of carbon cycling and carbon budgets in peatlands. While thorough in terms of surface-water runoff and water table fluctuation, they present little information regarding subsurface hydrogeology and carbon cycling of peatlands.

Hydrogeology and carbon cycling are connected by solute transport, the process by which dissolved substances are transported via advection, dispersion, and diffusion. The peat matrix presents particular challenges for studying solute transport because of its dual porosity nature and changing bulk density in response to water table changes [*Hoag and Price*, 1997; *Price*, 1997]. Decomposed peat has a high total porosity (>90%) but a low effective porosity and permeability [*Ours et al.*, 1997]. Advection is the primary process that moves solutes vertically through peat profiles, which is easily confirmed by looking at pore water chemistry profiles. For example, in the Red Lake bog, increasing total dissolved solids is observed near the base of profiles in conjunction with upward hydraulic gradients. These are interpreted to be the result of upward advection of minerotrophic waters from underlying inorganic sediments. Work by *Reeve et al.* [2001] has shown that transverse dispersion along long horizontal flow paths can also move solutes upward from underlying, solute-rich waters in mineral soils into peat profiles. Although dispersive mixing is important, the low flux of water through some peatland systems, especially in non-domed systems, creates situations where the diffusive flux of solutes may be more important than advection. At millennial scales, changing degrees of diffusion and advection of solutes from underlying sediments, coupled to the accumulation of peat, can significantly impact how water and solutes move in peat. *McKenzie et al.* [2002] found that the vertical accumulation of 3.5 m of peat in ~7000 C^{14} years in an alpine bog in Switzerland created downward hydraulic head gradients and forced a change from diffusive to advective vertical solute transport as the bog grew.

Although many of these processes happen on decadal or longer timescales, there are short, potentially more significant processes that control the flux of carbon into and out of the peatland system. Although dissolved organic carbon is removed from both the catotelm and acrotelm by runoff, and moved to deeper systems through groundwater recharge, the aerobic and anaerobic decomposition of organic carbon also leads to the loss of carbon to the atmosphere by diffusion of dissolved CO_2 and CH_4. Some gas emissions are sufficient to alter the elevation and horizontal position of the peat surface [*Glaser et al.*, 2004b]. CO_2 and CH_4 can also escape to the atmosphere through buoyancy-driven ebullition fluxes, potentially releasing large amounts of carbon [*Rosenberry et al.*, 2003; *Glaser et al.*, 2004b; *Baird et al.*, 2004; *Kellner et al.*, 2004; *Strack et al.*, 2006]. These episodic carbon emissions from boreal peatlands have largely gone undetected and have only recently begun to be quantified [*Rosenberry et al.*, 2006].

4. CHALLENGES

Previous and ongoing research has provided many key observations and hypotheses that lead to an improved understanding of many of the hydraulic and biogeochemical processes in peatlands. Research, both at the GLAP and at other northern peatlands, shows that the connections between hydrogeology and carbon cycling remain somewhat disjointed and site specific. Following are four suggested major scientific challenges for future research of large northern peatlands, in general. These challenges are by no means encompassing all research challenges, but are developed based on questions arising from GLAP research that are applicable to northern peatlands, in general. Generalized process-based solutions to these problems will help lead to a unified theory connecting hydrogeology and carbon dynamics.

4.1. The Spatial Extent of Climate-Driven Flushing; the Lost River Scenario

The GLAP Lost River peatland is notable for its climate-driven flushing of pore water [*Romanowicz et al.*, 1993; *Siegel and Glaser*, 1987; *Siegel et al.*, 1995], wherein short droughts lead to changes in the vertical direction of pore water flow at depth, and a multiyear drought leads to a complete upward flushing of the dilute peat system with minerotrophic groundwater. These results have important implications for the future ecology of peatlands within a climate change framework; specifically, changes in climate may substantially alter the solute transport regime via pore water flushing. Should extended droughts persist in peatlands, groundwater mounds under bogs will likely dissipate, and the combination of advective discharge from mineral soils, diffusion, and changes in pH will logically lead to more homogeneous landscapes where fen vegetation succeeds in bogs. This process would constitute a landscape-scale reversal of bog ecosystem succession over fens in patterned peatlands that has persisted for thousands of years, since deglaciation. While the Lost River conclusions are remarkable in demonstrating the potential dynamism and adaptability of flow in a large peatland, there are many research challenges.

The principle challenge is to determine and understand the controls of the geospatial frequency of occurrence of groundwater reversals within patterned peatlands, in general. Addressing this challenge will help generalize whether these types of flow systems should be expected in other peatlands and, if so, what their net effect would be in terms of carbon cycling.

For example, the contrast between the peatlands of the Hudson Bay Lowlands and the GLAP Lost River system is indicative of the difficulty in generalizing system dynamics. In the Hudson Bay Lowlands region, pore water flushing, driven by climate change, has not yet been observed; most pore water chemistry profiles are dominated by recharge with stable distributions of total dissolved solids with depth [*Reeve et al.*, 1996]. These differences lead to a general challenge of determining if the GLAP constitutes a unique hydrogeologic setting that is essentially the result of "ideal" conditions caused by more permeable underlying sediments, allowing for more upward discharge of groundwater. Similar studies need to be done in other patterned peatlands to come to valid generalizations.

There are a range of possible climate change impacts on northern peatland systems with the common research focus being on drier conditions leading to increased methane generation at depth [*Ise et al.*, 2008; *Moore et al.*, 1998]. If, in the Lost River peatland, drought conditions are assumed without a significant lowering of hydraulic head in regional groundwater systems, there would potentially be less methane generated because labile carbon delivered by advection downward to methanogens would cease [*Chanton et al.*, 1995; *Siegel et al.*, 1995]. On a broader scientific scale, these results present a significant challenge to understand and quantify the interaction of intermediate to regional scale groundwater-flow systems on the hydrogeology of northern peatlands and the associated transport of carbon [*Siegel*, 1993].

4.2. Hydrogeologic and Isotopic Complexities and Contradictions

The measurement of physical parameters, such as permeability, in peatlands is inherently difficult because of easily deformable peat materials, scale dependency of permeability, and hydrologic heterogeneity [e.g., *Chason and Siegel*, 1986; *Beckwith et al.*, 2003; *Rosa and Larocque*, 2007]. Isotopic tracers can be used to constrain uncertainties in the source, direction, and velocity of peatland flow systems [*Clark and Fritz*, 1997]. Research using pore water isotope data in the GLAP has produced results that may contradict some of the hydrogeologic hypotheses and point to previously unobserved heterogeneities and unexplained errors in the water budget.

At the Red Lake bog, *Hogan et al.* [2000] used strontium isotopes and observed that 90–100% of pore water in the upper 2 m of peat was meteorically derived, but the lower 2+ m of pore water was a mixture of precipitation and groundwater. At both the Red Lake bog and adjacent fen sites, the chemical concentrations and isotopic values showed extensive variability, potentially indicating macropore flow as an important hydrochemical transport pathway [*Beven and Germann*, 1982; *Holden et al.*, 2001; *Holden*, 2005]. The strontium isotopic results suggest there is a source of water that does not match the isotopic signature of groundwater or precipitation and could only partially be explained by organic matter mineralization. The results also showed that the groundwater discharging into the base of the bog and the fen were isotopically distinct, indicating multiple distinct flow systems. *Sarkar et al.* [2008] found water with an evaporative isotopic signal in deep fen water, suggesting groundwater flowing across the major southern watershed divide defining the GLAP watershed. *Chasar et al.* [2000] used carbon isotopes to assess the source of mineralized carbon at depth within the Red Lake peatland. The results showed that methane produced at depth within the peat profile was a mixture of both older, more recalcitrant, carbon and modern labile carbon sourced from the active top layer of peat. The carbon isotopic results also indicate that dissolved inorganic carbon in upwelling groundwater was a major source of subsequently reduced carbon.

These types of isotopic studies show that the linkages of regional hydrogeology (i.e., areas larger than the GLAP), peatlands hydrogeology, and carbon transport within the GLAP are potentially contradictory and far more complex than previously thought [e.g., *Siegel*, 1993], even after decades of research. For example, in the Red Lake bog, the main zone of deep methane generation (approximately 2-m depth) is hydraulically overpressured with respect to the pore water above and below, yet to explain the observed methane generation, based on the isotopic observations, there needs to be a significant downward flux of labile carbon to this zone [*Hogan et al.*, 2000]. Much of this complexity is likely related to subsurface heterogeneities in the mineral soils and within the peat.

These contradictory results present an interesting research challenge. Generalization to all northern peatlands may be impossible as there may be no unique model of dissolved organic carbon (DOC) transport, hydrogeologic setting, and carbon mineralization. Fundamental to the research challenge discussed in this section is determining if the complexities in the hydrogeologic heterogeneities observed at the GLAP are common to other northern peatlands, and if so, quantifying the hydrogeologic and biogeochemical processes.

4.3. Methane Production, Entrapment, and Release From Peatlands

A current peatland research topic is the characterization and quantification of the processes that control subsurface anaerobic production, entrapment, and release. *Weber* [1906] first reported on the phenomenon of peat land surfaces fluctuating in elevation over diurnal or longer time spans. *Fechner-Levy and Hemond* [1996] suggested that this vertical movement of a peat surface is caused by the interaction of free-phase methane gases trapped at depth within the peat column and atmospheric pressure, a now established hypothesis. At the GLAP Red Lake bog, high precision differential GPS were used to observe changes in bog surface elevation of up to 30 cm [*Glaser et al.*, 2004b] and that most of those elevation fluctuations closely correlated to daily and subweekly changes in barometric pressure [*Rosenberry et al.*, 2003]. Within these surface-elevation fluctuations, there are pronounced rapid drops and slow rises in the surface of the peat that are unrelated to barometric pressure, but are very closely related to rapid drops in pressure at 2-m depth within the peat profile. It is hypothesized that methane gas forms at or below 2 m at high enough concentrations to form free-phase methane gas and that changes in atmospheric barometric pressure compress this zone, allowing the surface of the peat to fluctuate up and down [*Glaser et al.*, 2004b]. The rapid drops in the peat surface are caused by large ebullition events when methane gas rapidly escapes to the surface, and the peat surface collapses similar to a soufflé deflating [*Glaser et al.*, 2004b].

Rosenberry et al. [2006] provide an in-depth review of much of the literature related to biogenic gas formation and escape. The ebullition pathways for methane release can be enigmatic and poses numerous research challenges including to understand the rates and mechanisms that control methane production, entrapment, and release. The primary controls on methane generation are temperature [*Hulzen et al.*, 1999; *Updegraff et al.*, 1998], carbon supply [*Chanton et al.*, 1995], and water table position [*Ise et al.*, 2008; *Moore and Dalva*, 1993]. The mechanisms that control production, entrapment, dissolution, and release of methane are not yet all understood, although researchers are using a variety of innovative techniques to image and understand the controlling processes, including geophysical methods [*Comas et al.*, 2007], time-domain reflectometry [*Beckwith and Baird*, 2001], coring [*Glaser et al.*, 2004b], differential GPS [*Glaser et al.*, 2004b], direct measurement [*Almendinger et al.*, 1986; *Roulet*, 1991], and ex situ imaging such as X-ray tomography [*Kettridge and Binley*, 2008]. Linking and scaling these disparate data to wider areas remains a major research challenge. Problems associated with scaling hydrologic data are well known, such as the scale dependency of hydraulic conductivity [*Freeze and Cherry*, 1979; *Surridge et al.*, 2005], and similar scale-dependent challenges will exist in expanding these detailed studies to the entire peatland scale, in particular, to expand to the peatland three-dimensional (3-D) landscape [*Rosenberry et al.*, 2006].

Evaluating how overpressured in situ methane at depth affects the internal hydrodynamics of bogs and the extent to which this overpressuring occurs in peatlands, in general, is another question for future research. The question of hydrodynamic trapping of overpressured methane still needs to be better quantified to determine why methane traps are being formed at depth. Why are methane traps being formed at depth? Local free-phase gas will obviously occlude pores and decrease hydraulic conductivity, but the observed cycles of trap and release is not fully understood. A possible model to explain these cycles would be elastic deformation [*Ingebritsen et al.*, 2006], though more research is required to understand the physical properties of the peat and the threshold required for accumulating and trapping gas. Although this research challenge focuses on ebullition from northern peatlands, seasonal freezing and permafrost are a dominant factor in controlling the atmospheric flux of carbon in the arctic and sub-Arctic [e.g., *Christensen et al.*, 2004; *Rivkina et al.*, 1998].

4.4. The Warming and Subsequent Thawing of the Arctic and Peatlands

As is often pointed out by research in high latitudes and in the popular press, continued climate change and warming of the arctic is leading to thawing permafrost in northern peatlands and the rapid release of carbon [e.g., *Kolbert*, 2005; *Serreze et al.*, 2000]. Observed hydrologic response to arctic warming include increasing groundwater contributions to surface-water systems [*Walvoord and Striegl*, 2007], disappearance of lakes sitting on top of permafrost [*Smith et al.*, 2005], and increases in the flux of carbon to the atmosphere [*Ise et al.*, 2008]. The net response of northern peatlands to warming is a balance between increased biological productivity and increased carbon decomposition rates, two factors that are closely controlled by soil temperature, meteorology, and surface hydrology [*Davidson and Janssens*, 2006].

A major concern with ongoing warming is that decreased moisture inputs will lead to a lowering of the water table that will consequently generate increases in methane emissions and DOC runoff due to additional aerobic peat mineralization [e.g., *Ise et al.*, 2008; *Moore and Dalva*, 1993]. An increase in surface temperatures will increase temperatures throughout the peat column [*McKenzie et al.*, 2006] leading to increased carbon decomposition rates within the catotelm [*Moore and Dalva*, 1993]. A long-term lowering of the water table will also increase vegetative activity and cause increased DOC production and runoff [*Strack et al.*, 2008]. All of these factors, on a global scale, have the potential to act as a warming feedback, reinforcing concerns regarding the warming of northern peatlands.

Increased temperatures with decreased moisture is the most commonly predicted scenario that is analyzed in terms of the peat-carbon dynamics. Increased temperatures should increase the size of the active zone. With a decrease in moisture and a thinner snowpack, greater frost penetration or a thicker accumulation of ice at the peat surface could trap free-phase gas within the peat, causing large ebullition events during spring thaw [*Rosenberry et al.*, 2006]. In areas where there would be increased moisture and a thicker snow cover, increased insulation could result in higher subsurface temperatures.

Within this context of a warming arctic, there are many challenges to understanding the impact on peatlands including the impact of thawing permafrost in arctic and subarctic regions, thickening of the active zone above permafrost, and changes in the moisture.

Heat transport through peatlands is primarily dominated by conduction as opposed to advection due to low pore water flow velocities [*McKenzie et al.*, 2006; *Moore*, 1987]. The impact of freezing on hydrogeology is commonly minimized, but it can have a strong control on pore water flow by decreasing permeability and effective porosity, while increasing heat capacity and thermal conductivity [*McKenzie et al.*, 2007]. Freezing can also exert a strong control on hydrogeology through cryogenic suction, ice lensing, and chemical segregation effects [*Williams and Smith*, 1989].

These potential future changes in temperature and hydrogeology inherently will lead to feedback mechanisms with implications for carbon cycling. Higher air temperatures will increase the size of the active zone in permafrost areas, leading to greater shallow methane production through longer growing seasons and warmer temperatures. In the Siberian peatlands, these feedbacks are already observed in permafrost lakes [*Walter et al.*, 2006] and with increasing DOC runoff [*Frey and Smith*, 2005].

Much of the focus on arctic warming has used anthropogenic peat dewatering as an analog for warming scenarios and enhanced methane generation, but it is unclear that in all areas, there will be a net drying of peatlands. The prediction of peatland moisture regimes is complicated [*Koutsoyiannis et al.*, 2008], and with increased moisture, it would actually be expected to decrease decomposition rates, a situation that may occur through increased precipitation or permafrost thawing [*Davidson and Janssens*, 2006]. Additionally, the response of the peat system to drying is likely a function of the rate of drying with time, which is much faster for dewatering than for climatic scale changes to precipitation rates.

Additional impacts of permafrost melting include effects of seasonal freeze-thaw cycles on transport of solutes and organic carbon. Current research and numerical models of these processes are primarily focused in 1-D vertical profiles [*McKenzie et al.*, 2007], which need to be scaled to 2- and 3-D experiments because it is known that carbon movement through these systems is strongly controlled by regional flow patterns [*Reeve et al.*, 2001].

5. SUMMARY AND CONCLUSIONS

There are many challenges remaining for a comprehensive understanding of the impact climate change has on northern peatlands. Within this context, this chapter identifies research questions related to the hydrogeology of peatlands and its connection to carbon transport and cycling. Although the research presented here is primarily focused on the GLAP, the identified water and carbon flux challenges are applicable to other large peatland areas such as the Hudson Bay Lowland [*Roulet et al.*, 1994] and the Siberian Peatlands [*Frey and Smith*, 2005]. This transfer ability assumes that the GLAP is representative of northern peatlands, an argument that is difficult to prove. This particular issue is itself a research challenge, how to generalize and apply the more site-specific

phenomena observed in the GLAP to other peatlands across the very broad northern-peatland scale.

This chapter is focused on the deeper groundwater processes within peatlands and, as such, does not give adequate treatment to methane generation or consumption near the peat surface. There are many complex feedbacks associated with water table position both seasonally and over climate timescales. These fluctuations will govern whether a given peatland is a net sink or source of carbon [*Strack et al.*, 2004], a situation with obvious implications for methane feedbacks to the atmosphere.

At the global scale, the impact of peatlands and their feedback on future climate is not well understood. There are still many unknowns regarding the rates of arctic warming and permafrost thawing, and their impact on peatlands. Global climate simulations do not include peatlands at this point [*Limpens et al.*, 2008], an omission that has important implications considering the potential feedback of peatlands for increasing carbon dioxide and methane levels in the atmosphere.

REFERENCES

Almendinger, J. C., J. E. Almendinger, and P. H. Glaser (1986), Topographic fluctuations across a spring fen and raised bog in the Lost River Peatland, Northern Minnesota, *J. Ecol.*, *74*, 393–401.

Baird, A. J., C. W. Beckwith, S. Waldron, and J. M. Waddington (2004), Ebullition of methane-containing gas bubbles from near-surface *Sphagnum* peat, *Geophys. Res. Lett.*, *31*, L21505, doi:10.1029/2004GL021157.

Beckwith, C. W., and A. J. Baird (2001), Effect of biogenic gas bubbles on water flow through poorly decomposed blanket peat, *Water Resour. Res.*, *37*(3), 551–558.

Beckwith, C. W., A. J. Baird, and A. L. Heathwaite (2003), Anisotropy and depth-related heterogeneity of hydraulic conductivity in a bog peat. I: Laboratory measurements, *Hydrol. Processes*, *17*, 89–101.

Belyea, L. R., and A. J. Baird (2006), Beyond "the limits to peat bog growth": Cross-scale feedback in peatland development, *Ecol. Monogr.*, *76*(3), 299–322.

Beven, K., and P. Germann (1982), Macropores and water flow in soils, *Water Resour. Res.*, *18*(5), 1311–1325.

Blodau, C. (2002), Carbon cycling in peatlands—A review of processes and controls, *Environ. Rev.*, *10*(2), 111–134.

Boelter, D., and E. S. Verry (1977), Peatand and water, *Gen. Tech. Rep. NC-31*, 22 pp., U.S. Dep. of Agric. For. Serv.

Chanton, J. P., J. E. Bauer, P. A. Glaser, D. I. Siegel, C. A. Kelley, S. C. Tyler, E. H. Romanowicz, and A. Lazrus (1995), Radiocarbon evidence for the substrates supporting methane formation within northern Minnesota peatlands, *Geochim. Cosmochim. Acta*, *59*(17), 3663–3668.

Chasar, L. S., J. P. Chanton, P. H. Glaser, D. I. Siegel, and J. S. Rivers (2000), Radiocarbon and stable carbon isotopic evidence for transport and transformation of dissolved organic carbon, dissolved inorganic carbon, and CH_4 in a northern Minnesota Peatland, *Global Biogeochem. Cycles*, *14*(4), 1095–1108.

Chason, D., and D. Siegel (1986), Hydraulic conductivity and related physical properties of peat, Lost River Peatland, northern Minnesota, *Soil Sci.*, *142*(2), 91–99.

Christensen, T. R., T. Johansson, H. J. Åkerman, M. Mastepanov, N. Malmer, T. Friborg, P. Crill, and B. H. Svensson (2004), Thawing sub-arctic permafrost: Effects on vegetation and methane emissions, *Geophys. Res. Lett.*, *31*, L04501, doi:10.1029/2003GL018680.

Clark, I. D., and P. Fritz (1997), *Environmental Isotopes in Hydrogeology*, 328 pp., CRC Press Lewis Publishers, Boca Raton, FL.

Clymo, R. S. (1984), The limits to peat bog growth, *Philos. Trans. R. Soc. London, B, Biol. Sci.*, *303*(1117), 605–654.

Comas, X., L. Slater, and A. S. Reeve (2007), In situ monitoring of free-phase gas accumulation and release in peatlands using ground penetrating radar (GPR), *Geophys. Res. Lett.*, *34*, L06402, doi:10.1029/2006GL029014.

Davidson, E. A., and I. A. Janssens (2006), Temperature sensitivity of soil carbon decomposition and feedbacks to climate change, *Nature*, *440*, 165–173.

Devito, K. J., J. M. Waddington, and B. A. Branfireun (1997), Flow reversals in peatlands influenced by local groundwater systems, *Hydrol. Processes*, *11*(1), 103–110.

Fechner-Levy, E. J., and H. F. Hemond (1996), Trapped methane volume and potential effects on methane ebullition in a northern peatland, *Limnol. Oceanogr.*, *41*, 1375–1383.

Freeze, R. A., and J. A. Cherry (1979), *Groundwater*, 604 pp., Prentice-Hall, Englewood Cliffs, N. J.

Frey, K. E., and L. C. Smith (2005), Amplified carbon release from vast West Siberian peatlands by 2100, *Geophys. Res. Lett.*, *32*, L09401, doi:10.1029/2004GL022025.

Glaser, P. H. (1992), Peat Landforms, in *Patterned Peatlands of Northern Minnesota*, edited by H. E. J. Wright et al., p. 327, Univ. of Minn., Minneapolis.

Glaser, P. H., G. A. Wheeler, E. Gorham, and H. E. Wright, Jr. (1981), The patterned mires of the Red Lake peatland, northern Minnesota: Vegetation, water chemistry and landforms, *J. Ecol.*, *69*, 575–599.

Glaser, P. H., D. I. Siegel, E. A. Romanowicz, and Y. P. Shen (1997), Regional linkages between raised bogs and the climate, groundwater, and landscape features of northwestern Minnesota, *J. Ecol.*, *85*, 3–16.

Glaser, P. H., B. C. S. Hansen, D. I. Siegel, A. S. Reeve, and P. J. Morin (2004a), Rates, pathways and drivers for peatland development in the Hudson Bay Lowlands, northern Ontario, Canada, *J. Ecol.*, *92*, 1036–1053.

Glaser, P. H., J. P. Chanton, P. Morin, D. O. Rosenberry, D. I. Siegel, O. Ruud, L. I. Chasar, and A. S. Reeve (2004b), Surface deformations as indicators of deep ebullition fluxes in a large northern peatland, *Global Biogeochem. Cycles*, *18*, GB1003, doi:10.1029/2003GB002069.

Gorham, E. (1991), Northern peatlands: Role in the carbon cycle and probable responses to climatic warming, *Ecol. Appl.*, *1*, 182–193.

Gorham, E., J. A. Janssens, and P. H. Glaser (2003), Rates of peat accumulation during the postglacial period in 32 sites from Alaska to Newfoundland, with special emphasis on northern Minnesota, *Can. J. Bot.*, *81*, 429–438.

Heinselman, M. L. (1970), Landscape evolution, peatland types, and the environment in the Lake Agassiz Peatlands Natural Area, Minnesota, *Ecol. Monogr.*, *40*, 235–261.

Hoag, R. S., and Price, J. S. (1997), The effects of matrix diffusion on solute transport and retardation in undisturbed peat in laboratory columns, *J. Contam. Hydrol.*, *28*, 193–205.

Hogan, J. F., J. D. Blum, D. I. Siegel, and P. H. Glaser (2000), $^{87}Sr/ ^{86}Sr$ as a tracer of groundwater discharge and precipitation recharge in the glacial Lake Agassiz peatlands, northern Minnesota, *Water Resour. Res.*, *36*(12), 3701–3710.

Holden, J. (2005), Piping and woody plants in peatlands: Cause or effect?, *Water Resour. Res.*, *41*, W06009, doi:10.1029/2004WR003909.

Holden, J., T. P. Burt, and N. J. Cox (2001), Macroporosity and infiltration in blanket peat: The implications of tension disc infiltrometer measurements, *Hydrol. Processes*, *15*, 289–303.

Hulzen, J. B. V., R. Segers, P. M. V. Bodegom, and P. A. Leffelaar (1999), Temperature effects on soil methane production: An explanation for observed variability, *Soil Biol. Biochem.*, *31*, 1919–1929.

Ingebritsen, S., W. Sanford, and C. Neuzil (2006), *Groundwater in Geologic Processes*, 2nd ed., 536 pp., Cambridge Univ. Press, New York.

Ingram, H. A. P. (1982), Size and shape in raised mire ecosystems: A geophysical model, *Nature*, *297*(5864), 300–303.

Ise, T., A. L. Dunn, S. C. Wofsy, and P. R. Moorcroft (2008), High sensitivity of peat decomposition to climate change through water-table feedback, *Nat. Geosci.*, *1*, 763–766, doi:10.1038/ngeo331.

Ivanov, K. (1975), *Water Movement in Mirelands*, translated from Russian by A. Thomson and H. A. P. Ingram, Academic, New York.

Janssen, C. R., (1968), Myrtle Lake: A late- and post-glacial pollen diagram from northern Minnesota, *Can. J. Bot.*, *46*, 1397–1408.

Kellner, E., J. S. Price, and J. M. Waddington (2004), Pressure variations in peat as a result of gas bubble dynamics, *Hydrol. Processes*, *18*, 2599–2605.

Kettridge, N., and A. Binley (2008), X-ray computed tomography of peat soils: Measuring gas content and peat structure, *Hydrol. Processes*, *22*(25), 4827–4837, doi:10.1002/hyp.7097.

Kolbert, E. (2005) The climate of man—I: Disappearing islands, thawing permafrost, melting polar ice. How the earth is changing, *New Yorker*, *81*, 56–71.

Koutsoyiannis, D., A. Efstratiadis, N. Mamassis, and A. Christofides (2008), On the credibility of climate predictions, *Hydrol. Sci. J*, *53*(4), 671–684.

Limpens, J., F. Berendse, C. Blodau, J. G. Canadell, C. Freeman, J. Holden, N. Roulet, H. Rydin, and G. Schaepman-Strub (2008), Peatlands and the carbon cycle: From local processes to global implications—A synthesis, *Biogeosci. Discuss.*, *5*, 1379–1419.

McKenzie, J. M., D. I. Siegel, W. Shotyk, P. Steinmann, and G. Pfunder (2002), Heuristic numerical and analytical models of the hydrologic controls over vertical solute transport in a domed peat bog, Jura Mountains, Switzerland, *Hydrol. Processes*, *16*(5), 1047–1064.

McKenzie, J. M., D. I. Siegel, D. O. Rosenberry, P. H. Glaser, and C. I. Voss (2006), Heat Transport in the Red Lake Bog, Glacial Lake Agassiz Peatlands, *Hydrol. Processes*, *21*(3), 369–378, doi:10.1002/hyp.6239.

McKenzie, J. M., C. I. Voss, and D. I. Siegel (2007), Groundwater flow with energy transport and water-ice phase change: Numerical simulations, benchmarks and application to freezing in peat bogs, *Adv. Water Resour.*, *30*, 966–983, doi:10.1016/j.advwatres.2006.08.008.

Moore, T. (1987), Thermal regime of peatlands in subarctic eastern Canada, *Can. J. Earth Sci.*, *24*, 1352–1359.

Moore, T., N. T. Roulet, and J. Waddington (1998), Uncertainty in predicting the effect of climatic change on the carbon cycling of Canadian peatlands, *Clim. Change*, *40*, 229–245.

Moore, T. R., and M. Dalva (1993), The influence of temperature and water table position on carbon dioxide and methane emissions from laboratory columns of peatland soils, *J. Soil Sci.*, *44*, 651–664.

National Wetland Working Group (1997), *The Canadian Wetland Classification System*, 2nd ed., 69 pp., Wetland Res. Cent. Publ., Waterloo, Ont., Canada.

Ours, D. P., D. I. Siegel, and P. H. Glaser (1997), Chemical dilation and the dual porosity of humified bog peat, *J. Hydrol.*, *196*(1–4), 348–360.

Price, J. S. (1997), Soil moisture, water tension, and water table relationships in a managed cutover bog, *J. Hydrol.*, *202*, 1579–1589.

Reeve, A. S., D. I. Siegel, and P. H. Glaser (1996), Geochemical controls on peatland pore water from the Hudson Bay Lowland: A multivariate statistical approach, *J. Hydrol.*, *181*(1–4), 285–304.

Reeve, A. S., D. I. Siegel, and P. H. Glaser (2000), Simulating vertical flow in large peatlands, *J. Hydrol.*, *227*, 207–217.

Reeve, A. S., D. I. Siegel, and P. H. Glaser (2001), Simulating dispersive mixing in large peatlands, *J. Hydrol.*, *242*, 103–114.

Reeve, A. S., R. Evensen, P. H. Glaser, D. I. Siegel, and D. Rosenberry (2006), Flow path oscillations in transient ground-water simulations of large peatland systems, *J. Hydrol.*, *316*, 313–324.

Rivers, J. S., D. I. Siegel, L. S. Chasar, J. P. Chanton, P. H. Glaser, N. T. Roulet, and J. M. McKenzie (1998), A stochastic appraisal of the annual carbon budget of a large circumboreal peatland, Rapid River watershed, northern Minnesota, *Global Biogeochem. Cycles*, *12*(4), 715–727.

Rivkina, E., D. Gilichinsky, S. Wagener, J. Tiedje, and J. McGrath (1998), Biogeochemical activity of anaerobic microorganisms from buried permafrost sediments, *Geomicrobiol. J.*, *15*, 187–193.

Romanowicz, E. A., D. I. Siegel, and P. H. Glaser (1993), Hydraulic reversals and episodic methane emissions during drought cycles in mires, *Geology*, *21*(3), 231–234.

Rosa, E., and M. Larocque (2007), Investigating peat hydrological properties using field and laboratory methods: Application to the Lanoraie peatland complex (southern Quebec, Canada), *Hydrol. Processes*, *22*(12), 1866–1875.

Rosenberry, D. O., P. H. Glaser, D. I. Siegel, and E. P. Weeks (2003), Use of hydraulic head to estimate volumetric gas content and ebullition flux in northern peatlands, *Water Resour. Res.*, *39*(3), 1066, doi:10.1029/2002WR001377.

Rosenberry, D. O., P. H. Glaser, and D. I. Siegel (2006), The hydrology of northern peatlands as affected by biogenic gas: Current developments and research needs, *Hydrol. Processes*, *20*, 3601–3610.

Roulet, N. T. (1991), Surface level and water table fluctuations in a subarctic fen, *Arct. Alp. Res.*, *23*, 303–310.

Roulet, N. T., A. Jano, C. Kelly, L. Klinger, T. Moore, R. Protz, J. Ritter, and W. Rouse (1994), Role of the Hudson Bay lowlands as a source of atmospheric methane, *J. Geophys. Res.*, *99*, 1439–1454.

Sarkar, S., D. I. Siegel, P. H. Glaser, and J. Chanton (2008), Deep Ground Water through Stable Isotopic Analysis in a Large Circumboreal Peatland, Abstract 147296, paper presented at Joint Meeting of the Geological Society of America, Houston, Tex.

Serreze, M. C., J. E. Walsh, F. S. Chapin III, T. Osterkamp, M. Dyurgerov, V. Romanovsky, W. C. Oechel, J. Morison, T. Zhang, and Barry, G. (2000), Observational evidence of recent change in the northern high-latitude environment, *Clim. Change*, *46*, 159–207.

Siegel, D. I. (1983), Ground water and the evolution of patterned mires, Glacial Lake Agassiz Peatlands, northern Minnesota, *J. Ecol.*, *71*(3), 913–921.

Siegel, D. I. (1993), Groundwater Hydrology, Chapter 11, in *The Patterned Peatlands of Northern Minnesota*, edited by H. E. Wright, Jr., pp. 163–173, Univ. of Minn. Press, Minneapolis.

Siegel, D. I., and P. H. Glaser (1987), Groundwater flow in a bog-fen complex, Lost River peatland, northern Minnesota, *J. Ecol.*, *75*, 743–754.

Siegel, D. I., A. S. Reeve, P. H. Glaser, and E. A. Romanowicz (1995), Climate-driven flushing of pore water in peatlands, *Nature*, *374*(6522), 531–533.

Siegel, D. I., J. P. Chanton, P. H. Glaser, L. S. Chasar, and D. O. Rosenberry (2001), Estimating methane production rates in bogs and landfills by deuterium enrichment of pore water, *Global Biogeochem. Cycles*, *15*(4), 967–975.

Smith, L., C. Y. Sheng, G. M. MacDonald, and L. D. Hinzman (2005), Disappearing Arctic Lakes, *Science*, *308*, 1429.

Strack, M., J. M. Waddington, and E.-S. Tuittila (2004), Effect of water table drawdown on northern peatland methane dynamics: Implications for climate change, *Global Biogeochem. Cycles*, *18*, GB4003, doi:10.1029/2003GB002209.

Strack, M., E. Kellner, and J. M. Waddington (2006), Effect of entrapped gas on peatland surface level fluctuations, *Hydrol. Processes*, *20*, 3611–3622.

Strack, M., J. M. Waddington, R. A. Bourbonniere, E. L. Buckton, K. Shaw, P. Whittington, and J. S. Price (2008), Effect of water table drawdown on peatland dissolved organic carbon export and dynamics, *Hydrol. Processes*, *22*, 3373–3385.

Surridge, B., A. J. Baird, and A. L. Heathwaite (2005), Evaluating the quality of hydraulic conductivity estimates from piezometer slug tests in peat, *Hydrol. Processes*, *19*, 1227–1244.

Updegraff, K., S. D. Bridgham, J. Pastor, and P. Weishampel (1998), Hysteresis in the temperature response of carbon dioxide and methane production in peat soils, *Biogeochemistry*, *43*, 253–272.

Verry, E. S., and D. Boelter (1978), *Wetland Functions and Values: The State of Our Understanding*, pp. 389–402, Am. Water Resour. Assoc., Middleburg, VA.

Waddington, J., T. Griffis, and W. Rouse (1998), Northern Canadian wetlands: Net ecosystem CO_2 exchange and climatic change, *Clim. Change*, *40*, 267–275.

Walter, K. M., S. A. Zimov, J. P. Chanton, D. Verbyla, and F. S. Chapin III (2006), Methane bubbling from Siberian thaw lakes as a positive feedback to climate warming, *Nature*, *443*, 71–75, doi:10.1038/nature05040.

Walvoord, M. A., and R. G. Striegl (2007), Increased groundwater to stream discharge from permafrost thawing in the Yukon River basin: Potential impacts on lateral export of carbon and nitrogen, *Geophys. Res. Lett.*, *34*, L12402, doi:10.1029/2007GL030216.

Weber, C. A. (1906), *Uber die Vegetation und Entstehung des Hochmors von Augstumal im Memeldelta*, 252 pp., Velagsbuchhandlung, Berlin.

Williams, P. J., and M. W. Smith (1989), *The Frozen Earth: Fundamentals of Geocryology*, 306 pp., Cambridge Univ. Press, Cambridge, U. K.

J. M. McKenzie, Earth and Planetary Sciences, McGill University, Montreal, Quebec, Canada H3A 2A7. (jeffrey.mckenzie@mcgill.ca)

D. O. Rosenberry, U.S. Geological Survey, MS 413, Building 53, Lakewood, CO 80225, USA. (rosenber@usgs.gov)

D. I. Siegel, Earth Sciences, Syracuse University, Syracuse, NY 13244, USA. (disiegel@syr.edu)

Water Relations in Cutover Peatlands

Jonathan S. Price and Scott J. Ketcheson

Department of Geography and Environmental Management, University of Waterloo, Waterloo, Ontario, Canada

Sphagnum mosses, the dominant peat-forming plant in many northern peatlands, generally do not regenerate spontaneously in mined peatlands because water transfer between the cutover peat and incipient moss diaspores cannot overcome the capillary barrier effect between the two hydraulically distinct layers. Artificial drainage networks established throughout peatlands, coupled with the removal of the acrotelm during the peat extraction process, drastically alter the natural system function through the exposure of more decomposed catotelm peat and increased compression, oxidation, and shrinkage, subsequently decreasing average pore diameter and enhancing this capillary barrier effect. Water table (WT) fluctuations, constrained within the reduced specific yield of the altered catotelm, exhibit increased variability and rapid decline. The increased effective stress caused by a declining WT can result in seasonal surface subsidence of 8 to 10 cm, thereby reducing saturated hydraulic conductivity by three orders of magnitude. Restoration efforts aim to alter the disturbed hydrological regime, creating conditions more favorable for the recolonization of *Sphagnum* mosses and the ultimate reestablishment of an upper acrotelm layer. Due to the large areal coverage and high organic carbon content, the response of peatlands to disturbances caused by resource extraction, and their return to functioning ecosystems, must be thoroughly addressed. This paper integrates both published and unpublished work to facilitate an overview of our understanding of the hydrological impact of peat cutting and its implications for restoration.

1. INTRODUCTION

Cutover peatlands are those exploited for their peat resource and sometimes for agriculture. The common feature is that the surface layer, previously the most biologically and hydrologically active zone [*Ingram*, 1983], has been removed. The hydraulic structure of this upper layer of living, dead, and poorly decomposed plant material is essential to maintaining the storage and fluxes of water that define its ecosystem functions including its hydrologic function, native plant community, geochemical attributes, and carbon balance. These functions are either lost or profoundly altered with peat cutting such that their ecohydrological trajectory may be forever altered.

Peat cutting is widespread in parts of Scandinavia, British Isles, Russia, and North America. Peatlands cover over 1 million km^2 in North America, with most (97%) of the total peatland-covered area located within the boreal and subarctic wetland regions [*Tarnocai*, 2006]. The peatlands located at temperate latitudes, though proportionately less, are more susceptible to disturbances due to anthropogenic activities as a consequence of their proximity to markets; hence, their economic value. For example, approximately

Carbon Cycling in Northern Peatlands
Geophysical Monograph Series 184
Copyright 2009 by the American Geophysical Union.
10.1029/2008GM000827

16,000 ha of Canada's peatlands have been or are currently being exploited for peat resources [*Bergeron*, 1994; *Keys*, 1992]. While peatlands include bogs, fens, and some swamps [*NWWG*, 1997], the primary target of peat extraction activities is on bogs for the production of *Sphagnum* peat because of its superior water-holding properties and resistance to decay [*Read et al.*, 2004]. There is a need to understand the impacts of peat cutting on the system ecohydrology in order to better manage the industrial exploitation, particularly in devising appropriate restoration strategies. Thus, the objective of this paper is to establish our current state of understanding of the hydrological impact of peat cutting and its implications for restoration. *Price et al.* [2003] provided a review of hydrological processes in abandoned and restored peatlands, and provide details of restoration attempts in Europe and North America. This paper includes some of the essential aspects outlined by *Price et al.* [2003] but includes more recent work, both published and unpublished, to highlight our current state of understanding.

2. ECOHYDROLOGICAL FUNCTIONS IN UNDISTURBED PEATLANDS

Peatlands are fundamentally an ecohydrological construct. Their ecological function is both an outcome and a determinant of their hydrogeomorphic setting, since plant materials contribute to peat development and accumulation. Undisturbed peatlands are typically characterized by an upper layer of living, dead, and poorly decomposed plant material (acrotelm) that is defined as the zone that exists above the average minimum annual WT [*Ingram*, 1978]. While the existence, development, and vertical extent of this layer are variable depending on peatland form and location, it has a disproportionately large influence on the ecohydrological and biogeochemical function. Toward the surface, there is an increase in porosity, hydraulic conductivity, and specific yield which regulates WT variability, infiltration capacity and water storage, groundwater and surface water runoff (RO), capillary rise, and evapotranspiration (ET) [*Ingram*, 1983]. Removal of the acrotelm exposes the more decomposed catotelm peat [*Ingram*, 1978] that previously existed in an anoxic state below the WT. This layer is characterized by lower hydraulic conductivity and lower porosity that imparts a low specific yield [*Price*, 1996] when it is drained. Consequently, the water regime no longer supports the original ecology and carbon regulation function of the peatland.

The structure of the acrotelm that arises from the vegetation community that formed it and the decomposition processes that transform it eventually into peat provide the ideal medium for sustaining that plant community. Near the surface the hydraulic matrix is characterized by relatively large pores derived from growing and dead undecomposed plant material as yet uncompressed by the weight of overlying materials over time. The large pores have a high saturated hydraulic conductivity, K_s [*Boelter*, 1968; *Hayward and Clymo*, 1982; *Hobbs*, 1986], which can effectively transmit water when a hydraulic gradient is present. Thus, natural drainage following large storms or snowmelt reduces the duration of surface flooding [*Spieksma*, 1999], conditions which are less than ideal for many plants even in wetlands [*Rochefort et al.*, 2002]. As drainage occurs and the WT falls, the effective K_s decreases sharply below the surface, and the ability to shed groundwater diminishes accordingly, a mechanism which ensures sufficient wetness is sustained. *Hoag and Price* [1995] noted that K_s of a bog acrotelm can decrease five orders of magnitude from the surface to 50-cm depth.

The poorly decomposed organic material in the upper acrotelm has low water retention capacity [*Boelter*, 1969]; thus, a high specific yield (S_y). Specific yield is the ratio of the volume of water drained from a soil by gravity (after being saturated) to the total volume of the soil. Large pore sizes dominate the upper acrotelm, resulting in the drainage of a large proportion of the pore space by gravity, consequently increasing the specific yield. On the one hand, this high storativity dampens the WT drawdown response to drainage and, thus, sustains the WT relatively close to the surface (a definitive factor for wetlands and especially peatlands where anoxic conditions must prevail), but the sharp decrease in volumetric water content (θ) with pore drainage rapidly reduces the unsaturated hydraulic conductivity of the moss [*Price et al.*, 2008], since hydraulic conductivity is a function of water content, $K(\theta)$. Thus, water flow in the matrix above the WT, especially near the surface of moss-dominated peatlands, is suppressed when the WT falls. Various studies have shown, for example, that evaporation from moss-dominated systems is sharply reduced when the WT declines [*Lafleur and Roulet*, 1992; *Price*, 1991; *Romanov*, 1968], since $K(\theta)$ becomes very low and upward capillary flows cannot effectively replenish water at a rate that meets the evaporative demand. One consequence of the reduced evaporation due to the low $K(\theta)$ is that the system retains a relatively high WT. Thus, the ability to readily shed water when the WT is high, while also preventing a large WT drawdown and reducing the groundwater efflux when the WT does drop, is an essential self-regulating feature of undisturbed bog peatlands. Another self-regulating feature of many peatlands is the ability to shrink and swell to adjust to changes in seasonal wetness [*Ingram*, 1983] or drainage [*Whittington and Price*, 2006].

The intimate relationship between ecology and hydrology of peatlands is a vital feature that both defines and regulates their function. Drainage and removal of the acrotelm obliterates the ecology and severely impairs the hydrological function. The characteristic hydrology of drained cutover peatlands is described below and then examined in the context of restoration.

3. NATURE AND EXTENT OF HYDROLOGICAL CHANGE EXPERIENCED BY CUTOVER PEATLANDS

Peat harvesting typically begins with the installation of a drainage network designed to dewater the peat. The dewatering is necessary to increase the bearing capacity for extraction operations [Hobbs, 1986] and additionally removes water that must otherwise be shed in the drying process for the commercial product. Drainage causes a short-term pulse in water flow as water is released from storage, but can also lead to an increase in base flow, at least on the short-term [Conway and Millar, 1960]. Drainage ditches are typically spaced ~30 m apart [Mulqueen, 1989]. Following the initial drainage and WT lowering, drainage must occur through catotelm peat where the low hydraulic conductivity impedes drainage. Moreover, with drainage, the effective stress on the peat increases causing peat compression and consolidation [Price, 2003], reducing the average pore size and further reducing the hydraulic conductivity of the peat and its drainage [Prevost et al., 1997]. Boelter [1972] showed ditches have little effect beyond 50 m in a forested peatland and that drainage was most effective within 5 m of the ditch. This was supported by a simple model [Price, 2003] that demonstrated the drainage of uniform cutover peat was rapid near the ditch, but largely ineffective beyond 10 m.

Most peat cutting directly removes the acrotelm. Once the surface of a bog is stripped and peat is harvested, the acrotelm/catotelm divisions no longer apply. What remains is a peat matrix that initially has the relatively uniform (with depth) hydraulic attributes of the catotelm peat. The relatively small pore size of the older, more decomposed peat has a high water retention capacity, low specific yield, and low hydraulic conductivity [Price, 1996]. Consequently, with drainage and evaporative water loss, the WT falls quickly [Price, 1996], unregulated as in the former acrotelm. The consequent alterations to the cutover peat matrix has a positive feedback loop, since the lower WT facilitates aerobic decomposition to greater depths, which can increase the efflux of carbon dioxide (CO_2) by 400% [Waddington and Price, 2000]. This further reduces pore size, with the consequent effect on water retention, specific yield, hydraulic conductivity, etc. Price [1996] reported the S_y of acrotelm mosses as 0.6, diminishing to 0.2 by 50-cm depth within the acrotelm peat. The exposed catotelm peat underwent compression and oxidation, and S_y declined to 0.06 in approximately 5 years. The consequence was a more variable

WT [Price, 1996]. The higher water retention capacity of the smaller pores results in higher volumetric water content at a given pressure $\theta(\psi)$, since the smaller pore radii are able hold more water at a higher tension (more negative pressure). Peat with a higher bulk density (hence smaller pore size) has a greater soil water retention [Boelter, 1968], resulting in the typical shape of the soil moisture-pressure characteristic curve [see Price, 1997]. The impact of the increased $\theta(\psi)$ is that more of the pores contain water, thus increasing the number and thickness of water films that conduct water flow, resulting in a higher unsaturated hydraulic conductivity at a given pressure, $K(\psi)$ [Price et al., 2008]. This increase in $K(\psi)$, coupled with stronger capillary rise, results in increased availability of water at the cutover peat surface, consequently increasing ET losses and enhancing the WT decline.

The increase in WT variability, and especially the accentuated drawdown during dry periods, has a number of important hydrological and ecological implications. As noted above, lowering the WT reduces the pressure (ψ) throughout the profile. This increases the effective stress (σ_e) at a given point below the WT, since

$$\sigma_e = \rho_T g h - \psi, \tag{1}$$

where ρ_T is the total density of the column of air, water, and peat of height, h, overlying that point, and where g is gravitational acceleration. The increase in σ_e results in "normal consolidation" of peat [McLay et al., 1992; Pyatt and John, 1989], where the volumetric change is equivalent to the amount of water lost [Terzaghi, 1943], resulting in initially rapid peat subsidence [Whittington and Price, 2006]. Price and Schlotzhauer [1999] mapped the subsidence to account for water storage change by normal consolidation (volume change equivalent to the volume of water lost) in an abandoned section of the cutover Lac St. Jean, Québec (LSJ) peatland. They found water storage change (ΔS_{tot}) was a function of pore drainage (S_y) and specific storage (S_s) times the saturated thickness of the aquifer (b) such that $\Delta S_{tot} = \Delta h (S_y + bS_s)$, where bS_s is the slope of the relationship between surface subsidence and WT, and Δh is the change in head (WT). At this site, bS_s (0.13) was greater than specific yield (0.05), underscoring the importance of including the storage change associated with peat compression (S_s) in estimates of total water storage change if the water balance is to be correctly specified [Price and Schlotzhauer, 1999]. The ability of the peatland to subside and dilate seasonally diminishes after prolonged drainage, but this mechanism is partly restored with rewetting [Shantz and Price, 2006a]. They also found the water storage change due to peat volume change over the post-snowmelt period in a rewetted site was equivalent to the volume of water lost as RO, while in the drained site, the volume change was nil.

Peat volume change affects the hydraulic properties of the soil. *Schlotzhauer and Price* [1999] and *Price* [2003] showed at LSJ that the seasonal subsidence of 8 to 10 cm caused up to a three order of magnitude decline in K_s, attributing this to collapse of the larger pores as the WT declined. This seasonal variation in K_s diminishes over time in synch with the reduced volume change over time [*Kennedy and Price*, 2004]. Long-term decrease in K_s was also noted by *Van Seters and Price* [2002] as a consequence of peat degradation. Despite the diminishing response with time, an abandoned block-cut bog near Cacouna, Québec (the Cacouna bog) exhibited surface subsidence of up to 5 cm at some locations within a single season 30 years after abandonment.

One important consequence of higher $K(\psi)$ and water retention, $\theta(\psi)$, in cutover peatlands is that evaporation remains relatively high when the WT drops, even in the absence of plants because of the greater availability of water, and the relatively effective capillary flow in the unsaturated zone near the surface. For example, *Price* [1996] found summer ET losses averaging 2.7 and 2.9 mm d^{-1} from an undisturbed and an adjacent unvegetated cutover bog (LSJ), respectively. *Shantz and Price* [2006a] found evaporation from bare cutover peat was about 20% greater than in a relatively wet restored section of the same peatland, although this was partly attributed to the effect of mulch applied to the surface of the restored section [*Petrone et al.*, 2004]. ET from the Cacouna bog increased during the summer of 2005, averaging 2.5 and 3.0 mm d^{-1} for the time periods 19 May to 1 July and 2 July to 16 August, respectively. This increase came despite an average WT decline of 19 cm over the same time period (S. Ketcheson, unpublished data, 2008). Increased atmospheric demands over the latter portion of the summer provide the explanation, as a 26% increase in daily net radiation flux was coupled with a 5°C increase in average air temperature. Further, with the exception of the last day of measurement, the average site WT remained within the upper 60 cm of the peat. This is within the lower limit of the shrub rooting structure [*Lafleur et al.*, 2005], indicating that transpiration through vascular vegetation was not substantially restricted due to physiological responses to water stress. Prolonged atmospheric demands for ET results in an eventual disconnect of the WT from the atmosphere [*Price*, 1997], at which point water is lost from soil storage in the unsaturated zone, and the WT becomes relatively stable.

4. POSTHARVEST CHANGES

As noted above, there are various mechanistic changes in the hydrological character of peatlands on the short-term that are a direct consequence of drainage and cutting. Over the longer-term, change occurs as a consequence of soil creep [*Price*, 2003], decomposition [*Van Seters and Price*, 2002], changes to RO due to drainage ditch collapse and infilling [*Van Seters and Price*, 2001], as well as macropore activity [*Holden*, 2005], and because of changes in vegetation [*Girard et al.*, 2002]. Following the abandonment of peat-cutting operations, cutover peatlands were often left without action to facilitate restoration. While this is less common nowadays, there are numerous abandoned peatlands that have solely spontaneously regenerated vegetation [*Lavoie et al.*, 2003]. In addition to invasive species, those plants common in peatlands return (e.g., *Sphagnum* mosses and Ericaceae), although not in the original proportion [*Lavoie and Rochefort*, 1996]. The hydrological conditions are typically too harsh to allow substantial recolonization of *Sphagnum*, which is the dominant peat-forming plant in undisturbed bogs [*Kuhry and Vitt*, 1996]. For example, at the Cacouna bog, a manually block-cut peatland, less than 10% of the area supported *Sphagnum* mosses 25 years after abandonment [*Lavoie and Rochefort*, 1996]. This recolonization was limited to the relatively low areas adjacent to occluded ditches and where cutting was greatest, which had soil water pressure >-100 mb and $\theta > 50\%$ in the upper 5 cm [*Price and Whitehead*, 2001]. In peatlands exploited with modern (vacuum harvesting) machines, spontaneous revegetation is much less successful than in block-cut sites [*Lavoie et al.*, 2005b]. *Price et al.* [2003] attributed this to the relative lack of hydrological variability in vacuum harvested sites, being uniformly poor for vegetation regeneration.

The recolonization of *Sphagnum* is related to moisture relations associated with position in the cutover landscape, typically relative elevation [*Price and Whitehead*, 2001]. The mode of recolonization is initially as isolated cushions, which in the wetter areas coalesce into a carpet. In areas less suitable, isolated cushions develop in a hemispherical form and exist as a microcosm closely tied to the moisture regime of the substrate and the influence of the Ericaceae canopy [*Price and Whitehead*, 2004]. The presence of the cushion itself affects the substrate, which compared to bare peat immediately adjacent, has a higher ψ and θ [*Price and Whitehead*, 2004]. Still, little is known about the water transfer between the peat substrate and moss cushion.

While much work has been done characterizing hydraulic properties of undisturbed peat [*Kellner et al.*, 2005; *Kennedy and Van Geel*, 2000; *Silins and Rothwell*, 1998], little is known of the hydraulic properties of the mosses themselves. Water retention characteristics [*Hayward and Clymo*, 1982] and, more recently, measurements of the unsaturated hydraulic conductivity function [*Price et al.*, 2008] for *Sphagnum* mosses have been published, providing the foundation for more in-depth characterization and modeling of water fluxes within the living and poorly decomposed mosses. Access to water is essential for the nonvascular *Sphagnum* mosses that are limited to relying on typically weak capillary pressure in its hyaline cells and interstitial spaces [*Hayward and Clymo*, 1982] to maintain a water supply to the growing part of the plant (capitula). *Sphagnum* is relatively intolerant to desiccation [*Sagot and Rochefort*, 1996]; thus, a lack of available water has implications for plant metabolic processes, such as photosynthesis and soil respiration (plant matter decomposition) [*McNeil and Waddington*, 2003]. *Strack and Price* [2009] have shown that photosynthesis is strongly coupled to the wetness of the moss surface, such that plant metabolic activity increases significantly with small additions of water (e.g., dew) that are insufficient to cause a change in volumetric water content 5 cm below the surface.

During the summer of 2005, 11 naturally recolonized *Sphagnum* cushions sitting atop the cutover peat substrate at the Cacouna bog were instrumented with time domain reflectometry (TDR) probes to measure moisture content at 5-cm intervals, centered at 2.5, 7.5, 12.5, and 17.5 cm below the cushion surface. The moisture content was fairly consistent near the surface of the cushions throughout the summer months, with variability increasing substantially with depth (Figure 1). The seasonal volumetric water content (θ) fluctuated within a small range (±10%) in the uppermost portion of the cushion, nearest to the capitula, where the cushions were driest. In contrast, the greatest range of θ (±60%) was exhibited across the 15- to 20-cm interval, where water was the most abundant, a trend evident in most of the cushions investigated (Table 1). A similar moisture profile was observed by *McNeil and Waddington* [2003].

An investigation of cushion size and its influence on moisture dynamics was conducted, lumping the cushions

Table 1. Seasonal Moisture Variability Across Four Depths Within Naturally Regenerated *Sphagnum* Cushions at the Cacouna Bog[a]

Depth Interval[b] (cm)	Mean θ	Standard Deviation	Maximum θ	Minimum θ	Range
0–5	0.19	0.012	0.28	0.18	0.10
5–10	0.23	0.025	0.38	0.16	0.22
10–15	0.33	0.071	0.65	0.22	0.43
15–20	0.48	0.134	0.79	0.19	0.60

[a]Data are for time period 8 June to 18 August 2005.
[b]For each depth interval, $n = 72$.

(*Sphagnum capillifolium*) according to approximate cushion volume (mean cushion size shown in brackets), resulting in three groups: largest (~125,000 cm³), mid-sized (~36,000 cm³), and smallest (~13,000 cm³). On average, the smallest-sized cushions had the highest moisture content (34%) compared to the larger sized (30%) and mid-sized (30%) cushions, though no direct relationship between cushion size and θ could be established. The least amount of moisture variability occurred in the uppermost portion of the cushion from each group, with increasing variability with depth. The smallest cushions exhibited the most variability overall, reflected in a comparatively large standard deviation with depth (Figure 2). The maximum and minimum θ values observed (79 and 16%, respectively) occurred within the smallest- and mid-sized cushion groupings, respectively.

Hummocks within natural peatland (i.e., hummock-hollow) landscapes can become isolated from the surrounding hydrologic system once more than 8 cm of peat has been deposited beneath the hummock, as this weakens the capillary rise to mosses near the top of the hummock [*Bellamy and Rieley*, 1967]. A similar restriction may apply to

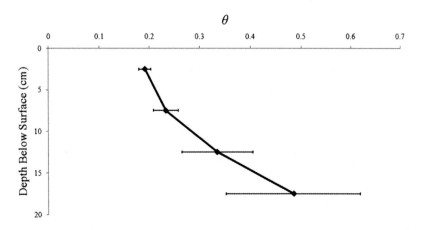

Figure 1. Average θ values within 11 naturally regenerated cushions at the Cacouna bog for the time period 8 June to 18 August 2005. Error bars indicate standard deviation ($n = 72$ for each depth).

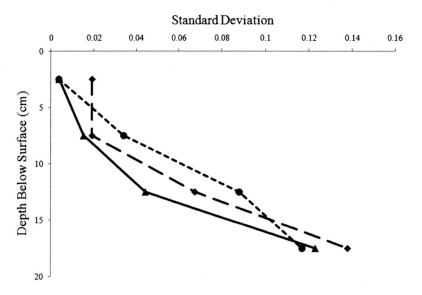

Figure 2. Variability in θ with depth for the largest (solid line; *n* = 25), mid-sized (large dashed line; *n* = 26), and smallest (small dashed line; *n* = 21) size cushion groupings in the Cacouna bog for the time period 8 June to 18 August 2005.

recolonized *Sphagnum* cushions. Considering the very low values of unsaturated hydraulic conductivity within living and poorly decomposed mosses [*Price et al.*, 2008], such as those comprising the newly formed cushions, regular water exchanges between the cutover peat substrate and the overlying moss cushions may occur only periodically. The abrupt change in pore size at the substrate-moss interface may create a capillary barrier effect [cf. *Kennedy and Van Geel*, 2000], where water exchange cannot occur until the pressure in the cutover peat is at or near zero (i.e., near saturation). Given the highly variable moisture content at depth within the recolonized cushions at the Cacouna bog and the small fluctuations of θ near the moss capitula, the upper and lower parts of the cushion may be only weakly coupled, so vertical extension (growth) is limited, and lateral spreading and coalescing of cushions is more favorable.

Schouwenaars and Gossen [2007] used a model to demonstrate that a thin layer of recolonized *Sphagnum* can be well-supplied by water, but once it grows to 5–15 cm thick, it is more susceptible to water stress because of the ineffective delivery of water to the moss from the substrate. However, at greater thicknesses, the water storage capacity of the thicker moss is sufficient to maintain an adequate water supply. Price (unpublished results) used Hydrus 1-D to simulate water flux in a moss layer subject to evaporation with a WT depth of 30 cm, using the hydraulic parameters presented by *Price et al.* [2008]. The results indicate that the mosses were unable to sustain evaporation at the (imposed) potential rate of 5 mm d^{-1} for more than 1 h, eventually equilibrating

at about half that rate. The five order of magnitude drop in hydraulic conductivity with drainage results in resistance to water flow in the upper moss layers. In addition to the liquid flux, there is a vapor flow, although this amounts to <2% of the former [*Price et al.*, 2009].

Vascular plants, especially ericaceous shrubs, more readily recolonize cutover sites, thriving in areas where ψ is much lower than −100 mb [*Farrick*, 2008]. Further, shrubs cleared from the original peatland surface were typically piled in the center of trenches during the abandonment process in block-cut peatlands, forming large seed banks and enhancing Ericaceae-dominated recolonization of manually cutover peat surfaces [*Girard et al.*, 2002]. *Girard et al.* [2002] reported that the spontaneous revegetation of the Cacouna bog by vascular plants resulted in 90–100% surface cover, while *Sphagnum* mosses recolonized less than 10% of the surface area.

Water availability and fluxes within cutover peatlands are influenced by the high abundance and distribution of shrubs [*Farrick*, 2008]. The presence of the shrub canopy and subsequent development of a leaf litter layer results in increased interception, transpiration, and altered peat surface evaporation dynamics [*Crockford and Richardson*, 2000; *Dingman*, 2002], consequently impacting ψ and θ in the upper portion of the peat substrate. Ψ below −100 mb results in the desiccation of *Sphagnum* mosses as water is removed from storage within hyaline cells [*Hayward and Clymo*, 1982]; however, the vascular shrubs are not only able to extract water and thrive under conditions with ψ much lower than this

[*Farrick*, 2008], but their presence might prevent extremely low ψ from occurring. In a lab investigation using peat monoliths extracted from the Cacouna bog, *Farrick* [2008] found that, under a falling WT, monoliths with shrubs present maintained ψ > –100 mb throughout the sampling period (85 days), while monoliths of bare peat (also from the Cacouna bog) subjected to similar conditions exceeded the ψ < –100 mb threshold after only 57 days. Lower net radiation under a shrub canopy cover and the presence of a thin litter layer (consequently reducing ET) accounted for the reduced water losses (thus higher ψ) from the shrub-covered peat surface. These reductions in water losses were able to offset the interception (*I*) from both the canopy (*I* = 33% of precipitation) and the leaf litter (*I* = 7% of precipitation), and Ericaceae were deemed beneficial, from a hydrological perspective, in the successful reestablishment of a *Sphagnum* cover [*Farrick*, 2008]. The benefits of an ericaceous cover, however, may be partially offset due to the potential interference of the leaf litter on the establishment of the incipient moss diaspores.

RO has been found to represent a substantial loss of water from disturbed peatlands. *Van Seters and Price* [2001] found in an old abandoned bog (the Cacouna bog) that RO losses corresponded to 18% of precipitation during the study period (2-year average; snow excluded). A snow survey and snowmelt RO study also at the Cacouna bog (March/April 2006) found average snow depths of 67–81 cm, with snowmelt waters accounting for approximately 109 mm of RO over a 29-day melt period (S. Ketcheson, unpublished data, 2008). During the following summer (19 May to 16 August 2006), RO from the same site was 73 mm (S. Ketcheson, unpublished data, 2008). The relatively short (29 days) snowmelt period resulted in 50% more RO in comparison to the 89-day summer study period. Thus, the snowmelt period represents a substantial proportion of the annual RO from a disturbed system and should be included in such estimations.

Typically, increases in RO, peak flows, and base flow relative to natural conditions are observed following drainage [*Price et al.*, 2003]. However, contradictory hydrological responses, such as decreased peak flow due to increased storage capacity in drained soils between storms, have also been observed [*Burke*, 1975]. Over the long-term, permanent structural changes can take place following WT lowering including soil-pipe formation and macropore development, which increases throughflow and correspondingly reduces overland flow [*Holden and Burt*, 2002]. Further, in a study comparing data from the 1950s to the early 2000s, *Holden et al.* [2006] identified long-term changes to the hydrology of disturbed (drained) catchments that were not apparent in the few years immediately following drainage [*Conway and Millar*, 1960].

5. HYDROLOGICAL PROCESSES RELATED TO SYSTEM RESTORATION

Given the desire to recolonize cutover peatlands with peat-forming *Sphagnum* moss, a variety of techniques including ditch-blocking, surface scarification, shallow excavation, low bunds, etc. are used to increase surface water detention [*Price et al.*, 2003]. The application of straw mulch has been shown to reduce evaporation [*Petrone et al.*, 2004] and thereby increase the surface wetness and consequently the survival and growth of reintroduced *Sphagnum* mosses [*Price et al.*, 1998].

An essential first step is the blockage of drainage ditches. This provides wetter conditions more suitable for nonvascular mosses. Blocking the drainage network at the Cacouna bog (October 2006) resulted in a rise in the average seasonal WT level by nearly 30 cm and made the substrate in previously marginal areas more favorable to *Sphagnum* mosses through increased ψ and θ at the cutover surface. However, previously established *Sphagnum* cushions in low areas were flooded (S. Ketcheson, unpublished data, 2008). Following the blockage there were substantial reductions in RO efficiency (percentage of precipitation produced as stream discharge). The average RO efficiency during the summer months of 2007 (following ditch blocking) was reduced to 10% from a 2-year average (2005 and 2006) preblockage efficiency of 23%.

No comparative studies of water budgets of undisturbed and drained sites including snowmelt RO are available, to our knowledge. However, snowmelt RO from both drained and restored sections of cutover peatland at Bois des Bel (BdB) peatland (also near Cacouna, Québec) amounted to 79% of the annual RO, but in the post-snowmelt period, the restored section lost only 25% of that experienced by the drained site [*Shantz and Price*, 2006b]. During snowmelt, the RO at both sites was dominated by surface water inputs (~85%), but surface water input declined to about 60% later in the summer [*Shantz and Price*, 2006b]. During the summer period, the restored site had wetter antecedent conditions; thus, RO was more responsive to rainfall, being larger and having a shorter time lag.

Rewetting at BdB significantly raised the WT, resulting in θ and ψ conditions suitable for the regeneration of a plant cover, notably with *Sphagnum* mosses. Rewetting restored some of the previous peat volume within 1 year, as the peat dilated in accordance with the higher pore water pressure, and consequently, the ability to change volume seasonally increased in response to WT variations [*Shantz and Price*, 2006a]. The greatest volume change within each layer occurred in the upper 30 cm (4.8–9.6%), while the 30- to 150-cm layer experienced 1.1–2.4% change over 3 years of monitoring. The unrestored section had <0.3% total volume change. At the restored site, the volume change was accompanied by a seasonal reduction of K_s of nearly one order of magnitude. Despite the growing layer of poorly decomposed mosses and nonvascular plants, the WT regularly falls below the cutover peat surface [*Shantz and Price*, 2006a], the CO_2 efflux continues to be high, and the essential water regulation function of an acrotelm are not yet present.

While site reconfiguration (blocking ditches, bunds, etc.) is effective when used in combination with straw mulch application, the use of companion species to nurse *Sphagnum* also has a benefit [*Lavoie et al.*, 2005a], as they provide shading. *McNeil and Waddington* [2003] found *Sphagnum* cushions (at the Cacouna bog) with the Ericaceae cover removed performed poorly, and it was assumed that shading was responsible. Price (unpublished data) used Hydrus 1-D (as noted previously) to model unsaturated flow in *Sphagnum* and found that splitting the potential evaporation demand between soil evaporation and transpiration resulted in higher ET losses, but accessed water from deeper within the cushion and resulted in a higher θ at the surface than if plants were absent (no transpiration).

Kennedy and Price [2004] developed a numerical model (FLOCOPS) to simulate surface elevation, WT, and soil moisture and pressure (at 5 cm below the surface) of a partly restored peatland (LSJ). They found the simulations to be most sensitive to the water retention and consolidation characteristics. A good fit was made to surface elevation, WT, and soil moisture, but only a reasonable fit to soil-water pressure, likely because of the difficulty in accounting for its high spatial variability [*Shantz and Price*, 2006a]. The analysis also revealed that the reduction in K_s due to compression resulted in less efficient transport of water to the surface in response to evaporation. This can reduce the evaporation loss, but increase the variability in θ. More variable θ increases soil respiration, thus carbon efflux [*Waddington et al.*, 2002]. However, in general, the compressibility of peatlands, including (but to a lesser extent) cutover peatlands, results in synchronous rise and fall of the surface and WT, thus maintaining higher levels of saturation than would otherwise be the case [*Whittington and Price*, 2006], with a consequent reduction in CO_2 loss [*Strack and Waddington*, 2007].

6. SUMMARY AND CONCLUSION

The changes in WT, ET, RO, and net carbon exchange caused by drainage, abandonment, and restoration are summarized in a conceptual model (Figure 3). In the undisturbed state, the WT is relatively high and stable, since it is regulated by the acrotelm. Consequently, methane (CH_4) fluxes

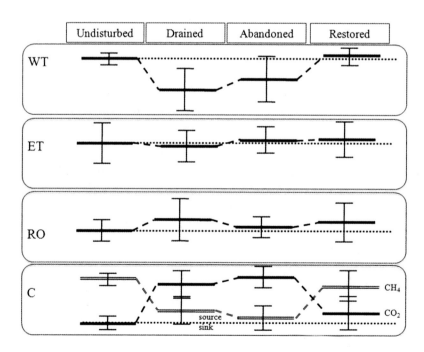

Figure 3. Changes in water table (WT), evapotranspiration (ET), runoff (RO), and net carbon exchange (C) relative to the undisturbed state. Vertical bar represents relative variability. Small dashed line represents natural levels, while larger dashed lines represent the transition between disturbance stages (slope not reflective of time for transition). Adapted from *Van Seters and Price* [2002] and *Waddington and Price* [2000].

to the atmosphere are high, while decomposition of the peat is low, resulting in a net atmospheric CO_2 sink [*Gorham*, 1991]. RO from bogs is quite variable, being highly dependent on the antecedent condition. Similarly, ET is highly variable, since *Sphagnum* relies solely on capillary transport of water to the surface, which is poor when the WT is low. Drainage lowers WT, and the decrease in S_y caused by peat consolidation increases WT variability. The drainage network is efficient, and water flows are transmitted quickly, increasing the flashiness of the hydrograph (variability). ET may drop slightly with the loss of vascular plants, but higher water retention strengthens the capillary transport of water, so bare-soil evaporation satisfies the potential ET demand; thus, the relative ET rate is similar to the undisturbed condition. ET variability is lower because capillary transport maintains water availability at the evaporating surface. Plant production (net ecosystem productivity) is eliminated completely (i.e., zero) following the removal of vegetation at mined sites [*Waddington and Price*, 2000]. The lower WT increases the depth of the aerobic zone and decreases soil moisture content, subsequently increasing soil respiration and decreasing CH_4 production, resulting in a large increase in net CO_2 loss to the atmosphere, and reduced CH_4 fluxes [*Waddington and Price*, 2000]. Over longer time periods in abandoned cutover bogs, WT will rise as the drainage network loses efficiency due to infilling and occlusion, which also reduces RO. However, WT variability is still high, since ET is the primary water loss, and it (ET) remains high, perhaps even increasing because of the spontaneous regeneration of ericaceous plants. Restoration measures to raise the WT often result in flooding, which is highly spatially variable but temporally less variable than in the drained and abandoned state because of the increased water retention by dams and bunds. The higher WT results in wetter antecedent conditions, which can increase RO and will increase RO variability. The RO efficiency, however, will be dependant upon antecedent conditions (capacity to retain additional water on-site) and event-based precipitation dynamics. CH_4 production increases in the wetter conditions following restoration, and recolonization of vegetation increases plant production and reduces net CO_2 losses to the atmosphere. ET in this wetter system may change little as the capillary transport in the drained and abandoned state was relatively efficient, although partial flooding of lower-lying areas will consequently increase ET, as open water is a freely evaporating surface. The return to a moss-dominated system may increase ET variability, as the newly emerging mosses with their huge range of $K(\psi)$ begin to control water delivery to the evaporating surface. In addition, the much wetter environment will likely cause a vegetation shift away from vascular vegetation, reducing transpiration losses from the

system, which is typically quite consistent. Over the long-term, once the WT resides in the newly developed moss layer, other system functions will return to a state similar to the undisturbed condition.

Cutover peatlands function very differently than undisturbed ones because stripping of the acrotelm during peat-cutting operations removes the essential self-regulating mechanisms characteristic of peatlands, including the ability to regulate WT and the water supply to vascular and nonvascular plants. Drainage ditches that lower the WT increase peat compression by normal consolidation and enhanced decomposition. This increases the WT variability and the water retention of cutover peat so that artificially or spontaneously introduced *Sphagnum* mosses have a limited ability to access water stored in the cutover peat. The presence of mosses does, however, increase the wetness of the substrate, which, if sufficient, will support the initiation and expansion of *Sphagnum* cushions that may eventually coalesce into a carpet. Restoration measures to detain water on-site for longer duration are effective and, coupled with reintroduction of plant materials, can begin to return ecohydrological functions to the peatland. Few, if any, of such restoration attempts have returned all such functions; presumably, this will have to wait until a layer with properties similar to the acrotelm is developed.

REFERENCES

Bellamy, D. J., and J. Rieley (1967), Some ecological statistics of a 'miniature bog,' *Oikos*, *18*, 33–40.

Bergeron, M. (1994), *Peat*, Nat. Resour. Can., Ottawa, Ont., Canada.

Boelter, D. H. (1968), Important physical properties of peat, in *3rd International Peat Congress, Québec City 1968*, pp. 150–156, Int. Peat Soc., Jyväskylä, Finland.

Boelter, D. H. (1969), Physical properties of peats as related to degree of decomposition, *Soil Sci. Soc. Am. J.*, *33*, 606–609.

Boelter, D. H. (1972), Water table drawdown around an open ditch in organic soils, *J. Hydrol.*, *15*, 329–340.

Burke, W. (1975), Aspects of the hydrology of blanket peat in Ireland, in *Hydrology of Marsh-Ridden Areas: Proceedings of the Minsk Symposium, June 1972: A Contribution to the International Hydrological Decade*, pp. 171–182, UNESCO, Paris.

Conway, V. M., and A. Millar (1960), The hydrology of some small peat-covered catchments in the northern Pennies, *J. Inst. Water Eng.*, *14*, 415–424.

Crockford, R. H., and D. P. Richardson (2000), Partitioning of rainfall into throughfall, stemflow and interception: Effect of forest type, ground cover and climate, *Hydrol. Processes*, *14*, 2903–2920.

Dingman, S. L. (2002), *Physical Hydrology*, 2nd ed., Waveland, Long Grove, Ill.

Farrick, K. K. (2008), The role of ericaceous shrubs in the surface water balance and soil water availability of a cutover peatland, Québec, M.Sc. thesis, Univ. of Waterloo, Waterloo, Ont., Canada.

Girard, M., C. Lavoie, and M. Thériault (2002), The regeneration of a highly disturbed ecosystem: A mined peatland in southern Québec, *Ecosystems*, *5*, 274–288.

Gorham, E. (1991), Northern peatlands: Role in the carbon cycle and probable responses to climatic warming, *Ecol. Appl.*, *1*(2), 182–195.

Hayward, P. M., and R. S. Clymo (1982), Profiles of water content and pore size in *Sphagnum* peat and their relation to peat bog ecology, *Proc. R. Soc. London, Ser. B*, *215*, 299–325.

Hoag, R. S., and J. S. Price (1995), A field-scale, natural gradient solute transport experiment in peat at a Newfoundland blanket bog, *J. Hydrol.*, *172*, 171–184.

Hobbs, N. B. (1986), Mire morphology and the properties and behaviour of some British and foreign peats, *Q. J. Eng. Geol. Hydrogeol.*, *19*(1), 7–80.

Holden, J. (2005), Peatland hydrology and carbon release: Why small-scale process matters, *Philos. Trans. R. Soc.*, *363*, 2891–2913.

Holden, J., and T. P. Burt (2002), Laboratory experiments on drought and runoff in blanket peat, *Eur. J. Soil Sci.*, *53*, 675–689.

Holden, J., M. G. Evans, T. P. Burt, and M. Horton (2006), Impact of land drainage on peatland hydrology, *J. Environ. Qual.*, *35*, 1764–1778.

Ingram, H. A. P. (1978), Soil layers in mires: Function and terminology, *J. Soil Sci.*, *29*, 224–227.

Ingram, H. A. P. (1983), Hydrology, in *Ecosystems of the World 4A, Mires: Swamp, Bog, Fen and Moor*, edited by A. J. P. Gore, pp. 67–158, Elsevier, Amsterdam.

Kellner, E., J. M. Waddington, and J. S. Price (2005), Dynamics of biogenic gas bubbles in peat: Potential effects on water storage and peat deformation, *Water Resour. Res.*, *41*, W08417, doi:10.1029/2004WR003732.

Kennedy, G. W., and J. S. Price (2004), Simulating soil water dynamics in a cutover bog, *Water Resour. Res.*, *40*, W12410, doi:10.1029/2004WR003099.

Kennedy, G. W., and P. J. Van Geel (2000), Hydraulics of peat filters treating septic tank effluent, *Transp. Porous Media*, *41*, 47–60.

Keys, D. (1992), Canadian peat harvesting and the environment, sustaining wetlands issues paper, North Am. Wetlands Conserv. Counc., Ottawa, Ont., Canada.

Kuhry, P., and D. H. Vitt (1996), Fossil carbon/nitrogen ratios as a measure of peat decomposition, *Ecology*, *77*(1), 271–275.

Lafleur, P. M., and N. T. Roulet (1992), A comparison of evaporation rates from two fens of the Hudson Bay Lowland, *Aquat. Bot.*, *44*, 55–69.

Lafleur, P. M., R. A. Hember, S. W. Admiral, and N. T. Roulet (2005), Annual and seasonal variability in evapotranspiration and water table at a shrub-covered bog in southern Ontario, Canada, *Hydrol. Processes*, *19*, 3533–3550.

Lavoie, C., and L. Rochefort (1996), The natural revegetation of a harvested peatland in Southern Québec: A spatial and dendroecological analysis, *Ecoscience*, *3*(1), 10.

Lavoie, C., P. Grosvernier, M. Girard, and K. Marcoux (2003), Spontaneous revegetation of mined peatlands: A useful restoration tool?, *Wetl. Ecol. Manage.*, *11*, 97–107.

Lavoie, C., K. Marcoux, A. Saint-Louis, and J. S. Price (2005a), The dynamics of a cotton-grass (*Eriophorum vaginatum L.*) cover expansion in a vacuum-mined peatland, southern Québec, *Wetlands*, *25*, 64–75.

Lavoie, C., A. Saint-Louis, and D. Lachance (2005b), Vegetation dynamics on an abandoned vacuum-mined peatland: 5 years of monitoring, *Wetl. Ecol. Manage.*, *13*, 621–633, doi:10.1007/s11273-005-0126-1.

McLay, C. D. A., R. F. Allbrook, and K. Thompson (1992), Effect of development and cultivation on physical properties of peat soils in New Zealand, *Geoderma*, *54*, 23–37.

McNeil, P., and J. M. Waddington (2003), Moisture controls on *Sphagnum* growth and CO_2 exchange on a cutover bog, *J. Appl. Ecol.*, *40*, 354–367.

Mulqueen, J. (1989), Hydrology and drainage of peatland, *Environ. Geol. Water Sci.*, *9*, 15–22.

National Wetlands Working Group (1997), *The Canadian Wetland Classification System*, 2nd ed., edited by W. R. Centre, B. G. Warner, and C. D. A. Rubec, Univ. of Waterloo, Waterloo, Ont., Canada.

Petrone, R. M., J. S. Price, J. M. Waddington, and H. von Waldow (2004), Surface moisture and energy exchange from a restored peatland, Québec, Canada, *J. Hydrol.*, *295*, 198–210.

Prevost, M., P. Belleau, and A. P. Plamondon (1997), Substrate conditions in a treed peatland: Responses to drainage, *Ecoscience*, *4*, 543–544.

Price, J. S. (1991), Evaporation from a blanket bog in a foggy coastal environment, *Boundary Layer Meteorol.*, *57*, 391–406.

Price, J. S. (1996), Hydrology and microclimate of a partly restored cutover bog, Quebec, *Hydrol. Processes*, *10*, 1263–1272.

Price, J. S. (1997), Soil moisture, water tension, and water table relationships in a managed cutover bog, *J. Hydrol.*, *202*, 21–32.

Price, J. S. (2003), Role and character of seasonal peat soil deformation on the hydrology of undisturbed and cutover peatlands, *Water Resour. Res.*, *39*(9), 1241, doi:10.1029/2002WR001302.

Price, J. S., and S. M. Schlotzhauer (1999), Importance of shrinkage and compression in determining water storage changes in peat: The case of a mined peatland, *Hydrol. Processes*, *13*, 2591–2601.

Price, J. S., and G. S. Whitehead (2001), Developing hydrological thresholds for *Sphagnum* recolonization on an abandoned cutover bog, *Wetlands*, *21*(1), 32–40.

Price, J. S., and G. S. Whitehead (2004), The influence of past and present hydrological conditions on *Sphagnum* recolonization and succession in a block-cut bog, Quebec, *Hydrol. Processes*, *18*, 315–328.

Price, J. S., L. Rochefort, and F. Quinty (1998), Energy and moisture considerations on cutover peatlands: Surface microtopography, mulch cover, and *Sphagnum* regeneration, *Ecol. Eng.*, *10*, 293–312.

Price, J. S., A. L. Heathwaite, and A. J. Baird (2003), Hydrological processes in abandoned and restored peatlands: An overview of management approaches, *Wetl. Ecol. Manage.*, *11*, 65–83.

Price, J. S., P. N. Whittington, D. E. Elrick, M. Strack, N. Brunet, and E. Faux (2008), A method to determine unsaturated hydraulic conductivity in living and decomposed *Sphagnum* moss, *Soil Sci. Soc. Am. J.*, *72*, 487–491.

Price, J. S., T. W. D. Edwards, Y. Yi, and P. N. Whittington (2009), Physical and isotopic characterization of evaporation from *Sphagnum* moss, *J. Hydrol.*, *369*, 175–182.

Pyatt, D. G., and A. L. John (1989), Modelling volume changes in peat under conifer plantations, *Eur. J. Soil Sci.*, *40*(4), 695–706.

Read, D. J., J. R. Leake, and J. Perez-Moreno (2004), Mycorrhizal fungi as drivers of ecosystem processes in heathland and boreal forest biomes, *Can. J. Bot.*, *82*(8), 1243–1263.

Rochefort, L., S. Campeau, and J.-L. Bugnon (2002), Does prolonged flooding prevent or enhance regeneration and growth of *Sphagnum*?, *Aquat. Bot.*, *74*, 327–341.

Romanov, V. V. (1968), *Hydrophysics of Bogs*, translated from Russian by N. Kaner, Isr. Program for Sci. Transl., Jerusalem.

Sagot, C., and L. Rochefort (1996), Tolérance des Sphaignes à la déssication, *Crytogamie Bryol. Lichenol.*, *17*(3), 171–183.

Schlotzhauer, S. M., and J. S. Price (1999), Soil water flow dynamics in a managed cutover peat field, Quebec: Field and laboratory investigations, *Water Resour. Res.*, *35*(12), 3675–3683.

Schouwenaars, J. M., and A. M. Gosen (2007), The sensitivity of *Sphagnum* to surface layer conditions in a re-wetted bog: A simulation study of water stress, *Mires Peat*, *2*, 1–19.

Shantz, M. A., and J. S. Price (2006a), Hydrological changes following restoration of the Bois-des-Bel Peatland, Québec, 1999–2002, *J. Hydrol.*, *331*, 543–553.

Shantz, M. A., and J. S. Price (2006b), Characterization of surface storage and runoff patterns following peatland restoration, Quebec, Canada, *Hydrol. Processes*, *30*, 3799–3814.

Silins, U., and R. L. Rothwell (1998), Spatial patterns of aerobic limit depth and oxygen diffusion rate at two peatlands drained for forestry in Alberta, *Can. J. For. Res.*, *29*(1), 53–61.

Spieksma, J. F. M. (1999), Changes in the discharge pattern of a cutover raised bog during rewetting, *Hydrol. Processes*, *13*, 13.

Strack, M., and J. S. Price (2009), Moisture controls on carbon dioxide dynamics of peat-*Sphagnum* monoliths, *Ecohydrology*, *2*(1), 34–41.

Strack, M., and J. M. Waddington (2007), Response of peatland carbon dioxide and methane fluxes to a water table drawdown experiment, *Global Biogeochem. Cycles*, *21*, GB1007, doi:10.1029/2006GB002715.

Tarnocai, C. (2006), The effect of climate change on carbon in Canadian peatlands, *Global Planet. Change*, *53*, 222–232.

Terzaghi, K. (1943), *Theoretical Soil Mechanics*, John Wiley, New York.

Van Seters, T. E., and J. S. Price (2001), The impact of peat harvesting and natural regeneration on the water balance of an abandoned cutover bog, Quebec, *Hydrol. Processes*, *15*, 233–248.

Van Seters, T. E., and J. S. Price (2002), Towards a conceptual model of hydrological change on an abandoned cutover bog, Quebec, *Hydrol. Processes*, *16*, 1965–1981.

Waddington, J. M., and J. S. Price (2000), Effect of peatland drainage, harvesting, and restoration on atmospheric water and carbon exchange, *Phys. Geogr.*, *21*, 433–451.

Waddington, J. M., K. D. Warner, and G. W. Kennedy (2002), Cutover peatlands: A persistent source of atmospheric CO_2, *Global Biogeochem. Cycles*, *16*(1), 1002, doi:10.1029/2001GB001398.

Whittington, P. N., and J. S. Price (2006), The effects of water table draw-down (as a surrogate for climate change) on the hydrology of a fen peatland, Canada, *Hydrol. Processes*, *20*, 3589–3600.

S. J. Ketcheson and J. S. Price, Department of Geography and Environmental Management, University of Waterloo, Waterloo, ON N2L 3G1, Canada. (jsprice@envmail.uwaterloo.ca)

The Influence of Permeable Mineral Lenses on Peatland Hydrology

A. S. Reeve and Z. D. Tyczka

Department of Earth Sciences, University of Maine, Orono, Maine, USA

X. Comas

Department of Geosciences, Florida Atlantic University, Boca Raton, Florida, USA

L. D. Slater

Department of Earth and Environmental Sciences, Rutgers, State University of New Jersey, Newark, New Jersey, USA

Cross-sectional computer models were created that incorporated different high permeability zones to explore the potential role of eskers and similar geologic units associated with peatlands on the hydrology of these systems. These computer simulations indicate that small isolated lenses of high permeability material will locally distort the flow field, shifting flow patterns and creating new discharge and/ or recharge zones. The simulation with the greatest hydraulic connectivity between the peat dome and peatland lagg displays widespread and continuous-with-depth downward flow beneath the bog dome, consistent with field observations. These simulations were compared with field data collected from a Maine (USA) peatland, and they suggest an esker identified within this peatland distorts the hydraulic gradients and resulting flow patterns within the peat. Similar subsurface features may be important in other peatland systems.

1. INTRODUCTION

Hydrology exerts an important influence on the vegetation patterns and the geomorphology of peatland systems by controlling the saturation state and the transport of nutrients within the peat column. The movement of water within the peat is controlled by the water table position and the permeability field within the peat and geologic materials underlying the peat. Several conceptual models, summarized by

Carbon Cycling in Northern Peatlands
Geophysical Monograph Series 184
Copyright 2009 by the American Geophysical Union.
10.1029/2008GM000825

Reeve et al. [2000, 2001b], have been proposed to describe the hydrology and water chemistry of peatland systems.

Reeve et al. [2000] described the impact of mineral sediment permeability underlying peat on groundwater flow patterns, particularly vertical flow, within the peat column. They noted a reduction in the vertical penetration of groundwater flow cells beneath a bog dome when the mineral sediment permeability underlying peat deposits decreased. These results have important implications for identifying where vertical flow is negligible in peatland systems and under what conditions vertical flow will be an important component of groundwater flow within peat deposits. In this paper, we expand on this work and assess the impact of isolated lenses of sand and/or gravel underlying peatland systems on flow patterns within the peat. We hypothesize that the assumption of homogeneous stratigraphy beneath peat deposits, as has been made in many studies [*Ingram*, 1982; *Winston*, 1994; *Reeve et al.*, 2000; *Glaser et al.*, 2004], may oversimplify flow patterns within peat and obscure important hydrogeologic characteristics of peatlands.

Many researchers have explored the influence of heterogeneity within peat [*Beckwith et al.*, 2003] or within geologic materials [*Freeze and Witherspoon*, 1967; *Winter*, 1976], but we are unaware of any systematic study that explores the influence of permeable lenses of mineral sediments on the flow patterns within peatlands. The importance of subsurface conditions on peatland hydrology has been widely recognized. The importance of low permeability units beneath a Newfoundland peat complex was assessed using cross-sectional computer models. That analysis identified the permeability of peat at the margin of the peatlands as particularly important in regulating water levels within a peatland [*Lapen et al.*, 2005]. The structure and stratigraphy of geologic materials may also direct groundwater discharge into a peatland [*Glaser et al.*, 1997; *Siegel*, 1992; *Boldt*, 1986], supplying solute-rich water to the peatland system. Geologic features may also hydraulically isolate peatland systems from nearby surface hydrology features [*Bradley*, 2002] or provide pathways for nutrient loading [*Drexler and Bedford*, 2002]. *Comas et al.* [2004, 2005a] used ground penetrating radar to image a subsurface feature beneath the central unit of Caribou Bog, Maine (USA) interpreted as an esker deposit. They speculate that the presence of this esker influenced the genesis of pools now present in this peatland.

2. METHODOLOGY

In this paper, we present results from several computer models that incorporate different permeable units beneath a peat deposit. The results from computer simulations are compared with field data collected from Caribou Bog, a peatland system associated with an esker deposit.

2.1. Computer Simulations

Computer simulations were run using FiPy [*Wheeler et al.*, 2007], a modeling package developed at the National Institute of Standards and Technology (USA). This software package is a general partial differential equation solver based on the finite-volume method. Each model simulates flow in a cross-section through an idealized peat basin. The cross-sectional model is 3000 m long with a maximum thickness of 25 m. This area represents a peat basin with upland hills on either end of the modeled area.

All sides of the model domain, except the top, were assigned no-flow boundary conditions (Figure 1). Constant

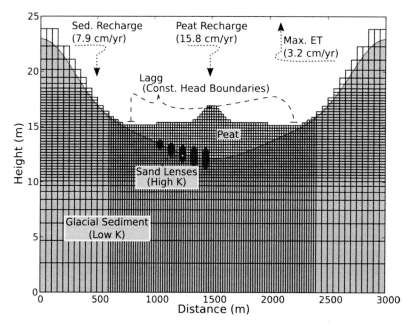

Figure 1. Configuration of computer models, including the grid used to simulate groundwater flow. The model domain is 23-m thick (maximum) by 3000-m wide. The areas for the peat and sediment lenses indicated in the upper figure were used to select cells to assign different hydraulic parameters by determining if a cell's center was within the different layers or lenses.

head boundaries were assigned along 100-m strips at the top of the peatland where the peat intersects the mineral soil (head equal to surface elevation), to simulate the standing water in a lagg. Recharge (R) assigned to the top cells in each simulation varied with the material present at the surface, with higher recharge rates assigned to peat (15.7 cm a^{-1}) and lower recharge rates assigned to the lower permeability glacial sediments (7.9 cm a^{-1}). Evapotranspiration rates (ET) varied with water table depth, changing linearly from a maximum value (3.2 cm a^{-1}) to 0 as the water table drops from the surface to a meter below the surface. The ET term is intended to simulate the removal of water through both surface runoff and ET, preventing the water table from rising far above the land surface.

Hydraulic conductivities assigned to the geologic units in the computer model are based on values measured at Caribou Bog. Peat deposits within the basin are subdivided into an upper permeable layer (1-m thick) and a lower less permeable layer (remaining thickness of peat). The hydraulic conductivity assigned to the upper peat layer decreased exponentially with depth from 1×10^{-3} to 1×10^{-4} m s^{-1}. The deep peat was assigned a hydraulic conductivity of 5×10^{-6} m s^{-1} based on measurements in Caribou Bog [*Stevens*, 2006]. Glacial sediments in the simulations were assigned

conductivities of 1×10^{-7} m s^{-1}, based on measurements within till that outcrops adjacent to Caribou Bog at a landfill site [*Cole*, 1992]. Five areas were predefined within the grid to represent heterogeneity within the geological materials beneath the peat. These sediment lenses were assigned a hydraulic conductivity of 1×10^{-4}, representing sand and/or gravel lenses within the till.

Three steady state computer models of a generalized peatland system were constructed to evaluate the potential impact of isolated permeable deposits on long-term flow patterns. Simulations were run assuming a ratio of horizontal (K_{xx}) to vertical (K_{zz}) hydraulic conductivity of 1 (Figure 2) and 10 (Figure 3). For each set of simulation scenarios, additional permeable features in the glacial sediment were intended to enhance the degree of lateral hydrogeologic communication within the subsurface. All models assume Darcy's law adequately describes the flow of water through the peat [*Hemond and Goldman*, 1985]. A diffusion equation was used within FiPy to calculate the hydraulic head (h) distribution in the model domain:

$$0 = \frac{\partial}{\partial x}\left(K_{xx}\frac{\partial h}{\partial x}\right) + \frac{\partial}{\partial z}\left(K_{zz}\frac{\partial h}{\partial z}\right) + R - \text{ET}. \qquad (1)$$

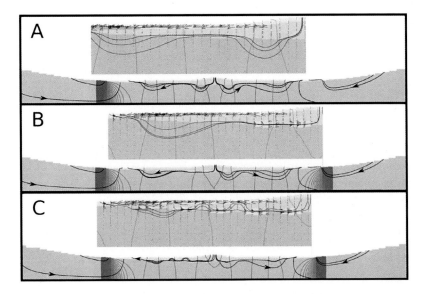

Figure 2. Results from computer simulations with isotropic hydraulic conductivity incorporating: (a) no heterogeneity in the mineral sediments, (b) two high hydraulic conductivity lenses in the glacial sediment, and (c) five high hydraulic conductivity lenses in the glacial sediment. For each of these scenarios pictured, the flow domain (3000-m long and 24-m high) is displayed with clipped left and right edges below an image focusing on the left portion of the peatland. The gray scale of the image is related to the hydraulic conductivity of the material, with lighter colors indicating higher hydraulic conductivity. Streamlines are indicated with thick black lines, and equipotentials are indicated by thin dark gray lines at a contour interval of 5 cm (equipotentials are not shown in the mineral sediment). Vectors indicating the magnitude and direction of flow are included in the detailed images.

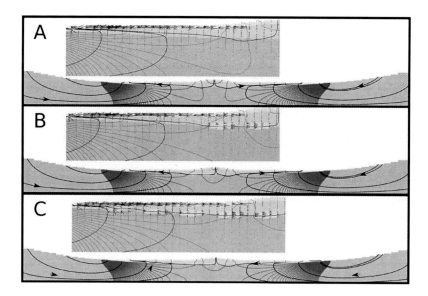

Figure 3. Results from computer simulations with a tenfold horizontal to vertical hydraulic conductivity anisotropy incorporating: (a) no heterogeneity in the mineral sediments, (b) two high hydraulic conductivity lenses in the glacial sediment, and (c) five high hydraulic conductivity lenses in the glacial sediment. For each of these scenarios pictured, the flow domain (3000-m long and 24-m high) is displayed with clipped left and right edges below an image focusing on the left portion of the peatland. The gray scale of the image is related to the hydraulic conductivity of the material, with lighter colors indicating higher hydraulic conductivity. Streamlines are indicated with thick black lines and equipotentials are indicated by thin dark gray lines at a contour interval of 5 cm (equipotentials are not shown in the mineral sediment). Vectors indicating the magnitude and direction of flow are included in the detailed images.

To account for desaturation of cells within the model, the methodology described by *Doherty* [2001] was used. This method involves smoothly decreasing the horizontal hydraulic conductivity until it reaches a negligible value and smoothly increasing the vertical hydraulic conductivity to some large value as the water level in a cell drops below the cell bottom. This results in negligible amount of horizontal flow in "dry" cells, while increasing the vertical flow to allow hydraulic source and sink terms assigned to the top cells to communicate with the active "wet" cells within the model.

The described simulation approach has a number of shortcomings. Cross-sectional simulations of peatland ecosystems assume there is no flow perpendicular to the cross section. As bog domes in peatlands produce radial flow patterns, this assumption is violated. Steady state simulations fail to account for variation in hydraulic stress controlled by seasonal and shorter term changes in precipitation and evapotranspiration. These temporal changes in hydraulic stress produce important seasonal or event-driven changes in flow within peatland systems [*Devito et al.*, 1997; *Reeve et al.*, 2006] that are not simulated in steady state models. Within peatland systems, biogenic gas [*Rosenberry et al.*, 2006;

Kellner et al., 2005; *Romanowicz et al.*, 1995] and changing water table position [*Price and Schlotzhauer*, 1999] may alter the hydraulic conductivity of peat over time by occluding pore space and changing the stress field within the easily deformed peat. Despite the false assumptions used in our computer models, they do allow the variation in mineral sediment permeability to be isolated and evaluated through numerical experiments.

2.2. Field Measurements

Hydraulic heads were measured along two transects (T1 and T2, Figure 4) across Caribou Bog in 1999 and 2000 to evaluate seasonal variability in hydraulic head and flow patterns. A short transect (T3) was installed in 2006 to evaluate the hydrologic influence of an esker deposit [*Comas et al.*, 2004, 2005a] buried beneath the peat. Caribou Bog is a multiunit peatland located in central Maine between the cities of Orono and Bangor and covers an area of about 2200 ha [*Davis and Anderson*, 1999] with peat thickness of up to 18 m (Figure 4). The central portion of Caribou Bog contains a raised bog surrounded by fen. A fen water track, a peat landform with ribbed pools, is located north of the central unit's

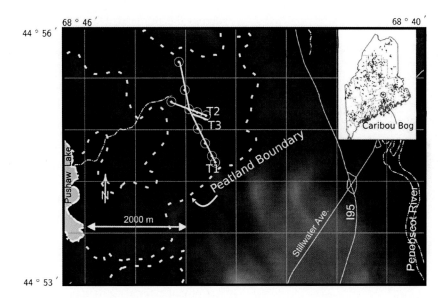

Figure 4. Map of area near Caribou Bog, Maine. Darker colors indicate topographically lower areas. The location of three monitoring well transects are indicated by solid white lines. Circles mark the location of wells along transects 1 and 2.

main raised bog unit [*Davis and Anderson*, 1999]. An intricate complex of narrow pools up to 200-m long occurs on the bog dome near the intersection of transects 1 and 2.

Monitoring wells installed along these transects were made from 1.91- or 2.54-cm nominal diameter PVC pipe. These flush-threaded pipes were installed to depths ranging from 1 to 18 m below the peat surface by manually pushing the well into the peat. Monitoring wells were installed in clusters of three or more wells fitted with 30-cm long machine slotted screens. Before installing wells, the mineral soil depth was estimated using either a 2.54-cm diameter piston corer or by probing the peat with a 2.54-cm diameter rod.

The top of a monitoring well at each well cluster was surveyed using an Ashtech dual frequency GPS. GPS data was postprocessed to differentially correct the data relative to a National Geodetic Survey reference station. Standard errors calculated by postprocessing software were less than 1 cm. Resurveying a monitoring well at one station yielded a difference of less than 1 cm, supporting the low calculated standard errors and the accuracy of dual frequency GPS. Elevation differences between wells within each cluster were measured periodically using a carpenter's level to assess movement over time within each cluster, and little change was observed. Before measuring water levels, all wells were developed by pumping water from them until organic sediment was no longer evident in the purged water. Water levels recovered in these wells following development in about 30 min to several hours. Water levels in monitoring wells

were measured using an electrical water-level indicator and converted to elevation relative to mean sea level using the surveying data.

3. RESULTS AND DISCUSSION

3.1. Computer Simulations

The baseline computer simulations (Figures 2a and 3a) are similar to previously published computer simulations of pristine bog-fen peatland complexes [*Siegel and Glaser*, 1987; *Reeve et al.*, 2001a]. In these simulations, groundwater flow cells develop under the peat dome, with downward flow near the center of this dome and groundwater discharge from the mineral sediments into the peat along the flanks of the peat dome. Little groundwater that has come in contact with the mineral sediments is advected into the upper portion of the peat column. There is little vertical flow through the peat column in the flat (fen) regions surrounding the bog dome, except near the peatland lagg where groundwater discharge occurs. The vector field in Figures 2a and 3a insets indicate that the majority of groundwater flux occurs in the upper, more permeable, peat, with much slower groundwater flux rates in the deeper peat. The limited flow (indicated by the very short vectors) near the bog dome is due to desaturation of the upper portion of the peat column near the crest of dome. Assigning tenfold anisotropy to the hydraulic conductivity of the different units decreases the amount of

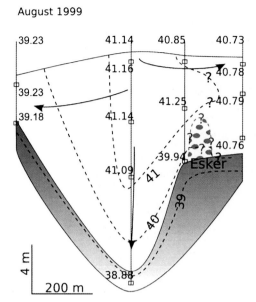

Figure 5. Cross-section along transect 2 with southeastern end on right side. Hydraulic head measurements (meters) collected in August 1999 are indicated next to well screen locations (boxes). Water table elevations are indicated above each well cluster. Equipotentials (dashed lines) are plotted at a one meter contour interval and inferred flow directions are indicated by solid lines with arrows. Gray shading indicates the location of mineral sediments underlying the peat.

vertical flow while increasing the vertical hydraulic gradient. A second flow cell near the edge of the peat unit is present when hydraulic conductivity is isotropic, but is replaced by groundwater discharge originating from the upland under anisotropic conditions.

Introducing two lenses of permeable material near the bog dome (Figures 2b and 3b) distorts the flow field, increasing downflow through the peat but reducing the flux of water from the peat into the underlying till. These results are clearly dependent on the location of the high permeability field within the mineral sediment. Under anisotropic conditions, this distortion is less pronounced but still present. The addition of high permeability units near the bog dome decreases the hydraulic head and lowers the water table position within the peat. In the isotropic case, the flux vectors in the peat have upward and downward vertical flow components on the left and right sides of each permeable lens, respectively. In the anisotropic scenario, increase in vertical flow is less pronounced, whereas vertical hydraulic gradients, illustrated by changes in equipotentials, are more apparent.

Increasing the number of permeable lenses from two to five (Figures 2c and 3c), extending them across the bottom of the peatland, results in cyclical shifts in vertical flow

across the peatland. Once again, shifts in the streamlines are more pronounced in the anisotropic scenario, and shifts in the equipotentials are more pronounced in the scenario with tenfold horizontal to vertical anisotropy in permeability. While these simulations alter the permeability within the mineral sediments, a similar insertion of high-permeability lenses in the peat will produce vertically converging and diverging flow on the hydraulically upgradient and downgradient side, respectively. These results suggest recharge and discharge function of peatlands can be influenced by subsurface shifts in permeability. We speculate that these shifts in vertical flow will produce variation in solute concentrations over short spatial scales, influencing vegetation communities and geochemical processes within the peat.

3.2. Comparison to Field Data

Computer simulations are compared with hydrologic data collected in the central portion of Caribou Bog, a peatland that overlies isolated sandy deposits interpreted by *Comas et al.* [2005a] to be esker beads that roughly follow transect 1. Seasonal changes in water levels measured in the central por-

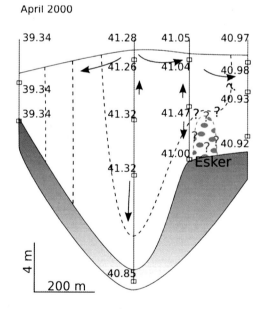

Figure 6. Cross-section along transect 2 with southeastern end on right side. Hydraulic head measurements (meters) collected in April 2000 are indicated next to well screen locations (boxes). Water table elevations are indicated above each well cluster. Equipotentials (dashed lines) are plotted at a 1-m contour interval, and inferred flow directions are indicated by solid lines with arrows. Gray shading indicates the location of mineral sediments underlying the peat.

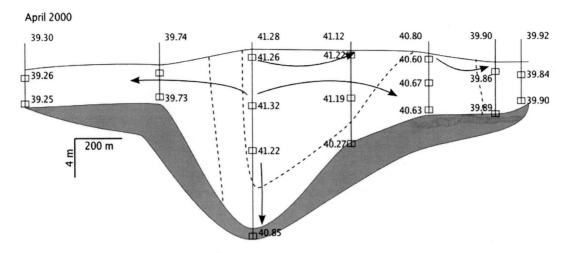

Figure 7. Cross-section along transect 1 with southern end on right side. Hydraulic head measurements (meters) collected in April 2000 are indicated next to well screen locations (boxes). Water table elevations are indicated above each well cluster. Equipotentials (dashed lines) are plotted at a 1-m contour interval, and inferred flow directions are indicated by solid lines with arrows. Gray shading indicates the location of mineral sediments underlying the peat.

tion of Caribou Bog mimic the hydrographs for water bodies in temperate regions, with decreasing water levels through the late spring and summer and a recovery in water levels in the late fall and winter. Rapid increases in water level, sustained over weeks, are associated with snow melt in the spring (data not shown). Hydraulic head measurements indicate that the water table rises about 1.3 m from the lagg to the bog dome throughout the growing season, and these water elevations decrease by about 20 cm from late spring to late summer (Figures 5–7).

Hydraulic gradients indicate water generally moves radially away from and downward within the peat dome. These flow patterns are driven by the water table mound beneath the peat dome as indicated by the groundwater flow simulations. The persistent downward hydraulic gradients in the deeper portions of this peatland, and the shifts from vertically divergent to convergent hydraulic gradients along transects, (Figure 7) suggest additional factors are influencing the hydraulic head in this peatland. Many researchers have invoked biogenic gas as the cause for anomalous hydraulic head measurements in peat deposits, and overpressuring observed in the central portion may be explained by the formation of gas. While gas production is clearly occurring within Caribou Bog [*Comas et al.*, 2005b, 2007], computer simulations in this paper coupled with the observation of esker-like sediment mounds beneath the peat [*Comas et*

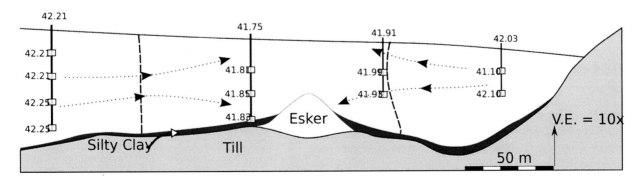

Figure 8. Cross-section along transect 3. Mineral sediment locations are based on GPR profiles collected in this area [*Comas et al.*, 2004]. Hydraulic head measurements (meters) collected in July 2006 are indicated next to well screen locations (boxes). Water table elevations are indicated above each well cluster. Equipotentials (dashed lines) are plotted at a 1-m contour interval, and inferred flow directions are indicated by solid lines with arrows.

al., 2004, 2005a], suggest an alternative explanation. Small local shifts in vertical flow may be controlled by high permeability zones in the mineral sediment and, perhaps, within the peat. Monitoring well transects that cross the esker ridge (Figures 5–6) show there are downward hydraulic gradients in the peat that overlies the esker. Hydraulic head data collected along a more detailed transect (Figure 8) support this interpretation. This pattern could be envisioned by traversing a simulated high-permeability unit on its hydraulically upgradient end perpendicular to the two-dimensional simulation. Groundwater flow driven by these hydraulic gradients will alter the geochemistry within the peat, potentially creating zones favorable for (or inhibiting) microbial activity and associated gas production.

4. CONCLUSION

Computer simulations of an idealized peat basin indicate that small isolated and permeable deposits will alter the flow patterns within a peatland, producing isolated recharge and discharge zones on the hydraulically up- and downgradient ends of these sediment lenses, respectively. Hydrologic data collected from a Maine (USA) peatland contain similar patterns, with vertical hydraulic gradients locally shifting from upward to downward. Downward hydraulic gradients were persistently measured over an esker-like feature identified by *Comas et al.* [2005a] beneath the peat. Computer simulations provide one possible explanation for shifts in vertical hydraulic gradients across peatlands, with these variations controlled by changes in the permeability of subsurface features beneath (and within) the peat. Identifying these features and interpreting their influence on the peatland hydrology will improve the broader understanding of these systems and may provide insight on vegetation patterning, solute fluxes, and gas production within peatland systems.

Computer simulation at the scale presented in this paper is a useful tool for evaluating the hydrology of peatland systems. These models provide an experimental framework to evaluate the importance of processes and the parameters controlling those processes. Complex relationships that are difficult to isolate in laboratory or field studies, such as feedback between biogenic gas production or carbon accumulation and hydrology, can be isolated in computer simulations.

Future computer simulation of peatland systems should attempt to link hydrologic, geochemical and biologic processes to better understand: (1) how these factors influence one another, (2) how peatlands will respond in various climate change scenarios, and (3) the hydrologic role peatlands play within the watersheds. Other peatland hydrology research amenable to computer simulation-based studies include the assessment of peatland heterogeneity and the unique hydrologic properties of peat on the hydrology and solute transport within peatland systems.

Acknowledgments. J. Rhoades provided valuable field assistance and provided feedback on the computer modeling component of this project. The National Institute of Standards and Technology hosted Reeve for 1 month during a sabbatical. This work was supported by the National Science Foundation (EAR-0510004) and the University of Maine.

REFERENCES

Beckwith, C., A. Baird, and A. Heathwaite (2003), Anisotropy and depth-related heterogeneity of hydraulic conductivity in a peat bog. II: Modelling the effect on groundwater flow, *Hydrol. Processes*, *17*, 103–113.

Boldt, D. (1986), Computer simulations of groundwater flow in a raised bog system, Glacial Lake Agassiz peatlands, northern Minnesota, Master's thesis, Syracuse Univ., Syracuse, N. Y.

Bradley, C. (2002), Simulation of the annual water table dynamics of a floodplain wetland, Narborough Bog, UK, *J. Hydrol.*, *261*, 150–172.

Cole, S. (1992), Supplemental hydrogeologic investigation, Orono Municipal Landfill, *Tech. Rep.*, S.W. Cole Eng., Inc. Bangor, Maine.

Comas, X., L. Slater, and A. Reeve (2004), Geophysical evidence for peat basin morphology and stratigraphic controls on vegetation observed in a northern peatland, *J. Hydrol.*, *295*, 173–184.

Comas, X., L. Slater, and A. Reeve (2005a), Stratigraphic controls on pool formation in a domed bog inferred from ground penetrating radar (gpr), *J. Hydrol.*, 40–51.

Comas, X., L. Slater, and A. Reeve (2005b), Geophysical and hydrological evaluation of two bog complexes in a northern peatland: Implications for the distribution of biogenic gases at the basin scale, *Global Biogeochem. Cycles*, *19*, GB4023, doi:10.1029/2005GB002582.

Comas, X., L. Slater, and A. Reeve (2007), In situ monitoring of free-phase gas accumulation and release in peatlands using ground penetrating radar (GPR), *Geophys. Res. Lett.*, *34*, L06402, doi:10.1029/2006GL029014.

Davis, R., and D. Anderson (1999), A numeric method and supporting database for evaluation of Maine peatlands as candidate natural areas, *Tech. Rep. 175*, Maine Agric. and For. Exp. Stn., Orono, Maine.

Devito, K., J. Waddington, and B. Branfireun (1997), Flow reversals in peatlands influenced by local groundwater systems, *Hydrol. Processes*, *11*, 103–110.

Doherty, J. (2001), Improved calculations for dewatered cells in MODFLOW, *Ground Water*, *39*, 863–869.

Drexler, J., and B. Bedford (2002), Pathways of nutrient loading and impacts on plant diversity in a New York peatland, *Wetlands*, *2*, 263–281.

Freeze, R. A., and P. A. Witherspoon (1967), Theoretical analysis of regional groundwater flow: 2. Effect of water-table configuration and subsurface permeability variation, *Water Resour. Res.*, *3*, 623–635.

Glaser, P., D. Siegel, E. Romanowicz, and Y. Shen (1997), Regional linkages between raised bogs and the climate, groundwater, and landscape of north-western Minnesota, *J. Ecol.*, *85*, 3–16.

Glaser, P., D. Siegel, A. Reeve, J. Janssens, and D. Janecky (2004), Tectonic drivers for vegetation patterning and landscape evolution in the Albany River region of the Hudson Bay Lowlands, *J. Ecol.*, 1054–1070.

Hemond, H., and J. Goldman (1985), On non-Darcian water flow in peat, *J. Ecol.*, *73*, 579–584.

Ingram, H. (1982), Size and shape in raised mire ecosystems: A geophysical model, *Nature*, *297*, 300–303.

Kellner, E., J. Waddington, and J. Price (2005), Dynamics of biogenic gas bubbles in peat: Potential effects on water storage and peat deformation, *Water Resour. Res.*, *41*, WO8417, doi:10.1029/2004WR003732.

Lapen, D., J. Price, and R. Gilbert (2005), Modelling two-dimensional steady-state groundwater flow and flow sensitivity to boundary conditions in blanket peat complexes, *Hydrol. Processes*, *19*, 371–386.

Price, J., and S. Schlotzhauer (1999), Importance of shrinkage and compression in determining water storage changes in peat: the case of a mined peatland, *Hydrol. Processes*, *13*, 2591–2601.

Reeve, A., D. Siegel, and P. Glaser (2000), Simulating vertical flow in large peatlands, *J. Hydrol.*, *227*, 207–217.

Reeve, A., D. Siegel, and P. Glaser (2001a), Simulating dispersive mixing in large peatlands, *J. Hydrol.*, *242*, 103–114.

Reeve, A., J. Warzocha, P. Glaser, and D. Siegel (2001b), Regional ground-water flow modeling of the Glacial Lake Agassiz Peatlands, Minnesota, *J. Hydrol.*, *243*, 91–100.

Reeve, A. S., R. Evensen, P. Glaser, D. Siegel, and D. Rosenberry (2006), Flow path oscillations in transient ground-water simulations of large peatland systems, *J. Hydrol.*, *316*, 313–324.

Romanowicz, E. A., D. I. Siegel, J. P. Chanton, and P. H. Glaser (1995), Temporal variations in dissolved methane deep in the Lake Agassiz Peatlands, Minnesota, *Global Biogeochem. Cycles*, *9*(2), 197–212.

Rosenberry, D., P. Glaser, and D. Siegel (2006), The hydrology of northern peatlands as affected by biogenic gas: Current developments and research needs, *Hydrol. Processes*, *20*, 3601–3610.

Siegel, D. (1992), Groundwater hydrology, in *The Patterned Peatlands of Minnesota*, edited by J. H. E. Wright, B. A. Coffin, and N. E. Aaseng, Univ. of Minnesota Press, Minneapolis, Minn.

Siegel, D., and P. Glaser (1987), Groundwater flow in a bog-fen complex, Lost River Peatland, northern Minnesota, *J. Ecol.*, *75*, 743–754.

Stevens, N. (2006), Characterizing solute transport processes through laboratory experiments and numerical simulation, Master's thesis, Univ. of Maine, Orono.

Wheeler, D., J. Guyer, and J. Warren (2007), Fipy user's guide, *Tech. Rep. version 1.2*, Nat. Inst. of Stand. and Tech., Gaithersburg, Md.

Winston, R. (1994), Models of the geomorphology, hydrology, and development of domed peat bodies, *Geol. Soc. Am. Bull.*, *106*, 1594–1604.

Winter, T. C. (1976), Numerical simulation analysis of the interaction of lakes and ground water, *U.S. Geol. Surv. Prof. Pap.*, *1001*.

X. Comas, Department of Geosciences, Florida Atlantic University, Boca Raton, FL 33431, USA.

A. S. Reeve and Z. D. Tyczka, Department of Earth Sciences, University of Maine, Orono, ME 04469, USA. (asreeve@maine.edu)

L. D. Slater, Department of Earth and Environmental Sciences, Rutgers, State University of New Jersey, Newark, NJ 07102, USA.

Index